DATE DUE

APR 0 7 2008			

GAYLORD PRINTED IN U.S.A.

MEDICINE ACROSS CULTURES

SCIENCE ACROSS CULTURES: THE HISTORY OF NON-WESTERN SCIENCE

VOLUME 3
MEDICINE ACROSS CULTURES

Editor

HELAINE SELIN, *Hampshire College, Amherst, Massachusetts USA*

MEDICINE ACROSS CULTURES

History and Practice of Medicine in Non-Western Cultures

Editor

HELAINE SELIN

Hampshire College, Amherst, Massachusetts, USA

Advisory Editor

HUGH SHAPIRO

University of Nevada, Reno, USA

KLUWER ACADEMIC PUBLISHERS

DORDRECHT/BOSTON/LONDON

Library of Congress Cataloging-in-Publication Data

Medicine across cultures : history and practice of medicine in non-Western cultures / Helaine Selin ; Hugh Shapiro, advisory editor.
p. cm. – (Science across cultures)
Includes bibliographical references and index.
ISBN 1-4020-1166-0 (alk. paper)
1. Medicine – Developing countries. I. Selin, Helaine, 1946– II. Shapiro, Hugh. III. Series.

2003041605

ISBN 1-4020-1166-0

Published by Kluwer Academic Publishers,
PO Box 17, 3300 AA Dordrecht, The Netherlands.

Sold and distributed in North, Central and South America
by Kluwer Academic Publishers,
101 Philip Drive, Norwell, MA 02018, USA

In all other countries, sold and distributed
by Kluwer Academic Publishers,
PO Box 322, 3300 AH Dordrecht, The Netherlands

Printed on acid-free paper

Printed and bound in Great Britain by Antony Rowe Ltd.

INTRODUCTION TO THE SERIES

SCIENCE ACROSS CULTURES: THE HISTORY OF NON-WESTERN SCIENCE

In 1997, Kluwer Academic Publishers published the *Encyclopaedia of the History of Science, Technology, and Medicine in Non-Western Cultures.* The encyclopedia, a collection of almost 600 articles by almost 300 contributors, covered a range of topics from Aztec science and Chinese medicine to Tibetan astronomy and Indian ethnobotany. For some cultures, specific individuals could be identified, and their biographies were included. Since the study of non-Western science is not just a study of facts, but a study of culture and philosophy, we included essays on subjects such as Colonialism and Science, Magic and Science, The Transmission of Knowledge from East to West, Technology and Culture, Science as a Western Phenomenon, Values and Science, and Rationality, Objectivity, and Method.

Because the encyclopedia was received with critical acclaim, and because the nature of an encyclopedia is such that articles must be concise and compact, the editors at Kluwer and I felt that there was a need to expand on its success. We thought that the breadth of the encyclopedia could be complemented by a series of books that explored the topics in greater depth. We had an opportunity, without such space limitations, to include more illustrations and much longer bibliographies. We shifted the focus from the general educated audience that the encyclopedia targeted to a more scholarly one, although we have been careful to keep the articles readable and keep jargon to a minimum.

Before we can talk about the field of non-Western science, we have to define both non-Western and science. The term non-Western is not a geographical designation; it is a cultural one. We use it to describe people outside of the Euro-American sphere, including the native cultures of the Americas. The power of European and American colonialism is evident in the fact that the majority of the world's population is defined by what they are not. And in fact, for most of our recorded history the flow of knowledge, art, and power went the other way. In this series, we hope to rectify the lack of scholarly attention paid to most of the world's science.

As for defining science, if we wish to study science in non-Western cultures, we need to take several intellectual steps. First, we must accept that every culture has a science, that is, a way of defining, controlling, and predicting events in the natural world. Then we must accept that every science is legitimate in terms of the culture from which it grew. The transformation of the word science as a distinct rationality valued above magic is uniquely European. It is not common to most non-Western societies, where magic and science and religion can easily co-exist. The empirical, scientific realm of understanding and inquiry is not readily separable from a more abstract, religious realm.

Medicine Across Cultures is the third book in the series. It includes about 20 chapters. Most deal with medical systems as they are perceived and practiced by different cultures: Australian Aboriginal people, Native Americans, Samoans, Indians, etc. The book also contains a variety of essays on related subjects, such as Religion and Medicine or Chinese and Western Medicine.

We hope the series will be used to provide both factual information about the practices and practitioners of the sciences as well as insights into the worldviews and philosophies of the cultures that produced them. We hope that readers will achieve a new respect for the accomplishments of ancient civilizations and a deeper understanding of the relationship between science and culture.

TABLE OF CONTENTS

ACKNOWLEDGMENTS

I would like to thank the contributors to this volume; these are the easiest, most reliable people I've ever worked with. I am so impressed with how committed they are to their subjects, while at the same time keeping their academic distance. It was a pleasure to read and work with their writing; no one was angry at me for changing their words. Thanks to Hugh Shapiro, the Advisory Editor, who read and commented on all the articles in addition to writing his own piece. Thanks to Dedie King, my acupuncturist, who makes it possible for me to maintain my equanimity and hard work at the same time. And thanks, always and again, to my loving family, Bob and Lisa and Lisa and Tim, my sources of balance, harmony, and wellbeing.

ACKNOWLEDGMENTS

ABOUT THE CONTRIBUTORS

HELAINE SELIN (Editor) is the editor of the *Encyclopaedia of the History of Science, Technology, and Medicine in Non-Western Cultures* (Kluwer Academic Publishers, 1997) and Science Librarian and Faculty Associate at Hampshire College in Amherst, Massachusetts. In addition to editing the new series, *Science Across Cultures*, she has been teaching a course on the Science and History of Alternative and Complementary Medicine.

HUGH SHAPIRO (Advisory Editor; Chinese and Western Medicine) is Associate Professor of History at the University of Nevada, Reno. He received his B.A. and M.A. from Stanford, and his Ph.D. in 1995 from Harvard University. His area of research is the history of disease in China, and his experience in East Asia includes extensive research in Beijing, Shanghai, and Taipei, and visiting appointments at the International Research Center for Japanese Studies (Nichibunken), Kyoto, Japan, and the National Taiwan Normal University, Taipei, Taiwan. In 1999, he received his university's highest teaching award, and in 2000, the Liqing Prize for the History of Chinese Science (Taiwan). For 2002–2003, he is at the School of Historical Studies, Institute for Advanced Study, in Princeton, New Jersey.

GUY ATTEWELL (Islamic Medicines) is studying for a Ph.D. at the School of Oriental and African Studies, University of London, in the social history of Unani medicine in 19th and 20th century Hyderabad (India). His research is funded by a studentship from the Wellcome Trust. He studied for his first degree at the SOAS in Arabic, and did an M.A. in European Renaissance history at the Warburg Institute, London, focusing on Arabic medical learning in Europe. His research interests also include Portuguese accounts of medicine in 16th century Goa, and patient experience in contemporary medical practice.

DON BAKER (Oriental Medicine in Korea) is an associate professor in the Department of Asian Studies at the University of British Columbia in Vancouver, Canada. His research has focused on the cultural history of Korea, particularly the history of traditional science, philosophy, and religion. He was one of the editors of *Sourcebook of Korean Civilization* (New York: Columbia University Press, 1993–96) and has also published *Joseon hugi yugyo wa cheonju-gyo ui daerip* [Confucians Confront Catholicism in the Latter Half of the Joseon Dynasty] (Seoul: Iljogak Press, 1997). In addition, he has published numerous articles in English on Korean religion, history, philosophy, and traditional science. He received his Ph.D. from the University of Washington in 1983.

KAREN McCARTHY BROWN (Caribbean Medicine), Drew University Professor of Anthropology of Religion, is the author of *Mama Lola: A Vodou Priestess in Brooklyn* (Berkeley: University of California Press, 1991), winner of the Victor Turner Prize in Ethnographic Writing (American Anthropological Society), and The Best First Book in the History of Religions (American Academy of Religion) as well as *Tracing the Spirit; Ethnographic Essays on Haitian Art* (Seattle: University of Washington Press, 1995). She has conducted fieldwork in Haiti and on the Haitian diaspora community in Brooklyn (New York), Ghana, Togo, Benin and Nigeria. Founder and Director of Drew's Newark Project, she has led an ethnographic exercise in mapping the religious life of the city of Newark, New Jersey.

VIGGO BRUN (Traditional Thai Medicine), Associate Professor in the Department of Asian Studies at the University of Copenhagen, has been teaching Thai and related courses in the Thai studies program since 1976. He has published *Sug, the Trickster who Fooled the Monk. A Northern Thai Tale with Vocabulary* (Lund: CINA Monograph Series 27, 1976) and, with Trond Schumacher, *Traditional Herbal Medicine in Northern Thailand* (Los Angeles: University of California Press, 1987). He has also published a Thai-Danish Dictionary, *Thai-Dansk Ordbog* (Bangkok: Craftsman Press, 1994). Brun is also editor of *The Copenhagen Journal of Asian Studies.*

CAI JINGFENG (Medicine in Ancient China) is Professor and Research Fellow at the China Institute for History of Medicine and Medical Literature, China Academy of Traditional Chinese Medicine, Beijing, China. His main interests include Traditional Chinese Medicine and its history, especially the history of Chinese minorities, and the history of Tibetan medicine and social medicine. He has published over 20 monographs, including *On the Origin of rGyud-bzi, Tibetan Medicine*, and over 100 academic papers in both Chinese and English in journals in China and abroad. Some of these specialized in the history of Tibetan medicine, especially in Tibetan medical painted scrolls (*Thangka*).

ANANDA SAMIR CHOPRA (Āyurveda) was born and brought up in Germany to Indian parents. He studied Medicine and Indology at the University of Heidelberg, Germany. After becoming a registered physician in Germany, he studied ayurvedic medicine for a year at the Shyamadas Vaidyashastrapith in Calcutta. Since 1996 he has headed the Āyurveda Section at the Habichtswaldklinik Āyurveda in Kassel, Germany. This clinic is part of a 350-bed German hospital. In a unique manner authentic ayurvedic medicine is practised as a complement with modern western medicine. In recognition of his services in Āyurveda, he was presented with a medal and shawl by the Āyurvedic Board of Karnataka (South India) in 1998. Since his childhood, Chopra has been learning classical Indian (Hindustani) music and has served as honorary director of studies at the Tagore-Institute in Bonn, Germany.

DAYALAN DEVANESEN (Traditional Aboriginal Health Practice in Australia) is the Director for Community Partnerships in the Department of Health and Community Services of the Northern Territory of Australia. After graduating as a doctor in India, Dr Devanesen commenced work as a Flying Doctor in the remote town of Alice Springs, Central Australia in 1974. Here he pioneered the development of the Aboriginal Health Worker Training Program and a bi-cultural approach to health service delivery for Aboriginal people. In 1985 he moved to Darwin, the capital of the Northern Territory, to become the Director of Aboriginal Health. From 1991–2001 he was the Director for Primary Health Care. In 1997 he was appointed as a Member of the Order of Australia for his contribution to the development of Aboriginal Primary Health Care services.

WASEEM ROSHDY ELSEESY (Medicine in Ancient Egypt) is Consultant Urologist at el Sahel Teaching Hospital at Heliopolis University in Cairo, Egypt. He is the inventor of a patented instrument used in urology and the author of a book, *Critical Medical Survey of Love and Sex* (Cairo: Engineering Center for Printing and Publishing, 3rd. ed., 1999). He is also a researcher in ancient Egyptian history and has a weekly column on issues in Egyptology in *Rose-el-Youssef*.

EDWARD C. (TED) GREEN (African Medicine) is an applied medical anthropologist who consults for a number of international development organizations. He has worked for many years in Africa, with shorter periods of research and applied work in Asia, the Middle East, Eastern Europe, Latin America, and the Caribbean. He worked as an advisor to the ministries of health in both Swaziland (1981–83) and Mozambique (1994–95). He is the author or editor of five books as well as over 200 journal articles, book chapters, and commissioned reports. He is currently a Takemi Fellow at the Harvard School of Public Health (2001–2002).

ÅKE HULTKRANTZ (Medical States and Soul Beliefs Among Tribal Peoples) holds a Ph.D. from the University of Stockholm and an Honorary Doctor of Divinity from the University of Helsinki. He is an emeritus Professor in Comparative Religion at Stockholm University, where he taught from 1958–1986. He has been a Visiting Professor and occasional lecturer at many European, African, North American and Japanese universities. He won prizes for scientific accomplishments from the Americanist Congress in Winnipeg (1980) and the International Society for Shamanistic Research, Budapest (1993). He has served as Chairman of the Swedish Americanist Society. He has published some twenty-five books and almost 400 articles on American Indians, Lapps and other Arctic peoples, shamanism and general ethnographical and religio-historical problems. He has also given the Gifford Lectures at Aberdeen University (1981 and 1982), and the Foerster Lecture at the University of California Berkeley (1983).

JOHN M. JANZEN (African Medicine) is Professor of Anthropology and Director of the African Studies Center at the University of Kansas. His special area of interest is in Sub-Saharan African health and healing. He explores the relationship of African medicine to Western biomedicine in *The Quest for Therapy in Lower Zaire* (Berkeley: University of California Press, 1978); the history and focus of therapeutic associations in *Lemba 1650–1930: A Drum of Affliction in Africa and the New World* (New York: Garland, 1982), and *Ngoma: Discourses of Healing in Central and Southern Africa* (Berkeley: University of California Press, 1992). His most recent book is *The Social Fabric of Health: An Introduction to Medical Anthropology* (New York: McGraw-Hill, 2001).

SAMUEL KOTTEK (Medicine in Ancient Hebrew and Jewish Cultures) is a certified pediatrician and Emeritus Professor and former Chair of the History of Medicine Department at Hebrew University Medical School in Israel. His work has concentrated on the early history of pediatrics in 18th century Europe and various aspects of the study of Jews and medicine. He has authored *Medicine and Hygiene in the Works of Flavius Josephus* (Leiden: E.J. Brill, 1994) and edited several others: *Moses Maimonides* (Northvale, New Jersey: Aronson, 1993) *Medicine and Medical Ethics in Medieval and Early Modern Spain* (Jerusalem: Magnes Press, 1996), *Health and Disease in the Holy Land* (Lewiston, New York: Edwin Mellen Press, 1996), and *From Athens to Jerusalem: Medicine in Hellenized Jewish Lore and Early Christian Literature* (Rotterdam: Erasmus, 2000).

MARGARET LOCK (Globalization and Cultures of Biomedicine in Japan and North America) is Professor in the Department of Social Studies of Medicine and the Department of Anthropology at McGill University, Montreal, Canada. She is the author of *East Asian Medicine in Urban Japan: Varieties of Medical Experience* (Berkeley: University of California Press, 1980) and *Encounters with Aging: Mythologies of Menopause in Japan and North America* (Berkeley: University of California Press, 1993), which won many scholarly prizes. Lock has edited seven other books and written over 150 scholarly articles. Lock's particular interest is the relationship among culture, technology, and the body in health and illness. She has done research in Japan into the revival of traditional medicine and into life cycle transitions. She has also completed a comparative study in Japan and North America about knowledge and practices associated with the introduction of new biomedical technologies that facilitate reproduction, and in connection with death and organ transplants.

CLUNY MACPHERSON (Pacific Medicine) teaches in the Department of Sociology at the University of Auckland, New Zealand. His interest in medicine across cultures stems from early work in Samoa with agriculturalists who wished to protect certain medicinal plants and the healers who wanted to use them. This developed into a ten-year study of the history of relations between indigenous and introduced medical beliefs and practices in Samoa. Cluny is a co-author of *Samoan Medical Belief and Practice* (Auckland University Press, 1991).

LA'AVASA MACPHERSON (Pacific Medicine) is a nurse. Her interest in medicine across cultures comes in part from her childhood experiences as an assistant and observer of her grandmother's traditional practice in Samoa and more recently from observations of the overlap between the two medical systems in her professional practice among migrants in Auckland. La'avasa is a co-author of *Samoan Medical Belief and Practice* (Auckland University Press, 1991).

PATRICK MAHER (Traditional Aboriginal Health Practice in Australia) has worked as a physiotherapist for the last eight years in the Northern Territory, Australia. He has undertaken qualitative research into the delivery of rehabilitation services to Aboriginal people, published articles on Aboriginal traditional health beliefs and disability in the Aboriginal population, and been part of a group which established a website to assist health professionals working with people with disabilities in remote indigenous communities. He is currently a Senior Disability Policy Officer with the Department of Health and Community Services in the Northern Territory, Australia.

LEWIS MEHL-MADRONA (Native American Medicine) attended Stanford University School of Medicine and completed residencies in family practice and psychiatry at the University of Vermont. He is certified in those specialties with added qualifications in geriatrics. He has also practiced emergency medicine since 1976. He currently is affiliated with the Program in Integrative Medicine and with the Human Energy Systems Laboratory, both at the University of Arizona. He teaches at the Southwest College of Naturopathic Medicine and has a private practice for intensive healing experiences in mind-body-spirit medicine, alternating between Arizona and Vermont. He is the author of *Coyote Medicine* (New York: Scribner, 1997).

RUBEN G. MENDOZA (Medicine in Ancient Peru and South America) is a founding faculty member of the California State University, Monterey Bay, and Professor in the Social and Behavioral Sciences Center. He is also founding Director of the Institute for Archaeological Science, Technology, and Visualization. Dr. Mendoza received his Ph.D. (1992) from the University of Arizona, Tucson. He has studied and published on the archaeology and ethnohistory of Mesoamerica and Peru. His fieldwork includes collaborative ventures in the excavation and mapping of the main pyramid complex at Cholula, Mexico; the 16th century Ex-convento de la Concepcion, Puebla, Mexico; the civic-ceremonial complex of Cañada de la Virgen, Guanajuato, Mexico; and the Crescent Rock Shelter of Colorado. He is principal investigator at Mission San Juan Bautista, California. Recent works include chapters in *Cultural Diversity in the United States: A Critical Reader* (Arlington, Virginia: American Anthropological Association, 2001), *U.S. Latino Literatures and Cultures:*

Transnational Perspectives (Heidelberg, Germany: Carl Winter-Verlag, 2000), and *Emerging Technologies in Teaching Languages and Cultures* (San Diego, California: LARC Press, 2001).

AHMED SHAFIK (Medicine in Ancient Egypt) is Professor and Chairman of the Department of Surgery and Experimental Research at Cairo University, Egypt. A surgeon by profession, he has developed many new surgical and laparoscopic procedures, invented instruments, designed drugs, and made so many medical contributions as to be a nominee for the Nobel Prize in Medicine. At the same time, he has written more than 500 pieces in the fields of anatomy, physiology, internal medicine and pathology. He is a member of many associations and organizations, and president or honorary president of many others.

ROBERT WILLIAM PRASAAD STEINER (Tibetan Medicine) is Professor and Director of the Division of Community Health in the Department of Family and Community Medicine at the University of Louisville School of Medicine in Louisville, Kentucky, USA. He received the M.D. degree from University of Louisville, certificate from New England School of Acupuncture, and the Ph.D. in Epidemiology from the University of North Carolina. He has studied many traditional health care systems from Asia, including Tibetan Medicine, traditional Chinese Medicine, Integral Yoga, Ayurveda, traditional Japanese Macrobiotics, and Sikh Dharma. He chaired the Society of Teachers of Family Medicine Task Force on Cross Cultural Health Experiences that developed national guidelines for introducing medical anthropology into family practice residency training programs. He currently serves on the Board of Directors of Health Care Excel, the Quality Improvement Organization for Indiana and Kentucky. He is a member of KTD Monastery in Woodstock, New York.

JIM B. TUCKER, M.D. (Religion and Medicine) is an Assistant Professor in the Department of Psychiatric Medicine at the University of Virginia Health System. He works in the Division of Personality Studies, where he conducts research, as well as the Division of Child and Family Psychiatry, where he served as Acting Director in 2000–2001.

CARLOS VIESCA received his M.D. in General Surgery at the Universidad Nacional Autónoma de México (UNAM) and his Ph.D. in History of Science at Universidad Autónoma Metropolitana, México. He is Head and Chairman of the Department at UNAM's Faculty of Medicine, where he has taught History and Philosophy of Medicine since 1967. He is a member and former President of the Mexican Society of History and Philosophy of Medicine and a member of the National Academies of Medicine and Surgery. He is the author of 12 books and some 200 chapters and papers about the history of medicine, mainly Mexican prehispanic and colonial medicine. Among them are *Medicina Prehispánica de México* (México, D.F.: Panorama Editorial, 1986),

Ticiotl. Conceptos médicos de los antiguos mexicanos (México, D.F.: UNAM, 1997), and *Cuerpo y Espíritu: Medicina y Plasticas en México* (2001).

ZHEN YAN (Medicine in Ancient China) holds a degree of Master of Medical History and is a doctoral candidate in Tibetan medical history at the China Institute for History of Medicine and Medical Literature, China Academy of Traditional Chinese Medicine, Beijing, China. Her main interests include Traditional Chinese medicine and its history, history of Tibetan traditional medicine, and history of Chinese minorities. Her publications include: *Comparative Studies on the Early Sphygmology of Tibetan Medicine and Traditional Chinese Medicine*, *On the Origin of Tibetan Medicine: a Preliminary Perspective*, and a review of papers presented at the 2000 Lhasa International Conference on Tibetan Medicine.

INTRODUCTION

Medicine Across Cultures explores the medical beliefs and practices of cultures around the world. Since the term "non-Western" is a cultural description, not a geographic one, this volume includes studies of the native cultures of the Americas. Ideas about health and illness are central to every culture. There are no universals regarding what it means to be well, or how to prevent, diagnose, or treat illness. Medical knowledge and practices in all societies are closely associated with other widely held values. Even though these values change with new knowledge and new technologies, we can still speak with confidence about culturally specific medical systems. A society's views on medicine arise from and reflect its cultural beliefs and customs.

In Western biomedicine, there is usually one cure for one disease, regardless of the person who has it. If a patient appears with all the symptoms for bronchitis, and the laboratory work confirms the diagnosis, then a standard treatment is prescribed. That treatment is prescribed to young and old, men and women. In many other cultures, diagnosis and therapy are person-specific rather than disease-specific. The treatment that I might receive, presenting to the physician with a cough and a fever, might be extremely different for someone with different personality traits. The diagnosis might be a complicated amalgamation of deciding factors: the way a person looks and sounds, his or her family history and place in the community, the patterns of personal relations, the climate and time of year, or diet. The treatment would grow out of that diagnosis, often touching on many of the factors involved in the diagnosis.

Different cultures have distinct ways of dealing with illness and with ill people; in many non-Western cultures the focus is not just on the content of the disease, but also on its context (Beinfeld and Korngold, 1998). It is important for a diagnostician to know about the person's situation and family to prescribe effectively.

In the last 500 years, the development of increasingly sophisticated medical instruments, such as the microscope, has led to a focus on smaller and smaller entities. Genetics and molecular biology are the latest manifestation of this trend. What is lost is the direct connection between patient and healer. Today in the West, if a person requires an x-ray, there is a fragmented process illustrating the gap between the two. First, a practitioner takes the x-ray, then a radiologist interprets it and reports to the physician. The physician then passes the results to the patient. Neither the doctor nor the patient will have seen the x-ray.

Recent years have witnessed a host of scholarly articles concerning the sciences in non-Western cultures. However, for the most part, whenever refer-

H. Selin (ed.), Medicine Across Cultures: History and Practice of Medicine in Non-Western Cultures, xix–xxiv.

ence is made to another culture's way of doing mathematics or astronomy, we usually see "science" in quotation marks, implying that it is not quite science in the superior Western sense. Interestingly, this disparagement is increasingly less so in the case of medicine, at least at the level of the patients rather than the physicians and scientists. Our anxieties about globalization do not apply in this case. Westerners, even with their techno-medicine, are increasingly drawn to medical systems from around the world. It is also important to note the global migration of people to the United States and Europe. As people migrate, they bring with them traditional knowledge and practices and help expose Westerners to these new ideas.

What do the world's medical beliefs offer that biomedicine lacks? The first factor seems to be personalization. The Cartesian model of the body as a machine that has dominated Western thought since the 18th century becomes increasingly alienating as many other parts of life become mechanized. Many of the medical practices that are discussed in this book involve the whole person and not just some affected parts. The Chinese means of diagnosis, by looking, listening or smelling, questioning, and touching, by knowing the family and the family's history, by factoring in personality types and food likes and dislikes, has increasing appeal in an increasingly impersonal world. There is an interesting irony here, since in many societies that value the collective good, medicine is very individual-oriented. In places like the United States, which places such a high regard on individuality, medicine largely ignores the person who has the disease (Beinfeld and Korngold, 1998).

Second, it seems that Western medicine can offer answers to the question "Why?", but it cannot answer "Why me?" The purely mechanistic explanation of causation does not necessarily offer solace to the sick person. Other ways of looking at disease offer reasons that the person became afflicted. These may not be entirely acceptable to a Western naturalistic view, especially when they have to do with malevolent spirits or past lives, but they are to the people who have the disease, and they fit comfortably within the worldview of their own culture. And even if Westerners might not believe that illness could result from transgressing a taboo, they can accept that something other than a microbe could cause someone to fall ill.

Third, although it is clear that biomedicine is the best choice for acute illness, for surgery, and for infectious diseases, most people in fact are more likely to have chronic conditions, especially as the population ages. It is another irony that the advances made by Western medicine in treating infectious diseases result in more people living longer, and often with chronic and disabling conditions (Martin, 2001). An approach that searches for a drug that will relieve the symptoms, and which someone might be expected to take for years and years, is not necessarily superior to the approach that looks for a gentler, less invasive, more user-friendly means of dealing with pain and disability.

There are also caveats to be aware of. After years of disparaging the scientific and medical ideas of non-Western cultures, the pendulum is swinging the other way, and there is often now a tendency to over-romanticize. Although it is true that biomedicine cannot always provide the psychological, spiritual, and personal support that people sometimes crave, it is the case that people with access to Western medicine, for the most part, live longer than people without (Information Please, 2002). At the same time, a medical system such as Chinese Medicine has been developing for more than 4000 years; it survives because it works.

It is also important to distinguish between what indigenous people say for themselves and what others write about them. Native people speak with an understanding of cultural reality an outsider can never hope to achieve. At the same time, they often have difficulty achieving critical distance from their own cultures and are sometimes constrained about raising controversial issues. Non-natives may not have the same constraints and have more freedom to be critical, but they will almost always lack that essential grasp of a group of people and the way they think that a cultural resident will have (Booth, in press). This volume presents the views of both native and non-native authors, who are both critics and believers, and, in some cases, both.

Another theme that runs through this book is the interconnection between religion and medicine. These are considered very separate spheres in the West, at least in recent times. Although some hospitals will acknowledge the need for religion by having a priest who makes visits to very sick patients, the medical establishment does not recognize an intimate connection between the two. Jim Tucker provides some insight into this issue in his chapter on Religion and Medicine, but in fact what Westerners call religion is present in all the medical systems in this volume. Non-Western cultures, and not just tribal cultures, do not see the mind and body and soul as separate, barely related entities; they know that curing just one part will not necessarily rid a person of a disease. No great intellectual leap is needed to realize that we are made up of many separate aspects, all of which might need attention.

Most of the chapters in this book focus, in one way or another, on healing. Some describe means of healing, sometimes using statues and images, or feeding and honoring one's ancestral and spiritual protectors, that may seem strange to a Western reader. But are these any stranger than chemotherapy or radiation, procedures that will certainly seem barbaric to people generations from now? Even perceptions of what it means to be ill vary from culture to culture. In many traditional healing philosophies, going back and forth between balance and imbalance, harmony and disharmony, are not perceived as illness, but just as part of the process of being alive.

In *Medicine Across Cultures*, the reader will find descriptions of many medical systems that are alive today (China, Tibet, India, Samoa, Australia) and many historical studies both of cultures that no longer exist (Aztec, Inca) and cultures that do (Hebrews, Islamic people). All focus on the history, but most bring the story up to date with information on how those cultures use their medicine today. All of us to some extent are becoming increasingly integrated in our medical approaches and views. As Cluny and La'avasa Macpherson say in their chapter on Samoa, why chose one system when you can choose the best parts of many?

A global kind of exchange is happening now in medicine. The West is willing to share antibiotics with the rest of the world, but in fact health care in many countries is not readily available, and drugs are often not affordable. In those countries, traditional medicine thrives, both because it is local and because it works for most of the diseases that people have. In fact, as Lock says in her chapter, the introduction and spread of medicine from Europe to Asia often stimulated a lively revival of indigenous medicine, commonly associated with nationalism and a celebration of the traditional.

Europe and America export their culture and cultural icons, but at the same time they find themselves dissatisfied with them. A study in 1998 showed that more than 40% of Americans used an "alternative" form of medicine, even though biomedicine was both accessible and paid for by their health insurance (Eisenberg, *et al.*, 1998). As tourism and global communications increase, so do our interests in trying to find what is better in other ways of seeing the world and incorporating them into our own lives, so that people can be healthier and happier as we live longer.

Medicine Across Cultures is divided roughly into three sections. The largest part of the book deals with different cultures and their medicine. Some are descriptive; some are more abstract and philosophical. John Janzen and Edward Green focus the attention of their article, "Continuity, Change and Challenge in African Medicine," on the ethnolinguistic group known as Niger-Congo. They discuss ethnobotany and divination as key factors in diagnosing and curing African diseases, as well as the paradigms that define and shape the system. They also show how the newest medical scourges in Africa are causing a re-legitimization of traditional medicine. Ahmad Shafik and Waseem Elseesy tackle the complex and ancient culture of Egypt, describing the medical papyri that have survived and how modern scientific analysis has show that the plant and other medications prescribed by the ancients have medicinal properties today.

Cai Jingfeng and Zhen Yan's article on medicine in ancient China describes the origin and formation of Traditional Chinese Medicine. The authors discuss the theory and the social organization of the system, as well as its clinical applications in fields from ophthalmology to pediatrics, and the long history of texts that continue to shape the beliefs and practices of Chinese medicine. Ananda Chopra describes the theory and applications of Ayurveda, Indian medicine, with detailed analyses and interpretations of the texts. Robert William Prasaad Steiner's chapter, "Cultural Perspectives on Tibetan Medicine", condenses both the complex history of medicine in Tibet and a detailed description of Buddhism and Buddhist practice and how the two interrelate. As in Chinese medicine, Tibetan medicine is based on carefully defined personality and social descriptions that determine diagnosis and treatment. Viggo Brun writes about Thailand, another Buddhist country, also describing the situation today and the government's role in keeping traditional medicine alive and vibrant.

Two other Asian countries are explored. Don Baker's "Oriental Medicine in Korea" describes the growth of Korean medicine, both in tandem with and as a separate entity from Chinese medicine. Baker shows how a medical system can be affected by colonialism, as invading countries suppress or encourage local beliefs. Margaret Lock considers globalization and cultures of biomedicine, when she compares ideas about menopause, informed consent, and recognition of brain death in contemporary Japan and North America.

Dayalan Devanesen and Patrick Maher write about traditional Aboriginal health practice in Australia, explaining the belief system, the Dreaming, that shapes Aboriginal views of life, health, and illness. They focus on the health system of the Warlpiri people, describing traditional healers and materia medica, as well as the integrated bicultural medical system that exists today. In "When Healing Cultures Collide: A Case from the Pacific", Cluny and La'avasa Macpherson also discuss the emergence of bicultural medical practice. Their focus is on Samoa, a society which has survived the colonial years without losing its own identity or its own medicine.

Lewis Mehl-Madrona's chapter on Native America describes herbal pharmacology, therapies, and elder care, both in ancient Native America and in the contemporary United States. In this case, the clash of cultures has resulted in a more difficult health situation for native Americans, and Mehl-Madrona offers some ideas about incorporating more traditional beliefs to relieve suffering. In "Lords of the Medicine Bag: Medical Science and Traditional Practice in Ancient Peru and South America", Ruben G. Mendoza compresses millennia of history and a wide geographic range in a chapter describing the detailed practices of Inca and other South American peoples. He pays special attention to the surgical skill of trephination, incorporating much archeological and paleopathological data into his study. Carlos Viesca Treviño adds to the chapters on the Americas with his study of medicine in ancient Mesoamerica, an area which includes the Aztec and the Maya people. He describes the connection between worldview and medicine, showing how the idea of humans as just another part of the cosmos shaped their medical customs. To complete the section on medicine in the Americas, Karen McCarthy Brown presents a case of Haitian Vodou in her chapter on healing relationships in the African Caribbean. This is another situation in which an indigenous tradition survived in spite of, or perhaps because of, suppression by colonial powers, and it is a tradition that survives today with Haitians around the world, including urban centers such as New York City. The connection between the contemporary African Caribbean and the Africa described by Green and Janzen is particularly striking.

The next two chapters deal specifically with a religious system and medicine. Samuel Kottek describes medicine in ancient Hebrew and Jewish cultures, based on the early Jewish texts, the Bible and the Talmud. He discusses the portrayal of physicians and surgeons, disease and diseases, dietary laws, anatomy and physiology, and surgery. Guy Attewell takes a different tack in his historiographical study of Islamic medicine. He studies both what he calls Greco-Islamic medicine and prophetic medicine, explaining why certain aspects of Islamic medicine have attracted much attention while others have not. He deals with specific themes pertaining to Islamic medical history that have been the subject of controversy, such as the translation movement.

The last contains three chapters. Hugh Shapiro compares Chinese and

Western medicine, using the notion of nerves to illustrate their basic philosophical differences. In tracing the introduction of the language of nerves into China, he raises questions about how the relationship between the two systems of thought has been understood. Jim Tucker briefly explores the connections between religious beliefs and medical states in Ayurveda and Buddhism and then describes the links between spirit and body and how they might be encouraged in Western medicine. He describes several interesting studies of the use of intercessory prayer in American intensive care units. Äke Hultkrantz writes about the role of the shaman as therapist and healer in many tribal cultures and how his extraordinary psychic states have a direct bearing on soul beliefs and medical procedures.

The chapters, as one can see from this brief overview, approach their subjects from descriptive, historiographical, linguistic, or historical viewpoints. The similarities, across huge distances of the globe, are remarkable, as are the differences in ways of seeing how the world works and why we become ill. What is clear is how much we have to offer each other, both from our combined and separate histories, and from the merging of Western and non-Western modes of medical practice.

BIBLIOGRAPHY

Beinfield, Harriet and Efrem Korngold. 'Eastern medicine for Western people.' *Alternative Therapies* 4(3): 80–87, 1998.

Booth, Annie. 'We are the land; Native American views of nature.' In *Nature Across Cultures: Views of Nature and the Environment in Non-Western Cultures*. Dordrecht: Kluwer Academic Publishers, in press 2003.

Eisenberg, David M. *et al.* 'Trends in alternative medicine use in the United States, 1990–1997: Results of a follow-up national survey.' *JAMA* 280: 1569–1575, 1998.

Information Please Online Almanac. <http://www.infoplease.com/ipa/A0774532.html>. Accessed August 2002.

Martin, Joseph B. 'Historical and professional perspectives of complementary and alternative medicine with a particular emphasis on rediscovering and embracing complementary and alternative medicine in contemporary Western society.' *The Journal of Alternative and Complementary Medicine* 7(Suppl. 1): S11–S18, 2001.

JOHN M. JANZEN AND EDWARD C. GREEN

CONTINUITY, CHANGE, AND CHALLENGE IN AFRICAN MEDICINE

Health and healing practices in sub-Saharan Africa have evolved over three millennia in constant interchange with those of other world regions. The medicine of Ancient Egypt shaped ideas of the civilizations around it, including the medicine of classical Greek and Roman Antiquity. This complex in turn spread to African regions, through the influence of Islamic medicine. Another dimension of Islam, "prophetic medicine", brought notions of health and healing to Africa from Persia and Arabia.[1] Christian faith healing, which spread first with early Christianity across North Africa and Ethiopia, later was part of European colonial expansion to sub-Saharan Africa. Post-Enlightenment scientific medicine, building upon ancient medicine, brought its ideas of public health and curative medicine. All these perspectives coexist in the early 21st century with African perspectives on health, sickness, and healing.

What then is African healing and medicine? Africa is of course a vast continent with a multitude of societies of great diversity, and we admit a certain hubris with the very idea of generalizing about it. We will focus our attention on the ethnolinguistic group known as Niger-Congo, within which we have ourselves had most of our experience.[2] This group of societies is geographically widespread and numerically large, covering sub-Saharan Africa from the Wolof of Senegal to the Swahili of coastal East Africa, and down to Southern Africa. The Niger-Congo grouping relates the Bantu-speaking peoples of the Congo basin and southeastern Africa historically and culturally to the dense population of West Africa. All of these societies share, in addition to historically related languages, "attitudes about God, religion, kinship, the nature of the world, and life" (Murphy, 1972: 179), and within them, health-related practices and beliefs.

TOOLS AND PERSPECTIVES FOR THE HISTORICAL STUDY OF CHANGING HEALING TRADITIONS

Ecologically distinctive zones of the rainforest, savanna, and desert have shaped both health and adaptive responses by human communities. The modes of

H. Selin (ed.), Medicine Across Cultures: History and Practice of Medicine in Non-Western Cultures, 1–26.

living – hunting and gathering, cultivation, herding, and then urban societies – also shaped the underlying determinants of health.

Hunter-gatherers, for example, such as the remaining Khoisan speakers of Southern Africa in the early 20th century, practiced infanticide for population control and birth spacing of up to four years between children. They also picked up camp whenever diseases broke out, in order to reduce deaths in the settlement. Given their small population concentrations, contagious diseases did not have a chance to take hold and become endemic. The health of the band was promoted through spirit healing ceremonies led by healer-singers (Katz, 1982; Katz, Biesele, and St. Denis, 1997).

Both the West African and the Bantu-speaking civilizations, defined primarily by sedentary agriculture, have also been cattle herders and pastoral nomads throughout their histories. Where the tsetse fly has been absent – as across the Sahel, the eastern Sudan, in the lake region of East-Central Africa, and into moderate Southern Africa – pastoralism has brought with it a distinctive set of ideas about health, sickness, and medicine.

As livestock herding spread southward about six millennia ago, it skirted the rainforest area. This created at the center of the African continent a vast population without domestic large animals. As a result, this population is unable to digest animal milk – a condition known as lactose intolerance. The boundary between pastoralist and non-pastoralist societies has historically been that between rainforest and wet savanna on the one hand and the dry savanna and the desert on the other. On the one hand, the pastoralists have had to manage their herds, concentrating on good breeding, learning the politics of being good neighbors (or superior raiders) on their annual transhumance treks to find seasonal pasture, and understanding the danger zones of the tsetse fly's habitat. On the other hand, the cultivators without large livestock have had to emphasize crop fertility, soil fallowing, irrigation and water management, and the importance of rainfall.

In West Africa, the domestication of plants and animals in sedentary settlements was well underway by 4000 BC. Urban centers and stratified societies emerged in the West African and the Sudanic savanna by the early centuries of the first millennium AD, and trade routes linked West Africa with the Mediterranean and Europe. By the early second millennium AD the influences of Islam and Arabia were felt in the savanna, but pre-Islamic healing rituals or therapeutic practices were not fully supplanted.

The spread of food cultivation and sedentary social modes southward through and around the equatorial rainforest has come to be associated with the spread of the Bantu, Cushitic, and Nilotic cultures and languages. Perhaps as early as 1000 BC the Bantu languages had begun to spread from the region that is now Cameroon and Nigeria. These languages ultimately came to be spoken throughout the whole of Central, Eastern, and Southern Africa, facilitating exchange of ideas and practices, including those related to healing. Food production and iron working spread rapidly through this area during the first millennium AD. The sedentarization of community life in sub-Saharan Africa and the domestication of food plants and livestock provided a moving threshold that also affected health and healing. With the transition to cultivation and larger, sedentary communities, new diseases appeared. There was sleeping sickness (endemic in the rainforests, a major threat to pastoralists), malaria (endemic in rainy forested areas), and smallpox (endemic once population concentrations emerged). Together with widespread environmental risks such as poisonous vipers, these diseases offer examples of health threats that encouraged the creation of "medicines" directed at them. They also show the vocabulary of health-related terms and concepts in a common linguistic and cultural background.

Language history – along with archeology and the study of the distribution of cultural practices including uses of plants, animals, and other natural substances – offers one of the most promising avenues for the study of the history of African medicine and health related practices (Ehret, 2000; Janzen, 1992; Schoenbrun, 1996; Vansina, 1990). Analysis of the words and their meanings and uses by language family permits the determination of which concepts, practices, terms, and phrases are part of the institutional infrastructure of these varied adaptations. Language analysis assists in determining which are inventions along the way, and which are borrowed from elsewhere. Language history also allows the reasonable dating of the origin and spread of specialized institutions, practitioners, techniques, concepts and ideas, and materia medica.[3]

Terms for suffering (*-*duaad*-), healer (*-*ganga*-), medicinal plant (*-*ti*-), the power of words and will to affect health in social relationships (*-*dog*-), and song-dance ceremonies of trance and healing (*-*goma*-) are part of the common vocabulary of the Bantu expansion from about 1500 BC. They are found throughout Equatorial, Central, East and Southern Africa (Janzen, 1992; Vansina, 1990).[4] Such a constellation also characterizes medicine on the Guinea coast and West Africa. As this cultural complex moved eastward to the Great Lakes region, the basic term for healer became bifurcated and the root term for "big man" or chief (*-*kumu*-) came to be applied to diviner (*-*mufuumu*-), suggesting the importance of the diviner in social control (Schoenbrun, 1997).

Profound transformations were brought to African medicine by the mercantile trade of the 16th to the 19th centuries, and then by 19th and 20th century European colonialism. Foreign trade, technology, ideologies, and social forms were often imposed by force. New ideas about health were part of this colonialism, ideas that discredited African medical systems. Missionaries and colonial regimes came to evangelize Africa. Just as Islamic crusaders had attacked "pagan" African forms of healing and religion, so Western Christian missions discredited the basis of knowledge as the overall approach to ritual healing. At a time when early positivist science was analyzing the causes of contagious diseases and public health campaigns were being waged to make Africa safe for "progress", assumptions that social dynamics could cause sickness were dismissed as witchcraft. Since they had cures for diseases such as yaws, leprosy, and later malaria and dysentery, Christian missions and their hospitals contributed to the conversions of many Africans. Although Christianity gained widespread following in sub-Saharan Africa, many of the marks of the African worldview of misfortune have been reincorporated or persisted quietly in private. In the post-colonial era – generally from 1960 on – the process continues of sorting out what indeed were pre-colonial health codes and realities and evaluating what in them is of importance and might be endorsed, revived, and further developed.

The latest economic and health crises lent an immediacy to these debates. African traditional medicine continues to be widespread, as biomedicine is expensive and often locally unavailable. The epidemic crisis of HIV/AIDS has revived the question of whether the African pharmacopoeia has something to offer. Such pressing issues have again raised questions about the fundamental character of African medicine and its ideas of health.

INTERPRETING PRACTICAL APPLICATIONS OF MATERIA MEDICA

Examples of pragmatic and empirical solutions to health threats include diagnoses and interventions for

> bone-setting, midwifery, and a host of specific interventions for such ailments as fever, rheumatism, intestinal disorders, parasites, lactation deficiency, earache, toothache, headache, epilepsy, menstrual disorders, and more. Most of the African pharmacopoeia under discussion here is derived from plants, but medicines from animal parts, sea shells, coral, soils and other substances of natural origin may also be used. Medicines may be drunk as herbal decoctions, or they may be ingested through inhalation, vaccination (dermal incisions), enemas, vaginal infusions, massage, bathing, or fumigation, in forms that include powders, porridges, soups, ointments, smoke, fumes, or eye drops (Bibeau, 1980).

These medications variously reflect the desert, savanna and rainforest ecologies and pastoral and farming ways of life. Following are two examples of pastoral knowledge of sleeping sickness and its vector the tsetse fly. As the early 19th century Nguni king Mzila was expanding his territory in Southeastern Africa, he concentrated his people in large settlements and had them garden close-in territories. Then he sent his hunters out to kill all large wildlife in the savanna forests at a greater distance before allowing his herdsmen to take the cattle into those areas (Swynnerton, in Ford, 1979). The Turkana pastoral nomads of northern Kenya are well aware of the dangers posed by the sleeping sickness bearing tsetse fly to their cattle and themselves during the rainy season, especially in the lush grasses near streams. Therefore they send out young herders with the least valuable animals to graze in those sites to establish their safety from the tsetse before bringing in fertile cows (Gray, personal communication, 1999, based on fieldwork with the Turkana in Kenya).

Malaria is an endemic disease over much of Africa. It is usually that variety of malaria for which the Anopheles mosquito serves as vector. It became a problem for West and Equatorial African cultivators at the time they settled into sedentary communities and began to clear forests for crops. A genetic adaptation to the high death rates occurred soon after, in the form of the blood cell sickling that in its heterozygous form created immunity to malaria (although proving lethal in its homozygous form). Clearly no one was aware of the genetic structure of this adaptation to malaria. However, settlers preferred to build villages and towns on breezy hillsides rather than in the quiet thickets. Thus, before quinine and the late 19th century association of mosquitoes with malaria, breezy hilltops were the preferred settlement site, provided they were near sources of good water.

Smallpox too has been a scourge in sub-Saharan Africa since the advent of cultivation and large concentrations of people (Dawson, 1992: 90). The central placement of Ipoona, the god of smallpox, in the pantheon of Yoruba (and other West African) societies, suggests that it has a history of millennia rather than centuries. In addition to the sacrifices made to the angry god Ipoona, who could kill, a pattern of actions during smallpox epidemics suggests pragmatic public health consciousness as well. Examples include the separate burial of victims, the abstinence from mourning in close proximity to the victims, and quarantining infected households or settlements. Most intriguingly, they refer to attempts to immunize those not yet infected by taking pus from the poxes of infected individuals and introducing it into scratches in their skin.[5]

Immunization as seen in the case of smallpox may be part of a broader principle of confronting the evil of disease or threat. Several kinds of poisonous vipers have posed a serious problem in the lands of the Nyamwezi and Sukuma peoples of Western Tanzania. Organizations of snake handling experts actively promote the encounter with these poisonous vipers and other snakes through public dance performances in which they demonstrate that they can come to terms with the threat. The demonstration includes allowing the otherwise venomous snakes to bite them. However, the snake handlers have been inoculated with small doses of venom that they have milked from the vipers. This understanding of immunization is mainly available to those who have been appropriately initiated to the Snake Handling Order. These examples of immunization to smallpox and snake venom are part of the much wider notion of the need to incorporate or confront the threat in order to overcome it.

Curative plants are central to African medicine. Two examples illustrate the ways in which specific problems were, and are, approached with the curative powers of plants and techniques. Mirau, a herbalist of the Meru people studied by Finnish ethnographer Harjula (1980), who records many of the healer's herbal treatments as one-disease, one-plant related, practices a widespread treatment for intestinal micro-organisms. Although this conveys a somewhat simplistic picture of African herbalism, it does permit a clear examination of Mirau's work from the outside. One of Mirau's 200 single-plant treatments is for children's diarrhea, a serious problem in many regions of the continent where infant mortality often reaches more than 100 per 1000 births. Using the plant known locally as *mamiso* (*Bidens pilosa L.*), Mirau takes 15–20 flowers and boils them to obtain one dose, which is given twice daily as oral medicine. According to Watt and Breyer-Brandwijk's (1962) massive compilation of East and Southern African medicinal and poisonous plants, this plant carries antibacterial substances against microorganisms, including five enteric pathogens. The same plant is reported in use against dysentery and colic in other regions of East and Southern Africa.

In another well-documented study from the national Zairian Research Institute (Bibeau, 1980), the work of six healers in Kinshasa was examined for effectiveness in twenty-two diabetes cases. Independent examination of blood sugar levels revealed an average glycene rate of 500 mg/108 ml of blood. After treatment, which lasted a week or longer, seventeen of the cases experienced a decline of glycene levels to an average of 100 mg/108 ml of blood. Although a surprisingly large diversity of plants was used in preparation of the oral medications, several plants stand out for their repetition from healer to healer, including *Crossopteryx febrifuga* (Rubiaceae), *Nauclea latifolia* (Rubiaceae), *Anchonames difformis* (Arceae), and *Bridelia feruginea* (Euphorbiaceae), the latter of which is used in Ghana for diabetes therapy. Some of the Kinshasa treatments were accompanied by dietary proscriptions for salt, ripe mangoes, pepper, beer, manioc and mushrooms.

Thus far these treatments have appeared to be of the empirical type that scientifically oriented Westerners would recognize. However, in classical African medical thought the added dimensions that Westerners might call "symbolic" or "social" are added without a conceptual break. In the next two examples of treatment with materia medica the fusion of the natural and the human dimension is evident; they deal with two types of bodily swelling.

Mama Mankomba of Mbemba village in the Luozi region of Lower Zaire was well known for her treatment for bodily swelling. She distinguished between two types of swelling, the first thought to be due to heart congestion, the second to poisoning, the result of anger growing out of animosity. Simple swelling was dealt with by an initial emetic from the drops of sap of the finger cactus (*diza kia nlembo*, *Euphorbia tirucalli L.*) with a soapy base to keep the poison from harming the body. This was followed by a potion made from the roots of six savanna plants taken three times daily (*Nlolo, Annona arenaria Thonn, Annonaceae; Mumpala-mbaki, Crossopteryx febrifuga* (*Afz. ex G. Don*) *Benth.* = Rubiaceae; *Nkizu, Syzygium guineense* (*Willd.*) DC. = Myrtaceae; *Votila, Psorospermum febrifugum Spachh.* = Guteriferaceae; *Luvete, Hymenocardia acida Tul.* = Euphorbiaceae; *Kinsangula, Maprounea africana Muell. Arg.* = Euphorbiaceae). Dietary restrictions against sugar, salt, and pepper were also imposed. Poisoning cases received the same initial purge, but were followed by a second purge of the bark scrapings of only the kinsangula plant with salt and palm oil to provoke diarrhea and vomiting. Although Mama Mankomba treated the physical manifestations of anger illness, she refused to become involved with the deeper causes of anger which required conflict resolution and judicial action (Janzen, 1978).

The use of multiple plants and techniques not only introduces a degree of complexity from the botanical and chemotherapeutic compound involved, but also adds to these symbolic classifications, names, songs and other ritual connotations and devices, and above all, the human dimension in health and disease. Studies of plant uses in circumstances where both chemotherapeutic and consciously exercised symbolic principles and human issues are at work need to be examined further.

DIVINATION: DIFFERENTIATING WHAT "JUST HAPPENS" FROM "AGENCY-CAUSED" MISFORTUNE

As important as practical medicines is the pervasive concern that Kongo therapy managers spoke of as "something else going on" (Janzen, 1978). There

Figure 1 Kongo healer-herbalist Kitembo of Balari commune in North Manianga stands in a forest clearing with a handful of leaves collected from wild plants, as is most of his pharmacopoeia. However, other Kongo healer-herbalists tend botanical gardens for their most frequently used herbaria (Photograph by J. Janzen).

is a shift from pragmatic to ritualized therapy that occurs because the misfortune or affliction is perceived to be fraught with anxiety and fear of pollution by both human and superhuman conflict (Prins, 1981). This shift amounts to a purposeful amplification in practical care with affective symbols referring to the human dimension, to spirits, and to efforts to manipulate them. Usually only consecrated persons are considered capable of handling such powerful therapies as the purification of polluted persons and settings, making sacrifices to ancestors or neutralizing menacing spirits.

The pervasiveness of divination in treating African sickness and misfortune attests to the importance of causation, especially the suspected shift in cause from a mundane to a highly charged cause in the human or spirit realm. Usually consultation with a diviner is not undertaken until there is sufficient reason in the kin group of the sufferer to suspect causes other than natural ones. Such a precipitating factor may be the worsening turn of a sick person, a sudden and mysterious death, the coincidence of a sickness with a conflict in the close social environment of the sufferer, or the paradoxical occurrence of

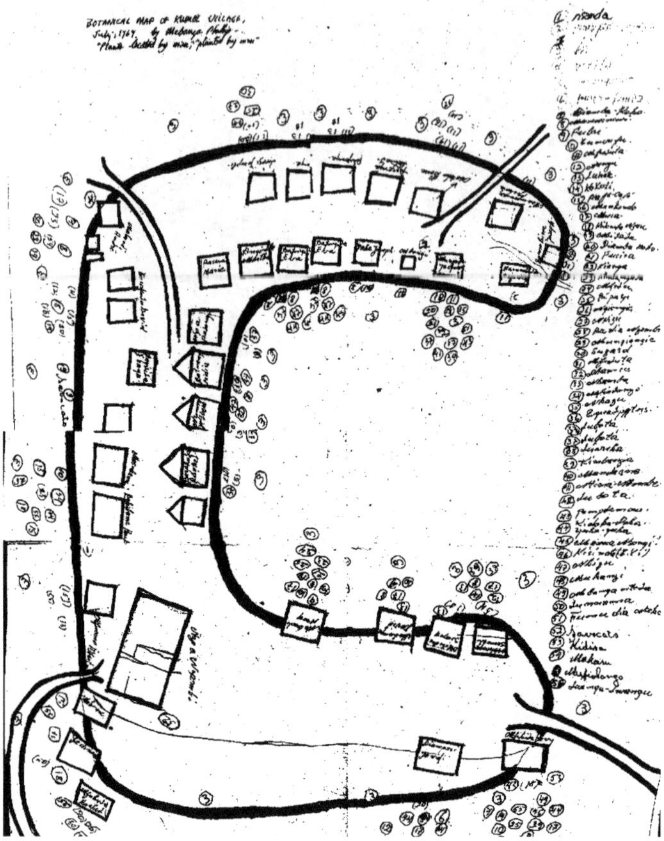

Figure 2 A 1969 botanical survey map by Mabanza Philip of "plants planted by man" in Kumbi village in North Manianga, Lower Congo, Democratic Republic of Congo. The map shows 56 varieties of shrubs and trees for food, medicinal, ritual and other purposes. The *ba*, oil palm (3) part of the ancient West African domestic plant set, is used for cooking oil, palm wine, and raffia fibre. Others, such as *payi-payi* (25) and avocado (52) are much cherished edible plants. *Kienga* (22) and *lubota* (42) are potent medicinal plants; *lemba-lemba* (6) is a sedative and a symbolic plant of the entrances; *mpese-mpese*, a poplar, used to outline chief's courtyards, here marks the men's lodge. *Kidiza* (53) is a cactus whose poisonous sap is used to catch fish and for emetics. The *nsanda* fig tree (*Ficus bubu*, sp. = Moraceae) (1), a harbinger of water, thus of a good village site, is said to be the first planted to determine the quality of the site. A botanical map such as this reveals the 2000-year history of plant domestication and agriculture, just as the illustration of healer Kitembo with wild plants in his hand suggests the continuing history of foraging. Both modes of making a living are joined in Kongo culture and medicine.

a disease on only one side of a family. In such cases the clients are looking for answers to questions not only of "Why did it happen?" but "Why did it happen to us?" and possibly "Who caused it?" and "What should we do about it?"

Scientific explanations of health may not necessarily lay to rest these questions, which are of a different order from the ideas in natural causation. A community may know very well that the spirochete transmitted by the bite of an anopheles mosquito causes malaria in the blood of a human. But the diviner may shed light on the question of why some people are infected and not others, or why some died when all were infected. Divination may also clarify the human causes behind accidents or provide a pattern with which to explain them. Western medicine is often good at answering "why?" but not "why me?"

In the broad West Africa belt from Central Nigeria to Ghana, the prevailing mode of divination is known as Ifa. A cup bearing a set of usually 16 cowries or pods is thrown out into a tray. The combination of "ups" and "downs" is coded to indicate a set of verses, numbering in the thousands, which illuminate the life situation involved in the affair before the diviner. The tray or the cup usually bears the image of Eshu Elegba, the trickster, who is believed somehow to hold in his hands individuals' and families' fortunes. At least he attempts ceaselessly to surprise humans with contradictory and unintended turns of events, often for the worse. Thus he and his character of trickery, deceit, and surprise embody the essence of what divination seeks to illuminate (Pemberton, in Pelton, 1980: 136).

On the Southern Savanna, from the Atlantic coast southeastward to the Copper Belt, the Ngombo basket mode of divination is common. Its thorough integration into the societies suggests that the genre may be a thousand years old. Carved figurines and natural objects, representing human situations and predicaments, lie together in the basket. As the basket is shaken, one of the objects emerges at the basket's rim between two lumps of clay, one red, and the other white. This "gateway" of white and red suggests the liminality of the threshold between the visible and the invisible spirit world. The diviner reads the case before him in the light of the emergent object or the constellation of objects in the basket (Turner, 1975: 315–316).

In Southern Africa a common mode of divination is a bag of animal bones and perhaps seashells (brought or traded from coastal areas) which are shaken out and thrown onto a mat before the diviner and the client (Figure 3). The bones, whose constellation represents issues in human life, relationships, and the world of spirits, may be combined with trances to indicate a complex hierarchy of causation behind the surface realities of a misfortune.

These and many other types of divination in sub-Saharan Africa are predicated on the assumption that sickness or other misfortunes may be caused by an untoward turn of events in the human or related spirit world. The immediate cause or agent such as the sign or symptom of disease is thought to require interpretation in the light of ultimate natural, human or spirit agents. Thus, despite widespread acceptance of modern science, divination continues to be a common method for discerning the dividing line between that which "just happens" and the human or mystical factor that is seen as important in the pattern of misfortune.

SCIENCE OR MAGIC? A HISTORY OF SCHOLARLY DEBATES ABOUT ETIOLOGY IN AFRICAN MEDICINE

Anthropologists and other scholars have debated the nature of African medicine and thought ever since David Livingstone published his debate with an

Figure 3 Divination by means of throwing "bones" in Swaziland (Photo by E. Green).

Mbundu rain maker (Janzen, 1978: 38–40), designed to show that the latter, although rational, was arguing with false premises. All participants in the anthropological debates accept empirically effective medicines, as mentioned earlier. The debate focuses on the nature of the other logic – the human and spirit logic – and the relationship between these and the empirical or "natural" realm of causes and cures. The arguments range along a spectrum from prioritizing the empirical treatments to charging that witchcraft overrules other causalities in African medical thought. A range of terms has been put forward to identify the several logics that work together in African healing thought, as for example naturalistic, personalistic, God-caused, or human-caused.

Many scholars of African medicine today would not be likely to use Foster's global distinction between "personalistic" and "naturalistic" treatments (1976: 775), because it does not fit well. Illnesses that "just happen" – we would say naturally – are attributed to God, a personalistic force. Nor would this dichotomy very readily do justice to impersonal ideas of pollution brought about by exposure to the dead, to certain diseases, and brushing one's feet against polluting substances. Similarly, few would accept Murdock's global survey of theories of illness, in which African societies, based on the survey's reading of available ethnographies, demonstrated a prevalence of supernatural (including fate, ominous sensations, contagion, mystical retribution, soul loss, spirit aggression, sorcery, and witchcraft) over natural (including infection, stress, deterioration, and accident) etiologies (1980: 48). Nevertheless, British anthropologist Robert Pool, who has studied Cameroonian societies (1994), has joined Murdock and scholars of other disciplines, missionaries, travelers, government administrators (pre- and post-colonial, foreign and African), doctors and health officials, and economic development professionals, who have taken these simplistic dichotomies and characterized African health beliefs as operating primarily, or solely, in the domain of "personalistic" or "supernatural" shaped witchcraft, sorcery and/or spirits.

E.E. Evans-Pritchard's classic on Azande ideas of misfortune (1937: 67) would seem to endorse their views. Yet Evans-Pritchard has been much misrepresented by his excerpt on the Azande granary's collapse being attributed to witchcraft rather than natural causes. A careful reading of Evans-Pritchard reveals that he describes a "hierarchy of resorts ranging from simple to serious, with recourse first to empirical treatments, then to magical interventions" (Janzen, 1981: 188–189). Yet many scholars of African healing and religion, who have experienced attributions to witchcraft for events Westerners would say were caused by gravity, germs, or sheer coincidence, prioritize this explanation over one in which events merely occur because they occur (e.g., see Turner on the Ndembu, 1967: 300–301).

A third group of anthropologists began to find evidence of empiricism and rational, logical thought in African ethnomedicine (Horton, 1967; Fortes, 1976; Yoder, 1982; Morris, 1997). Horton in particular sought parallels between African and Western thought, including in the domain of health and illness. Anthropological opinion has changed considerably since the 1970s, in part because of the involvement of anthropologists in applied research of infectious diseases such as child diarrhea and sexually transmitted diseases (Inhorn and Brown, 1998; Green, 1999).

For example, recent research suggests that while magico-religious or supernatural ideas may often be associated with mental illness and certain other conditions, naturalistic etiologic notions rooted in empiricism are often found to underlie the infectious and contagious diseases that have always accounted for the greatest morbidity and mortality (Green, 1999). Diseases such as malaria, tuberculosis, schistosomiasis, cholera, amoebic dysentery, AIDS and other sexually transmitted diseases, typhoid, acute respiratory infections including pneumonia, yellow fever, leprosy and dengue tend to be understood within a framework that may be called indigenous contagion theory. In this analytic framework, one becomes ill because of impersonal exposure. One comes into contact with something that anyone could come into contact with, not because an avenging spirit or an ill-intended person singles one out for misfortune in the form of sickness (Green, 1999).

A fourth and final group of scholars has sought to formulate the relationship between disparate types of logic and misfortune causation in African thought. Morris notes that Chewa medicine includes an "empirical herbalist tradition, based on a belief in the intrinsic efficacy of certain plant and animal substances." Yet it also includes "a cosmological tradition, which sees the human subject as a microcosm of the world and in which health was seen as restoring a balance or mix between certain vital 'humors' or principles, and a tradition that focused on 'communal rites of affliction', and involved spirit healing" (Morris, 1997: 86). How do these multiple realms of African healing relate to each other?

The late Rwandan scholar and physician Pierre-Claver Rwangabo offers an insight into contemporary African thinking on the question. Even though not all aspects of the Rwandan medicine system are amenable to modern science, Rwangabo believes that it is a part of modern reality rather than a fossil. He divides the causal domains of Rwandan medicine into "physical" and "mystical" causes. Diseases range across a variety of types which may be attributed to either causal category or to both. Rwangabo's medical training is evident in his listing of disease classes that include: parasitic diseases, microbial diseases, systemic diseases and bodily accidents, gynecological and obstetrical diseases, and psycho-mental and behavioral diseases. But under the latter group he identifies current psychopathologies that entail abnormal behavior as understood in traditional thought and diseases believed to be caused by broken prohibitions and beliefs about ancestors (*abazimu*) and other spirits (*ibitega, amahembe, nyabingi, amashitani, amajini*) which often are identified in relation to mental illnesses. "Poisoning", the result of human aggression, is a major aspect of the human source of misfortune. Misfortunes brought on by the breach of social rules also have a mystical though not necessarily mysterious causal character. Rwangabo's insight into the character of traditional medicine lies in the observation that most pathologies may have both a physical and a mystical dimension. This affects the way therapy will be arranged. The decision to seek physical or other therapy has to do with the context in which it occurs, its severity, the suspected human etiology, and response to treatment.

This emphasis on the context of the causal attribution makes all the difference in how sufferers, their therapy managers, diviners, healers and medical practitioners will treat illness. If the misfortune is considered to be ordinary and predictable, it will be seen along the lines of the material world. If catastrophic forces or circumstances have precipitated it, or if it seems to be the result of the chaos of underlying affairs in the human and mystical realm, it must be handled differently. Thus the same condition may need to be treated with different medicines. The first realm we might term "natural", the second "unnatural". But this dichotomy requires closer examination so as not simply to read into it influences of Western thinking.

In widespread sub-Saharan African parlance in the 20th century, the natural realm is associated with God, or caused by God – the God of the created universe (Turner, 1969: 52; Orley, 1970: 137; Swantz, 1970; Ngubane, 1977: 22–24; Gilles, 1976: 358–369; Janzen, 1978: 44–49; Davis, 2000: 94–95), not the God of a mechanistic Enlightenment world nor the God of puritanical retribution for human sin, as in the African Islamic or Christian view. Rather, God-caused misfortune is widely seen to be the created order of things such as the seasons and rhythms of birth and death in society and in the surrounding world. The death of an elderly person would be "in the created order", whereas the death of a child-bearing mother, for example, would be seen as "unnatural", or caused by some other human or spirit force. The unnatural source of misfortune – which may be manifested in an otherwise scientifically understood disease – may be attributed to human error or malicious motive, arising from the many inconsistencies inherent in human society, or the deceptive, anti-social nature of some individuals. This view of humanity often includes the ancestors or demigods who have a vested interest in the outcome of human affairs, especially in their clans and localities.

PARADIGMS IN AFRICAN MEDICAL THINKING

Sub-Saharan African understandings of health, sickness and healing are often couched ideas about the nature of the world and life within it, ideas or images which offer powerful metaphors with which to make sense of suffering and uncertainty. These ideas are discernible in verbal concepts that have a deep history and broad geographical and cultural distribution and a continuing use in diagnosis, the formulation of the sickness experience, and in therapies.[6]

The first organizing idea defines an ordered structure of the body. Any disruption, negation, or distortion of this ideal suggests sickness, as expressed in a verbal cognate that extends from West to Equatorial Africa – *eela* or *ele* in Yoruba of Nigeria (Buckley, 1985) and *beela* in Kongo of Western Zaire (Janzen, 1978 and 1992). This notion often relates to a non-verbal code of three or more colors and related substances used in medicine and ritual to situate the body within a wider cosmology. Chalk or kaolin from stream banks is used to represent purity and wholeness; camwood or other red is used to represent transition and danger. Charcoal or a substitute represents human chaos. White is associated with goodness and the legitimacy of the created order. Chalk is the most widespread of these colors used in healing and ritual. For example, chalk is smeared onto the face and body to represent the presence of ancestors, white beads are strung around the body or head to represent spirit possession, flour or another white powder is used to trace the outlines of sacred space, and white sap may allude to milk or semen as sources of life. Whiteness represents the realm of the water, the spirits, and the beyond. It usually stands for the clarity and goodness of the spiritual world where human life is rooted. Usually, the colors are used in combination. Redness, smeared on the face and body, dyed into cloth, or covering a ritual object, appears alongside whiteness to suggest the juxtaposition of "the white" with transition and danger. The combination suggests power that can build up the created order but can also destroy it. Hierarchies of spirits may appear represented in these two colors. Charcoal, the remains of fire, represents the chaos and destructiveness of human life left to itself. In association with white, this chaos and energy is balanced by the clarity, legitimacy, and order of spiritual truth. Victor Turner wrote the classic study of this color code in sub-Saharan African ritual about the Ndembu of northern Zambia (1967: 59–92).

In a second idea, balance or harmony is necessary to a state of health in the relationship between an individual and the people surrounding him or her, as well as between the human community and the natural and spiritual environment. The verbal concept *lunga* refers to a principle of health in the Zulu society of South Africa (Ngubane, 1977) and the Kongo society of Western Congo (Janzen, 1978) and to a type of healer in the Luba-influenced Tabwa culture of the western shores of Lake Tanganyika (Davis, 2000). In regions influenced by Galenic humoral theory via Islamic medicine, balance may have the connotation of an equilibrium between the humors and between heat and cold. Balance between opposing humors or fluids leads to health, imbalance to disease.

The idea of balance or harmony often extends to relations with kinsmen, neighbors and others with whom people come into regular social contact. Social *dis*harmony leads to resentment, anger, envy, evil wishing, cursing, and gossip, and often finds expression in sorcery and witchcraft. Cases of illness or misfortune that seem particularly dramatic or ambiguous are commonly a cause for divination.

A third concept is that of "coolness" as grace and health, in contrast to the "heat" of conflict and ill-health (widely distributed in sub-Saharan Africa under the term *-*pod*- in connection with cooling down, being cured). Conflict and anger are often associated with heat and fire that, like energy out of control, can cause great destruction, disease, and death (among the Ndembu of northern Zambia: Turner, 1969: 27–31; among the Tabwa of eastern Congo: Davis, 2000: 58–60). Medicines that extinguish the fire or cool the situation or sufferer are administered as an antidote to the dangerous condition. A balance between the extremes of fire and the cold of death are regarded as important for health and life (in the widespread *ngoma* ritual of healing, Janzen, 1992; among the Hemba of eastern Congo: Blakeley and Blakeley, 1994). The sub-Saharan African concept of the "cool" extends an aesthetic notion widely seen in the arts and in human relations into a definition of health.

Fourth, the widespread concept of "flow and blockage" is the closest to a classical African anatomy that inspires clear ideas about health and infuses numerous therapeutic practices. The prevalence of purgatives and emetics, fertility medicines, and herbal drinks in the African tool kit reflects this conceptual scheme (Janzen, 1978: 170). However, the physiological coordinates are usually linked to the wider world of a person's relationships in society, if not to society itself as a body. In fact, an outside observer sees a clear homology between the physical realm of the body and exchanges in society. Both are seen as needing to flow openly to thrive. Just as food and fluids need to be ingested for the physical body to be healthy, so the body social needs to be fed with reciprocal gifts and gestures of good will. Grudges, envy, and ill will in the social body are seen to cause blockage in the physical body (Taylor, 1992).

A fifth idea expresses purity, a ritual state in which the dimensions of the human world are in order; its opposite is a state in which these affairs are out of order, causing ritual pollution or sickness. Several Africanists have noted the importance of the antonymous concepts of pollution and purity and have noted practices related to these ideas in widely distributed Bantu-speaking societies (e.g., among the Kongo of Western Equatorial Africa: Janzen, 1982; the Lele of southern Congo: Douglas, 1966; among Nguni-societies of southern Africa: Hammond-Tooke, 1981a, 1981b; Ngubane, 1977; and in Mozambique and elsewhere: Green, 1999). Pollution commonly results from mystical contamination, which in turn is caused by death (including abortion) and women's reproductive processes (birth, menstrual blood, breast milk). The distinguishing feature in all of these examples seems to be an absence of spirits, witches or other malevolent humans as the cause of distress. Pollution is an impersonal condition that can be righted by purification rituals or curative interventions. In the course of years of applied ethnomedical research in Africa, Edward Green came to believe that purity and pollution concepts represent a traditional set of natural contrasts that may have served in the past as a foundation for ideas of health and disease prevention.

Finally, sixth, there are the related notions of contagion and protection, or immunization. That microorganisms can spread diseases is in fact a relatively old idea that goes back well before the theory of microbial vectors of disease. It is a widely held notion in African thinking about health, although that which is believed to spread and infect may include ill will, poison, malefic medicines, and a variety of forces which may cause harm. A well-known example of this idea is the case of pre-colonial Kenya, where smallpox-infected communities were quarantined as a health measure and the healthy were immunized with a bit of fluid from the pustule of an infected individual (Dawson, 1992: 96).

In most societies where this theory is found, the agents of infection are described as worms or tiny insects, e.g., *kadoyo* among the Bemba (Zambia), *iciwane* among the Zulu (South Africa), *liciwane* among the Swazi (Swaziland), *atchi-koko* among the Macua (northern Mozambique), *khoma* among the Shona (central Mozambique), and *kokoro* among the Yoruba of Nigeria (Green, 1999; Foster, Osunwole, and Wahab, 1996). Sexually transmitted diseases are often thought to involve such infection agents.

Some scholars speak of African contagion as mystical, comparing it to pollution (see Murdock, 1980). Yet it is not so mystical when closely examined. People become ill as a result of contact with, or contamination by, a substance or essence considered dangerous because it is unclean or impure. Africans considered in an unclean or polluted state are often kept apart from other people, since they are considered contagious until ritually purified. In central Mozambique, people believe that several kinds of child diarrhea and/or dehydration are caused by contact with polluting essences (or by eating bad or spoiled food). One source of pollution that may appear mystical is unfaithful behavior on the part of a parent: if a mother or father commits adultery, he or she acquires a contaminating essence that makes the child sick. The immediate cause is physical contact with the child, or drinking "hot", "spoiled" or "contaminated" breast milk. This belief reinforces the importance of fidelity in marriage (Green, 1999: 13).

A third component of contagion theory, environmental danger, is based on the belief that elements in the physical environment can cause or spread illness.

One expression of this is the notion that contagious illness can be carried in the air or wind. For example, the Bemba of Zambia believe that tuberculosis is an "illness in the air", spread by inhalation of unclean dust carried by the wind. The Bambara of Mali classify smallpox, measles, and other contagious illness as "wind illness", because only wind has sufficiently widespread contact with the body to cause outbreaks (Imperato, 1974: 15). *Tifo temoya* (illness in the air) is a general Swazi term denoting illnesses that are contracted through inhalation. Colds, influenza, tuberculosis, severe headaches (probably malarial) and some types of contagious child diarrhea are examples (Green, 1999: 189–190).

Recognition of this sort of contagious disease, along with sound preventive practices, can also be found in indigenous African veterinary practices. McCorkle and Mathias-Mundy (1992: 67) have noted:

> African herding strategies often reflect a highly sophisticated understanding of contagion and immune responses. For example, Fulani may move upwind of herds infected with hoof and mouth disease in order to avoid contagion; or they may move downwind so as to expose their animals to the disease, knowing that a mild case confers immunity.

HEALTH IN THE SOCIAL FABRIC: SHRINE COMMUNITIES AND CULTS OF AFFLICTION

The texture of society is extremely rich in sub-Saharan Africa. Social codes and the power of words are considered important to shaping health. A widespread verbal concept whose root is reconstructed as **-gidu-* refers to the role of social prohibitions, taboos, and the consequences of their violation. This is mentioned with reference to the restriction on eating or killing one's clan or individual totems and familiars. Other observers note that these prohibitions help individuals adhere to social codes in general, including health promoting restrictions on such things as over-consumption of alcohol, overeating, or health destroying excesses of any kind (Rwangabo, 1993).

Another aspect of the social dimension of health is bound up in the role of human agency in sickness or misfortune, in the action of anger or ill will in a relationship, and the action or gesture to harm the other, be it an injurious word, a blow to the head, or a bit of poison in the drink or food. The single root that most frequently refers to this "human cause" of misfortune is **-dog-* or **-dok-* (part of the Proto-Bantu lexica of at least 3,000 years ago), modern derivations of which are found from Cameroon and the Kongo coast in the west (KiKongo: *kuloka*), Central Africa (Kinyarwanda: *kuroga*), to the Nguni speakers in South Africa (Zulu: *kuthaka*).

In contemporary diagnoses of misfortune, victims will often identify a string of misfortunes and try to recall the exact words spoken by others prior to or in association with the events, drawing the logical inference that these utterances had caused, or could have led to, the bad luck. Words of warning or injurious words spoken in anger are especially suspect. Therefore, in divination, these moments are recalled so that the individuals or the relationships may be repaired. Without treating the root cause, the surface signs and symptoms cannot be permanently overcome.

There have been collective healing rites based on common afflictions or groups of devotees identified with gods or spirits considered to be the cause or medium of these afflictions. Such communities or orders have arisen and declined over time as particular constellations of afflictions have occurred. Often they have been a kind of barometer of the major dislocations and diseases in a region.

In West Africa, shrine communities and cult memberships were often associated with major shrines and cults to the earth, to the water, to nature, and to the sky, as well as to the rulers of local cities and states. Some shrines were addressed to specific conditions, such as twinship (as in the Ibeji of Yorubaland), or particular diseases, such as the very widespread shrine complex to the god of smallpox (variously Ipoona, Shapanna, Shapata). On the Guinea coast forms of gender-divided societies developed around the female Sande association and the male Poro association. These addressed many aspects of health and the public good such as instruction of youth, midwifery, social control, and hygiene. These shrine societies and cults of West Africa may well have been part of very early agrarian society, having become an integral feature of the cycles of celebrations and sacrifices.

Across Equatorial Africa this type of therapeutic ritual assembly often centers around particular issues as well, such as fertility, twinship, women's reproductive issues, the health and wellbeing of infants and children (Turner, 1968 and 1975; Devisch, 1993; Spring, 1978), debilitating chronic conditions, fortune and misfortune for men in hunting, mental illnesses, the survival of kin groups (Janzen, 1992; Nisula, 2000), and a range of social and environmental issues (Janzen, 1992; van Dijk, 2000). Membership was usually made up of the afflicted and formerly afflicted, who underwent a therapeutic initiation with stages from sufferer-novice to healer-priest. These "drums of affliction", as Victor Turner (1968) dubbed them, are often associated with the voice of the ancestors and spirits which inhabit the celebrants and are expressed in the song-dances at the core of the ritual performances. Sometimes the mark of growth or healing in the sufferer-novice is the creation of a personal song based on the ordeal of suffering, a dream/vision, or other moving experience. Such a song constitutes a unique set of powerful words, recalling the cognate *dok*, that offset and overcame the destructive forces of disintegration, misfortune, sickness and chaos of the previous period of the individual's life. Where such a "drum of affliction" addresses community issues, the healing ritual may be directed to the community, and society becomes the body that is cured.

These cults and shrines have related to Christianity and Islam for centuries. Sometimes the African institution has absorbed the outside idea or symbol; in other cases Christian and Islamic institutions have recreated the African forms and substance. Especially widespread in sub-Saharan Africa are the Independent African Churches, many of which encourage healing, exorcisms, and various kinds of incorporating rites of purification, protection, and sanctification (Sundkler, 1976; Jonker, 2000). Prophet-founders play the role of ancestor-mediators, while prominent or talented members assume the diagnostic role of diviners.

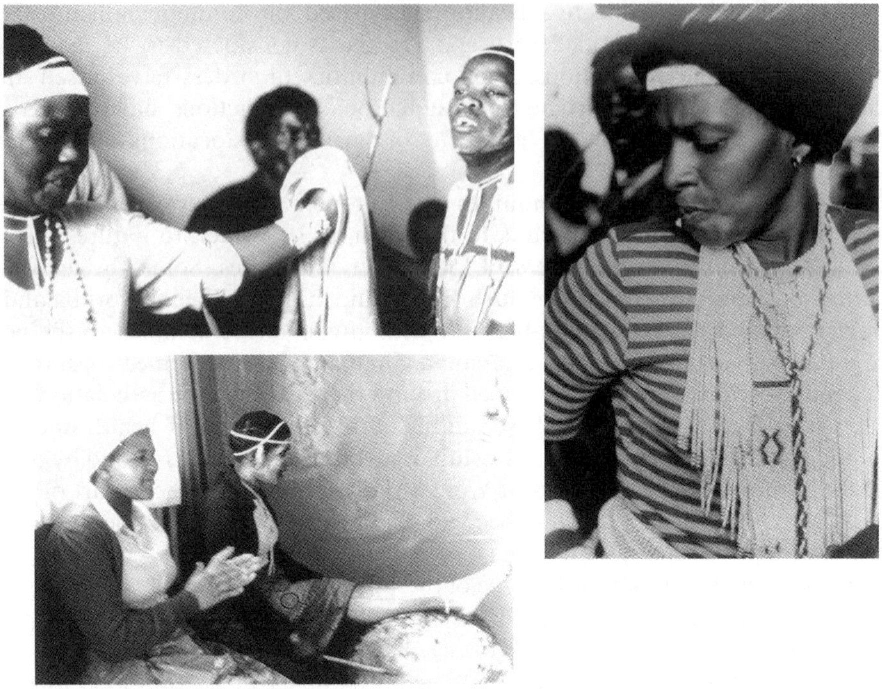

Figure 4 "Doing *ngoma*" in Guguleto Township, Capetown, South Africa, 1982. This composite shows the essential elements of a very widespread Central and Southern African therapeutic ritual. Lower left, two novices provide the drum rhythm; upper left, other novices sing-dance and "confess their dreams" to novices and a few senior healers; right, a trained graduate of *ngoma* leads this particular event with her bold sure step and her colorful clothing and beadwork, which represents her well-developed sense of self in contrast to the novice-sufferers, who are entirely "white" (Photograph by J.M. Janzen).

Although orthodox Muslims frown on blending Islam with African indigenous religion, the interpenetration of Islam and African ritual healing is extremely common (among Hausas of Northern Nigeria, see Abdullah, 1992 and Wall, 1988). *Jin* and *amasheitani* spirits widely cohabit the spirit worldviews of *ngoma* associations in Eastern and Central Africa. Muslims healers of the Swahili coast have long practiced ngoma as part of their medicine kit, along with reading the Koran; the purification symbolism of African healing merges with that of the ritual ablutions of Islam in connection with prayer. In northern Swahili towns such as Lamu, early 20th century ngoma Maulidi was introduced for performance in the mosque; its songs celebrated the prophet Mohammed, much to the chagrin of fundamentalist Muslims.

NEW HEALTH CRISES AND THE RE-LEGITIMIZATION OF AFRICAN MEDICINE

The financial crises of African societies, and the search for an infrastructure of health, has led planners to take a second look at African institutions such as cults of affliction and the education of healers. By 1998, findings from a survey by the World Health Organization's Regional Office for Africa showed that a national management or coordination body for traditional medicine activities existed in 17 of 30 countries surveyed. Twenty-two indicated that associations of traditional healers had been established, and 10 said that a national directory existed. Four countries reported that a training program for healers existed (the actual number would surely be higher) and 17 countries had such programs for traditional birth attendants. Twenty countries indicated that institutions in their country were carrying out research related to traditional medicine. Fifteen reported that there was local production of indigenous medicines, and 17 countries reported having botanical gardens or arboreta for cultivating medicinal plants (WHO, 2000).

Research related to the medicinal value of plants in the African materia medica is important for several reasons related to public health: (1) achieving community or national self-reliance in health by promoting locally-available and already-accepted herbal medicines; (2) developing an indigenous pharmacological industry, based on local plants, that reduces national dependence on expensive, imported drugs; (3) promoting natural health care and reducing the iatrogenic effects of modern medicine; (4) finding out what traditional healers are using and, if medicines are found to be dangerous and highly toxic, trying to persuade healers to substitute safer plants or at least reduce dosages; or (5) finding an effective-seeming indigenous medicine for the symptoms of a high-priority illness such as child diarrhea, then finding ways to promote more widespread use of the medicine (Green, Jurg, and Djedje, 1994: 44).

Preliminary pharmacological research is showing that phytomedicines (medicines derived from plants) used by African healers may indeed be effective for diseases in two of the most severe current public health crises: schistosomiasis and childhood diarrhea.

Ndamba *et al.* (1994) analyzed the most commonly used plants used to treat schistosomiasis by 286 traditional healers in Zimbabwe, administering the crude extracts orally to hamsters infected with *S. haematobium cercariae.* It was found that plant extracts from *Abrus precatorius* (Leguminosae), *Pterocarpus angolensis* (Leguminosae) and *Ozoroa insignis* (Anacardiaceae) were lethal to adult schistosomes. In a study of HIV-positive Ugandans with Herpes zoster, herbal mixtures used by healers were found to be at least as effective as biomedical treatments, including the antiviral drug acyclovir, in treating symptoms (Homsy *et al.*, 1999).

Recent phytochemical research has shown that the roots of *Mirabilis jalapa*, used in South Africa as a purgative to treat some child diarrheas, in fact exhibit antibacterial activity against an impressive range of diarrhea-causing pathogens: *Staphylococcus aureus, Streptococcus pyogenes, Escherichia coli, Enterobacter sp., Vibrio cholerae, Shigella flexneri* and *Salmonella typhi* (Chifundera, Kizungu and Mbuyi, 1991).

At the World Health Organization Forum on Traditional Medicine in Health Systems held in Harare in 2000, the African Regional WHO office expressed keen interest in the mass production of phytomedicines for the treatment of malaria, AIDS, and other diseases identified as priority diseases by member states. It is a strategic objective of the WHO to develop a framework for the integration of traditional medicine into national health systems. The idea is to encourage local industry to invest in the local production of indigenous medicines and make them commercially viable. Governments were urged to create policies related to conservation, safety and toxicity, and regulation in order to assist a local production industry (Green, 2000).

Mass production, promotion and distribution of African phytomedicines have begun. For example, Nigeria has developed medicines for ulcers, anemia, contraception, malaria and HIV, and it now holds patents for some of these in several countries. In late 2000, Nigeria became the first African nation to officially promote a plant medicine for the treatment of HIV/AIDS.

The ratio of traditional healers to the general population of various African countries seems to be in the range between 1:200 and 1:800, based on surveys and censuses (see Green, 1994: 19 for a review of studies). In Mozambique, a

Figure 5 An example of African medicine given a modern scientific basis and commercialized. Dr. Byamungu Lufungula, left, a French-trained pharmacist, stands before instruments in his laboratory in Bukavu, Kivu, Eastern Congo. In 1994 his enterprise, SODIPHAR, employed 25 workers. Pharmacies in Goma, Bukavu, and Uvira sold about 20 products, all based on laboratory tested traditional medicines obtained in collaboration with herbalist healers. Byamungu is the author of *Les Plantes Médicinales, Les Rites Therapeutiques, et Autres Connaissances en Médicine des Guérisseurs au Kivu*, 1982.

doctor theoretically serves about 10,000 people. In practice, coverage is even less. Most doctors live in larger cities, while most rural Africans are lucky if they live within 5–10 kilometers of a clinic staffed by a minimally trained nurse, where medicines may or may not be available.

It is widely accepted that at least 80% of Africans rely on traditional healers for much or all of their health care. This had led some that work in public health to think that healers ought to play a role in curbing the spread of infectious disease. From a public health viewpoint, this would seem to make sense. Healers are found everywhere; they are culturally acceptable; they explain illness and misfortune in terms that are familiar. In the 1970s there were several collaborative programs involving traditional healers in areas such as child diarrheal disease and family planning (e.g., Green, 1987 and 1996; Good, 1987; Warren, 1989).

With the explosion of HIV/AIDS in East and Southern Africa, an even greater number of collaborative programs developed. In Southern Africa, once they had participated in workshops on AIDS and STDs (sexually transmitted diseases), healers proved willing to promote condoms and safe sex. Once it was accepted that standard STDs facilitate the transmission of HIV, it occurred to some that healers ought to be involved in STD treatment programs – at least through referrals to clinics – since patients with STDs so often consult healers. Moreover, avoiding AIDS by sticking to one partner usually made sense to healers because they already interpret locally recognized sexually transmitted illness as resulting (at one level at least) from violations of the codes that govern proper sexual behavior and exposure to essences believed to be polluting

Figure 6 Indigenous South African healers collaborating in public in AIDS prevention program, 1993 (Photograph by E. Green).

Figure 7 South African diviner-mediums (traditional healers) teaching each other biomedical ideas about HIV/AIDS, in a USAID-sponsored workshop in Tsitsikama, South Africa, 1993 (Photo by E. Green).

(reproductive fluids, death). They feel encouraged to learn that their own governments as well as the international community *also* wish to warn people against having sex with "just anyone", with too many people, with strangers, with prostitutes, or with someone other than one's spouse. Finally, with drugs and hospitals in short supply, healers are already caring for a large proportion of those already infected with HIV. UNAIDS, the United Nations agency for coordinating HIV/AIDS programs, recently published an official "Best Practices" paper summarizing the role of healers in HIV/AIDS prevention and support programs, which concluded that they have made a substantial contribution (King, 2000), although the situation regarding HIV in Africa remains grim.

NOTES

1 The term "Islamic medicine" encompasses the traditions of medical theory, practice and literature that have been developed in Islamic cultural contexts and expressed most commonly in the languages of Islam, principally Arabic, Persian, Turkish and Urdu. There are two medical traditions that developed in Islamic contexts. One ultimately derives its authority, and many of its theoretical and practical components, from ancient Greek and Hellenist sources, called in this essay Islamic medicine; the other encompasses traditions associated with the Prophet Muhammad, or prophetic medicine [Adapted from Guy Attewell's article on Islamic Medicine, in this volume].

2 For Janzen: Central Africa (specifically, the Western southern savanna and Kongo coastal regions, the Great Lakes region, coastal Tanzania), and Southern Africa (Swaziland and Capetown); for Green: Swaziland, Mozambique, Southern Africa, East and West Africa.

3 One of the best methods for reaching back through contemporary and recent African experience to find the faint images of ancient thought and practice is to follow the widespread vocabulary

having to do with health, the nature of disease, and the ideas and techniques of healing. Sub-Saharan healing is not codified in written texts, but it is transmitted through rich oral instructions from generation to generation. These texts and their vocabulary are as persistent and more durable – in the tropics – than papyrus or parchment. Historical linguistics is based on the premise that core vocabulary in language remains relatively constant and can be traced by comparing languages with a historical affinity. Core vocabulary shows both lexical and phonetic change at a rate of about 20% per millennium. Thus, if two languages are shown to have a 60% common core vocabulary and related phonetic structure, they are about two thousand years apart in their history. Newer techniques, ideas and related terms will show a different distribution, whose origins can sometimes be traced by examining the distribution and the phonetic transformations they have undergone. Phonetic changes occur in certain directions within the basic human sound-making potential. This "historical linguistic" methodology permits scholars to study concepts at a depth of up to five thousand years.

[4] A term headed by an asterisk (e.g. **-ganga-*) indicates that it is a widespread cognate in either the proto-Bantu or proto- regional reconstructed core of verbal concepts.

[5] See Green, 1999: 69–70 for such evidence in southern Africa.

[6] This is the most recent of a series of formulations of such broad ideas in sub-Saharan African medicine. The present essay is particularly informed by "Sub-Saharan healing" (Janzen, 1997).

BIBLIOGRAPHY

Abdullah, Ishmael. 'Diffusion of Islamic medicine into Hausaland.' In *The Social Basis of Health and Healing in Africa*, Steven Feierman and John Janzen, eds. Berkeley: University of California Press, 1992, pp. 177–194.

Balick, M.J., E. Elisabetsky and S.A. Laird, eds. *Medicinal Resources of the Tropical Forest*. New York: Columbia University Press, 1996.

Bibeau, Gilles *et al. Traditional Medicine in Zaire*. Ottawa: International Development Research Centre; Kinshasa: Centre de Médecine des Guérisseurs, 1980.

Blakeley, Pamela and Thomas D. Blakeley. 'Ancestors, "witchcraft", and foregrounding the poetic: men's oratory and women's song-dance in Hemba funerary performance.' In *Religion in Africa*, Thomas D. Blakeley, Walter van Beek and Dennis L. Thomson, eds. London: James Currey, 1994, pp. 399–442.

Buxton, Jean. *Religion and Healing in Mandari*. Oxford: Clarendon Press, 1973.

Chavunduka, Gordon. *Traditional Healers and the Shona Patient*. Gwelo: Mambo Press, 1978.

Davis, Christopher O. *Death in Abeyance: Illness and Therapy among the Tabwa of Central Africa*. Edinburgh: Edinburgh University Press, 2000.

Dawson, Marc. 'Socioeconomic change and disease: smallpox in colonial Kenya, 1880–1920.' In *The Social Basis of Health and Healing in Africa*, Steven Feierman and John Janzen, eds. Berkeley: University of California Press, 1992, pp. 90–103.

Devisch, Rene. *Weaving the Threads of Life: The Khita Gyn-Eco-Logical Healing Cult among the Yaka*. Chicago: University of Chicago Press, 1993.

de Zoysa *et al.* 'Perceptions of childhood diarrhoea and its treatment in rural Zimbabwe.' *Social Science and Medicine* 19(7): 727–734, 1984.

Douglas, Mary. *Purity and Danger: An Analysis of Concepts of Pollution and Taboo*. New York: Praeger, 1966.

Ehret, Christopher. *An African Classical Age: Eastern and Southern Africa in World History, 1000 B.C. to A.D. 400*. Charlottesville: University Press of Virginia; Oxford: James Currey, 1998.

Evans-Pritchard, E.E. *Witchcraft, Oracles and Magic among the Azande*. Oxford: Clarendon Press, 1937.

Feierman, Steven and John M. Janzen, eds. *The Social Basis of Health and Healing in Africa*. Berkeley: University of California Press, 1992.

Ford, John. 'Ideas which have influenced attempts to solve the problems of African trypanosomiasis.' In *The Social History of Disease and Medicine in Africa*, John M. Janzen and Steven Feierman, eds. Special Issue of *Social Science and Medicine* 13B(4), December 1979, pp. 269–276.

Fortes, M. 'Foreword.' In *Social Anthropology and Medicine*, J. Loudon, ed. London: Academic Press, 1976, pp. ix–xx.

Foster, G. 'Introduction to ethnomedicine.' In *Traditional Medicine and Health Care Coverage*, R. Bannerman, J. Burton and Ch'en Wen-Chieh, eds. Geneva: World Health Organization, 1983, pp. 17–24.

Foster, L.M., S.A. Osunwole and B.W. Wahab. 'Imototo: indigenous Yoruba sanitation knowledge systems and their implication for Nigerian health policy.' In *Alaafia: Studies of Yoruba Concepts of Health and Well-Being in Nigeria*, F. Fairfax, B.W. Wahab, L. Egunjobi and D.M. Warren, eds. Studies in Technology and Social Change, No. 25. Center for Indigenous Knowledge for Agriculture and Rural Development. Ames: Iowa State University, 1996, pp. 26–38.

Gilles, Eva. 'Causal criteria in African classifications of diseases.' *In Social Anthropology and Medicine*, Joe B. Loudon, ed. London: Academic Press, 1976, pp. 358–395.

Good, C.M. *Ethnomedical Systems in Africa. Patterns of Traditional Medicine in Rural and Urban Kenya.* New York: Guilford Press, 1987.

Gray, Sandra. Personal communications, 1997.

Green, Edward C. 'Collaborative programs for traditional healers in primary health care and family planning in Africa.' In *African Medicine in the Modern World*, C. Fyfe and U. Maclean, eds. Edinburgh: Centre of African Studies, University of Edinburgh, 1987, pp. 115–144.

Green, E.C. Annemarie Jurg and Armando Djedje. 'The snake in the stomach: child diarrhea in Central Mozambique.' *Medical Anthropology Quarterly* 8(1): 4–24, 1994.

Green, E.C. 'The WHO forum on traditional medicine in health systems.' *Journal of Alternative and Complementary Medicine* 6(5): 379–382, 2000.

Green, E.C. *Indigenous Theories of Contagious Disease.* Walnut Creek, California: Altamira Press, 1999.

Green, E.C. *Indigenous Healers and the African State.* New York: Pact Publications, 1996.

Green, E.C. *AIDS and STDs in Africa: Bridging the Gap Between Traditional Healers and Modern Medicine.* Boulder, Colorado: Westview Press, 1994.

Harjula, R. *Mirau and his Practice: A Study of the Ethnomedicinal Repertoire of a Tanzanian Herbalist.* London: Tri-Med Books, Ltd., 1980.

Hammond-Tooke, W.D. *Boundaries and Belief: The Structure of a Sotho Worldview.* Johannesburg: Witwatersrand University Press, 1981.

Homsy, J. *et al.* 'Evaluating herbal medicine for the management of Herpes zoster in human immunodeficiency virus-infected patients in Kampala, Uganda.' *Journal of Alternative and Complementary Medicine* 5(6): 553–565, 1999.

Imperato, P.J. 'Traditional medical practitioners among the Bambara of Mali and their role in the modern health-care delivery system.' *Rural Africana* 26: 41–54, 1974.

Inhorn, Marcia C. and Peter J. Brown. 'The anthropology of infectious disease.' *Annual Review of Anthropology* 19: 89–117, 1990.

Janzen, John M. *The Quest for Therapy in Lower Zaire.* Berkeley: University of California Press, 1978.

Janzen, John M. *Lemba 1650–1930: A Drum of Affliction in Africa and the New World.* New York: Garland Publishers, 1982.

Janzen, John M. *Ngoma: Discourses of Healing in Central and Southern Africa.* Berkeley: University of California Press, 1992.

Janzen, John M. 'Sub-Saharan healing.' In *Ancient Healing: Unlocking the Mysteries of Health and Healing Through the Ages.* Lincolnwood, Illinois: Publications International, 1997, pp. 322–338.

Janzen, John M. 'Afterword.' In *The Quest for Fruition through Ngoma*, Rijk van Dijk, Ria Reis and Marja Spierenburg, eds. Oxford: James Currey, 2000, pp. 155–168.

Janzen, John M. and Steven Feierman, eds. *The Social History of Disease and Medicine in Africa.* Special Issue of *Social Science and Medicine* 13B(4), December 1979.

Janzen, John M. and G. Prins, eds. *Causality and Classification in African Medicine and Health.* Special Issue of *Social Science and Medicine* 15B(3), 1981.

Katz, Richard. *Boiling Energy: Community Healing among the Kalahari !Kung.* Cambridge, Massachusetts: Harvard University Press, 1982.

Katz, Richard, Megan Biesele and Verna St. Denis. *Healing Makes our Hearts Happy: Spirituality and Cultural Transformation among the Kalahari Ju!'hoansi.* Rochester, Vermont: Inner Traditions International, 1997.

Kendall, C. *et al.* 'Urbanization, dengue, and the health transition: anthropological contributions to international health.' *Medical Anthropology Quarterly* 5(3): 257–268, 1991.

King, Rachel. *Collaboration with Traditional Healers in HIV/AIDS Prevention and Care in Sub-Saharan Africa: A Literature Review.* Geneva: UNAIDS, 2000.

Last, Murray and Gordon Chavunduka, eds. *The Professionalisation of African Medicine.* Manchester: Manchester University Press, 1986.

MacGaffey, Wyatt. *Art and Healing of the BaKongo.* Stockholm: Folkens Museum-Etnografiska. 1991.

MacGaffey, Wyatt. *Astonishment and Power. The Eyes of Understanding: Kongo Minkisi; Resonance, Transformation, and Rhyme – The Art of Renee Stout.* Washington, DC: Smithsonian Institution Press, 1993.

McCorkle, Constance M. 'Ethnoveterinary medicine in Africa.' *Africa* 62(1): 59–93, 1992.

Mendonsa, Eugene L. *The Politics of Divination: A Processual View of Reactions to Illness and Deviance among the Sisala of Northern Ghana.* Berkeley: University of California Press, 1982.

Morris, Brian. 'The powers of nature.' *Anthropology and Medicine* 5(1): 81–102, 1998.

Murdock, G.P. *Theories of Illness.* Pittsburgh: University of Pittsburgh Press, 1980.

Murphy, E. *History of African Civilization.* New York: Dell Publishing, 1972.

Ngubane, Harriet. *Body and Mind in Zulu Medicine: An Ethnography of Health and Disease in Nyuswa-Zulu Thought and Practice.* London: Academic Press, 1977.

Nisula, Tapio. *Everyday Spirits and Medical Interventions: Ethnographic and Historical Notes on Therapeutic Conventions in Zanzibar Town.* Helsinki: Transactions of the Finnish Anthropological Society NRO 43, 1999.

Orley, John. 'African medical taxonomy: with special reference to mental illness.' *Journal of the Anthropological Society of Oxford* 1: 137–150, 1970.

Peek, Philip M. *African Divination Systems: Ways of Knowing.* Bloomington: Indiana University Press, 1991.

Pelton, Robert D. *The Trickster in West Africa: A Study of Mythic Irony and Sacred Delight.* Berkeley: University of California Press, 1980.

Pool, R. 'On the creation and dissolution of ethnomedical systems in the medical ethnography of Africa.' *Africa* 64(1): 1–20, 1994.

Reid, W.V. *et al. Biodiversity Prospecting: Using Genetic Resources for Sustainable Development.* Washington, DC: World Resources Institute, 1993.

Rwangabo, Pierre Claver. *La médecine traditionnelle au Rwanda.* Paris: Karthala, 1993.

Schoenbrun, David. *A Green Place, A Good Place: A Social History of the Great Lakes Region, Earliest Times to the 15th Century.* Portsmouth: Heinemann, 1998.

Spring, Anita. 'Epidemiology of spirit possession among the Luvale of Zambia.' In *Women in Ritual and Symbolic Roles,* Judy Hoch-Smith and Anita Spring, eds. New York: Plenum, 1978, pp. 165–190.

Sundkler, Bengt. *Zulu Zion and Some Swazi Zionists.* London: Oxford University Press, 1976.

Swantz, Marja-Lisa. *Ritual and Symbol in Transitional Zaramo Society.* Lund: Gleerup, 1970.

Taylor, Christopher. *Milk, Honey, and Money: Changing Concepts in Rwandan Healing.* Washington, DC: Smithsonian Institution Press, 1992.

Turner, Victor, *The Forest of Symbols.* Ithaca, New York: Cornell University Press, 1967.

Turner, Victor. *The Drums of Affliction: Religious Processes among the Ndembu of Zambia.* Oxford: Clarendon Press, 1968.

Turner, Victor. *The Ritual Process: Structure and Anti-Structure.* Chicago: Aldine Publishing Company, 1969.

Turner, Victor W. *Revelation and Divination in Ndembu Ritual.* Ithaca: Cornell University Press, 1975.

Twumasi, Patrice. *Medical Systems in Ghana.* Accra: Ghana Publishing Corp., 1975.

Van Dijk, Rijk, Ria Reis and Marja Spierenburg, eds. *The Quest for Fruition through Ngoma: The Political Aspects of Healing in Southern Africa.* Oxford: James Currey, 2000.

Vaughan, Megan. *Curing their Ills: Colonial Power and African Illness*. Cambridge: Polity Press, 1991.

Wall, Lewis. *Hausa Medicine: Illness and Well-Being in a West African Culture*. Durham, North Carolina: Duke University Press, 1988.

Warren, D.M. *Disease, Medicine, and Religion among the Bono of Ghana: A Study in Culture Change*. Ph.D. Dissertation, Indiana University, 1974.

Warren, D.M. 'Utilizing indigenous healers in national health delivery systems: the Ghanaian experiment.' In *Making our Research Useful: Case Studies in the Utilization of Anthropological Knowledge*, J. van Willigen, B. Rylko-Bauer and A. McElroy, eds. Boulder, Colorado: Westview Press, 1989, pp. 159–178.

Watt, J.M. and M.G. Breyer-Brandwijk. *The Medicinal and Poisonous Plants of Southern and Eastern Africa*. Edinburgh: Livingstone, 1962.

World Health Organization. *Traditional Medicine in the African Region: An Initial Situation Analysis (1998–1999)*. African Forum on the Role of Traditional Medicine in Health Systems, 16–18 February, 2000. Harare: WHO Regional Office for Africa, 2000.

Yoder, P. Stanley, ed. *African Health and Healing Systems*. Los Angeles: Crossroads Press, 1982.

AHMED SHAFIK AND WASEEM R. ELSEESY

MEDICINE IN ANCIENT EGYPT

Ancient Egyptians had a widespread reputation for their medical knowledge. Homer's *Odyssey* mentions that the physicians of Egypt were skilled beyond all others, and Warren R. Dawson (1924) wrote, "There is no doubt that ancient Egyptians were a highly gifted people with a great capacity for practical achievement." Herodotus said that each of the physicians of Egypt was a specialist who had committed himself to the study of one particular branch of medicine. Many of the foundations of medical science were established in Egypt more than fifty centuries ago. We can document this with the rich archival material left behind by the ancient Egyptians. Time has spared for our admiration a mass of documentary evidence, the so-called medical papyri (Nunn, 1997: 30–41).[1]

PRACTICED MEDICAL SPECIALITIES

There were many Egyptian physicians, since they specialized in the treatment of one specific disease. Reportedly, there was specialization in the following fields.

Gynecology and obstetrics

According to the Kahun and Carlsberg VIII (Copenhagen) papyri, determining whether a woman would or would not be able to bear children was accomplished by keeping a bulb of onion or garlic in her *iuf* (vagina) over night until dawn. If the specific odor of either appeared in her mouth, she would be able to bear children. The scientific foundation of this fertility test is that onion and garlic contain volatile oils which pass from the cervix through the uterus to the fallopian tubes and, if these are unobstructed, reach the peritoneal cavity which has a very high absorbency to circulation. The route of excretion of these volatile oils is the respiratory tract.

For a pregnancy test, *emmer* (wheat) and *it* (barley) were moistened with the woman's urine daily. If they both grew, she would bear a child. If the barley grew, the child would be a male; if the wheat grew, it would be female. If neither grew, the woman was not pregnant. Ghalioungui, Khalil, and Ammar

H. Selin (ed.), Medicine Across Cultures: History and Practice of Medicine in Non-Western Cultures, 27–47.

put this technique to the test in 1963: there was no growth of either seed when watered with male or non-pregnant female urine. With 40 specimens from pregnant women there was growth of one or both species in 28 cases. Thus, growth seemed to be a good indicator of pregnancy. The prediction of sex was correct in 7 cases and incorrect in 16.

As they were very observant, the ancient Egyptians recognized dilated veins over the breasts as an important physical sign for early indication of pregnancy. In the Kahun Papyrus we read:

> The woman lies down while you smear her breast and her two arms and her two shoulders with new oil. You rise early in the morning to examine her. If you find her blood vessels, metu, fresh and good and none being collapsed [literally: sunken], bearing children will be hetep, [i.e. happy or satisfactory]. If you find them collapsed, this means bened [meaning unknown]. If you find them green and dark, she will bear children late.

Strangely enough this early physical sign is still mentioned in our latest obstetrics and gynecology books. The urine wheat and barley test was used in Germany in the 18th century AD (Westendorf, 1992).

Prolapse of the uterus was described in the Ebers Papyrus. The components of the remedies to be drunk, or smeared on the pubic region, or applied to the umbilicus, or rubbed on the limbs, or to fumigate the vagina, give no clue to the meaning of their use. There are many remedies for "cooling the uterus and driving out heat". Whether this heat means inflammation or pain is not clear, nor is its pathology. Again, whether *setet* of the uterus means mucus coming from the uterus or pain as suggested by Dawson (1934) is not known, and even the treatment does not give any clue to the condition.

Amenorrhea had a clear definition in the Ebers Papyrus. Not clearly defined was excessive bleeding (menorrhagia), but there were drugs to draw out the blood from a woman. Discharge from the uterus, *khaau*, was also described.

Breast diseases were included in the gynecological section of the Ebers Papyrus. The most important description was that of breast cancer. "If you examine a breast swelling with the flat of the hand and you feel it like an unripe pear, tell the relatives of the patient: this is a disease for which I cannot do anything. It is humanitarian to tell the relative, not the patient". It speaks highly of the achievements and ethics of ancient medicine that our medical students are taught the same description of clinical examination (Shafik, 1998).

Giving birth

Women were delivered while squatting on two large bricks or while sitting in a chair from which the center of the seat had been removed. Two or three women, not *SWNW* (doctors), assisted women in labor, at least among the nobility. Most of the few Egyptian pictures of birth show the infant emerging head first, as is normal. It is important to mention that labor in a squatting position or on a bottomless chair is considered far less painful than in the conventional supine position in bed (Fayad, 1996).

The first specialized hospitals for antenatal care were established in Egypt 4000 years ago. A separate room called *Mameze* was built in the house garden or upper story to isolate the mother for two weeks after giving birth and to protect her from puerperal sepsis.

Contraceptives

The ancient Egyptians practiced family planning. For example, contraceptive devices of different shapes and sizes were inserted into the uterus. They believed that semen was formed in a man's heart and stored in the holy bone called the sacrum. They said, "He gave her some of his heart" and called the role of man in pregnancy the "beautiful role". There is a temple wall showing a sperm beside an ovum, then the division of a fertilized ovum into two, then four cells! (Imam, 1999).

General surgery

Ancient Egyptians practiced topical anesthesia, a necessity for minor operations. By putting vinegar (acetic acid) in a certain concentration over marble stone from Memphis, a cooling effect of carbon dioxide resulted from the interaction with acetic acid. The marble stone thus had an anesthetizing effect.

Within the first six hours after an injury, wounds were sutured with a material manufactured from catgut. If a wound became infected, healers placed fermented pulp of bread inside the wound (acting much as the antibiotic penicillin), or honey, which can kill bacteria by its hygroscopic power, readily absorbing moisture from the wound. If the wound was bleeding, they resorted to packing the wound with flesh (muscle), because they had found that muscle fibers of fresh meat contained coagulants which helped to stop bleeding. This procedure is described in modern textbooks such as Farqharsson's *Operative Surgery* and practiced up to the present day (Elseesy, 1999).

Circumcision

The procedure of male circumcision is very clearly shown in the famous relief on the east doorway to the tomb of Ankh-ma-hor (King Teti, Old Kingdom, 6th Dynasty, 2345 BC) (Figure 1). This relief carries several inscriptions:

(Priest to assistant): Hold him fast! Do not let him fall!
(Assistant to operator, i.e., priest): I shall act for your praise!
(Operator to patient): I will make it comfortable!

Circumcision in Egypt belonged to the domain of the priest rather than of the surgeon. The hygienic aspects of circumcision have been used as arguments to understand its origins. Another possible reason for circumcision is the concept of bisexuality, originating in priestly speculations over the creation of the world by specific gods whom they considered both father and mother of mankind. Accordingly every human being was believed to have both male and female souls. The male soul in girls was located in the vestigial phallus, i.e. the clitoridial eminence, and the female soul in boys was placed in the prepuce, taken to represent the labia. Hence young individual adolescents had to shed their heterosexual outfit before qualifying as integral members of their sex

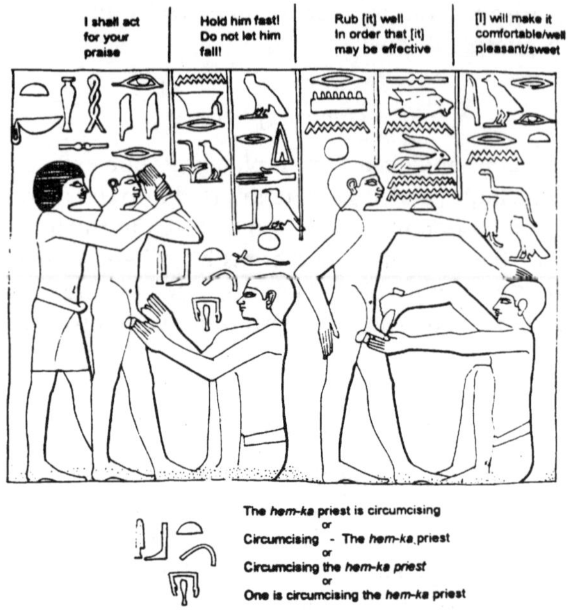

Figure 1 The circumcision in the tomb of Ankh-ma-hor, 6th Dynasty, Saqqara. (From Nunn, John F. *Ancient Egyptian Medicine*. London: British Museum Press, 1997, p. 170.) Drawn by Richard Parkinson of the British Museum; reproduced with the kind permission of Dr. John Nunn.

community. Herodotus reported that the Egyptians were the first to circumcise children, well before the Syrians and the Phoenicians, and that the Hebrews acquired this custom from them (Shafik, 1998). Female circumcision, or female genital mutilation as it is better known, was not practiced in ancient Egypt (Elseesy, 1999).

From the early Dynastic Period, ancient Egyptians were capable of superb needlework, the oldest surviving needles being Predynastic gold needles dating to the 1st Dynasty and on display in the Manchester Museum (Petrie, 1917; Petrie, 1907). Stitching of wounds was recommended in the Edwin Smith Papyrus in the following seven cases: gaping wounds of the head, gaping wounds above the eyebrows, wounds of the nostril, flesh wounds of the ear, wounds of the upper lip, gaping wounds of the throat, and gaping wounds of the shoulder. Breasted (1930) advanced the view that adhesive bandage was used on minor lacerations as is still widely used today. No stitching was used in infected wounds. Treatment was confined to local applications of various herbal and mineral remedies, which included "green pigment". It is believed that "green pigment" stands for green malachite, which, since it contains copper salts, could be expected to have some antibacterial effect, especially against *staphylococcus aureus*. Barley bread was recommended to dry the wound.

Ancient Egyptians practiced trephining, the process of perforating the skull with a surgical instrument. Breasted (1930) and Ghaliounghui (1973) reported skulls of mummies with well-healed edges, which means that patients lived after a trephine operation. This operation required great skill in not injuring the dura mater (covering of the brain), because exposure of the brain would be fatal. The instruments used for trephinations were hammers and chisels or a convex scraper with a wide radius.

Ancient Egyptians described hernias as "that which comes into being when he coughs". They also described swellings of vessels, i.e. aneurysms: "The swelling dilates with speech of the heart [pulse]", as well as the treatment of the feeding vessel. Healers removed lipomas and other benign swellings.

Egyptian surgeons almost certainly did not participate in legal amputations, but public executioners did perform punitive amputations for different crimes. They amputated the tongue for treason, the hand for stealing, nose and ears for committing adultery, and phallus and testes for the crime of rape.

A wall in the temple at Kom Ombo depicts 40 assorted instruments of the Roman period (2nd century AD), probably used by Egyptian surgeons, including knives, spatulas or spoons, two small bags for drugs, vessels for holding liquids or oils, and pan scales similar to those used for measuring drugs. Four objects may represent forceps or even specula for dilating and examining the cervix or rectum (Figure 2).

The ancient Egyptians practiced venesection or bloodletting. Physicians used this procedure to treat scorpion stings or snakebites. "Crocodile bites should be bandaged with fresh meat on the first day" (Hearst, 239). As mentioned earlier, this procedure was a standard Egyptian remedy for major flesh wounds.

Diseases of bones and joints

Stet is the ancient Egyptian term for rheumatic diseases. The Egyptians also described deformities of the spine such as cervical and lumbar spondylosis and spondylitis deformans. Cervical spondylosis was described in this way: "If you examine a patient suffering from pain in his neck and who cannot look at his tummy because of this pain, say: this patient has Stet at his neck". Treatment involved anointing the neck with saffron. Smith (1908) noticed a prevalence of rheumatic diseases among ancient Egyptians. Pictures on temple walls showing cases of rickets, spinal deformity, rib deformity, and bow legs are ascribed to the 12th Dynasty, 2000 BC. Osteoarthritis was represented in the form of an old man. The Egyptians also noticed the fusion of the 5th lumbar vertebra with the 4th one, or with the sacrum.

Injuries to bones are well described in the Edwin Smith Papyrus. There are descriptions of simple closed fractures without a wound, in addition to comminuted, compound, and impacted fractures. Perforation of the skull is classically shown in the mummy of the Great Pharaoh Seqenenra (17th Dynasty, 1600 BC) who fought against the Hyksos; new growth of bone around his uppermost injury indicates that he lived for few months after this head injury (Fleming and Fishman, 1980).

The ancient Egyptian physicians resorted to a valuable clinical sign, the crepitation between the two fractured ends of bone, as a diagnostic technique for differentiating between sprain and fracture. A classical sign of a fractured

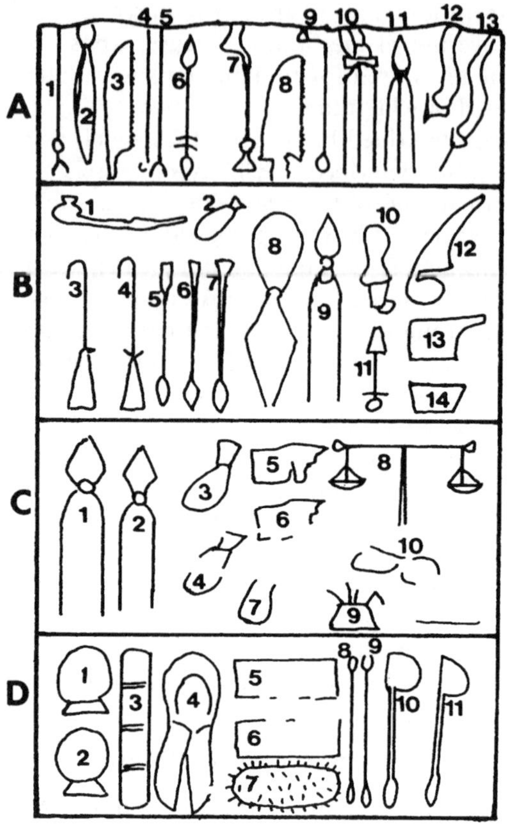

Figure 2 (A1) bifurcated sharp hook, (A8) saw blade, (B2) flask or clyster, (B3 and 4) hooks, (B5, 6 and 7) scoop-probe, (B8) cranioclast/craniotribe, (C1 and 2) tooth forceps or bone forceps, (C3 and 4) flask or clyster, (C8) balance, (D1 and 2) cupping vessels, (D3) instrument case, (D4) shears, (D5 and 6) bandages or bandage boxes, (D7) sponge, (D8 and 9) double-ended probes, (D10 and 11) scalpels. From Nunn, John F. *Ancient Egyptian Medicine*. London: British Museum Press, 1997, p. 164. Drawing by Dr. John Nunn; used with his permission.

skull base is bleeding from the ears and nostrils. This is reported in 7 patients. In 4 patients it was remarked that they could not flex their necks to look at their breasts. This is a well-known sign of meningeal irritation from blood or infection; the prognosis was pronounced unfavorable. The Egyptians described cervical fracture (fracture of cervical vertebra) with its full-blown clinical picture, namely quadriplegia and erect penis, and even reported the post mortem findings in this kind of injury. "You find the upper vertebra is sunken into the lower one, the same as if your foot is immersed in clay".

Treatment

Egyptian physicians used wooden splints padded with linen (Smith, 1908). They used *imru*, the exact meaning of which is unknown but seems to resemble plaster of Paris. Healers used *adope*, clay casts which they molded to the shape of the fractured limb and then left to dry. Egg white was used as splint: they soaked linen with egg white, wrapped it around the fractured limb after its reduction, and then left it to dry.

Egyptians described joint dislocations, for example mandible and shoulder dislocation.

> If you examine a man having a dislocation in his mandible and you find his mouth is open and does not close, then place your fingers on the back of the two 'rami' of the mandible inside the mouth and your two claws [i.e. thumbs] under his chin and cause them [i.e. the two mandibles] to fall, so they lie in their correct place.

The tomb of Ipwy in Deir-el-medina contained a remarkable wall painting showing stages of reduction of a dislocated shoulder, according to the classical method of Kocher. The arm is first externally rotated to stretch the pectoral muscles, the elbow is then brought forward in front of the chest and, finally, the hand is rapidly swept across to the opposite shoulder, when the shoulder should fall into place. The procedures described in Bailey and Love's surgical textbook confirm the opinion of M.K. Hussein (Cairo University) that modern medicine has to date no better method for reduction of a dislocated shoulder or dislocated mandible than those used by the ancient Egyptians. Even the 8-shaped bandage for a fractured clavicle was used in ancient procedures thousands of years ago. Moreover, physicians described the neurological complications of trauma, especially after skull injury. "His eye is askew on the side of injury; he walks limping with his sole on his side with that injury which is in his skull!" The explanation of this paradox in case 8 (Edwin Smith) is the contra-coup injury, i.e. the brain is injured on the side opposite to the trauma.

Diseases of the eyes

The eye plays a major part in Egyptian mythology. Seth pulled out Horus' eye, which was magically restored by Thoth. The restored eye became the most powerful of the protective amulets (Andrews, 1994).[2]

The ancient Egyptian physicians were renowned for their skill in the treatment of eye problems. Herodotus wrote that the Persian Emperor, Cyrus, sent to Ahmose II (26th Dynasty, 664–525 BC) to ask for the services of the best ophthalmologist in Egypt, an incident which, according to Herodotus, played a part in the Persian invasion of Egypt in 525 BC. *SWNW irty* (doctors of the eyes) in ancient Egypt described many eye diseases and their treatment.

Night blindness was treated using the liver of an ass or ox. Night blindness, we know now, is caused by a deficiency in vitamin A, a vitamin abundantly contained in ox liver. The Ebers Papyrus, 351, moreover, suggests ox liver, roasted and pressed, as treatment for *sharu* disease (blindness) of both eyes.

Cloudiness and darkness described opacities in the cornea or the lens (cataract). As to surgical treatment of cataracts, neither the ancient Egyptian literature nor the oculists' instruments handed down to us indicated that cataract operations were performed. Yet we have a wall painting showing a very fine tube between the oculist's mouth and the patient's eye. Some ophthalmologists interpret this as showing that the opaque lens (cataract) is being sucked from the eye. Other wall paintings show the lens pushed to fall away from the light pathway.

Descriptions of burning, heating and redness in the eyes are found, which were probably expressions for inflammation. Blood, hemorrhages and blood vessels in the eyes were mentioned. *SHDW* disease probably means corneal scars or leucoma.

The Egyptian eye disease, trachoma, which is caused by *Chlamydia trachomatis* or conjunctival unevenness, were mentioned in the Ebers Papyrus, 346. Trichiasis is one of the trachoma complications which was treated by removal of the cilia and various ointments to prevent recurrence.

Contraction of the pupil or iris, perhaps a symptom of iridocyclitis, is mentioned in the Ebers Papyrus, 345. The treatment was application of ointments. The papyrus also described pterygium (flesh in the eyes) as well as blindness resulting from "a small ball in the eyes" (possibly the nucleus of a Morgagnian cataract). Leca (1988) suggested gonorrhea as a possible cause of iritis – which would explain a case mentioned in the Kahun Papyrus.

> A woman was recommended to eat raw donkey liver for the treatment of a severe bilateral eye disease and pain in the neck related to a disease of the uterus. Her vulva was to be treated with resin from terebinth in new oil and the eyes bathed with a solution of bones from a bird.

The existence of malignant tumors in the eyes or orbit has never been proved in Egyptian mummies or skeletons.

Topical treatment was frequently prescribed for eye diseases. As a rule, ointments called *collyria* were applied directly to the exterior of the eyes. More rarely, eye drops were inserted with eye droppers made of goose feathers (Elseesy, 1999). Honey was added to many ointments, partly in order to obtain the proper consistency, partly because of its bactericidal effect, and because honey was thought to have an effect upon demons. The ointments often included extracts from plants such as resin from terebinth, gum from *Cistus ladanum*, leaves from papyrus and other plants, castor oil, seeds, wheat and juice from dates. Grated minerals were often added, e.g. green copper carbonate, red natron, granite, flint, ochre, lead, alum, or chalcedony (a variety of quartz). Animal materials consisted of extracts from crocodiles, galls from fish, mussels, pigs, tortoises, vultures, and goats. They also used bone marrow and extracts from donkey teeth.

An extract from healthy pig eyes mixed with black eye make up, and minerals such as antimony, manganese, copper, and lead were also found in make up. Epilation was performed with tweezers.

A wall painting in the tomb of the master builder IPWY (1200 BC) depicts an oculist treating the eye of a craftsman; a basket of drugs behind him suggests the high standards of the oculist. A typical representation of blindness is a blind harpist in the tomb of Pa-aten-em-heb in Saqqara which belongs to the Amarna period of the 18th Dynasty.

Many Egyptian words apparently denoting eye diseases remain untranslated. Consequently, there are eye diseases which we suspect might have existed in pharaonic times for which we do not know the Egyptian words (Nunn, 1997: 202).

Dentistry

Hesy-Ra, chief of dentists and doctors (3rd Dynasty, *ca.* 3686–2613 BC) was the first recorded doctor and dentist (Figure 3).

Teeth and jaws are highly resistant to decay. Most mature adult teeth from the earliest times show attrition due to hard particulate matter in food, partly as a result of corn grinding with stone and partly from contamination of grain with wind-blown sand (Nunn, 1997: 203). A modest decrease in incidence between 4000 BC and 1000 AD, first reported by Ruffer (1921) and confirmed

Figure 3 One of several wooden panels from the Dynasty III tomb of Hesy-Ra. This one shows him sitting before an offering table piled with bread with a scribe's ink palette hanging over his right shoulder. The hieroglyphs list his many titles; the three hieroglyphs in the upper right corner identify him as a "chief tooth physician". Drawing by Yvonne Markowitz, from Estes, J. Worth. *The Medical Skills of Ancient Egypt* revised edition. Canton: Science History Publications, 1993, p. 22.

by Thornton (1990), was probably due to improved techniques of corn grinding. Because sugar was absent from the ancient Egyptian diet, all investigators have reported a low incidence of caries extending from the earliest times until the first millennium BC (Smith and Jones, 1910; Leek, 1972; Grilletto, 1973). Tetracycline in food or beer may have been a factor in the prevention of caries (Rose *et al.*, 1993). The incidence of dental caries increased from 3% in the Predynastic Period to 20% in the early Christian period (Leek, 1979). It seems likely that increasing foreign influence led to dietary changes which encouraged caries formation (Nunn, 1997: 203). Ancient Egyptians knew how to fill the holes in dental caries with copper sulphate mixed with gum and honey. The filling was sometimes made of pure 24 carat gold (Elseesy, 1999).

Dental abscesses

Leek (1979) found 25 abscesses in six mummies. The infections arose from caries or through the pulp exposed by attrition (Smith, 1986). A root abscess is difficult to treat, and we do not know whether the hole shown in the bone of a mandible is a natural sinus or was drilled to let pus out. Both Leek and Smith reported widespread periodontal disease, leading to loss of alveolar supporting bone and consequently to loose teeth which either fell out or could have been easily removed.

Extraction

Although the method of removal was known and practised (Nunn, 1997: 204), it was the policy of the ancient Egyptian dentist not to remove a tooth except in a hopeless case. Leek (1979), however, reached the conclusion that some teeth must have been removed from strong and well-developed jaws. We do not know the instrument of extraction.

Tooth transplant

Ancient Egyptians were superb technicians (Cope, 1994: 265). They used the cavity of a recently extracted tooth or prepared a cavity in the jawbone, then placed a healthy tooth inside the cavity and fixed it to an adjacent tooth with a fine gold wire. They noticed that the transplant was not rejected if the tooth came from a twin but might be rejected if the tooth came from a person unrelated to the family of the patient or even from a relative.

Remedies

There were five remedies to fix or strengthen the teeth. It is unclear, however, whether these remedies were used to treat a tooth which was loose from periodontal disease or was crumbling from caries; a loose tooth is the more likely interpretation. The remedies included scrapings from a millstone and using malachite which might have a favorable effect on sepsis. Malachite ($2CuO.CO_2.H_2O$) is a bright green copper ore, characteristically layered in cross-section, and composed chiefly of cupric carbonate. It is used for eye diseases and as a wound dressing but never for disorders of the chest or applied directly to the female organs. The author of the Smith Papyrus used what he called a "powdered green pigment". Copper sulphate was also used as a home remedy for infections in the 20th century. Two remedies, carob and honey, were applied locally. Carob is a bean of the evergreen tree *Ceratonica siliqua.* Apart from the mentioned application, it is used chiefly for disorders of the urinary tract and gut and ailments of the anus. There were more plants in use, but they remain unknown to us because of the language barrier (Ebers, 553).

Internal medicine

Heart and vessels

The ancient Egyptians called the pulse "speech of the heart through peripheral vessels". Although the *SWNW* could palpate the pulse at several sites, he did not yet know that the blood circulates. He only seems to have recognized that the pulse was an important diagnostic sign linked to cardiac activity. The author of the Ebers Papyrus noted that the pulse could be expected to disappear when his patient fainted. Ancient physicians also seem to have been able to detect several possible abnormalities of the heart beat, for example, extra beats (extrasystoles), identifiable by what they called "forgotten beats". They recognized left ventricular enlargement, because the *SWNW* reported that the heart impulse on the chest wall was displaced to the left of its usual position, and they noted the weak thready pulse that would later be associated with heart failure.

Henry Sigerist (1951) took the explanation of primordial medical systems another step further when he concluded that physiology began when man tried to correlate the actions of blood, air, and food. The Egyptians found the focal point of this correlation in the heart, which they regarded as the central organ of the body, essential to both the living and the dead. They reasoned that all senses report directly to the heart and that it was the seat of thought, emotion, and intelligence. Lacking a word for "mind", they used "heart" to represent the same idea. Most of what we know about Egyptian physiological concepts comes from the Ebers Papyrus, taken from the Smith Papyrus and called the "physician's secret": knowledge of the heart's movement and knowledge of the heart (Estes, 1993: 78). The original word for heart was *ib*, later supplemented by *haty* which had a far more precise anatomical meaning; the later term often replaced the earlier term (Westendorf, 1992). The Egyptians said, "If you want to know how the heart is, put your hand on any organ, because the heart speaks through its vessels". They called the large vessel beside the heart, most probably the aorta, the "vessel".

Atherosclerosis and calcification of large arteries have been found and are well demonstrated by Isherwood, Jarvis and Fawcitt (1979).

Aneurysms were described, and treatment was to ligate the feeding vessel. The description of the clinical picture of aneurysm read this way:

> If you examine a swelling of a vessel in an extremity, and this swelling is hemispherical and dilates under your hand, but if you isolate it from the body, it does not pulsate, and so it does not dilate or shrink, say: it is a swelling of a vessel, and I will cure this disease (Ghaliounghui, 1967).

They wrote about angina pectoris: "If you examine a patient complaining of pain in his chest, left arm, a side of his or her stomach, say: 'Something (spirit) entered from his mouth, and death is very near'."

The Ebers Papyrus notes that, "a liver or spleen that is too full of blood can produce all diseases" (organs' congestion from heart failure). Atherosclerosis was not mentioned although it was discovered in some mummies.

Alimentary tract

There is a piece of advice in Ebers Papyrus, saying, "Do not eat unless you have an appetite for food". Herodotus mentioned that Egyptians were accustomed to have purgatives on 3 days every month to clean their intestines. They used castor oil as a purgative and also prescribed castor oil for cases of diarrhea! This may appear strange but seems reasonable if the goal of therapy was to hasten expulsion of the disease that was causing diarrhea. They designed the enema when they observed the bird Ibis filling its beak with water and then injecting the water through its anus to wash its intestine.

The usual translation "shepherd of the anus" should probably be altered to reflect the fact that the word commonly translated as anus really refers to the entire lower gastrointestinal tract below the stomach. All the papyri describe conditions such as "hot anus", "warm anus", rectal prolapse, and probably ordinary hemorrhoids. Generalized abdominal discomfort was related to incomplete digestion especially after overeating. Similar dyspeptic symptoms were also attributed to the presence of the ever-dangerous "blood-eater" in the stomach. Senna was used for constipation, calcium carbonate for acidity, pepper, cardamom, cumin, anise, almond, camomile, helba, and many other remedies for different gastrointestinal tract disorders.

Urology

The urinary tract is mentioned in the Ebers, Berlin, and Hearst papyri. Egyptian words such as *wesesh* or *khaa* have the general meaning of evacuation or voiding of waste products.

The ancient physicians knew about cystitis and retention of urine. The Ebers Papyrus, 265, tells of a remedy to eliminate *tau* (heat) in the bladder or outflow obstruction due perhaps to urethral stricture or an enlarged prostate. Elliot Smith and Dawson (1924) reported only 3 cases of renal and 2 of vesical calculus in 30,000 bodies for which they had accurate records.

The ancient Egyptians used the term *aaa* for bilharziasis (schistosomiasis). According to Ghalioungui (1967), *aaa* was mentioned 28 times in the Ebers, Berlin, Hearst, and London papyri. Eppell was of the view that the term stands for bilharziasis because of its clinical picture, i.e., abdominal pain, escape of a heartbeat (extrasystole), and blood in the urine or stools. More details on schistosomiasis are reported in the section on Parasitic Infections.

Hydrocele, which is effusion of fluid in the tunica vaginalis (covering the testis), is shown on a relief from the tomb of Ankh-ma-hor (6th Dynasty, *ca.* 2345–2181 BC) at Saqqara. It is a hydrocele, not an inguinal hernia extending to the scrotum, because there is a distinct upper boundary to the swelling.

Nutritional, metabolic and endocrine disorders

Obesity

It is difficult to detect obesity after the process of mummification. The most important evidence is in statues, reliefs, and paintings (Nunn, 1997: 81). Ancient Egyptians scorned obese people. The pharaohs are not usually shown in a state of obesity but rather as young men with a build corresponding to the classical canon of proportions. Yet many tomb owners are shown displaying rolls of fat and often pendulous breasts, as for example the legendary healer, Amenhotep-son-of-Hapu (Cairo Museum, No. 4486). The Queen of Punt presents a difficult diagnostic problem in this connection: a relief from Queen Hatshepsut's temple at Deir el Bahri, 18th Dynasty (Cairo Museum, 4276), shows her as if suffering from filariasis, a disease transmitted by mosquitoes which blocks the lymph vessels and leads to swelling and edema of any part of the body, especially the lower and upper limbs. Also Akhenaten, a pharaoh of the 18th Dynasty, is represented with a massive lower jaw, well-developed breasts and feminine hips; he was often shown with six daughters. Elliot Smith proposed the diagnosis of Frölich's syndrome (dystrophia adiposa genitalis), but Charles Edmonds (Nunn, 1997: 84) assumes that the likeliest diagnosis would be Klinefelter's syndrome. This is a congenital chromosomal abnormality with doubling of the female X strand normally present in the male XY chromosome, which then has an XXY pattern. There is often mild obesity but the subjects are usually quite tall with long legs. Muscularity tends to be underdeveloped, with small testes and gynaecomastia. Although the sexual drive may be normal, infertility is the rule. Akhenaten's six daughters, if they are his own, are hardly compatible with this diagnosis (Nunn, 1997: 84).

Nutritional diseases

The Egyptians knew about night blindness and treated it with roasted ox liver. As established by modern medicine, ox liver is very rich in vitamin A, and a deficiency of vitamin A causes night blindness. Ox liver, which we know today contains an abundant amount of vitamin B_{12}, was given to anemic patients. Some held the view that Queen Cleopatra had thyroid enlargement, but Ghaliounghui studied her painting at the Dendera Temple and came to the conclusion that what looks like thyroid enlargement is the excessively round neck which was specific for the carving technique of the Ptolemaic era. Reliefs were also found on temple walls showing protruded eyes, which is a sign of exophthalmic goitre, yet there was no enlargement of the thyroid gland. It seems that thyroid disease was rare because plenty of fish containing iodine was part of their diet.

Parasitic infections

Schistosomiasis or bilharziasis

The first archaeological evidence for the presence of bilharziasis in ancient Egypt was the finding of calcified ova. In addition, the worm *harrart* displayed on some temple walls could be the worm of bilharzia.

Beside the main symptom, hematuria (the presence of blood in the urine), the clinical picture also includes abdominal pain, escape of a heartbeat (extrasystole), and blood in the urine or stools.

Bilharziasis, or *aaa*, is mentioned 28 times in the Ebers, Berlin, Hearst, and London papyri. Among them, Ebers 62 says the disease is caused by harrart (cercaria). This is a parasitic worm with a complex life cycle alternating between two hosts, humans and fresh-water snails, *Bulinus* and *Planorbis*, which live on riverbanks. According to Ghaliounghui (1987), the male adult worm is 1 cm and the female double this length but much thinner than the male. In order to see the worms, it is necessary to dilute the blood in water before clotting. A magnifying lens is needed. Although we do not have any proof that such lenses existed at that time, Elseesy (1999) believes that the ancient Egyptians, who manufactured glass and fiberglass, also invented the magnifying lens. On the other hand, the details of miniature engravings on jewelry suggest that the services of people with gross short sight (myopia) were used for tasks requiring close vision.

It seems that Egyptians had recognized the urethra as an essential portal for the parasite to enter the human body. This would explain the sentence by someone who, aware of the mode of infection, said, "I have not waded in the water" (Faulkner, 1972), as is reported in the negative confession in Chapter 125 of the *Book of the Dead*. Elseesy opines that the penile sheaths shown in some tomb murals, whether they were intended to prevent urination in water or to block the access of the parasite through the urethra, also have the same hygienic measures and effect.

There is an association between schistosomiasis and bladder cancer, perhaps due to the liberation of nitrosamines which are known to be carcinogenic (Hicks, 1983). Schistosomiasis of the rectum can be painful and may explain the high percentage of ancient Egyptian remedies for the anus (Majno, personal communication with John Nunn, 1995).

It is noteworthy here that the ancient Egyptians treated *aaa* with antimony chloride and that modern medicine up to about 20 years ago treated schistosomiasis using antimony tartrate (Elseesy, 1999).

Filariasis

This disease, which is transmitted by mosquitoes, is defined by swelling and thickening of the skin. It is caused by the adult worm (*Filaria bancrofti*) which block the lymphatics. It is difficult to detect filariasis in mummies, but some tomb pictures of servants show enlarged male external genitalia (Weeks, 1970). The typical swollen legs of elephantiasis are almost never seen (Ghalioungui, 1949). Scrotal skin from the Leeds mummy, Natsef-Amun, was examined by Tapp and Wildsmith (1992), and they reported filaria worms.

Round worms

Adult worms (*Ascaris lumbricoides*) are quite large and can be seen in stools. It may be assumed that the ancient Egyptians would have known this, and Ebbell (1937) suggested that the cause was the still unidentified *hefat* worm.

Cockburn and his colleagues (1975) found evidence of Ascaris infection in the mummy *PUM II*, unwrapped in the United States. This very common intestinal parasite undergoes an internal cycle within the human host, in whom eggs laid in the intestine hatch into larvae to complete the cycle. If it is a heavy infestation, the patient may vomit round worms. Ascariasis can cause intestinal obstruction, as was reported by Elseesy (1985) who removed 233 worms by laparotomy in a single case of intestinal obstruction.

Tapeworms (Tenia saginata and Tenia solium)

Tapeworm eggs are ingested in undercooked meat and hatch into long, flat intestinal worms that attach themselves by their heads to the intestinal mucosa. Tapeworm eggs were found in the mummy *ROM (NAKHT)* examined in Toronto (Hart *et al.*, 1977).

Unidentified Egyptian worms

Ebbell interprets *hereret* as the schistosome parasite, but this has not been generally accepted in more recent studies. Herbal remedies were used for parasitic infestations, e.g. acacia leaves, barley, cumin, pine oil, and pomegranate roots. In addition to mineral remedies like natron, malachite, desert oil, red ochre and antimony chloride, animal remedies such as honey, white oil, ox fat, and goose fat were also used. For all these agents, beer or milk or wine provided the necessary vehicle. Pomegranate and wormwood are known vermifuges.

Malaria

Miller *et al.* (1994) applied the *Para*Sight™ F- test to mummies from the Predynastic Period to the New Kingdom, 25th Dynasty (716–656 BC). This test detected antigens produced by *Plasmodium falciparum* (causing tertian fever) in mummies from all periods investigated. The authors concluded that this finding suggests that all were suffering from malaria at the time of their death (Nunn, 1997: 73). Elseesy (1999) notes that the vast areas of land covered with river Nile water in the form of lakes and canals were indeed a good medium for the disease.

Bacterial and viral infections

Tuberculosis

Pott's disease affects the thoracic vertebra with tuberculosis, which consequently collapses and leads to a spinal deformity called kyphosis or, clinically, gibbous. Ruffer reported this disease in a priest of Amun of the 21st Dynasty (1085–945 BC). Derry (1938) reported 9 cases of TB spine. Morse's review (1964) included a single case of collapsed lung with pleural adhesions, which might have been pulmonary tuberculosis. According to Shafik, the collapsed vertebrae of Pott's disease distinguish it from Porter's hump.

Tetanus

Tetanus is caused by *Clostridium tetani*, an organism which may contaminate deeply lacerated wounds. Mummies do not usually show evidence of this disease. However, case 7 of the Edwin Smith Papyrus gives a clinical picture of lockjaw and distortion of the face which, according to Elseesy (1999), is called Risus sardonicus (the weeping laughter). The patient suffers stiffness of his neck and cannot bend it to look at his shoulders or his breast. His heart beats slowly and saliva emerges from his mouth but does not drop down.

Poliomyelitis

The funerary stela of a doorkeeper (18th Dynasty) by the name of Roma portrayed him with a grossly wasted and shortened leg, showing equinus deformity of the foot (Figure 4). The most favored view is that this is a case of poliomyelitis contracted in childhood, before completion of growth of the leg bones. The equinus deformity would then be a compensation for walking on the shortened leg. Roma appears with a stick which could be used as a crutch.

Figure 4 The grossly wasted and shortened leg belonged to Roma, a doorkeeper of the 18th or 19th Dynasty. The differential diagnosis includes club foot (talipes equinus) and poliomyelitis. Roma does not appear to be using his staff as a crutch. From Nunn, John F. *Ancient Egyptian Medicine*. London: British Museum Press, 1997, p. 77. Taken from The Egyptian Collection, Ny Carlsberg Glyptotek, Copenhagen, Denmark; used with their kind permission.

Osteomyelitis

A wide variety of organisms may infect bone. Sandison (1972) agreed with earlier observations that pyogenic osteomyelitis is surprisingly rare in skeletal remains, and there are many cases of healed compound fractures in which the bone shows no signs of infection. This could be explained on the basis of (1) low virulence of organisms, (2) high resistance of the immune system in the ancient Egyptians, or (3) use of fermented bread pulp (penicillin) in compound fractures which means exposure of bones to the outside atmosphere.

Smallpox

Ruffer and Ferguson (1911) reported a skin lesion resembling smallpox in a mummy of the 20th Dynasty (1200–1085 BC). The most distinguished case is that of Ramses V, described by Elliot Smith (1912). Nothing in the medical papyri can be related to smallpox.

THE MATERIA MEDICA OF PHARAONIC EGYPT

We have compiled some of the remedies that were used in pharaonic Egypt along with their clinical indications (Table 1).[3] It is important to remember that, because of the language barrier, only 20% of the prescribed remedies are understood.

THE INFLUENCE OF ANCIENT EGYPTIAN MEDICINE ON LATER CIVILIZATIONS

Homer and Herodotus remarked on the Egyptians' reputation for healing, especially for the remedies they made from plants that grew along the Nile. According to the *Odyssey* (8th century BC, contemporaneous with the 23rd and 24th Dynasties), Queen Helen of Sparta dosed her banquet guests one night with a drug that could alleviate their painful memories of the Trojan War. Egypt provided direct potential pathways for the transmission of medical knowledge to other parts of the ancient world. The Greeks derived several pathophysiological concepts from ancient Egyptian writings. Egyptians anticipated aspects of later medical thinking with their approach to ailments: first the clinical picture, second the diagnosis, and third the treatment, including the prognosis. In ancient Egypt, moreover, patients had the right to complain about the doctor if the treatment went wrong (Shafik, 1998).

The ancient Egyptians had two great medical schools: Heliopolis and Imhoteb. James Henry Breasted said, "Egyptians reached the secrets of physiology, anatomy and medicine 2000 years before Hippocrates". Hippocrates made no mention of the pulse, although the Egyptians had already termed it "speech of the heart in every periphery of the body".

Alexandria University produced Erasistratos, a pioneer in comparative anatomy, Eudimos in embryology, Apollodoros in toxicology, Philinos in experimental medicine, and Ammonios, an expert in diagnosing stones of the urinary tract and knowing how to get rid of them.

The Greco-Roman period between 31 BC and 642 AD was a time of both

Table 1 The materia medica of pharaonic Egypt

acacia	topical use for sore limbs; superficial skin conditions (especially swollen glands)
aloe	for eye diseases; laxative; astringent
alum	astringent (externally); an emetic if given in high doses
amm	a photosensitizer if eaten by patients suffering white skin patches, i.e. vitiligo. They become sensitive to sunlight.
anise	carminative (relieving flatulence)
antimony salts	trisulfide; used primarily for eye diseases; treatment of bilharzia (which was recently successfully treated by antimony tartrate, Elseesy, 1985)
barley	as dry grain or as paste, for skin conditions
bat blood	0.12 percent for eye disorders
bee wax	a binding agent
beer	from barley, wheat, or dates; 14% for urinary and musculoskeletal disease
bolti	symbol of fecundity because its young are hatched in the adult's mouth; 0.12% for women's disorders
calamine	0.59% zinc oxide, ointment, mild antiseptic and astringent
calcium carbonate	antacid
carob	for diseases of the anus (Chester Beatty)
castor oil	laxative
charcoal	0.12% for skin diseases
copper salts	externally; astringent and antiseptic
saffron	external use for rheumatic pains
date wine	2.73%: for superficial skin conditions; according to Elseesy could be used as an antiseptic.
eggs	chiefly of ostrich or goose (yolk only), for skin and hair conditions
ergot	(Hearst Papyrus) to stimulate uterine contractions and constrict small arteries (Leake)
hedgehog fat	to cure baldness
honey	all systemic disorders, especially wound dressing
hyocyamus	1.66% for ailments of limbs and gut. Probably it was and is still used for burning micturition (Elseesy, 2000)
kaolin	still used as antidiarrheal agent
lime	for blood eating, scurvy (Ebbell)
liver of ox	for night blindness (vitamin A) and anemia (Vitamin B_{12})
malachite	cupric carbonate, for eye diseases; wound dressing
marijuana	0.24% for sore limbs, female disorders, ingredient of anal remedy (Chester Beatty)
memphis stone	with vinegar gives local analgesia (Discorides and Pliny)
mint	carminative, stomach pains
opium	narcotic effects; sedative
papyrus strips	as bandage
pomegranate	for intestinal worms, still used in Egypt (Elseesy)
senna	1.07% laxative
willow buds	from the Egyptian willow tree, (*Salix safsaf*), was probably used as an analgesic since aspirin (acetyl salycilic acid) is from the willow tree (Elseesy, 2000)
wine (red or white)	5.23% was used as a vehicle or for dilution of other active drug materials, because wine appears only in remedies for diseases of the skin, the musculoskeletal system, gut, and female organs.

assimilating and increasing the ancient Egyptian civilization. When the Arabs invaded Egypt and settled there, they started translating and spreading this huge amount of knowledge. Muslim researchers, most of them non-Arabs, added to the knowledge and transmitted this to the world. Greco-Romans and Muslims were the bridges, but we believe that those who had ignited the spark were the ancient Egyptians.

NOTES

[1] Following is a list of the surviving papyri and their locations.

1. The Ebers Papyrus is preserved at the University of Leipzig, Germany and is in almost perfect condition. It is the longest and most famous among all papyri. Written in the beginning of the 18th Dynasty (1570–1320 BC) or even earlier, it deals with diseases of the stomach, action of the heart and its vessels, and the surgical treatment of cysts, boils, carbuncles and similar conditions.
2. The Hearst Papyrus is kept at the University of California. It has been assigned to the reign of Tuthmosis III in the 18th Dynasty. Its contents are similar to those of the Ebers Papyrus.
3. The Edwin Smith Papyrus is in the possession of the Historical Society of New York. In 13 medico-magical incantations and prescriptions, it deals with the surgical treatment of wounds and fractures.
4. The Chester Beatty Papyrus, at the British Museum (No. 10686), contains a series of prescriptions and remedies for diseases of the anus and rectum. It might be called an early treatise on proctology.
5. The Berlin Medical Papyrus, kept at the Berlin Museum (No. 3038), dates back to the 19th Dynasty (1320–1200 BC). It is similar in character to the Ebers and Hearst papyri.
6. The Kahun Papyrus, assigned to the 12th Dynasty (*ca.* 1991–1786 BC), was discovered at Lâhûn, Fayoum, Egypt, in 1889. It deals only with gynecology.
7. The London Medical Papyrus, from the 18th Dynasty, is preserved at the British Museum (No. 10059). It is badly written and entirely medico-magical in its contents.
8. Other documents. A considerable number of magical papyri exist in museums in Paris, Leiden, Turin, Berlin, Budapest, Copenhagen, and Rome; they are concerned with the treatment of diseases. An important magical papyrus, the London-Leiden Demotic Papyrus, which was written in the 3rd century AD in demotic script, contains a good deal of medical matter.

[2] Egyptian deities, Seth, Horus, and Thoth, were the reason that the eye has played a major part in Egyptian mythology. The God Seth, also known as Set, as the patron of the district of Upper Egypt, was worshipped at Avaris during the rule of the Hyksos invaders (1650–1567 BC). He is represented as a composite creature with a greyhound's body, slanting eyes, square-tipped ears, and a long forked tail. Seth has been connected with various animals (e.g. dog, pig, okapi, and anteater), which may have served as a prototype of this remarkable monster, the Typhonian animal, so called after the Greek demon Typhon. According to Egyptian mythology the God of the skies, Horus, became the opponent of Seth when Seth murdered his father, Osiris. The myth has it that Horus, usually depicted as a falcon whose eyes represent the sun and the moon, defeated Seth, thus avenging his father and assuming the rule. In his fight with Seth, however, his left eye (the moon) was damaged – this being the mythological explanation of the moon's phases – and was healed by the God Thoth. The depiction of the restored eye became a powerful protective amulet and the cult of Horus spread in prehistoric times over the whole of Egypt. Thoth was the god of reckoning and of learning in general in ancient Egypt; he weighed the hearts of the deceased at their judgment and reported the result to the presiding god, Osiris, and his fellow judges. He was also held to be the inventor of writing, the founder of the social order, the creator of languages, the scribe, interpreter and advisor of the gods, and the representative of the Sun God Re on earth. In the myth of Osiris, Thoth protected Isis, Osiris' wife, during her pregnancy and, as already mentioned, healed the injury inflicted on her son Horus by Osiris' adversary, Seth. Thoth's sacred animals were the ibis and the

baboon. He was usually represented as a human with an ibis head (*Encyclopaedia Britannica*, Micropaedia, vol. 1X).

[3] The list was compiled from the following sources: B. Ebbell, trans., *The Papyrus Ebers: The Greatest Egyptian Medical Document* (London: Oxford University Press, 1937); Chauncey D. Leake, *The Old Egyptian Medical Papyri* (Lawrence: University of Kansas Press, 1952); Frans Jonckheere, *Le Papyrus Médicale Chester Beatty* (Brussels: Fondation Égyptologique reine Elisabeth, 1947). Ida Hay (Boston) provided important assistance in ascertaining the most recent scientific names for many plant species. Loutfy Boulos' book, *Medicinal Plants of North Africa* (Algonac, Michigan: Reference Publications, 1983), provides a good deal of information about modern Egyptian uses of many ancient botanical drug ingredients.

BIBLIOGRAPHY

Andrews, C.A.R. *Amulets of Ancient Egypt*. London: British Museum Press, 1994.

Bailey, Hamilton *et al. Bailey & Love's Short Practice of Surgery*. 14th edition. London: H.K. Lewis, 1968.

Boulos, Loutfy. *Medicinal Plants of North Africa*. Algonac, Michigan: Reference Publications, 1983.

Breasted, J.H. *The Edwin Smith Surgical Papyrus*. 2 volumes. Chicago: University of Chicago Press, 1930.

Cockburn, A., R.A. Barraco, T.A. Reyman and W.H. Peck. 'Autopsy of an Egyptian mummy.' *Science* 187: 1155–1160, 1975.

Dawson, W.R. 'Studies in the Egyptian medical texts-III.' *Journal of Egyptian Archaeology* 20: 41–46, 1934a.

Dawson, W.R. 'Studies in the Egyptian medical texts-IV.' *Journal of Egyptian Archaeology* 20: 185–188, 1934b.

Derry, D.E. 'Pott's disease in ancient Egypt.' *Medical Press and Circular* 197: 196–199, 1938.

Ebbell, B. *The Papyrus Ebers*. London: Oxford University Press, 1937.

Elseesy, Waseem R. نظرة فاحصة في الحب و الجنس (*Critical Survey of Love and Sex; Nathra fahessa fi al hob wa al guinss*). 3rd edition. Cairo: Engineering Center for Printing and Publishing, 1999.

Elseesy, Waseem R. 'Drugs in ancient Egypt.' From a series of 100 articles on Egyptology, *Rose-el-Youssef Weekly Magazine*, No. 3706, June 19, 1999.

Estes, J. Worth. *The Medical Skills of Ancient Egypt*. Revised edition. Canton: Science History Publications, 1993.

Faulkner, R.O. *The Ancient Egyptian Book of the Dead*. London: British Museum Publications, 1972.

Fayad, Mohamed. فن الولادة في مصر القديمة (*Art of Obstetrics in Ancient Egypt; Fun al awlada fi misr al qadeema*). Cairo: Dar el Shorouq, 1996.

Fleming, S. and B. Fishman. *The Egyptian Mummy: Secrets and Science*. Philadelphia: University Museum, 1980.

Ghaliounghui, Paul, S. Khalil and A.R. Ammar. 'On an ancient Egyptian method of diagnosing pregnancy and determining foetal sex.' *Medical Historian* 7: 241–246, 1963.

Ghaliounghui, Paul. الطب عند القدماء المصريين (*Medicine in Ancient Egypt; Al tibb aend alqadama' al misri'een*). Cairo: Future Printing House, 1967.

Ghaliounghui, Paul. *Magic and Medical Science in Ancient Egypt*. 2nd edition. Amsterdam: B.M. Israël, 1973.

Ghaliounghui, Paul. *The Ebers Papyrus*. Cairo: Academy of Scientific Research and Technology, 1987.

Ghaliounghui, Paul. 'Sur deux formes d'obésité representée dans l'Égypte ancienne.' *Annales du Service des Antiquitées de l'Égypte* 49: 303–316, 1949.

Grilletto, R. 'Caries and dental attrition in the early Egyptians as seen in the Turin collections.' In *Population Biology of the Ancient Egyptians*, D.R. Brothwell and B.A. Chiarelli, eds. London: Academic Press, 1973, pp. 325–331.

Hart, G.D. *et al.* 'Autopsy of an Egyptian mummy (Nakht-Rom I).' *Canadian Medical Association Journal* 117: 461–476, 1977.

Hicks, R.M. 'The canopic worm: role of bilharziasis in the aetiology of human bladder cancer.' *Journal of the Royal Society of Medicine* 76: 16–22, 1983.

Imam, Adel. *Slide with ova and split cells.* 2nd Channel Program of Egyptian Television, 1999.

Isherwood, I, H. Jarvis and R.A. Fawcitt. 'Radiology of the Manchester mummies.' In *The Manchester Museum Mummy Project*, A.R. David, ed. Manchester: Manchester University Press, 1979, pp. 25–64.

Leake, Chauncey D. *The Old Egyptian Medical Papyri.* Lawrence: University of Kansas Press, 1952.

Leca, Ange-Pierre. *La Médecine Égyptienne au Temps des Pharaons.* Paris: Roger Dacosta, 1988.

Leek, F.F. 'Bite, attrition and associated oral conditions as seen in ancient Egyptian skulls.' *Journal of Human Evolution* 1: 289–295, 1972.

Leek, F.F. 'The dental history of the Manchester mummies.' In *The Manchester Museum Mummy Project*, A.R. David, ed. Manchester: Manchester University Press, 1979, pp. 65–78.

Miller, R.L. *et al.* 'Diagnosis of Plasmodium falciparum infection in mummies using the rapid manual *Para*Sight™-F test.' *Transactions of the Royal Society of Tropical Medicine and Hygiene* 88: 31–32, 1994.

Morse, D., D.R. Brothwell and P.J. Ucko. 'Tuberculosis in ancient Egypt.' *American Review of Respiratory Diseases* 90: 524–541, 1964.

Nunn, John F. *Ancient Egyptian Medicine.* London: British Museum Press, 1997.

Petrie, W.M.F. *Gizeh and Rifeh.* London: Bernard Quaritsch, 1907.

Petrie, W.M.F. *Tools and Weapons.* London: Constable, 1917.

Rose, J.C., G.J. Armelagos and L.S. Perry. 'Dental anthropology of the Nile valley.' In *Biological Anthropology and the Study of Ancient Egypt*, W.V. Davies and R. Walker, eds. London: British Museum Press, 1993, pp. 61–74.

Ruffer, M.A. and A.R. Ferguson. 'Note on an eruption resembling that of variola in the skin of a mummy of the XXth Dynasty.' *Journal of Pathology and Bacteriology* 15: 1–3, 1911.

Ruffer, M.A. *Studies in the Paleopathology of Egypt*, R.L. Moodie, ed. Chicago: University of Chicago Press, 1921.

Sandison, A.T. 'Evidence of infective diseases.' *Journal of Human Evolution* 1: 213–224, 1972.

Shafik, Ahmed. 'Medicine in ancient Egypt.' *Journal of Investigative Surgery* 11: 291–293, 1998.

Sigerist, Henry E. *A History of Medicine: Primitive and Archaic Medicine.* New York: Oxford University Press, 1951.

Smith G.E. 'The most ancient splints.' *British Medical Journal.* 1908, pp. 732–734.

Smith, G.E. and W.R. Dawson. *Egyptian Mummies.* London: George Allen and Unwin, 1924.

Smith, G.E. and F.W. Jones. *The Archaeological Survey of Nubia: Report for 1907–1908.* Vol. II: *Report on the Human Remains.* Cairo: National Printing Department, 1910.

Smith, N.J.D. 'Dental pathology in an ancient Egyptian population.' In *Science in Egyptology*, A.R. David, ed. Manchester: Manchester University Press, 1986, pp. 43–48.

Tapp, E. and K. Wildsmith. 'The autopsy and endoscopy of the Leeds mummy.' In *The Mummy's Tale*, A.R. David and E. Tapp, eds. London: Michael O'Mara Books, 1992, pp. 132–153.

Thornton, F. *Oral Pathological Comparison of Discrete Ancient Nile Valley and Concurrent Populations.* M.Phil. dissertation, University of Bradford, 1990.

Weeks, K.R. *The Anatomical Knowledge of the Ancient Egyptians.* Ph.D. Thesis, Yale University, 1970.

Westendorf, W. *Erwachen der Heilkunst: Die Medizin im Alten Aegypten.* Zürich: Artemis und Winkler, 1992.

CAI JINGFENG AND ZHEN YAN

MEDICINE IN ANCIENT CHINA

Traditional Chinese Medicine (TCM) is an integral part of ancient Chinese civilization. Diagnosis and treatment are based on knowledge acquired through long-term medical practice in combination with an assessment of natural conditions including climate, geography and phenology. The theoretical basis was laid down some 3000 years ago in the *Huang di nei jing (Huang di's Inner Canon of Medicine)*, in Zhang Zhongjing's *Shang han za bing lun (Essays on Cold and Miscellaneous Diseases)* of the Eastern Han dynasty (202 BC–AD 220) and other medical classics. The system includes the *yin-yang*, five evolutionary phases and other theories such as those on channel collaterals, viscera, pathogenesis, etiology, and treating principles (Editorial Committee, 1985). TCM contributes to the health of Chinese people and to some extent to other Asian nationalities as well and, increasingly to people in the rest of the world.

ORIGIN AND FORMATION OF TCM

Through the long evolutionary process, as knowledge and experience developed, prescriptions for fighting diseases were accumulated and enriched. The use of fire improved hygiene and health care. The change from raw to cooked food, combined with being able to keep warm, greatly reduced the incidence of disease. Earthenware appeared during the Neolithic age in China (5000–3000 BC), and this paved the way for the manufacture of wine. Legend has it that Yidi prepared wine during the Xia Yu period (2070–1600 BC), which demonstrates the ancient origin of fermented liquor (Li, 1999). The *Shuo wen jie zi (Dictionary for Words)* states, "the Chinese character *yi* (medicine) refers to a man who treats disease ... with 'liquor' which cures disorders" (Xu, 121/1979). According to the *Jiu shi (History of Wine)*, "during the end of the Western Han dynasty, Wang Mang's imperial edict mentioned that liquor was viewed as the most significant medicine of all".

Early in the Paleolithic age, stone knives were elaborately carved. By the Neolithic age, the knife had evolved into a stone needle (Figure 1). As early as 3000 years ago, bronze metallurgy enabled people to make bronze needles (Figure 2). About 2470 years ago, the Iron Age began and iron needles

H. Selin (ed.), Medicine Across Cultures: History and Practice of Medicine in Non-Western Cultures, 49–73.

Figure 1 Stone "needle" with bilateral blade, pointed end, 7 × 3 cm, used in the Paleolithic age for blood and pus letting, unearthed at Henan province. From Fu Weikang, *A General History of Chinese Medicine*, Vol. IV: *Illustration Album*. Used with the permission of the publisher.

Figure 2 Bronze needle from 3000 years ago (Western Zhou dynasty) unearthed at Shaanxi, China, 9.2 cm long, triangle in cross section, used for blood-letting and acupuncture. From Fu Weikang, *A General History of Chinese Medicine*, Vol. IV: *Illustration Album*. Used with the permission of the publisher.

appeared. The book *Ling shu (Miraculous Pivot)* states that there are nine needles in different dimensions, lengths, calibers, and shapes with round, blunt or sharp ends, for different kinds of treatment (Figure 3). For instance, the minute needle is for dredging the channels to regulate the *qi*[1] and blood. This demonstrates that, at the time when the *Spiritual Axis* was compiled about 2000 years ago, the Chinese art of acupuncture and moxibustion had already reached a high theoretical level.

There is also an ancient tradition for herbal medicine in China. The *Huai nan zi (Book of Master Huainan)* says, "in ancient times, people ate plants, fruits from trees and grasses, worms, and clams. But diseases are bound to happen. Shennong teaches people to grow grains [and] ... to taste all herbs to be able to differentiate the good from the bad. He himself has encountered seventy kinds of poisons in a single day". In ancient society, as agriculture developed, herbs were applied for treating diseases. Archeological materials show that in a Han tomb excavated in Gaocheng, Hebei province, there are apricot and peach stones from the Yin dynasty (1600–1046 BC). Many medicinal materials are recorded in ancient classics such as the *Yi jing (Classic of*

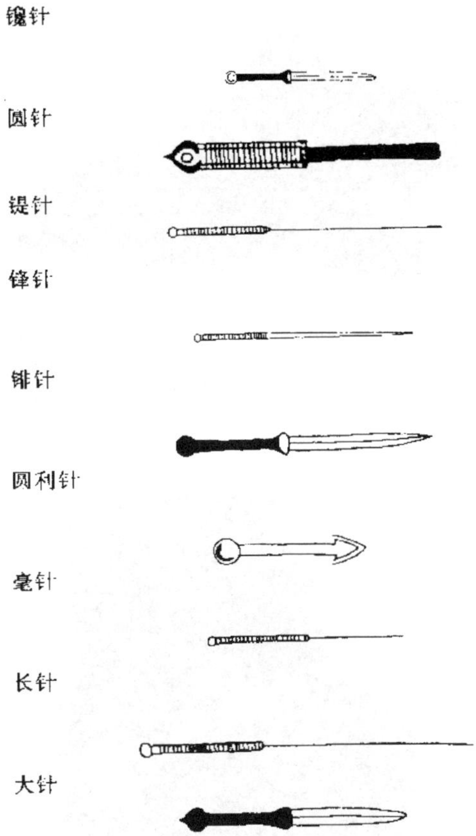

Figure 3 Nine ancient acupuncture needles in different forms and lengths for use in different therapeutic situations. From a xylographic edition of *Ling shu jing (Classic of Miraculous Pivot)*. Collection of the Library of the China Institute for the History of Medicine and Medical Literature. Used with the permission of the Institute.

Changes) and the *Shi jing (Classic of Poets)*. The *Shan hai jing (Classic of Seas and Mountains)* describes some 130 kinds of medicines, including grasses, woods, animals, fish, worms and birds. The *Shennong ben cao jing (Shennong's Classic of Materia Medica)* lays down the foundation of Chinese materia medica (Anonymous, *ca.* 2nd century AD/1963).

Archeological investigation also found that inscriptions on bones and tortoise shells included characters dealing with diseases like headache, abdominal pain, dizziness, common cold, tinnitus, deafness, eye disease, and ulcers, as well as children's and women's diseases and dental diseases such as caries (Figure 4). They also included infectious diseases and parasitosis (Figure 5). The *Zhou li (Ritual of the Zhou Dynasty)* mentions that "diseases can be seen in all seasons, viz. head disease in spring, scabies and other itching diseases in summer, malaria and cold diseases in autumn, and asthmatic and coughing diseases in

Figure 4 Ancient inscription of the Chinese character "caries" in an oracle of animal bone and tortoise shell, indicating an orifice in a tooth. From Fu Weikang, *A General History of Chinese Medicine*, Vol. IV: *Illustration Album.* Used with the permission of the publisher.

Figure 5 Ancient inscription of the Chinese character "parasite" in an oracle of animal bone and tortoise shell, indicating a worm within the interior of a container. From Fu Weikang, *A General History of Chinese Medicine*, Vol. IV: *Illustration Album.* Used with the permission of the publisher.

winter" (Anonymous, *ca.* 8th century BC/1929). This indicates the recognition of relationships between seasons and diseases, an ancient etiological idea.

The earliest extant medical classic is the *Huang di nei jing (Huang di's Inner Canon of Medicine)* (Figure 6). It includes two parts, the *Su wen (Plain Questions)* and *Ling shu (Miraculous Pivot)* (Anonymous, *ca.* 475 BC/1956). The *Su wen* deals with visceral manifestations, etiology, pathogenesis, treating principles, therapeutics, prophylaxis, health care, and theories like *yin-yang* and the five evolutionary phases; the *Spiritual Axis* deals mainly with the art of acupuncture and moxibustion. The *Shang han za bing lun (Essays on Cold and Miscellaneous Diseases)* (Zhang, 219/1976) is the origin of one of the special features of TCM, treatment based on differential diagnosis of syndromes. TCM

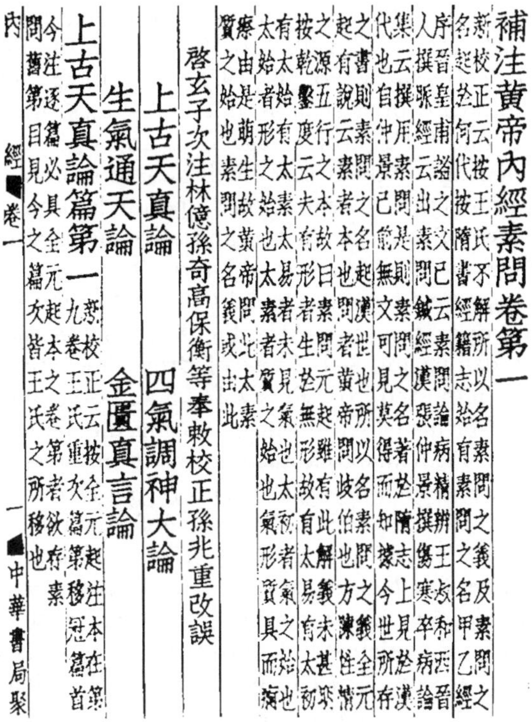

補注黃帝內經素問卷第一

啓玄子次注林億孫奇高保衡等奉敕校正孫兆重改誤

上古天真論 四氣調神大論

生氣通天論 金匱真言論

上古天真論篇第一

Figure 6 A page from the first chapter of the xylographic edition of *Huang di nei jing su wen (Plain Questions* of *Huang di's Inner Canon of Medicine).* Collection of the Library of the China Institute for the History of Medicine and Medical Literature. Used with permission of the Institute.

diagnosis is based on a differential system where several parameters are examined based on the four diagnoses – asking, observing, touching, and listening – and the predominant concurring findings are taken as the basis for the diagnosis. This is called "differential" and strongly contrasts with the earlier overly simplistic diagnostic systems based on one finding only. Many prescriptions in this book are still in use today.

BRIEF HISTORY OF TCM

The *Nan jing (Classic of Questioning)*, actually a supplement to the *Inner Canon* in the form of questions and answers, deals with physiology, anatomy, pulse diagnosis, physiology, visceral organs, etiology, pathogenesis and treating principles (Qin, *ca.* 5th century BC/1999). The *Zhen jiu jia yi jing (A-B Classic of Acu-moxibustion)* (Huangfu, 282/1956), written by Huangfu Mi, and the *Mai jing (Classic of Pulse Diagnosis)* (Wang, 280/1957) by Wang Shuhe, are representative medical classics of the Wei-Jin dynasties (220–420 AD), the latter mentioning 24 kinds of pulses. Ge Hong of the Eastern Jin dynasty (317–420 AD) wrote the book *Bao pu zi nei pian (Inner Chapter of Master*

Baopu), which is viewed as the beginning of Chinese pharmacological alchemy. In the Liang dynasty (502–557 AD), Tao Hongjing wrote *Ben cao jing ji zhu (Variorum of Classic Herbal)* (Tao, 500/1955), describing 730 kinds of herbal drugs organized into categories such as jade and stones, grasses, woods, birds and beasts, insects and fishes, fruits, vegetables and grains. This approach to the classification of drugs on the basis of their natural sources is more advanced than the classification of three categories of "superior, intermediate and inferior" in *Shennong's Classic of Materia Medica* and also has a more scientific basis. Lei Xiao's *Lei gong pao zhi lun (Master Lei on Drug Processing)* became the first book on drug processing with special Chinese features (Lei, 588/1985).

In the Sui-Tang dynasties (581–907), the *Xin xiu ben cao (Newly Revised Herbal)* provided details for 844 kinds of drugs, including many, such as benzoin, arura, etc., imported from Persia, India and other countries, reflecting the active exchange of medicinal products during this period (Su, 659/1957). The study of herbal medicine continued to make progress during the Song, Liao, Jin and Yuan periods (760–1368). The *Jiayou Herbal* was compiled on the basis of a general nationwide investigation; unfortunately, both these books were later lost. The *Jing shi zheng lei bei ji ben cao (Classified Herbal for Emergency Use Based on Classics and Historical Documents)*, written by Tang Shenwei, was compiled by consulting relevant works in Buddhism and other historical documents; it explains and illustrates 1746 kinds of herbal drugs. This is a work of great significance (Tang, 1082/1957). Other famous medical books include Wang Weiyi's *Tong ren shu xue zhen jiu tu jing (Illustrated Acu-moxibustion Classic of Acupoints in Bronze Models)* (Wang, 1026/1955), Hua Shou's *Shi si jing fa hui (Elaboration on 14 Channels)* (Hua, 1341/1956), Chen Wuze's *San yin ji yi bing zheng fang lun (Discourses on Recipes and Syndromes of Three Etiological Factors)*(Chen, 1174/1957), Chen Ziming's *Fu ren da quan liang fang (Complete Effective Recipes for Women's Disorders)* (Chen, 1237/1956) and Qian Yi's *Xiao er yao zheng zhi jue (Children's Medicines and Syndromes)* (Qian, 1119/1955).

Each of the so-called Four Schools of the Jin-Yuan dynasties, represented by Liu Wansu, Zhang Congzeng, Li Gao and Zhu Zhenheng, propounded various unique medical ideas. Liu advocated the application of cold medicines to cure diseases, which he believed were all caused by heat pathogens; Zhang suggested the use of drastic allopathic therapies such as emesis, catharsis and diaphoresis. Li recommended applying ascending tonics to strengthen the spleen and stomach, and Zhu advocated using yin-nourishing remedies as he believed a deficiency of yin was responsible for illness (Figure 7).

In the Ming-Qing dynasties (1368–1911) herbal medicine was further developed. Li Shizhen's *Ben cao gang mu (Compendium of Materia Medica)* (Li, 1590/1957) was the most influential work. In 52 volumes it describes 1892 kinds of drugs in 62 categories, appended with some 10,000 recipes. The author compiled the book on the basis of arduous field investigation and documentary research; it is the best model for research on historical documents in TCM and has been rendered into many foreign editions. Zhao Xuemin's *Ben cao gang mu shi yi (Supplement to the Compendium of Materia Medica)*, and Wu Qishun's

Figure 7 Woodcut Portrait of Zhu Zhenheng (1281–1358), one of the chief representatives of the four medical schools of the Jin-Yuan transition. From Fu Weikang, *A General History of Chinese Medicine*, Vol. IV: *Illustration Album*. Used with the permission of the publisher.

Zhi wu ming shi tu kao (Investigation on Plants with Illustrations) are both world famous. By the end of the Ming dynasty (1368–1644), Wu Youke, a physician known for his work on infectious diseases, held that the so-called Warm Epidemic diseases were caused by certain kinds of invisible *li qi* (malignant pathogens) invading the body through the nostrils and mouth (Wu, 1642/1928).

Medicine from Western countries spread into China at the turn of the Ming-Qing dynasties and was popularized gradually during the late Qing and Republican periods (Figure 8). There was a trend of trying to match TCM theories with Western physiology, anatomy and pathology, together with a tendency to seek correspondence of disease names in both systems based on clinical diagnosis and treatment. The modern concept of integrated medicine is derived from these trends and tendencies.

BASIC THEORY

Basic theory in TCM includes the *yin-yang* theory, the five evolutionary phase theory, and theories of viscera channels, etiology and pathogenesis, treatment, therapeutics, prevention and health care.

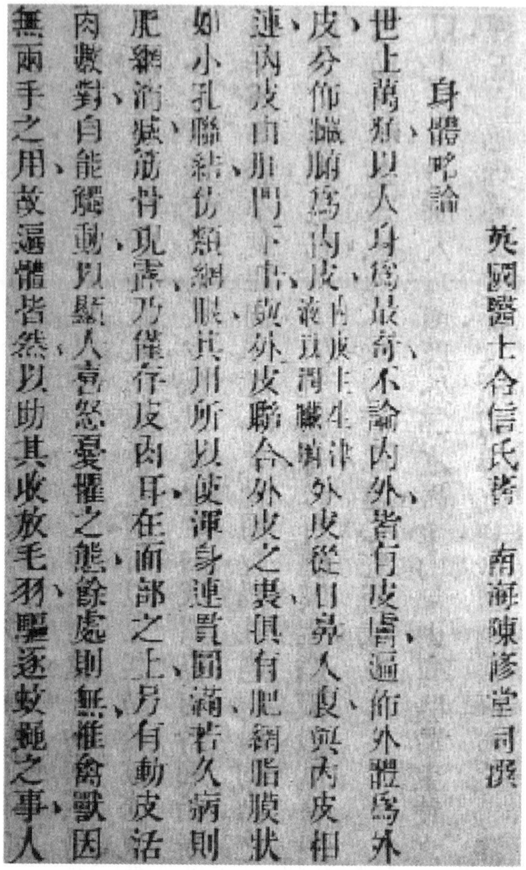

英國醫士合信氏著　南海陳修堂同撰

身體略論

世上萬類以人身為最奇不論內外皆有皮膚遍衛外體為外
皮分佈臟腑為內皮內皮上生津液以潤臟腑外皮從口鼻入處與內皮相
連內皮由肛門下出與外皮聯合外皮之裏俱有肥網脂膜狀
如小孔聯結彷類網眼其用所以使渾身連貫圓滿若久病則
肥網消減筋骨現露乃僅存皮肉耳在面部之上另有動皮活
肉數對自能攝動以顯人喜怒憂懼之態餘處則無惟禽獸因
無兩手之用故遍體皆然以助其收放毛羽驅逐蚊蠅之事人

Figure 8 A page from a xylographic edition of *Outline of Anatomy and Physiology* written by an English missionary, Doctor Hobson, carved in 1851. This is one of the earliest western medical works published in Chinese. From Fu Weikang, *A General History of Chinese Medicine*, Vol. IV: *Illustration Album*. Used with the permission of the publisher.

The yin-yang and five phase theories guide clinical practice in TCM. By applying the law of contradictory unity (yin/yang), TCM explains body structure, physiological and pathological changes and their mutual promotion, mutual antagonism and mutually dependent relations and the general law of diagnosis and treatment. The underlying basis of TCM is that all of creation is born from the interdependence of two opposite principles, yin and yang. These two opposites are in constant motion, creating a fluctuating balance in the healthy body. Disease results when either yin or yang is in a state of prolonged excess or deficiency.

The five element theory (*wuxing*), also called the five phase or five evolutionary phase theory, holds that everything in the universe, including our health, is governed by five natural elements: wood, fire, earth, metal and water. This theory underscores the Chinese belief that human beings, both physically and mentally, are intertwined with nature. According to the theory, each of the five elements has a season and particular organs and senses associated with it. The wood element, for example, is associated with spring, the liver, and the gall bladder. Fire is associated with early summer, the heart, and small intestines; the earth element corresponds to late summer, the stomach and spleen. Metal is associated with autumn, the lungs and large intestine, and water is associated with winter, the kidneys and the bladder.

In addition to the theory of the five elements, TCM practitioners employ the Eight Guiding Principles to analyze and differentiate the energetic imbalances in the body or the nature of a patient's condition. The eight guiding principles consist of four polar opposites: yin/yang, cold/heat, deficiency (*xu*)/excess (*shi*), and interior/exterior.

- The cold/heat principle is used to determine the patient's overall energy. A slow metabolism, chills, pale skin, and a low-grade fever mark a cold condition; a hot condition is characterized by a heightened metabolism, sensations of heat in the body, high fevers, and a flushed complexion.
- The interior/exterior principle describes symptoms in terms of the location of the patient's problem. Exterior conditions are caused by the invasion of the body by pathogens, and are usually acute and superficially located with a short duration. Exterior symptoms are those that affect the hair, skin, muscles, joints, peripheral nerves and blood vessels. Interior conditions result from pathogens that enter the interior of the body. Interior symptoms affect the organs, deep vessels and nerves, brain, spinal cord, and bones.
- The principle of deficiency/excess describes the strength of an illness and the status of the body's resistance. A deficient condition would be viewed as a lack of blood (such as in anemia), energy (*qi*), heat, or fluids. Chronic illness would fall in this category. An excess condition, by contrast, means that the body has pathogenic factors or pathologic conditions such as heat evil, yet the body is still strong enough to react to the pathogens. In TCM, an acute condition is seen as an excess condition.
- The yin/yang principles are the generalization and combination of the above principles, and a condition can be categorized in terms of the relative dominance of either yin or yang. In Chinese medicine, all organisms have both yin and yang qualities and a balance of the two is necessary for good health. In general, yin energy is associated with cold, female energy, and the solid organs; yang is associated with hot, male energy, and the hollow organs. Chronic illness is seen as yin, while acute illness is seen as yang.

The theory of visceral manifestation studies the physiological functions, pathological changes, and the relationship between the viscera and their outside environment. These form the basis for directing all aspects of treatment, for the transformation of spirit, qi, blood and body fluids and for their distribution and action.

In channel theory a system with passages runs throughout the whole body to connect the surface with the interior and tissues with organs. In the system of channel and collaterals, qi and blood circulate to form a complete system.

The theory is extensively applied to explain physiology, pathology, diagnosis and treatment.

TCM holds that disease occurs because of disharmony between the inner and outer environment. The body's insufficient status or weakness of resistance is the root cause of disease, and external pathogens can invade only if there is weakened inner resistance.

There are four diagnostic methods in TCM: inspection (looking), listening, interrogation (asking), and palpation (touching), with inspection of the tongue and pulse taking as the most important procedures. By objectively analyzing information obtained through these methods, combined with various combinations of the eight guiding principles, practitioners of TCM diagnose disease (Figure 9). For example, a patient might be diagnosed as having an "internal cold", "external heat", "deficient yang" or "exuberant yin" condition. Used in conjunction with the five element theory, the eight guiding principles give the TCM practitioner a more complete picture of a patient's energy imbalances and determine the treatment to be recommended.

Treating principles include proper arrangements for root causes and superficial manifestations (symptoms and signs), supporting the body's resistance and eliminating pathogens, orthodox and heterodox therapies and regulating yin-

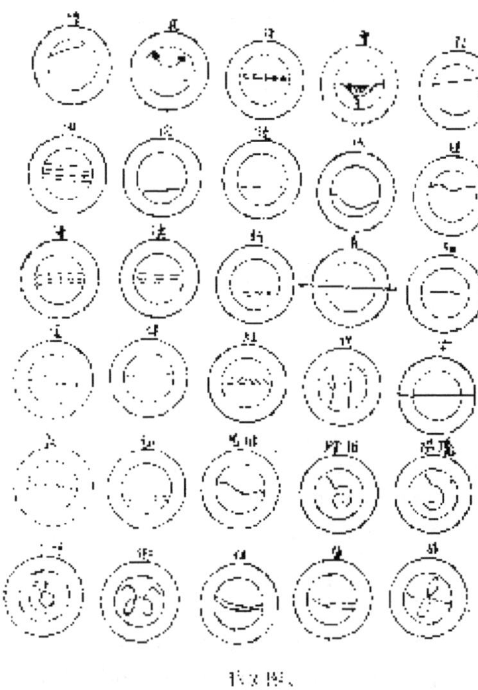

Figure 9 Diagram illustrating the imaginary figures of 30 kinds of pulses, each with a special name describing its properties. From *Cha bing zhi nan* (*Guidelines for Diagnosis of Diseases*) by Shi Fa, 1241. Used with the permission of the China Institute for the History of Medicine and Medical Literature.

yang harmony. Treatment varies according to individual conditions such as mental type and personal constitution and physique. Different recipes many be prescribed to different individuals with the same disease or with the same disease at different stages in the same person. Under the guidance of such principles, proper therapy is administered, which usually involves recipes and drugs.

In a TCM context, prophylaxis includes prevention of diseases before they actually occur and preventing new complications or aggravations of existing diseases. Care for the aged and the search for longevity are also important parts of the system (Cai, 1998).

CHINESE MATERIA MEDICA

Ben cao (herbal), the ancient term for Chinese materia medica, first emerged in the Han dynasty (221 BC–220 AD). This study of herbal medicinals included:

- Knowledge of the properties of Chinese natural drugs and principles for their utilization. The significant properties included taste, nature or property, channel affinity or channel tropism, toxicity, tendency towards tonification and reduction, compatibility when compounded, indications and contraindications, dosage, and methods of decoction and administration. All the knowledge was accumulated through clinical practice.[2]
- Indications and theoretical studies aimed at better understanding and application in clinical usage.
- Preparation of crude drugs, including the origin, classification, identification, cultivation, animal domestication for drugs derived from animals, crude drugs, the so-called "genuine producing locations of drugs" and drug processing for eliminating or minimizing drug toxicity and increasing their therapeutic potential.

The art of drug forms is a branch of science dealing with the compatibility of different ingredients when put together in a compound recipe and their clinical applications. As a basis for treatment in TCM, the art of drug forms investigates the classification, principles for compounding, compatibility among various drugs in a compound prescription, methods of administration, actions and indications, clinical applications, as well as therapeutic connections with modern research. Since it is an integral part of the TCM system of "rationale-principle-recipe-drug", the art of drug forms can be used to differentiate syndromes and establish principles and is therefore subordinate to treating principles. The latter form the basis for creating and applying new recipes. Recipes, a central part of clinical practice, are created from many individual ingredients according to strict compounding principles. These compound recipes satisfy the needs of complex disease conditions and yield synergic action among all ingredients while eliminating untoward actions of certain ingredients at the same time (Editorial Committee, 1985).

Herbs used in Chinese medicine are derived from plant, animal, and mineral substances. Although plant-derived herbs, such as ginseng and ginger, are the most common, minerals and animal parts such as oyster shells, deer antlers, and bear gall bladder are also prescribed.

CLINICAL SCIENCE

Internal medicine

Internal medicine is based on the general ideas that inner impairment will always have an outward manifestation and that superficial and external symptoms reflect an underlying impairment. In this sense, internal medicine is the basis of all other clinical departments. Generally it includes two aspects: seasonal infections (including cold pathogenic diseases and warm epidemic infections) and miscellaneous disorders of the viscera. Cold diseases include, in a broad sense, all kinds of disorders caused by exopathogens, while in a narrow sense they refer only to diseases caused by cold exopathogens. Exopathogens invade the body through the cutaneous layer and hit the viscera by way of channels. Manifestations appear at the superficies in the initial stage, but eventually other symptoms develop such as abdominal distension, diarrhea, and syncope, which are due to yang depletion and impairment of the body's resistance. Epidemic warm diseases are mostly caused by warm pathogens invading the body through the nostrils and mouth. Most miscellaneous diseases are caused by internal injures or emotional disturbances, overstrain, or improper diet.

Gynecology

Gynecology is a branch of medicine dealing with the prevention and treatment of women's diseases. TCM gynecology has a long history of several thousand years. As early as 3000 years ago, in inscriptions on oracles of tortoise shells and animal bones, there is mention of a "delivery disease", while in the *Classic of Seas and Mountains* (*ca.* 2000 years ago), there are medicines for fertilization and infertility. The *Yellow Emperor's Inner Canon of Medicine* lays down the foundation of gynecology, mentioning the uterus and discussing issues related to menstruation. There are also descriptions of amenorrhea, leucorrhea, and massive uterine hemorrhea.[3]

Pediatrics

Pediatrics is a branch of science dealing with the diagnosis, treatment, and prevention of diseases in children and with their growth and development. Pediatrics in TCM plays an important role in the flourishing of the Chinese population. Since children's bodies are particularly tender while they are growing, the manner in which their illnesses are diagnosed and treated is especially important. Children are by no means miniature adults, and their physiology and the pathogenesis of their diseases demands special knowledge. However, pediatrics is also derived, like other branches of medical science, from the *Inner Canon*, *Classic of Questioning*, *Shennong's Classic of Materia Medica*, and *Essays on Cold and Miscellaneous Diseases.* It has developed as an independent discipline since the Jin-Tang dynasties (255–907), and it flourished in the Song dynasty. There was a branch of pediatrics in the Imperial Medical Academy of the Tang dynasty (618–907); in the Song dynasty (960–1234), a Minor-Recipe-Pulse Department was set up in the Imperial Medical Academy, and a Benevolence Bureau for Children was established for foundlings, devoted exclusively to pediatrics.

External medicine

External medicine deals with the diagnosis and treatment of diseases such as ulcers or tumors, diseases of the thyroid gland, mammary gland or anus, and dermatosis. Formerly, injuries due to falling, external wounds, and injuries of the sense organs were also included under this category, but they all belong to independent disciplines now. Most diseases in external medicine are visible and treated by both external and internal therapies; of course, some can be cured by external therapy alone. The treatment of external diseases is based on the general principles of TCM, with different therapies assigned for different people, paying attention not just to the lesions but also to the body as a whole.

Osteology and traumatology

In the oracles on animal bones and tortoise shells there were records of diseases of the hands, arms, legs, feet and toes which were treated with wine preparations. At the same time, bronze instruments were used for treatment. As early as 2000–3000 years ago, there was a special discipline exclusively for fractures and dislocations of bones. Written during the Southern-Northern dynasties (420–589), *Liu juan zi gui yi fang (Master Liu's Recipes Bequeathed by Spirits)* was the first monograph dealing with ulceration and injuries and with reduction therapy for bone fractures. Other monographs appeared in the same period. In the Sui-Tang dynasties (581–907), hemostasis and suturing methods were used for bone fractures, and paralysis and circulatory disturbances were recognized in bone injuries. During this time, hot compresses and wax were used for injuries to bones and tendons and for dislocations, as subordinate therapies in osteology. Since the Tang dynasty dislocations of all kinds of joints were treated with reduction manipulation (Figures 10, 11).

Ophthalmology

Records on eye diseases may be traced far back to over 3000 years ago, when the King's eye diseases were recorded on oracle bones. More information on the anatomy, physiology and clinical manifestations of the eyes was recorded in the Qin-Han dynasties (221 BC–220 AD), laying the foundation for the establishment of a special discipline. Early in the Jin dynasty (265–420 AD), eye operations, including those for tumors of the eyes, were performed, and snow blindness was also mentioned. In the Tang dynasty (618–907), an independent department was established and monographs on ophthalmology appeared;

Figure 10 Diagram of treatment for dislocation of the vertebral bone by the "hanging-piled-brick method". From Wu Qian, *Yi zong jin jian (Golden Mirror of Medical Orthodoxy)* carved in 1742, collected at the Library of the China Institute for the History of Medicine and Medical Literature. Used with the permission of the Institute.

Figure 11 Diagram showing manipulation for the dislocation of the mandibular joint. From *Yi zong jin jian* (*Golden Mirror of Medical Orthodoxy*) carved in 1742. Used with the permission of the China Institute for the History of Medicine and Medical Literature.

operations on external eye disorders were common. The Five Orbiculi theory was described in *Mi chuan yan ke long mu lun* (*Nagajuna's Ophthalmic Theory Secretly Handed Down*), and the theory of Eight Ramparts was first proposed in works of the Southern Song dynasty (1127–1279).

Oto-rhino-laryngology and stomatology

Since all sense organs connect with the viscera through channels, symptoms of the sense organs reflect the conditions of the underlying viscera. In *Zhu bing yuan hou lun (Pathogenesis and Manifestations of All Diseases)* of the Sui dynasty (581–618), a full chapter specifying 70 conditions was devoted to diseases of these organs with detailed analyses of their pathogenesis and symptoms.

Acupuncture-moxibustion

Acu-moxibustion is extensively applied in all clinical situations. Today this field includes the ancient arts of acupuncture and moxibustion combined with other modern scientific techniques such as electro-acupuncture. Acupuncture is a method of applying needles, herbs or pressure at acupoints, while moxibustion is the application of burning moxa floss or other combustible materials at the acupoints or diseased sites. Acupoints are the specific sites through which the qi is transported to the body surface. The Chinese characters for an acupoint mean respectively "transportation" and "hole". Acupoints are not only the pathways for the circulation of qi and blood, but also the loci of response to diseases. As early as 10,000 years ago, stone knives were used. These stone instruments were excavated in Dolun County in Inner Mongolia and Rizhao County in Shandong Province. It is stated in the *Inner Canon* that acupuncture originated with fisherman of the eastern coast, while moxibustion came from northern nomadic people. In the 6th century, acu-moxibustion was transmitted to Paeche in ancient Korea and eventually to Japan, to which Zhi Cong, a Chinese native from Wu (now Jiangsu province), carried an *Atlas of Acupoints.* These traditions are still in use in some 60 countries. There is evidence that acu-moxibustion also spread to Uddyana in ancient India as early as the 6th century AD. But it only arrived in Europe in the 16th century, with France as its first site of practice. Bronze models for acupuncture were first cast in the Song dynasty and became more popular in later ages (Figure 12).

The basic mechanism of acupuncture has recently been described as operating in a similar fashion to a secondary neurological system or information system in the body, with the act of stimulating an acupuncture point seen as communicating information vital to basic physical functions. The central concept is that if the circulation of information is disrupted, optimum function is affected and this results in pain or illness.

Tuina or therapeutic massage

Tuina (literally, push and pull) is an effective measure for treating diseases by applying various kinds of manipulation which activate the body and extremities for the prevention and treatment of diseases. It can dredge the channels, smooth the joints, regulate qi and blood, and strengthen the body's resistance. By stimulating or subduing the energy in the body, practitioners help bring the patient's body back into balance. To determine what channels need work, tuina

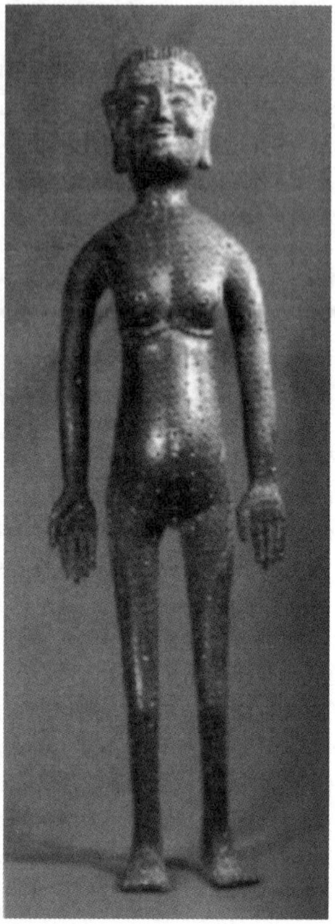

Figure 12 A bronze statue of acupuncture points, cast in 1744 during the Qing dynasty, 46 cm high. From Fu Weikang, *A General History of Chinese Medicine*, Vol. IV: *Illustration Album*. Used with the permission of the publisher.

practitioners feel the patient's wrist pulse. Because it is based on the same channel points as acupuncture, it is often called "acupuncture without needles". Tuina regulates the nervous system so that qi flows properly, boosts the immunological qi of the body, and flushes out metabolic waste. It is a popular part of the practice of TCM, because it is both comfortable and effective. This technique was originally applied over 2000 years ago. A special department was established in the Imperial Medical Academy in the Tang dynasty as one of four, the other three being internal medicine, acupuncture, and conjuration and incantation (see Table 1).

Health care

In a TCM context, health care refers to the discipline devoted to the study of birth, senility, illness and death, especially focusing on the prevention and treatment of senile disorders and on prolonging life. This branch of science began over 2500 years ago. It takes a comprehensive approach to wellbeing, including living a healthy lifestyle, regulating sexual activity, cultivating mental health, eating proper and healthy foods, performing physical labor and taking certain drugs and doing breathing exercises (*Qigong*). Qigong incorporates posture, movement, breathing, meditation, visualization and conscious intent in order to cleanse or purify the qi.

Nursing

Following the basic TCM principles, based on clinical results obtained through the four diagnostic methods and the eight diagnostic guidelines, in combination with psychology and nutrition therapy, nursing aims at the rehabilitation of diseases. The six external pathogenic factors (wind, cold, heat, moisture, dryness and summer heat) interacting with the seven emotions (anger, joy, sorrow, grief, pensiveness, fear and fright) form the theoretical foundation of disease pathology. The *Inner Canon* established the principles for nursing, which involves the concept of the unity of nature and the human body, observation of the course of illness and diagnostic and therapeutic approaches through channel theory. It developed in all successive dynasties.

Diet therapy

Diet therapy originated in remote antiquity and developed rapidly. It was founded as an independent discipline in the Zhou dynasty about 3000 years ago. In the *Rituals of the Zhou Dynasty*, nutrition is one of the four medical departments and is responsible for foods, drinks, and seasonings. Foods were viewed as medicines in early society and were differentiated into warm, hot, cool, and cold natures. They were also classified into five tastes: sweet, sour, salty, acrid, and bitter, and into properties of ascending, descending, floating and sinking. The Chinese believe that diet is one of the three sources (diet, heredity, and environment) of qi. The foods we eat directly influence the excesses and deficiencies in our bodies. The Chinese approach to diet is grounded in the five element and eight guiding principles theory. Foods are seen as having yin and yang, warming and cooling, drying and moistening properties. Certain foods are better for some people than others, depending on their body type and condition. A person with a "cold damp" condition should not eat a diet of raw fruits and vegetables (which are yin), because they would further exaggerate the loss of body heat and fluid secretion. Conversely, foods that are fried, broiled, high fat, or spicy are seen as warming (yang) because they generate heat and stimulate circulation. A person whose diagnosis is "hot dry" should avoid these foods. In general, the TCM approach to diet is to optimize digestion and increase qi, moisture, and blood. It is an important measure for treating disease; it is even the first choice for treatment in most cases.

SOCIAL BACKGROUND OF TRADITIONAL CHINESE MEDICINE

Traditional Chinese medicine is part of the indigenous knowledge of the Chinese people; it is a summary of years of experience in combating diseases and striving for health and longevity as well as the collective wisdom gathered after China became less isolated from the rest of the world.

Zhang Zhongjing, a famous physician of the Han dynasty (2nd century AD), exhorted people to take care of their health and value their lives. He emphasized that medicine was important not only for protecting oneself but also for treating those higher up on the social scale, including monarchs and parents, and for the poorer classes. In the preface of his book, *Shang han za bing lun (Essays on Disorders of Cold and Miscellaneous Diseases)*, Zhang noted that two thirds of more than 200 people in his gens died within a decade, simply because they did not pay attention to their health. 70% died of febrile diseases caused by cold exopathogens, according to his understanding of etiological factors. Diseases, especially epidemics, were the cause of enormous fatalities. Historical facts are very convincing in showing that TCM helps to keep the population flourishing.

Naturally, population changes are influenced by many factors, including warfare, natural disasters and the prevalence of epidemics. All the factors are mutually affected, related and restricted to each other. Most natural disasters are unavoidable, while warfare is a man-made adversity, the former often being one of the causes of the latter. Physical disorders and epidemics are the results of both, which can, in turn, cause further deterioration. Despite the many factors influencing changes in population, however, epidemics most certainly cause the population to drop abruptly.

The role of TCM in population changes can readily be seen on a macroscopic level. According to demographic statistics, the Chinese population always fluctuated by about fifty million during a period of about two thousand years up to the middle of the 18th century. The population reached one hundred million before the reign of Qianlong (1706–1795) of the Qing dynasty. After this, it increased quickly to above four hundred million during the reign of Daoguang (1821–1850). Though the factors influencing the increase of population are very complex, and one cannot exclusively attribute it to the merits of TCM, the significant role of TCM is by no means to be ignored.

The ancients in China mentioned the body, the country, the society and the universe in the same breath, and looked upon the body as a small world or universe. A physician's treating patients, eliminating diseases and taking care of health is analogous to a monarch's governing a nation.

As for the social status of medical professionals, in the early stages of feudal society, or before the Tang dynasty (10th century), the social status of medicine as a profession was still very low. In a book of the Song dynasty, *Chi shui xuan zhu (The Black Pearl from the Red River)*, there is a story about Wen Zhi, a famous physician of the Song Kingdom in the Spring and Autumn period (before the 5th century BC). When Wen Zhi was invited by the King of Qi to treat his severe pyogenic and ulcerous infections of the skin, he used a kind of psychotherapy that enraged the patient. Though he told the prince beforehand that he would die of the treating method when the king was cured, and the prince did guarantee to preserve his life, Wen Zhi eventually was put to death after the king's health was restored.

Certainly, some people engaged in medical work did have higher social positions, but these men were basically not professional physicians but doctors who used witchcraft to treat diseases while concurrently applying some medical measures. The social status of witch doctors was rather high during the period from the Warring States (475 BC–221 BC) to the Qin and Han dynasties (221 BC–220 AD). For instance, Xia Wuju, an attending physician, and Li Xi, an imperial physician to the King during the Qin Kingdom, were both high-ranking officials. They achieved their positions not because of their medical skill but because of their mastering of witchcraft.

Until before the Song dynasty (960–1279), doctors still were located in the middle and lower classes or sometimes even at the bottom of society. Han Yu, a famous writer of the Tang dynasty (618–907), in his book *Shi shuo (Discourses on Teachers)* held that "wizards, doctors, musicians and all kinds of physical laborers are all looked on with disgrace by gentlemen". Another medical writer, Zhang Ji of the Qing dynasty (1844–1911), said, "those who master a task to serve their superiors belong to humble professions. Doctors are all ignorant Philistines". [*Bian que cang gong lie zhuan bu zhu xu (Preface to Supplemental Annotations to the Biographies of Bian Que and Cang Gong)*, Editorial Committee, 1985].

In a feudal society, the monarch defined the nation. His personal desires and hobbies even influenced the development of medicine. Before the Tang dynasty, the loves and hates of the feudal lords had definite influences on medicine as a whole. Some medical organizations and officers were established in each dynasty, such as the Imperial Physician in the Qin dynasty, Aide to the Imperial Physician in the Han dynasty, Medical Scholar and Director of Palace Medication in the Three Kingdoms period, Western and Eastern Jin dynasties, Northern and Southern dynasties and so on. Starting in the Tang dynasty, monarchs paid more attention to medicine. The Imperial Medical Office, a large educational institution, was established for the training of medical personnel on a large scale. A unified national standard for materia medica was issued, namely the *Xin xiu ben cao (Newly Revised Herbal)*, the first pharmacopoeia in China. Some voluminous medical books, such as *Bei ji qian jin yao fang (Emergency Prescriptions Worth a Thousand Pieces of Gold)*, and *Wai tai mi yao (Medical Secrets of an Official)* achieved the emperor's favor, and some of the authors even assumed high ranking official positions.

The first (Taizu), second (Taizong), fourth (Renzong) and the last but one (Huizong) emperors of the Northern Song dynasty all showed a great interest in medicine. Some of them ordered testing, verifying and collecting various prescriptions, and some of them even wrote and compiled their own medical books. For example, Zhao Ji, the last but one emperor, compiled the *Sheng ji jing (Classic of Holy Benevolence)*. Thanks to the founding of the Jiao zheng yi shu ju (Office of Editing Medical Books), which arranged, systematized and collated important ancient medical classics, these books were saved from oblivion. There was a series of outstanding achievements in the Song dynasty, such as the casting of a bronze model for acupuncture, substantial developments in clinical medicine, and comprehensive collation and summarization of herbal classics.

Confucianism also played a role in the advancement of TCM. Under the guiding ideology of *San gang wu chang* (the three cardinal guides [ruler guides subject, father guides son, and husband guides wife] and the five constant virtues [benevolence, righteousness, propriety, wisdom, and fidelity]), men in the Song dynasty prepared to become government officials. If one failed the government's examinations, one would shift to the medical profession. Many learned TCM because of filial piety and to prove loyalty to the emperor. Fan Zhongyan, a famous figure of the Song dynasty who was once a prime minister, coined the aphorism, "Either be a good prime minister or a good doctor". The academic knowledge of medical scholars in the Jin and Yuan dynasties combined with the philosophy of Neo-Confucianism in the Ming and Qing dynasties, or the philosophical schools formed from the new stage of Confucianism, exerted significant influence on medicine. Physicians like Zhu Danxi, Zhang Jingyue, and Li Shizhen established several new schools of thought; a school that taught the doctrine of epidemic warm and seasonal febrile disorders and a school of therapy by warm tonification were also founded. At the same time there were tremendous developments in the field of herbal medicine.

However, Confucianism also had negative and restrictive actions on the medical system. The philosophy of the three cardinal guidelines and the five constant virtues badly trammeled some aspects of medicine. There was, for example, an anti-scientific view regarding the discipline of anatomy, an indispensable subject for the progress of medicine as a whole.

MEDICAL INSTITUTIONS, ORGANIZATIONS AND POLICY

Medical institutions and organizations and their related policies represent, to some extent, a nation's medical developmental level. Medical institutions in China have a long history. The *Zhou li (Rituals of the Zhou Dynasty)* describes how medicine was organized into four different offices: master physician, dietetic physician, massage therapist and veterinarian. These demonstrate the clear division of labor in the Zhou dynasty.

Medical institutions developed further in the Qin and Han dynasties. Imperial physicians and their aides were attached to Shao Yu (Chamberlain for Palace Revenues). In the Northern and Southern dynasties, there were other substantial changes. In the Southern dynasty, the imperial physicians and aides became attached to the *qi ju yuan* (Imperial Diary Office) and its commandant. In the Tang dynasty, there were medical scholars and assistants under each provincial governor divided into superior, intermediate, and inferior levels; in the Song, all medical organizations were under the command of the Medical Institute.

The most important medical organization in the Ming and Qing dynasties was the Imperial Academy of Medicine. Different positions were established for medical professionals, including *yuan shi* (Director of the Academy), administrative assistant, imperial physician, medical secretary, medical official, physician and student of general medicine. There were 13 departments in the

Imperial Academy of Medicine in the Ming dynasty and 11 (later changed to 9) in the Qing. Physicians and students of general medicine were selected from those who passed their examinations with excellent results and qualified. The Imperial Academy also had a storehouse for crude drugs collected from all corners of the country. In addition, in the Ming dynasty, there was an Imperial Dispensary in the service of the royal palace to which chief stewards and chief and apprentice pharmacists were appointed; in the Qing dynasty, clerk, master, assistant and other positions were established.

MEDICAL EDUCATION

China has a long history of medical education. In the beginning medical education was conducted by private teaching, mostly in a tutor-apprentice pattern.

The scale of the Imperial Medical Office of the Sui dynasty was enlarged and expanded in the subsequent Tang dynasty. There were 4 faculties: medicine, acupuncture, massage, and conjuration and incantation (Table 1). A pharmacological department was also set up, with a medical botanical garden attached. Table 1 is a summary of the organization, institutions, personnel, teaching curricula and subjects of the Office.

MEDICAL ETHICS

In China, the chapters *Da yi xi ye (Practices of an Excellent Physician)* and *Da yi jing cheng (Good Faith of an Excellent Physician)* in the book *Bei ji qian jin yao fang (Emergency Prescriptions Worth a Thousand Pieces of Gold)*, written by Sun Simiao of the Tang dynasty are popularly recognized as the codex setting out the general principles for medical morality in TCM. These chapters

		Personnel					
		Scholar	**Instructor**	**Tutor**	**Worker**	**Student**	**Curriculum**
Department	**Medicine** divided into disciplines of pediatrics, dermatopathy, otorhinolaryngology, ophthalmology, stomatology, etc.	1	1	20	100	40	*Ben cao jing (Classic Herbal), Zhen jiu jia yi jing (A-B Classic of Acupuncture and Moxibustion), Mai jing (Classic of Pulse Diagnosis), and others*
	Acupuncture nine forms of acupuncture needles and their usage	1	1	10	20	20	*Ming tang tu zhen jing (Channels and Acupoints Chart), Su wen (Plain Questions), and others.*
	Massage	1		4	16	15	
	Conjuration and Incantation	1		2	8	10	

Table 1 The organization of the Imperial Medical Office of the Tang dynasty (619–907 AD).

include all requirements for a physician in medical practice. New ideas were all based on the above codices, especially the *Da yi jing cheng* (Cai, 1984).

Medical ethics in TCM have a very wide range. They can be summarized as follows:

1. Be assiduous and love the profession.

   If you want to become an excellent doctor, you must read extensively to increase your medical knowledge. Ignorant and idle people without aspiration are not qualified to be doctors. Even a doctor with a great deal of knowledge should continuously improve and update his information and skill.

2. Do your best to save lives regardless of payment and rewards.

   Doctors of TCM must be humanitarians, making no distinction between the rich and the poor.

3. Work in a proper style.

   A doctor must treat all patients. He must visit them at their homes instead of waiting for patients to come to him. If he does not behave well, or if he has some immodest evil thoughts, he will suffer material and spiritual harm.

4. Adhere to science and fight against superstition.

   Medicine in China was intermingled with witchcraft to some extent at an early stage and both were involved in the development of TCM. However, it became clear that they were antagonistic to each other and eventually each had its own separate development. Real doctors must get rid of superstitions and insist on scientific behavior.

MEDICINE OF MINORITIES

China is a multinational country. Traditional Chinese medicine is created collectively by all nationalities, although each nationality also has its own unique medical experiences and achievements.

Because of differences in environment, diet, habits and customs, views about health care and religious beliefs, each nationality has distinct local characteristics. There is a wealth of rich minority medical practices in China directly connected to ethical beliefs and value systems, customs, and natural conditions, forming a multifarious medical community.

As a rule, theory in medicine is created by accumulating experiential knowledge which in turn guides clinical practice. In China, medical developments of the 55 minority groups are very uneven. Their level of development can be categorized into three types (Hong, 1996):

1. Medicine with a complete theoretical system. There are many recorded documents in different linguistic characters that describe basic theory, medical organizations, diagnostic methods and therapy and pharmacology. The medical systems of Mongolians (Figure 13), Uygurs, Koreans and the Dai people fall into this category.
2. Medicine with rich practical experiences and diversified therapeutic methods, forming some theories. However these groups have no linguistic script of their own and no documentary records. Their fragmentary theory and medical experiences are recorded using other writing forms, generally Han

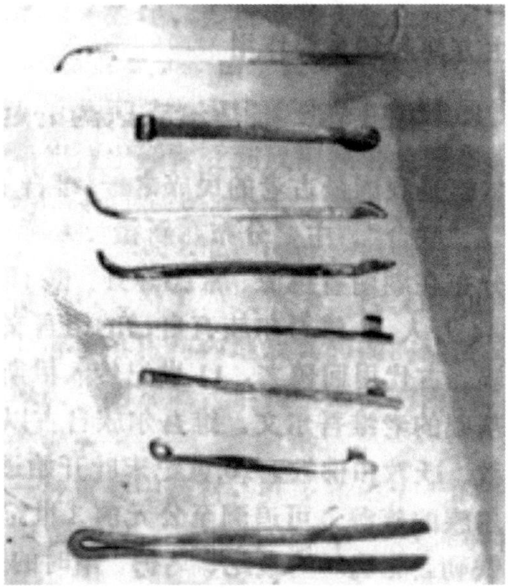

Figure 13 Medical instruments used in Mongolian medicine. From Hong Wuli's *A History of Medicine of Chinese Minorities*. Nanning: Guangxi Press of Science and Technology. Used with the permission of the publisher.

(commonly called Chinese) characters. Because of the scarcity of records, the theoretical contents have depended on oral transmission and are difficult to summarize. They cannot be further upgraded and systematized. Medical know-how is practiced mostly without systematic theoretical knowledge. Most of the Chinese minority medical systems fall into this group.

3. Medicine based on the customs and habits of the minority group and their own natural surroundings, medical techniques and folkloric skills. Local herbs are applied without exchanging knowledge with the outside world; no theoretical knowledge has been summarized. Minorities with shorter histories, without their own writing characters and literature, and with a small population always fall into this group.

All kinds of minorities' medical systems and practices are parts of traditional Chinese medicine as a whole, and contribute to some degree to the formation and development of TCM.

NOTES

[1] Qi is the energy that gives us the ability to move, think, feel, and work. Qi circulates along a system of conduits, the principal ones being channels or meridians. There are twelve main bilateral channels of qi, each intimately connected with one of the viscera of the body, and each manifesting its own characteristic qi (e.g. liver qi, gallbladder qi, etc.). When the flow of qi becomes unbalanced through physical, emotional, or environmental insults, illness may result.

[2] Herbs have four basic qualities and properties: nature, taste, affinity, and primary action.

- Nature: An herb's nature is often described as cooling or heating, but it can also be described as moistening, relaxing, and energizing. The peppermint herb, for example, has a cooling energy, and is used to lower the metabolism or reduce gas and bloating.

- Taste: Herbs are categorized by five tastes – sour, bitter, sweet or bland, spicy, and salty, and herbs representing different tastes are used to treat different conditions.
- Affinity: This property refers to the affinity that an herb has for a particular organ network.
- Primary action: An herb may be used to dispel (move), astringe (restrain), purge (expel), or tonify (strengthen).

[3] Amenorrhea is the absence or discontinuation or abnormal stoppage of the menstrual periods, Leucorrhea is excess formation and discharge of mucus in the female genital tract, and hemorrhea is excessive bleeding.

BIBLIOGRAPHY

Anonymous. *Zhou li (Ritual of the Zhou Dynasty), ca.* 8th century BC. Shanghai: The Commercial Press, reprinted 1929.

Anonymous. *Ling shu (Miraculous Pivot), ca.* 475 BC. Beijing: People's Health Publishing House, reprinted 1956.

Anonymous. *Shen nong ben cao jing (Shennong's Classic of Materia Medica), ca.* 2nd century AD. Beijing: People's Health Publishing House, reprinted 1963.

Cai, Jingfeng and Wuli Hong. 'A brief introduction to the development and characteristics of medical ethics in early Tibetan medicine.' In *Anthology of an Academic Symposium of Tibetan Nationality*. Lhasa: People's Publishing House of Tibet, 1984, pp. 516–531.

Cai, Jingfeng and Wuli Hong. *Zhong guo zang yi xue (China's Tibetan Medicine)*. Beijing: Science Press, 1995.

Cai, Jingfeng and Ying'ao Yu. *Yi yao xue zhi (Annals of Medical Science)*. Shanghai: Shanghai People's Publishing House, 1998.

Cai, Jingfeng. *A General History of Chinese Medicine*, Vol. 3: *Contemporary Period.* Beijing: People's Health Publishing House, 2000.

Chen, Wuze. *San yin ji yi bing zheng fang lun (Discourses on Recipes and Syndromes of Three Etiological Factors)*, 1174. Beijing: People's Health Publishing House, reprinted 1957.

Chen, Youbang. *Emergency Medicine in TCM*. Fuzhou: Fujian Press of Science and Technology, 1995.

Chen, Zhiming. *Fu ren da quan liang fang (Complete Effective Recipes for Women's Diseases)*, 1237. Shanghai: Shanghai Health Publishing House, reprinted 1956.

Editorial Committee. 'Traditional Chinese medicine.' In *Chinese Medical Encyclopedia*. Shanghai: Shanghai Press of Science and Technology, 1985, pp. 3–51.

Han, Jisheng. 'Acupuncture analgesia and the related neural-pathway and neurotransmitter.' *Progress in Physiology* 4: 284–290, 1984.

Hong, Wuli. *A History of Medicine of Chinese Minorities.* Nanning: Guangxi Press of Science and Technology, 1996.

Hua Shou. *Shi si jing fa hui (Elaboration on 14 Channels)*, 1341. Shanghai: Shanghai Health Publishing House, reprinted 1956.

Huangfu, Mi. *Zhen jiu jia yi jing (A-B Classic of Acu-moxibustion)*, 282. Beijing: People's Health Publishing House, reprinted 1956.

Lei Xiao. *Lei gong pao zhi lun (Master Lei on Drug Processing)*, 588. Nanjing: Jiangsu Press of Science and Technology, reprinted 1985.

Li, Jingwei. *A General History of Chinese Medicine*, volume 1: *Ancient Period.* Beijing: People's Health Publishing House, 1999.

Li, Shizhen. *Ben cao gang mu (Compendium of Materia Medica)*, 1590. Beijing: People's Health Publishing House, reprinted 1957.

Qian, Yi. *Xiao er yao zheng zhi jue (Key to Differentiation and Treatment of Children's Diseases)*, 1119. Shanghai: Shanghai Health Publishing House, reprinted 1955.

Qin, Yueren (Bian Que). *Nan jing (Classic of Questioning)*, ca. 5th century BC. Beijing: Press of Literature of Science and Technology, reprinted 1999.

Ren, Yingqiu. *Theories from Various TCM Schools.* Shanghai: Shanghai Press of Science and Technology, 1980.

Shen, Ziyin. 'Introduction to TCM basic theory.' *Journal of TCM* 10: 12–15, 1979.

Su, Jing. *Xin xiu ben cao (Newly Revised Herbal)*, 659. Shanghai: Shanghai Health Publishing House, reprinted 1957.

Tang, Shenwei. *Chong xiu zheng he jing shi zheng lei bei ji ben cao* (*Revised Classified Herbal of the Zhenghe Period for Emergency Use Based on Classics and Historical Documents*), 1082. Beijing: People's Health Publishing House, reprinted 1957.

Tao, Hongjing. *Ben cao jing ji zhu (Variorum of Classic Herbal)*, 500. Shanghai: Qun Lian Press, reprinted 1955.

Wang, Minghui. 'The history and present status of studies on theory of qi (vital energy).' *Journal of New TCM* 2: 1–5, 1980.

Wang, Shuhe. *Mai jing (Classic of Pulse Diagnosis)*, 280. Shanghai: Shanghai Health Publishing House, reprinted 1957.

Wang, Weiyi. *Tong ren shu xue zhen jiu tu jing (Illustrated Acu-moxibustion Classic of Acupoints in Bronze Models)*, 1026. Beijing: People's Health Publishing House, reprinted 1955.

Wu, Youxing. *Wen yi lun (On Epidemic Warm Diseases)*, 1642. Beijing: People's Health Publishing House, reprinted 1955.

Xu, Shen. *Shuo wen jie zi (Dictionary for Words)*, 121. Beijing: China Press, reprinted 1979.

Zhang, Zhongjing. *Shang han za bing lun (Essays on Cold and Miscellaneous Diseases)*, 219. Shanghai: Shanghai People's Publishing House, reprinted 1976.

ANANDA S. CHOPRA

ĀYURVEDA

The Sanskrit term Āyurveda may be translated as "science or wisdom (*veda*) of life (*āyus*)". Āyurveda denotes an indigenous living medical tradition on the Indian subcontinent. Āyurveda has a long textual history, and proponents of present-day Āyurveda see their science as the continuation of a living tradition spanning at least two millennia. This perspective tends to overlook historical evolution and changes of ideas and is fundamentally different from a historical-philological approach.

This article aims at giving a concise description of the fundamentals of the Ayurvedic science as it forms basis and background for Ayurvedic physicians. Following a popular practice of Ayurvedic physicians, we will be referring to classical texts in an eclectic manner. This practice reflects the observation that many basic concepts of Āyurveda have remained surprisingly constant over centuries, although medical practice underwent changes.

As the term "science of life" implies, Āyurveda is not only a mode of therapy but, as an old text says (CS, Sū, 30.26), its purpose is to "protect the health of the healthy and pacify the disease of the diseased". Thus a whole system of dietary recommendations also forms part of Ayurvedic science.

The doctrine of "five great elements" (*pañcamahābhūta-*, literally "five great exisiting entities") forms a point of departure for systematic Ayurvedic science. The elements – earth, water, fire, air and space – make up the entire phenomenal world including the human body. This idea, sometimes called the fundamental correspondence between the world and man, is central to Āyurveda, as it explains the interaction between humans and the world in many respects. A simple example may be cited to illustrate this point: in dietetics it is said that sweet foods are predominantly made up of the elements earth and water. In the human body fatty tissue (*medas*) is made up mainly of the same elements (see the verses quoted in Ḍalhaṇa's comment ad SuS, Sū. 15.14.) Thus one sees that excessive eating of sweet substances increases fatty tissue in the body. In practice the elementary composition of a substance is known by its properties or qualities (Sanskrit: *guṇa*). In Āyurveda a list of 20 qualities became widely accepted (see Table 1) and served as a framework for diagnosis, therapy and dietetics. The principle is to treat with opposite qualities.

H. Selin (ed.), Medicine Across Cultures: History and Practice of Medicine in Non-Western Cultures, 75–83.

Table 1 20 guṇa- (qualities) (see CS, Sū, 25. 36)

heavy (*guru-*)	light (*laghu-*)
cold (*śīta-*)	hot (uṣṇa-)
unctuous (*snigdha-*)	dry (*rūkṣa-*)
dull (*manda-*)	sharp (*tīkṣṇa-*)
stable (*sthira-*)	mobile (*sara-*)
soft (*mṛdu-*)	hard (*kaṭhina-*)
non-slimy (*visáda-*)	slimy (*picchila-*)
smooth (*ślakṣṇa-*)	coarse (*khara-*)
minute (*sūkṣma-*)	gross (*sthūla-*)
viscous (*sāndra-*)	liquid (*drava-*)

Humans are made of three pillars: the imperishable self (*ātmā-*), the mind (*sattva*) and the body (*śarīra*) (CS, Sū. 1.46). As the self is eternal and imperishable, health and disease affect body and mind. Turning to the body, one may roughly distinguish between structural and functional components. There is early anatomical data from ancient India because of the practice of Vedic sacrifice, where the anatomy of the sacrificial animal was carefully studied. A crude form of dissecting the human body to acquire anatomical knowledge was probably practised in early times (SuS, Śā. 5.49), but later the study of anatomy became less important to Ayurvedic physicians. The classical texts contain long enumerations of body parts, which may differ at times, as for example regarding the total number of bones in the body. The surgically-oriented *Suśruta-Saṃhitā* counted a total of three hundred bones in contrast to the opinion of "Vedists" (*vedavādinaḥ*, cf. SuS, S̄, 5.18) who counted 360. Various ducts or channels were also named (designated as *srotas-*, *sirā-*, *dhamanī*); however, the exact meaning as well as the possible anatomical correlates of these is poorly understood.

A special feature of Ayurvedic anatomy was the *marma-*s: a number[1] of vulnerable points which had to be specially protected, especially while performing surgical procedures. The body is further made up of seven constituents (in this context usually called *dhātu-* literally "essence"): *rasa-* (juice, which evolves after digestion of food), *rakta-* (blood), *māṃsa-* (flesh), *medas-* (fat), *asthi-* (bone), *majjā-* (marrow) and *śukra-* (semen). These are formed in a process of metabolism starting with the digestion of food (see CS, Sū. 28.4). If this physiological process functions well then *ojas* (vitality) is produced. Along with the seven *dhātus*, "secondary *dhātus*" (*upadhātu*, including breast milk, skin, and sinews [CS, Ci. 15.17]) and excretions (*mala-*, stool, urine, and bile [CS, Ci. 15.18–19, SuS, Sū. 46.527]) are also formed. This list, as well as the exact nature of the metabolic process, was a matter of controversy among Ayurvedic authors.[2] All these may roughly be classed as human structural components. There are also functional components; these comprise *agni* (literally "fire") and the three *doṣas* in the narrower technical sense of the word (see below). Transforming forces in general are called *agni* in Āyurveda. The most prominent one (and probably also the most accessible to examination) is the "digestive fire" (*jāṭharāgni*, *kāyāgni*, etc.) which "cooks" the food and renders it useful.

Other *agnis* are also distinguished, acting on the five elements or the seven constituents mentioned above.

One of the most important features of Āyurveda is the doctrine of three *doṣas*. In a narrow technical sense we speak of three *doṣas* called *vāta* (literally "wind"), *pitta* (bile) and *kapha* (phlegm). Blood was counted as the fourth *doṣa* in some old schools, but over time the idea of three *doṣas* prevailed. The word *doṣa*- may literally mean "vitiator", to corrupt or make ineffective; however, the three *doṣas* play important roles in many respects. For the sake of clarity we will treat some of these aspects separately, but it should be borne in mind that in the original texts these are intertwined.

Doṣa regulating physiological processes

Vāta regulates processes which are characterised by movement, such as respiration, intestinal peristalsis, and excretion. *Pitta* is responsible for transformatory processes such as digestion and metabolism as well as psychic processes. *Kapha* gives firmness to the body, regulates stabilising functions like lubricating the joints and refreshes the sense organs.

Doṣa-constitution

Every human being possesses an individual constellation of the three *doṣas* at birth. This is his or her unique constitution (*prakṛti*). The constitution accounts for many individual characteristics, both physical and psychological (Table 2). This is not the only theory of constitution we find in Āyurveda (see for example the "psychic constitutions" in CS Śā. 4 and SuS, Śā. 4.80–98 or the "essence/

Table 2 Some characteristics of *doṣa*-constitution (*prakṛti*)

VĀTA	PITTA	KAPHA
Underdeveloped physique, prominence of veins and tendons, prominence of bones and joints Dry, cracked, lustreless skin, cracked and rough nails Feels cold easily	Medium physique Moles, soft and wrinkled skin, reddish and soft nails Sweats easily	Well-developed physique, veins and tendons not visible Soft smooth skin, soft and thick hair
Thin eyelashes, restless eyes	Tendency to develop conjunctival infection, medium eyebrows	Dense eyebrows, moist eyes
Poor irregular appetite, flatulence and severe constipation Disturbed sleep (while going to sleep) Little will power Good mental grasp, short memory, moody	Strong appetite, good digestion, thirst Light sleep Medium memory Assertive	Regular and good appetite, moderate digestion, soft stool Sound and prolonged sleep Slow grasp, good memory Strong will power

constitution" in CS, Vi. 8.102–111), but it turned out to be the most influential one and has assumed great significance in contemporary "New Age Ayurveda".

The doṣa as morbific (disease-causing) factor

The *doṣa*s regulate physiological processes, but they may also initiate pathological processes. To explain this aspect, as well as the determination of *doṣa* constitution, the *doṣa*s are characterised by qualities: *vāta* is dry, cold, light and unstable; *pitta* is slightly viscous, hot, acrid and sour and *kapha* is viscous, cold, heavy and slimy (comp. CS, Sū. 1.59–61). Influences of the same quality as the respective *doṣa*s lead to an increase of the *doṣa*. For example, dry food and cold weather may lead to an increase of *vāta*, which has the qualities dry and cold; hot and sour food increases *pitta* and heavy and sweet food increases *kapha*. Although the *doṣa*s act on the whole body there are regions where a particular *doṣa* is predominantly found. *Kapha* is mainly in the head and the chest, *pitta* in the upper abdomen and *vāta* below the navel. The pathophysiological process starts usually in the region of the affected *doṣa* and then spreads to other parts of the body. At this stage of disease development *vāta* plays an important role, as it alone has the capacity to cause movement. If this process continues, the *doṣa* coalesces with other body constituents and leads to a localised disease. In this context the term *doṣa* is employed in a sense close to its literal meaning (i.e. vitiator); the affected body parts are then summarily termed *dūṣya* (the one to be affected/vitiated).

The qualities associated with the *doṣa*s are also important for dietetics and therapy as actions or drugs of the opposite quality treat the respective *doṣa* increase.

Doṣas as characterising biological rhythms

Day and night are respectively divided into three equal parts dominated sequentially by *kapha*, *pitta* and *vāta*. The same applies to the digestion process, which has three phases regulated by *kapha*, *pitta* and *vāta* respectively. Stages of life and seasons are also described by using the *doṣa*s.

DIETETICS

Ayurvedic dietetics comprise a host of recommendations, ranging from preparation and consumption of food, to healthy routines for day and night, sexual life, and rules for ethical conduct. In contrast to contemporary practitioners of New Age Ayurveda, older Ayurvedic authors tended to be religiously neutral. Even Buddhist authors refrained from trying to convert the patient to follow their particular religious ways.

DIAGNOSIS AND NOSOLOGY

In such a long tradition, change and variation in diagnostic strategies are to be expected. A remark in the *Suśruta-Saṃhitā* indicates that there may always have been different opinions regarding diagnostic strategy (SuS, Sū. 10.4 as read against AHS, Sā. 1.22ab). The patient should be questioned and the doctor use all five sense perceptions, including smell and sight. Other diagnostic concepts were more elaborate: the *Caraka-Saṃhitā* (Vi. 8.94) speaks of a tenfold examination including

- constitution (*prakṛti*),
- abnormality (*vikṛti*),
- "essence" (*sāra*),
- stability (*saṃhanana*),
- measurements (*pramāṇa*, here different body parts are measured in finger breadths),
- suitability (*sātmya*, suitability of food),
- psychic strength (*sattva*),
- digestive capacity (*āhāraśakti*),
- physical fitness (*vyāyāmaśakti*) and
- age (*vayas*).

From about the 13th–14th centuries onwards, pulse diagnosis (*nāḍīparīkṣā-*) was introduced to Āyurveda. Feeling the pulse enables the physician to assess the status of the three *doṣa*s and may also provide a host of other information. When the system of pulse diagnosis was mentioned for the first time in Ayurvedic literature, in the *Śārṅgadhara-Saṃhitā*, it had already acquired the systematic shape which is referred to in later literature. The question of where and when pulse diagnosis actually entered the Ayurvedic scientific world is still an unsolved riddle. In addition to looking for foreign influences (Chinese medicine or Arabic medicine, which was in close contact with Āyurveda up to the 19th century at least), I believe a remark by Th. Brosse (Filliozat, 1953, § 1671) indicating that influences of tantric Yoga may have helped in bringing pulse diagnosis to an important position in Āyurveda merits further research. Although pulse diagnosis found a firm place in Ayurvedic literature from the 14th century, the impression remains that it "... did not become fully integrated with Āyurveda ..." (Meulenbeld, 1995: 6). In diagnosing disease a concept of five criteria became most influential. The five criteria are: origin of the disease, prodrominal (precursory) symptoms, typical symptoms of the fully developed disease, observing the effect of therapeutic procedures and the pathological process. The most important nosological classification in Āyurveda is associated with the name of Mādhava, whose work, popularly known as *Mādhavanidāna* (*Mādhava's pathology*), is dated about 700 CE. With minor alterations, Mādhava's system served as a model almost up to the present time. In Mādhava's systematic classification of disease entities, the *doṣa*s are often employed in the differential diagnosis of single diseases. Among the many forms of fever, there are, for example, separate *vāta-*, *pitta-* and *kapha-*born fevers (cf. MN, 2.8–13).

THERAPY

Ayurvedic medicine largely relied on its own system of physiology and pathophysiology to explain the nature of humans and disease. Treatment was also decided on this basis, the aim being to restore the normal state. In classical times Āyurveda included eight special disciplines (cf. SuS, Sū. 1.6, comp. also CS, Sū. 30.28):

1. *śalyam* (surgery),
2. *śālākyam* (dealing with diseases above the clavicle),
3. *kāyacikitsā* (medicine),
4. *bhūtavidyā* (science of dealing with spiritual possession),
5. *kaumārabhṛtyam* (paediatrics),
6. *agadatantram* (toxicology),
7. *rasāyanam* (preventive and roborating [strengthening] measures) and
8. *vājīkaraṇam* (aphrodisiacs).

As this short enumeration may already show, the system of Ayurvedic medicine did not need a belief in supernatural (in this context we mean extraneous to its own system of explaining nature) influences to explain or treat disease. The field of *bhūtavidyā* (possession) forms a notable exception, and in the old Ayurvedic texts the use of magical spells, ritual offerings and the like is recommended in such diseases. However, a quote from the *Suśruta-Saṃhitā* reveals that even in these diseases pharmaceutical preparations were used or perhaps even deemed superior (Ut. 60.37cd–38ab): "If with the previously enjoined spells or offerings as taught by the science of dealing with possessions it is not possible to conquer (the diseases) they may be treated by (pharmaceutical) compounds".

Ayurvedic therapy comprised a large number of different techniques and strategies which are classified in diverse ways. Major surgery lost its importance over the centuries, at least among the respected Ayurvedic physicians. Among other groups of people surgical knowledge seems to have survived (Wujastyk, 1993). Herbal preparations remained an immensely popular mode of therapy, which grew with the inclusion of new kinds of herbs introduced to India by different groups of immigrants. For example, Muslim invaders brought opium to India which found its way into the Ayurvedic materia medica.

On the basis of the doctrine of five great elements, an elaborate system to classify plants based on their taste and action on the body evolved. However, as is to be observed in many other medical traditions, it seems that therapeutic recommendations were often made on the basis of empirical data rather than by relying on systematic theory (Meulenbeld, 1987). This applies also to the enormous number of compound formulations which are usually classed according to their therapeutic use for certain diseases. Under the name of *pañcakarman* (five actions) a special mode of therapy is described comprising therapeutic emesis, purgation, oily enemas, watery enemas and the use of errhines to cause sneezing. These measures should be undergone only after previous oleation (applying medicated oils externally and internally) and sudation (sweating). Therapeutic applications of oils, usually prepared with herbal decoctions, were widely used, and up to this day Ayurvedic medicine in southwestern India heavily relies on oil therapies. In the second millennium of the Common Era alchemy became more and more popular in Āyurveda, especially mercury and its preparations. In practical Ayurvedic therapeutics alchemical preparations assumed great importance, although theoretically the systems of alchemy and Āyurveda are not yet convincingly reconciled.

A typical dilemma of healers, not only in India, is reflected here: the physician treads a narrow path between his everyday practice largely dominated by empirical knowledge and his being a member of a learned community adhering to a scientific system. Āyurveda grew in a particular cultural environment, and for Ayurvedic physicians this cultural environment offered both an orientation and a point of departure for formulating a theoretical system. Ayurvedic authors thus used terminology developed in philosophical circles. A case in point is the doctrine of five great elements. This doctrine is to be found in several philosophical schools. Ayurvedic physicians probably borrowed it from philosophers but adapted it to suit their own purposes, as it was found to be useful in explaining medical phenomena happening in practice. The physician was content to have found a theory that served as a basis to his science. Questions beyond this, e.g. the question of evolution of the five great elements, were of minor importance although they are dealt with in old Ayurvedic texts.

In the *Suśruta-Saṃhitā* (Śā. 1.13), a concise description of the evolution of the five great elements in line with classical Sāmkhya philosophy is followed by the statement that "its use is mentioned in respect to therapy, whatever is beyond the elements is not a consideration in therapy". Erich Frauwallner (vol. II, 1956: 7 and 15) remarked that we may distinguish two currents of thought in ancient India. One is interested in metaphysics, as is apparent in the philosophical schools of Sāmkhya and Buddhism, whereas a different current – in general perception quite untypical of Indian culture – is interested in the objects of the external world as they present themselves to man. Frauwallner included materialistic schools of Indian philosophy as well as the classical Vaiśeṣika philosophy in this current of thought, and we may add that Āyurveda is also strongly linked to this worldview. The traditional Ayurvedic physician conveys the image of a pragmatic person in regard to the health and wellbeing of his patients, even if that means at times recommending the eating of beef or the judicious use of alcoholic drinks. In the course of time, as Āyurveda became a *śāstra* (traditional science) and part of the scientific lore of traditional Indian culture, it became more static without ever completely losing its ability to change.

Abbreviations used in the article:

AHS = *Aṣṭāṅgahṛdayasaṃhitā*
CS = *Caraka-Saṃhitā*
MN = *Mādhavanidāna*
Ni. = *Nidānasthāna*
Śā. = *Śārīrasthāna*
Sū. = *Sūtrasthāna*
SuS = *Suśruta-Saṃhitā*
Vi. = *Vimānasthāna*

NOTES

1 The classical number is 107; see SuS, Sā. 6.1.
2 See, for example, Cakrapāni's comments ad CS, Sū. 28.4 and Ci. 15.16 or Ḍalhana's commentary ad SuS, Sū. 14.10.
3 For a classical description of this pathophysiological process see SuS, Sū. 21.18–36.

BIBLIOGRAPHY

(An asterisk marks the editions of classical texts used for references in the text)

Aṣṭāṅgahṛdaya (A Compendium of the Ayurvedic System). Composed by Vāgbhaṭa with the Commentaries *Sarvāṅgasundarā* of Arunadatta and *Āyurvedarasāyana* of Hemādri, Annā Moreśwar Kunte and Kriṣṇa Rāmchandra Sātri Navre, collators; Bhiṣagāchārya Hariśāstri, and Parādkar Vaidya, eds. Varanasi: Krishnadas Academy, 1995. Reprint.

Aṣṭāṅgasaṅgraha Induvyākhyāsaṃhita, D.V. Paṇḍitarāva and Ayodhyā Pāṇḍeya, eds. 2 vols. New Delhi: Central Council of Research in Ayurveda and Siddha, 1991.

**Charakasaṃhitā* by Agniveśa. Revised by Charaka and Dṛidhabala with the Āyurveda-Dīpikā commentary of Chakrapāṇidatta, Vaidya Jādavaji Trikamji Āchārya, ed. 3rd ed. Bombay: Nirnaya Sāgar Press, 1941.

Chopra, Ananda Samir. 'Pañcakarma in Deutschland – Grundlagen, Praxis, Perspektiven.' *Erfahrungsheilkunde: Acta Medica Empirica* 47: 192–198, 1998.

Chopra, Ananda Samir. 'Die Lehre von den Fünf Elementen (*Pañcamahābhūta*) als wissenschaftliche Grundlage der äyurvedischen Therapie.' *Erfahrungsheilkunde: Acta Medica Empirica* 47: 179–182, 1998.

Comba, Antonella. 'Carakasaṃhitā, Śārīrasthāna I and Vaiśeṣika philosophy.' In *Studies on Indian Medical History. Papers presented at the International Workshop on the Study of Indian Medicine held at the Wellcome Institute for the History of Medicine 2–4 September 1985*, G. Jan Meulenbeld and Dominik Wujastyk, ed. Groningen: Egbert Forsten, 1987, pp. 43–61.

Comba, Antonella. 'Universal (*sāmānya*) and particular (*viśeṣa*) in Vaiśeṣika and Āyurveda'. *Journal of the European Ayurvedic Society* 1: 7–32, 1990.

Das, Rahul Peter. 'On the nature and development of "traditional Indian medicine." *Journal of the European Ayurvedic Society* 3: 56–71, 1993.

Dasgupta, Surendranath. *A History of Indian Philosophy*. 5 vols. Cambridge: Cambridge University Press, 1922; 1st Indian edition Delhi: Motilal Banarsidass, 1975, reprinted 1991/1992.

Emmerick, R.E. *The Siddhasāra of Ravigupta Volume 1: The Sanskrit Text* (Verzeichnis der orientalischen Handschriften in Deutschland, Supplementband 23,1). Wiesbaden: Franz Steiner Verlag, 1980.

Filliozat, Jean. 'Médecine.' In *L'Inde classique: Manuel des études indiennes*, Tome II, Louis Renou et Jean Filliozat, eds. Paris: École française d'Extrême-Orient, 1953.

Filliozat, Jean. *The Classical Doctrine of Indian Medicine: Its Origins and its Greek Parallels*, Dev Raj Chanana, trans. Delhi: Munshiram Manoharlal, 1964.

Filliozat, Pierre-Sylvain. '*Yukti, le quatrième pramāṇa des médecins* (Carakasaṃhitā Sūtrasthāna XI, 25).' *Journal of the European Ayurvedic Society* 1: 33–46, 1990.

Filliozat, Pierre-Sylvain. 'Caraka's proof of rebirth.' *Journal of the European Ayurvedic Society* 3: 94–111, 1993.

Frauwallner, Erich. *Geschichte der indischen Philosophie* [History of Indian Philosophy]. 2 vols. Salzburg: Otto Müller Verlag, 1953 (Vol. I) and 1956 (Vol. II).

Halbfass, Wilhelm. 'The therapeutic paradigm and the search for identity in Indian philosophy.' In *Tradition and Reflection: Explorations in Indian Thought*, Wilhelm Halbfass, ed. Albany: State University of New York Press, 1991, pp. 243–263.

Hilgenberg, Luise and Willibald Kirfel. *Vāgbhaṭa's Aṣṭāṅgahṛdayasaṃhitā, ein altindisches Lehrbuch der Heilkunde aus dem Sanskrit ins Deutsche übertragen mit Einleitung, Anmerkungen und Indices*. Leiden: E.J. Brill, 1941.

Jolly, Julius. *Indian Medicine*, C.G. Kashikar, trans. 3rd ed. New Delhi: Munshiram Manoharlal, 1994.

Kutumbiah, P. *Ancient Indian Medicine*. Madras: Orient Longmans, 1962.
Leslie, Charles. 'The ambiguities of medical revivalism in modern India.' In *Asian Medical Systems: A Comparative Study*, Charles Leslie, ed. Delhi: Motilal Banarsidass, 1998, pp. 356–367.
Leslie, Charles, ed. *Asian Medical Systems: A Comparative Study*. 1st Indian ed. Delhi: Motilal Banarsidass, 1998.
**Mādhavanidāna*, by Mādhavakara. With the commentary *Madhukośa* by Vijayarakṣita and Śrīkaṇṭhadatta and with extracts from Ātaṅkadarpaṇa by Vāchaspati Vaidya, Vaidya Jādavjī Tricumjī Āchārya, ed. Varanasi: Chaukhambha Orientalia, 1986.
Meulenbeld, Gerritt Jan. *The Mādhavanidāna and its Chief Commentary*. Leiden: E.J. Brill, 1974.
Meulenbeld, Gerrit Jan. 'Reflections on the basic concepts of Indian pharmacology.' In *Studies on Indian Medical History, Papers presented at the International Workshop on the Study of Indian Medicine held at the Wellcome Institute for the History of Medicine, 2–4 September 1985*, G. Jan Meulenbeld and Dominik Wujastyk, ed. Groningen: Egbert Forsten, 1987, pp. 1–17.
Meulenbeld, Gerrit Jan. 'The characteristics of a *doṣa*.' *Journal of the European Ayurvedic Society* 2: 1–5, 1992.
Meulenbeld, Gerrit Jan. 'The many faces of Āyurveda.' *Journal of the European Ayurvedic Society* 4: 1–10, 1995.
Meulenbeld, Gerritt Jan. *A History of Indian Medical Literature*. 2 vols. Groningen: Egbert Forsten, 1999 (vol. 1A/B); 2000 (vol. 2A/B).
Sharma, Priya Vrat. *Introduction to Dravyaguṇa* (Indian Pharmacology). Varanasi: Chaukhambha Orientalia, 1976.
Sharma, Priya Vrat. *Āyurveda kā vaijñānika itihāsa* (Scientific History of Āyurveda). 2nd ed. Varanasi/Delhi: Chaukhambha Orientalia, 1981.
Sharma, Priya Vrat, ed. *History of Medicine in India: From Antiquity to 1000 AD*. New Delhi: Indian National Science Academy, 1992.
**Sushrutasamhitā* of Sushruta. With the Nibandhasangraha commentary of Shri Dalhaṇāchārya, Jādavji Trikumji Āchārya, ed. Varanasi: Chaukhamba Surbharati Prakashan, reprinted 1994.
Suśruta-Samhitā. With English translation of text and Ḍalhaṇa commentary along with critical notes, Priya Vrat Sharma, ed. and trans. 3 vols. Varanasi: Chaukhamba Visvabharati, 1999–2001.
Udupa, K.N. and R.H. Singh, eds. *Science and Philosophy of Indian Medicine*. 2nd ed. Nagpur: Baidyanath Ayurved Bhawan Ltd., 1990.
Vogel, Claus. *Vāgbhaṭa's Aṣṭāṅgahṛdayasaṃhitā, The First Five Chapters of its Tibetan Version*, C. Vogel, ed. Wiesbaden: Franz Steiner Verlag, 1965.
Wezler, Albrecht. 'On the quadruple division of the Yogaśāstra, the Caturvyūhatva of the Cikitsāśāstra and the "four noble truths" of the Buddha. (Studies in the Pātañjalayogaśāstravivaraṇa II)'. *Indologica Taurinensia* 12: 289–337, 1984.
Wezler, Albrecht. 'On a prose passage in the Yuktidīpikā of some significance for the history of medicine.' *Journal of the European Ayurvedic Society* 3: 282–304, 1993.
Wujastyk, Dominik. 'Indian medicine.' In *Companion Encyclopedia of the History of Medicine*, W.F. Bynum and Roy Porter, eds. London and New York: Routledge, 1993, vol. 1, pp. 755–778.
Wujastyk, Dominik. *The Roots of Āyurveda: Selections from Sanskrit Medical Writings*. New Delhi: Penguin Classics, 1998.
Zimmermann, Francis. *The Jungle and the Aroma of Meats: An Ecological Theme in Hindu Medicine*. 1st Indian edition, Delhi: Motilal Banarsidass, 1999.
Zysk, Kenneth G. 'The science of respiration and the doctrine of the bodily winds in ancient India.' *Journal of the American Oriental Society* 113: 198–213, 1993.
Zysk, Kenneth G. *Medicine in the Veda: Religious Healing in the Veda*. Philadelphia: American Philosophical Society, 1985. 1st Indian edition, Delhi: Motilal Banarsidass, 1996.
Zysk, Kenneth G. *Asceticism and Healing in Ancient India: Medicine in the Buddhist Monastery*. New York: Oxford University Press, 1991. Corrected edition, Delhi: Motilal Banarsidass, 1998.
Zysk, Kenneth G. 'New Age Āyurveda or what happens to Indian medicine when it comes to America.' *Traditional South Asian Medicine* 6: 10–26, 2001.

ROBERT WILLIAM PRASAAD STEINER

CULTURAL PERSPECTIVES ON TRADITIONAL TIBETAN MEDICINE

In memory of Lucy Virginia Ashton Steiner
(November 17, 1913 – December 14, 2000)

His Holiness Dalai Lama XIV offered a statement for the First International Congress of Tibetan Medicine in 1998 (Dalai Lama, 1999):

> As an integrated system of health care, Tibetan medicine can offer allopathic medicine a different perspective on health. However, like other scientific systems, it must be understood in its own terms, as well as in the context of objective investigation. In practice it can also offer Western people another approach to achieving happiness through health and balance.

Tibetan medicine synthesizes components from the health traditions of neighboring regions. Elements from Ayurveda in India, Traditional Chinese Medicine, Unani medicine from the former Persian Empire, shamanic practices from Siberia and Mongolia, and Bön customs of pre-Buddhist Tibet are all present. Despite the presence of these components from other cultures, Tibetan medicine retains unique characteristics.

Traditional Tibetan medicine is a health care system that is intertwined with the teachings of Tibetan Buddhism. In part, it is based on the construct of reality derived from the oral and written traditions of Buddhist meditation masters and their respective practice lineages. Philosophically, there is no duality of body and mind; the focus is on the experiential verification of the insubstantiality (emptiness) of both the body-mind and the phenomenal world, also know as the relative world of appearances. To demonstrate the order, consistency, richness and complexity of this tradition, this article will discuss both Tibetan medicine and Tibetan Buddhism. This does not imply that only Buddhists can benefit from the techniques and methods of Tibetan medicine; rather, this approach provides a context for understanding its basic tenets.

Tibetan medicine emphasizes a sacred outlook on life. The Tibetan word for Buddhist is *nangpa*, literally one who is inner-directed. Tibetan Buddhists examine the nature of mind to find a fruitful approach to living (Karthar Rimpoche, 1992: 25). Practitioners of Tibetan medicine and Buddhism cultivate

H. Selin (ed.), Medicine Across Cultures: History and Practice of Medicine in Non-Western Cultures, 85–113.

attitudes and behaviors based on compassion and wisdom to benefit others. The attitudes toward the preciousness of human life provide a framework for understanding the root causes of suffering and opportunities for transforming suffering into enlightenment.

The political changes in Tibet and China in 1959 made Western societies more aware of Tibetan medicine. Health professionals in all cultures reify disease into recognizable patterns that are based on selective attention to objective signs and subjective symptoms expressed by patients. Clinical patterns of disease and illness are socially constituted constructs of reality (Kleinman, 1980: 38), known as patient explanatory models. The clinical realities of health care systems are mediated by symbolic reality within a culture. Relating to the symbols and meaning of different aspects of health, disease and illness can facilitate practical insights into Tibetan medicine for readers in other regions of the world.

A BRIEF HISTORY OF TIBETAN MEDICINE

Traditional Tibetan medicine is based on the teachings of the Sakyamuni Buddha. He lived as a Brahmin prince in India about 2,500 years ago. According to tradition, he witnessed the sufferings associated with birth, old age, sickness and death which inspired him to begin a spiritual search. He eventually attained Enlightenment – the realization of the true nature of Mind. Then he began to teach.

Sakyamuni Buddha presented teachings about Tibetan medicine during meditation in groves of medicinal plants in Oddiyana, probably in or near present day Afghanistan. He explained the teachings in ways that would be comprehensible to all listeners in accord with their level of understanding. These teachings became the *Oral Tradition Tantra*, known in Tibetan as the *bDud-rtzi Snying-po Yan-lag brGyad-pa gSan-ba Man-ngag gi rGyud-bzhi* (The Four Secret Oral Tantras on the Eight Branches of the Essence of Nectar), or more concisely the *rGyud-bzhi* (Dhonden, 1977). The eight branches refer to sections of the Ayurvedic medical tradition: anatomy, embryology, childhood diseases, women's diseases, diseases caused by evil spirits, trauma, poisoning and rejuvenation therapies designed to strengthen sexual potency and relieve problems of old age (Thakkur, 1974).

Vairochana, a Tibetan translator and designer of the Tibetan alphabet, went to India during the eighth century. He translated the original Sanskrit medical text *Amrta Astanga Guhyopadesa Tantra* into Tibetan; the Tibetan translation is the *rGyud-bzhi*. Upon his return to Tibet, Vairochana presented a copy of the text to Padmasambhava, the patron saint of Tibetan Buddhism. The text was hidden in the first monastery in Tibet at Samye during its construction and was later discovered by the terton[1] Trapa Ngonshe. Vairochana also gave the text to King Trisong Detsen (790–858 AD). The king passed it to the royal court physician, Elder Yuthog YonTan Gonpo (708–833 AD), and he in succession passed it to Yuthog the Younger (1112–1203) (Rechung Rimpoche, 1976).[2] The *rGyud-bzhi* became the main text for studying Tibetan medicine for centuries; it is still used today.

During the reign of the Fifth Dalai Lama (1671–1682), the regent Sangye Gyamtso founded the School of Medicine and Astrology on Chagpori Hill near Lhasa. Sangye Gyamtso wrote several medical works, including the most complete commentary of the *Four Tantras*, known as *Lapis Lazuli* (1688). *Lapis Lazuli* later became part of the Tibetan Buddhist canon, the *Tengyur*. It is complemented by the *Atlas of Tibetan Medicine* (1688), a series of paintings, graphics and classification schemes used in teaching at the monastic medical schools (Avelon, 1998: 42–50). The two texts represent a synthesis of classical Indian and Chinese medical knowledge in the context of popular customs and beliefs in Tibetan culture at that time (Bolsokhovea, 1998).

The Mendzekhang School of Medicine opened to both monks and lay people in Lhasa in 1916 under the reign of Dalai Lama XIII. During the domination of Tibet by China in 1959, Chagpori Medical College was destroyed. In 1961, the Dalai Lama XIV, living in exile in Dharamsala, India, requested his personal physician Dr. Yeshe Dhonden to begin a new Tibetan Medical and Astrological Institute for teaching and clinical services. Other traditional teaching sites for Tibetan medicine are in Ladakh, Nepal, Bhutan, Mongolia, Buryatia and within the Tibetan Autonomous Region (TAR). Schools and courses for the study of Tibetan Medicine are also available in Europe, North America and elsewhere.

Tibetan medicine survives today despite wide swings in the politics of its institutional legitimacy during the past few decades. In an attempt to survive domination by central Chinese authority beginning in 1959, traditional Tibetan medicine began to move away from the Buddhist orientation and toward biomedical therapies (Janes, 1995). In 1962, Tibetan medicine was sanctioned as a component of the public health system, only to be de-legitimized in 1966 with the onset of the Cultural Revolution. The rise of professionalism in Tibetan medical education in TAR was accompanied by a concurrent loss of practitioners in popular and folk medicine, especially in rural areas. By 1985, Tibetan medicine was again a legitimate part of governmental health programs, especially for primary care services in TAR (Janes, 1995).

Some critics of traditional medicine cite the movement toward biomedical principles and away from traditional methods as a sign that Tibetan medicine lacks validity and credibility. However, others contend that the dominant culture reshapes and transforms traditional medicine to adapt to the mainstream forces of political, economic and cultural authorities (Lock and Scheper-Hughes, 1990). The adaptation of traditional Tibetan medicine is most likely a symptom of a culture in forced transition; it is a means to assure the survival of a cultural heritage under siege.

Traditional texts of Tibetan medicine: the Four Tantras

The *rGyud-bzhi*, also known as the *Four Tantras*, is the main text for studying Tibetan medicine. It includes the *Root Tantra* (*rtsa rgyud*), the *Explanatory Tantra* (*shad-rgyud*), the *Instructional Tantra* (*man-ngag rgyud*) and the *Subsequent Tantra* (*phi ma-i rgyud*) (Dhonden, 1977; Tsarong, 1981; Clark, 1995; Finckh, 1988). Each volume has a different focus. The first lists the causes of disease and means of treatment. It includes a description of the Tree of Health and Disease, an organizational scheme described in more detail in the next section.

The main body of the Explanatory Tantra addresses theoretical issues of Tibetan pathophysiology and constitutional analysis in terms of the "elements" and "humors". These names are metaphors for somato-psycho-cosmologic aspects of the human condition. They are metaphors for qualities of relationship rather than anatomical constituents. Both elements and humors in Tibetan medicine are similar to the Ayurvedic tradition of *tridoshas* (Lad, 1984).[3] Indeed, some authors refer to traditional Tibetan medicine as "Tibetan Ayurveda" (Sachs, 1995). The second volume concludes with a section on the characteristics of a good physician. [Editor's note: See the chapter on Āyurveda in this volume.]

The Instructional Tantra is the third volume. It has fifteen divisions that describe practical aspects of diseases and treatment. There are special sections for diseases occurring during childhood and old age and for female conditions. Combinations of constitutions are described in this section. The Subsequent Tantra contains sections on the examination of the pulse and urine for diagnosis and preparation of medicines and their indications. There are special physical therapies, including both invasive approaches (e.g., acupuncture, moxibustion and surgery) and more gentle methods (e.g., baths and compresses). Major surgery is no longer practiced in traditional Tibetan medicine, and acupuncture is less commonly used.

Tibetan tree of health and disease

The Tree of Health and Disease described in the Root Tantra serves as an organizational scheme for the categories and concepts of Tibetan medicine (Avelon, 1998; Baker, 1997; Dhonden 2000: 21; Finckh, 1978 and 1988). The tree has three roots and nine trunks. The trunks support a total of forty-seven branches with 224 leaves (see Figure 1). The three roots correspond to the physical body, diagnosis and therapy. The first root supports two trunks, one for the healthy person, including embryology, physiology and anatomy, and another for the unhealthy organism, also known as pathology. The second root supports three trunks that represent each method of diagnostic assessment: observation, palpation and questioning. The third root has trunks for each major category of treatment: nutrition, behavior, medicine and external treatments. The root for therapy has the most trunks, branches and leaves; the trunk corresponding to therapies is the largest. It provides a classification for medicines according to origin, potency, application and qualities.

To provide an example of this organization of the text, one root, one trunk and the corresponding branches, leaves and flowers will be described. The first trunk represents the natural or unaltered state of health. It arises from the first root that represents the physical body. The trunk has three branches with 25 leaves. The 25 topics represented by the leaves explain the processes that occur during digestion. The lower branch shows the varieties of the afflictive aspects

Figure 1 Tibetan Tree of Health and Disease.

of the body (Tibetan: *nyes-pa*; Sanskrit:*doṣa* or *dosha*). The middle branch has seven leaves, each representing the bodily constituents (food, blood, flesh, fat, bone, marrow and regenerative fluids). The leaves for three excretory products (defecation, urination and perspiration) are located on the upper branch. Two flowers adorn the upper part of the trunk, representing the attainment of longevity and good health. The blossoms of the flowers represent the realization of spiritual practices, including the accumulation of wealth and happiness. The fruits are the symbols for the realization of Buddha-nature, or Enlightenment.

THEORY AND PRACTICE OF TIBETAN MEDICINE

Anatomy and physiology in Tibetan medicine: aggregates and elements

Maintaining balance among the three aspects of bodily constituents (Tibetan: *Nyipa sum*) is the goal of Tibetan medicine. The three aggregates are complexes of mind-body that include physical, personality and spiritual aspects. They are *rlūng, mKhris-pa* and *Bad-kan*, often collectively translated as "humors" and individually translated as "wind, bile and phlegm". However, these terms do not accurately portray their unified and multi-dimensional nature. The terms *rlūng*, *mKhris-pa* and *Bad-kan* can also be translated as "subtle energy, transforming energy and supportive energy" to convey their functional interactions. The term "aggregates" will be used in the remainder of this chapter, in place of the more common translation, "humor".

The elements that comprise the ordinary physical body are translated as space, air, fire, water and earth (*man mkha*, *rlūng*, *me*, *chu* and *sa*). The five sensory consciousnesses (consciousness, concept, perception, feeling and form) each correspond to the respective elements. Thus, each element is a metaphor or "sphere of influence" (Porkert, 1974), representing the many dimensions of life experiences. For example, space represents formless potentiality (e.g., consciousness) and air is associated with subtle movements (e.g., conduction of impulses in the nervous system, circulation, respiration, or intuition). Fire represents active transformation (e.g., digestion of food into nutrients or intellectual discourse), while water represents cohesion and communication (e.g., empathy, lubrication, or feelings), and earth is manifest form (e.g., anatomic structure or physical sensations). Each element is associated with a sense organ and sensory perception, emotions, sacred images and more (See Table 1).

The elements combine to form the aggregates that represent a person's constitution (Tibetan: *rang zhin*). Space and air together comprise *rlūng* (Sanskrit: *vāta*). *Rlūng* is traditionally considered the most important aggregate in Tibetan medicine, since movement is so basic to all vital functions. Consequently, *rlūng* conditions are highly prevalent. Fire is the primary element in *mKhris-pa* (Sanskrit: *pitta)* constitution, while water and earth combine to form *Bad-kan* (Sanskrit: *kapha*). The aggregates can also combine so that primary and secondary aspects are present. For example, primary *rlūng* and secondary *mKhris-pa* constitutions (and vice versa) allow for refined behavioral and medicinal prescriptions. When all aggregates are afflicted simultaneously, a condition known as *mukpo* is present. The Tibetan elements differ from the five-phase theory in the traditional Chinese medical model (Porkert, 1974). [Editor's note: See the article on Chinese medicine in this volume for a fuller explanation.]

An example of the explanatory model for Tibetan physiology may be useful. When we eat, the nutrients go to the stomach where *Bad-kan* helps to mix them (lubrication function); then the *mKhris-pa* helps to digest them (transformation activity), and the *rlūng* helps to separate the essential nutrients from the waste products (movement metaphor). The essences of the nutrients eventually form blood. In a cascade of events, the essence of blood forms muscle

Table 1 Multi-dimensional symbolism of the sacred and mundane merge within the aspects of the elements (Adapted and modified from Lauf, 1976: 130; Clifford, 1990: 74; Dummer, 1988: 22–25)

ELEMENT	SPACE	AIR	FIRE	WATER	EARTH
Taste	astringent	salty	pungent	bitter	sweet
Sense	taste	touch	sight	hearing	smell
Emotion	grief	fear	sympathy	anger	joy
Physical form		lung/colon	heart/sm. intestine	kidney/bladder	spleen – pancreas/ stomach
Awareness	consciousness	volition	perception	form	feeling
Defilement	ignorance	jealousy; envy	craving	hate; anger	pride
Tibetan:	*GTimug*	*Phrag-dog*	*Dod-chags*	*Zhe-sdang*	*Nga-ryal*
Wisdom	universal law	accomplishing works	discriminating insight	mirror-like wisdom	equality; equanimity
Tibetan:	*Chos-dbying yeshes*	*Bya-grub yeshes*	*Sor-rtogs yeshes*	*Me-long yeshes*	*Mnyam-nyid yeshes*
Dhyani budhha	Vairocana	Amoghasiddhi	Amitabha	Aksobhya	Ratnasambhava
Seed syllable	Om	Ah	Hrih	Hum	Sva; Tram

NOTE: Different perspectives are associated with alternative assignments.

tissue, the essence of muscle tissue forms fat, the essence of fat forms bones, the essence of bones forms marrow and the essence of bone marrow forms the regenerative fluid. The qualities of the regenerative fluids are responsible for the "radiant aura" or "glow" of good health (Sanskrit: *ojas*). When the three aggregates are balanced, then the seven bodily sustainers are also balanced.

Functional relations are emphasized more than physical anatomy (Rechung, 1976). Anatomic organs are classified into two groups: the five solid organs and the six hollow organs. The five solid organs – lung, heart, spleen, liver, kidney – are reservoirs for storage of various vital essences, while the six hollow organs transport and disperse the products of digestion and waste. The corresponding six hollow organs include the large intestine, small intestine, stomach, gall bladder, urinary bladder and a function without a well-defined anatomic substrate for sexual potency. This arrangement of organs into solid and hollow is similar to the traditional Chinese medicine system of *tsang* and *fu* based on complementary opposites of *yin* and *yang*.

The mind-body aspects in the Tibetan explanatory model are not easily recognized when translations name anatomic parts to denote multi-dimensional functions. For example, in traditional Tibetan physiology, both heart and small intestine are associated with the fire element. Consequently, they have physical, emotional and cognitive functions that are associated with the metaphor of transformation. They are both involved with transforming foods into nutrients that are made into blood and then circulated to the entire body. The function of the small intestine is to sort nutrients to be assimilated from those that are passed on as waste. By correspondence in the traditional model, the small intestine is then also associated with a cognitive function of discriminating insight. The other organs, their functions and elements are each related to a particular wisdom portrayed in sacred iconography (Lauf, 1976: 117–37). Body and mind, mundane and spiritual are fused within Tibetan medicine.

Clinical diagnosis in Tibetan medicine

A Tibetan physician employs his own senses to examine the patient's general balance of health. There are three main methods of diagnosis: questioning, observation and palpation. Objective technologies for clinical diagnosis are practically non-existent. There are clinical survey instruments to assess Ayurvedic constitution (Sachs, 1995; Chopra, 1991), but no psychometric studies are readily available to validate their clinical utility.

Diagnosis by questioning

Questioning is based on a series of twenty-nine items that help to identify the nature of the imbalance (Clark, 1995). Questioning is analogous to taking a clinical history in Western medicine. Asking about specific signs and symptoms assists the physician in understanding the imbalances among the three aggregates. The qualities that the patient exhibits in answering questions are as important as the content of the response.

Chronobiology is an integral part of Tibetan medicine that can be discerned by questioning. *Rlūng* arises at dawn and again in the evening; it is prominent in summer. *mKhris-pa* arises during the autumn and at midday and midnight. *Bad-kan* arises in the spring and in the morning and at dusk. The times of increasing influence from each aggregate correspond to symptoms of prototypical biomedical diseases that are associated with their respective aggregates. For example, the symptoms of major depressive episodes and anxiety are best explained by excessive *rlūng*. Thus, early morning awakening among many depressive patients is compatible with this explanatory model, since *rlūng* increases near dawn. Similarly, the symptoms of peptic ulcer disease typically worsen near midnight, corresponding with the natural rise of *mKhris-pa*. The most prevalent time for manifestation of symptoms among patients with western disease categories can be explained by Tibetan medicine.

Diagnosis by observation: urine and tongue diagnosis

The urine is examined to assess constitution, condition and response to treatments. The patient is instructed to refrain from eating spicy foods, drinking alcohol, performing unusual strenuous activity and having sexual intercourse on the day before the examination. The physician examines the first early morning urine sample by stirring it in a small bowl with a clean stick and then examining the color, vapor, odor, bubbles, sediments and protein.

The color of urine is determined by the intake of food and drink, seasons, diseases and imbalances. In general, a reddish color with a small volume of urine indicates a hot disorder, while clear urine with a large volume indicates a cold disorder. For the patient with *rlūng*, the urine is very clear, and it has many bubbles that may last only a short time. The urine from a person with *mKhris-pa* constitution is a reddish-yellow color, with a very strong odor and moderate-scant volume. The *Bad-kan* patient has urine that is clear, with little odor, large volume and large bubbles that often last a long time.

The Tibetan doctor may view the patient's tongue to provide additional clinical information. For example, the *rlūng* patient will probably have a red, dry and rough tongue. The patient with *mKhris-pa* constitution will likely have a tongue with a thick yellow coating. The tongue of a *Bad-kan* patient is white, smooth and wet. Tongue and pulse diagnosis may also prompt additional questioning to support or reject clinical assessments.

Diagnosis by palpation: pulse diagnosis

Examination by touching includes palpation of the body parts to assess temperature, areas of tenderness, tightness, weakness, or abnormal growths. Pulse diagnosis is a key method (Dhonden, 1986; Finckh, 1978; Steiner, 1989). The physician interprets the waveforms of the patients' radial arteries to assess ongoing physiologic processes and relationships between body parts and aggregates. As before, patients are instructed not to engage in sexual intercourse on the day prior to the pulse reading, not to consume alcohol, eat excessively or engage in strenuous activities. These recommendations are intended to normalize the three aggregates by minimizing extraordinary external influences. The assessments of constitutional pulses are typically performed during the early morning, while the patient's stomach is empty.

The physician palpates the radial pulses on both wrists, while the patient is in a comfortable and quiet environment (see Figure 2). The left hand is slightly flexed, so that the wrist creases are clearly visible. A special pillow may be used to rest the wrists during this part of the examination. The doctor places the fingers of his right hand on the left radial pulse of the patient, so that the right index finger is proximal to the wrist crease, at a distance about as wide as the patient's distal thumb. The second and third fingers then fall naturally along the artery. The three palpating fingers generally do not touch one another unless the patient is small in stature. The patient's right radial artery pulses are palpated in a similar manner. The physician may examine the pulses on one wrist at a time, or he may examine both wrists simultaneously. The radial pulse is palpated with increasing pressures at each site proximal to the wrist crease.

Each of the physician's palpating fingers can assess the functional characteristics of specific physiologic systems. It is again important to note that although the names for organ systems are used here, these terms are metaphors for the mind-body functions associated with the elements and aggregates. According to the text, heart and small intestine are assessed at the first position on the male patient's left hand, spleen and stomach with the second palpating finger and so on. Similarly, the pulses at the first position on the male patient's right radial artery correspond to the lungs and large intestine, liver and gall bladder pulses are located at the second position, and right kidney and urinary bladder are at the third position.

The pulses of women are reversed at the first position only, so that the heart and small intestine are assessed at the first position on the woman's right hand, and the pulse for lungs and large intestine is read on a woman's left radial pulse. The reason for the switch is due to slight differences between genders for the apex of entry for consciousness that enters the heart, although no

Figure 2 Pulse diagnosis.

physical differences in the anatomy of the heart is present. This is consistent with teachings of Tibetan embryology.

There is a complex system for interpreting the qualities of the pulses at each of the twelve sites on the radial arteries. For example, tight or twisting qualities may be present in the pulse in the presence of disorders of *rlūng*. This finding often indicates the presence of pain symptoms that may radiate or have a breaking quality in specific sites indicated by the pulse. Smooth pulses in the proximal sites are often present in early pregnancy, a finding attributed to the excess blood from the fetus (*Bad-kan* condition). Fast or jumping qualities indicates excess heat (*mKhris-pa*). The physician can discern underlying patterns within the pulses; this allows him to select various medicines in accord with their taste, potency and site of action.

Constitution analysis and treatment methods

The first priority in Tibetan medicine is to restore harmony to each person in the context of his or her life experiences. Symptoms are a result of an imbalance caused by disturbances in circumstances such as diet, lifestyle and seasonal and mental conditions. Treatment depends on the total situation rather than the individual symptom; two people with the same symptom may be given very different advice and treatment.

Tibetan medicine provides a systematic approach to treatment and prevention. There are four methods of intervention: diet, behavior modification, medication and physical therapies. The three constitutions (*rlūng*, *mKhris-pa* and *Bad-kan*) have patterns of symptoms, preferences, and tendencies that can be remedied with categoric therapies. If the illness is not serious, diet and behavioral advice may be sufficient; otherwise, prescriptive medicines may be needed.

However, the best use of medical treatment cannot provide good health through physical means alone; a healthy mind is needed as well (Tai Situ Rimpoche, 1992: 21). Tibetan medicine gives priority to psychological and spiritual factors in its definition of health. It teaches that we can make peace with the disease once we know and understand it. The literature is rich with specific meditation techniques used to restore balance (Akong Tulku Rimpoche, 1996; Sachs, 1995; Tarthang Tulku, 1979; Tulku Thondup, 1998; and many others). Each constitution and typical treatments are discussed below.

Constitution and conditions in Tibetan medicine: rlūng

The physical characteristics of *rlūng* constitution typically include an asthenic body, sensitivity to cold and a talkative and lively personality. *Rlūng* people are easily aroused from sleep, tend to eat small amounts of food frequently and work hard but sweat little. They are prone to worry and sorrow. Symptoms of excess *rlūng* tend to be worse during cold and windy climates, during summer, or during early morning and/or late afternoon. Aging tends to produce imbalances in *rlūng*, so that these conditions are more common among the elderly.

Egocentrism is the usual source for imbalances of *rlūng*. Relational issues about attachment, sometimes translated as passion, characterize people with *rlūng* constitutions (Sachs, 1995). Ego, in this context, means identifying with the false concept that each of us is separate and independent from the world in which we live. In Tibetan medicine and culture, ego is viewed as a limitation to be overcome, rather than a psychological resource to be enhanced. Ego is synonymous with self-clinging habitual patterns of mind, and so it is viewed as an obstruction to mental clarity.

A person with excess *rlūng* shows a tongue that is rough and dry. Pulses are irregular, empty or floating. Urine is clear and has many bubbles that last only a short time. Symptoms are worse with ingestion of foods with light or rough potency, and symptoms change quickly over time. Typically, symptoms worsen when the patient is hungry and they may be relieved with foods that are oily, hot or heavy. Symptoms tend to have a breaking quality, particularly in the bones and joints and in the lower body. Nervous symptoms including anxiety are common in this group.

Patients with excess *rlūng* are instructed to rest – both physically and mentally. Minimizing anxiety and worry is important, since this psychic discord is a basic manifestation of *rlūng*. They can be advised to stay in comfortable, quiet, dimly lit, warm places. Company with a few close friends is beneficial, as are small amounts of alcoholic drinks. Patients with *rlūng* disorders are asked to pay special attention to regularity of lifestyle (e.g., eating, sleeping and excretory function), to find time for calm activities and socializing and to exercise in ways that promote good overall circulation, using techniques such as hatha yoga exercises and pranayama (yoga breathing). External cautery, similar in principal to traditional Chinese or Mongolian moxibustion (the application of burning moxa floss or other combustible materials at the acu-points) may be used to heat the body.

Constitution and conditions in Tibetan medicine: mKhris-pa

People with this constitution may be very ambitious; they tend to be active and may try to control situations physically or through positions of power. They tend to be keen thinkers and may have extraordinary motor skills, yet control and aggression are sources of suffering for them. People with *mKhris-pa* constitutions can be characterized by aggressive behaviors sometimes translated either as "hatred" or "aversion".

Patients with *mKhris-pa* constitutions are typically medium size, muscular, with a yellowish tinge to the skin. They are frequently hungry and/or thirsty, sweat easily and they like activities and challenges. They are prone to annoyances and anger. Symptoms of excess *mKhris-pa* tend to be worse during hot and dry climates, during autumn, or during midnight and midday hours. Imbalances in *mKhris-pa* tend to occur among middle age groups.

A person with excess *mKhris-pa* shows a tongue that is coated with a thick, yellowish substance. Pulses are fast, strong and jumping. Urine is yellowish-red, with a strong odor and vapor and small bubbles. Symptoms are worse after eating, especially after foods with sharp or hot qualities. Symptoms may include a bitter taste in the mouth, a sense of heat or fever within the body and aches in the middle and upper parts of the body. Ingestion of cool foods may relieve the symptoms.

Therapy for excess *mKhris-pa* is directed to foods and medicines that are sweet, bitter and astringent in taste and cool, smooth and thin in potency. Patients with an imbalance in *mKhris-pa* should eat beef, vegetables, fresh butter, fresh low fat cheese, cow's yogurt and buttermilk and drink weak tea or spring water. A simple vegetarian diet is suitable. Foods that are heating in nature should be avoided, including peanut butter, mustard and hot spices, garlic, ginger, onion, alcohol, grilled meats [esp. lamb], oily and greasy foods and soups made with bones.

Syrups and powders are traditional types of prescribed Tibetan medicines for *mKhris-pa*. Purgatives and strong laxatives are used to reduce the excessive amount of heat in the body. Behavior change includes the use of cold baths and showers, sitting in shaded places, walking by the sea and cooling scents such as sandalwood. Individuals suffering from a *mKhris-pa* disorder should avoid situations causing conflict and direct, excessive exposure to the sun. Physical activities that are relaxing are encouraged; an environment that is conducive to calmness is beneficial. External treatments may target the production of sweat and may also include bloodletting.

Constitution and Conditions in Tibetan medicine: Bad-kan

Someone with a *Bad-kan* constitution typically has a large or tall body, can endure hunger and thirst well, and shows a pleasant, easy going and friendly personality. People in this group tend to sleep soundly, eat too much and too quickly and wear clothing that is insufficient to keep them warm. Excessive sleeping can worsen conditions. *Bad-kan* people tend to be slow, persistent and methodical in daily routines. Cultivation of good habits can be beneficial and enduring. Established routines are easily maintained, but changing habits can be difficult. Changing unhealthy habits may call for exertion that may initially make the person uncomfortable. This information is part of the prescription for health for this constitution. Babies and children tend to have imbalances in *Bad-kan*.

A person with excess *Bad-kan* shows a tongue that is moist, smooth and glistening with a thick gray lusterless coating. Pulses are slow, weak, sinking or deep. Urine is usually clear, with little odor and few bubbles that last a long time. Symptoms are worse after eating, especially after heavy foods and relieved after warm or hot foods. Symptoms include a sense of discomfort, heaviness or coldness in the upper body or a sense of heaviness in the mind.

Therapy is directed to foods and medicines that are pungent, bitter and astringent in taste, with potencies of sharp, rough and light. Patients should adopt a heating diet with respect to both the nature and the temperature of the food. For example, they are encouraged to consume a traditional diet to counter excess cold and retention of fluids, including hot water, cooked foods, mutton, fish, barley, pomegranates, sheep cheese, yogurt, ginger, radish, honey, garlic and wine. They should avoid cold drinks and raw foods such as salads, potatoes, tomatoes, eggplant, bell peppers and sugar.

Patients with *Bad-kan* disorders may be instructed to keep warm and perform vigorous exercise such as running or dancing. Swimming in cold water is not appropriate for *Bad-kan* conditions. Periodic emetics (e.g., induced vomiting) are recommended as part of a traditional health maintenance routine, since this reduces the excessive amount of cold and damp that tends to accumulate in the body. An environment that is conducive to heat producing activities is beneficial, including lots of exposure to warm sun and fires burning in the home. External treatments may target the cautery and other forms of heat.

Mukpo is the combined dysfunction of all three aggregates. This is a complex chronic condition in which each of the imbalances ultimately affects the digestive and metabolic functions. The diet includes fresh, cooked foods to facilitate the severely compromised digestion that is typical of this condition. Recommended foods include fresh meat, dairy products, vegetables, fruits and whole grains. Garlic, tomatoes, eggplant and bell peppers exacerbate this condition, as do foods that are aged, fermented, cured, smoked, sour, chilled, oily, or processed.

Pharmacology in Tibetan medicine

Natural herbal preparations are a large part of therapy in Tibetan medicine. The medicinal effects associated with Tibetan preparations are not due to the presence of a specific ingredient, as in western pharmaceuticals. Rather, the herbal combinations act on the body and mind aspects to restore balance. Remedies act on the specific imbalance, while other ingredients support the aggregates that are not maligned during the period of the intervention, although they may be challenged during the dynamics of therapy and healing.

Most of the ingredients in the Tibetan materia medica are vegetable in nature. There are eight categories of natural substances used in preparing prescribed therapeutic remedies: precious metals and gems, stones, earth medicines, trees, resins, plants, herbs and animal products (Clark, 1995; Clifford, 1990; Dash, 1976). The remedies can take ten forms: decoction, pills, powder, gruels, medicinal butters and calxes, concentrated extractions, medicinal wine, herbal medicine and gems as medicine. Enemas, especially if oily, mild and soothing, are useful for *rlūng* imbalances. Purgatives reduce the heat in *mKhris-pa* conditions. Emetics reduce water and mucous associated with excessive *Bad-kan*.

Tibetan herbal pills are often complex, with 3 to 150 herbs per formula. Each herbal formula is prescribed to fit the manifestation of the disease and the evolving condition of the individual patient. The physician often starts with less potent preparations and advances to stronger forms if necessary. Typically, two-four formulas are prescribed for each day at specific times. Morning remedies commonly include those for *Bad-kan* disorders or digestive disorders; afternoon remedies are typically used to treat *mKhris-pa* disorders. Remedies given in the late afternoon or evening are usually given to treat *rlūng* disorders. Herbal remedies are often modified at each visit.

The substance in which a medicine is dispensed is called the medicine horse (*men-ta*) (Burang, 1974). Typical carrier agents include water, alcohol, sugar, treacle or honey. Sugar is the men-ta for dispelling *mKhris-pa*, treacle is for reducing *rlūng*, and honey overcomes *Bad-kan*. The compounded remedies are sometimes coated with resins, so that they pass through the stomach and into the intestine where they are integrated into the body in accord with traditional concepts of physiology.

The art and science of preparing traditional Tibetan medicines is based on tastes and potencies. The tastes are associated with medicinal properties. Each taste is present as the result of interaction between two or more elements. Specific tastes can increase or decrease the activity of specific aggregates. The text says that sweet, sour and salty tastes reduce *rlūng*. Likewise, bitter, sweet and astringent abate excessive *mKhris-pa* and pungent, bitter and astringent tastes counter *Bad-kan* (see Table 2).

Potencies are the qualities of therapeutic actions associated with medicines and nutrients (Clifford, 1990; Lad, 1984). There are eight potencies, which are combined in pairs of complementary opposites: heavy/light, oily/rough, cool/hot and dull/sharp. Heavy and oily potencies reduce *rlūng*, cool and dull remove *mKhris-pa* and light, rough, hot and sharp counter *Bad-kan*. On the other hand, light, rough and cool potencies can increase *rlūng*, hot, sharp and oily can increase *mKhris-pa*, and heavy, oily, cool and dull potencies can enhance *Bad-kan*. There are complex interactions and hierarchies of order that determine the therapeutic actions of foods and medicines. Several texts provide extensive lists of the medicinal properties of foods (Sachs, 1995; Lad, 1984) and traditional medicines (Clark, 1995), with lists of dietary recommendations and foods to be avoided for each constitution (Tokar, 1998, 1999).

The refined sense perceptions of Tibetan doctors determine the properties of herbal remedies and traditional medicines. The tastes and actions of therapeutic agents on specific parts of the tongue correspond with medicinal properties at specific parts of the body. For example, medicines with astringent tastes that influence the sides of the tongue may be useful for treating imbalances that adversely influence the gallbladder organ and associated dimensions, including some migraine headaches. Knowledge about tastes and medicinal actions of therapeutic agents is embodied in a cultural tradition of health care, known as microsystems of anatomic correspondence (Porkert, 1974).

Typically the Tibetan pharmacist is also a physician. Preparation of botanical medicines involves collection of specimens, drying, cleaning, storage and preparation, detoxification and neutralization and compounding. Various parts of plants are gathered at different seasons – fruits in the autumn, leaves during the summer, branches and barks during the spring and roots during the winter. Mantras to invoke the Medicine Buddha and other deities are often recited during the time of collecting and during other stages in processing to imbue the remedies with additional potency.

Table 2 Influence of tastes and qualities on the aggregates (adapted and modified from Clifford, 1990: 120; Lad, 1984)

ELEMENTS	TASTE	QUALITIES	FOOD AND SPICES	AGGREGATE INCREASED	AGGREGATE DECREASED
Earth-Water	sweet	cooling	wheat, rice, peppermint	*Bad-kan*	*MKhris-pa; Rlūng*
Earth-Fire	sour	heating	yogurt. cheese, lemon, rose hips	*MKhris-pa; Bad-kan*	*Rlūng*
Water-Fire	salty	heating	salt, kelp	*MKhris-pa; Bad-kan*	*Rlūng*
Water-Space	bitter	cooling	dandelion, turmeric	*Rlūng*	*MKhris-pa; Bad-kan*
Fire-Wind	pungent	heating	onion, radish, ginger, garlic	*MKhris-pa; Rlūng*	*Bad-kan*
Wind-Earth	astringent	cooling	unripe banana, goldenseal	*Rlūng*	*MKhris-pa; Bad-kan*

TRADITIONS OF HEALING IN TIBETAN BUDDHISM

Traditionally, Buddhist teachings are viewed as a medicine to alleviate suffering, prevent illness and promote health and happiness. Sakyamuni Buddha taught in a series of three teaching cycles after his Enlightenment. The three cycles comprise the sutras, the original oral teachings of the Buddha; they are complemented by shastras, or commentaries. The sutras and shastras comprise the Tibetan canons, the Kangyur and Tengyur. The three cycles of teachings correspond to the vehicles of Buddhism – Hinayana, Mahayana and Vajrayana. Each vehicle has methods for people with various levels of understanding and willingness to make commitments to practice.

Vehicles of Tibetan Buddhism

The Hinayana tradition, also known as the "lesser vehicle", emphasizes disciplines for purification and moral codes of ethical conduct. The major teaching from the first cycle is the Four Noble Truths (Dalai Lama, 1998). The fruition of this path is self-realization (Sanskrit: *pratimoksha*; Tibetan: *so-sor-tharpa*), in which the individual is liberated from the ties to *samsara*[4] (Trungpa, 1981: 1–6). This form of Buddhism is also known as Theravadin and is common in South India, Sri Lanka, Thailand, Burma and other regions of Southeast Asia.

Tibetan physicians are usually trained in the Mahayana and Vajrayana traditions. The Mahayana Path, also known as the "Great Vehicle", emphasizes the altruistic motivation to benefit all sentient beings. The primary discipline of the Mahayana approach is to generate the enlightened attitude (*bodhicitta*) to benefit others through the development of genuine warmth (*maitri*) and compassion (*karuna*). This form of Buddhism spread to China, Japan and Korea.

The Vajrayana teachings, also known as the "Indestructible Vehicle", state that all beings are endowed with the enlightened essence, Buddha Nature (Thrangu Rimpoche, 1993). Vajrayana is known as the direct path to enlightenment, in contrast to the gradual path (Tibetan: *lamrim*) of the Mahayana. The realization of the true nature of mind is not a matter of learning more or gathering more credentials. Realization of Buddha Nature is a path of abandoning habitual, egocentric patterns of ordinary mind to experience directly "suchness" (Sanskrit: *tathatagarba*), things-as-they-are, or Ultimate Reality. It is a sacred view that requires a committed relationship with a spiritual teacher. This form of Buddhism spread to Mongolia and Siberia and it remains today in TAR.

Styles of meditation within each vehicle (Yana)

Different methods of meditation are emphasized in each vehicle (Gyamtso Tsultrim Rimpoche, 1986). Each serves as the foundation for the next stage of development in mind training. The Hinayana approach emphasizes concentration and focusing methods for taming the mind (Sanskrit: *shamata*; Tibetan: *zhi-gnas*; shine). Realizing peace of mind and purification of the body-mind are emphasized.

The Mahayana approach adds a quality of genuine warmth and compassion for others to the self-realization methods of Hinayana. The motivation for the Mahayana practices is the cultivation of a heartfelt, enlightened attitude called *bodhicitta*. The disciplines of the bodhisattva include practice in the six perfections (Sankrit: *paramitas*): generosity, moral discipline, patience, perseverance, meditation and wisdom awareness. These are attitudes that pervade the practice of meditation and daily activities.

Richness, directness and clarity characterize the meditation methods of the Vajrayana Path. This set of teachings has various names: fruition vehicle, tantrayana, secret mantrayana (Tibetan: *sang ngak*, "secret and profound"), direct path to enlightenment and others. Vajrayana methods for attaining Buddhahood have shamanic qualities (Samuel, 1993), including rituals and vivid, detailed visualizations. Varieties of suffering can be transformed into their sacred counterparts through the practices of mindfulness, linked with contemplative analysis about the insubstantiality of all phenomena, also known as emptiness (Sanskrit: *shunyata*). The sacred and the secular merge in this tradition, since samsara and nirvana are viewed as inseparable. The formless meditation traditions of Mahamudra and Dzogchen emphasize the spaciousness qualities of simply resting the mind in uncontrived naturalness.

Tibetan traditions of healing

The Tibetan tradition of healing begins with the conversion of suffering into the aspiration to attain enlightenment. The healing process is ultimately spiritual. Egocentric patterns limit human experiences to repeating cycles of existence known as samsara. Suffering is the pervasive mark of samsara. From the Mahayana perspective, there arises an unrivaled sense of compassion, such that if only one person suffers in samsara, then the whole of humankind is adversely influenced. Consequently, traditionally trained Tibetan physicians engender bodhicitta to benefit all beings.

According to Buddhist philosophy, there are six realms of samsaric existence, each with particular characteristics. Happiness and positive health appear in some measure within the three upper realms of gods, demi-gods and humans. Despite the pervasiveness of suffering, humans have the greatest potential for realizing nirvana. Nirvana is the direct experience of ultimate reality, the world as-it-is. Nirvana is the realization of our true nature that is beyond the limits of the discursive mind. It is beyond all concepts and dualities, beyond the limits of space and time, beyond all hopes and fears. Freedom from the ties to habitual patterns of suffering is nirvana.

Habitual patterns of mind are attributed to *karma* and *klesas*. Karma is the result of actions, including those from previous lives. Karma includes positive and negative impressions from intentions and volitional acts that continue to influence the qualities and content of our lives. Klesas are emotional defilements and mental obscurations, such as attachment and aversion, sometimes translated as passion and aggression, that are the root causes of suffering. Ignorance is the root cause of attachment, aversion and other forms of suffering.

Ignorance in this context is not a lack of knowledge; it is an incorrect view of our basic relationships in life, including ego-fixation and habits of dualistic thinking.

Healing involves removing the karmic veils and klesas that obscure proper relationships within the life experiences. Tibetan Buddhism posits that we are ignorant of our true nature, also known as Buddha Nature. Buddha Nature is not something to be created or achieved, for it is innate within all sentient beings. Enlightenment is a process of revelation, since there is nothing to be gained or rejected. The afflicted aggregates simply dissolve in mindfulness upon the realization of Ultimate Reality (Thrangu Rimpoche, 1993b). Through the disciplines of mindfulness, generosity and compassion, suffering can be transformed. Thus, healing is synonymous with complete awakening or enlightenment.

Mahayana meditations for healing: Tong len

Tong len is a healing meditation from the Mahayana tradition. It is based on compassion, whereby the meditator engenders bodhicitta, the awakened heart to benefit all beings. "*Bodhicitta* is the unconditional intention to help all sentient beings become free of suffering. It is the complete abandonment of any sort of personal territory, both in one's relationships with others and in one's understanding of the world as it is" (Kongtrul Rimpoche, 1987: 55). The cultivation of bodhicitta enables the practitioner to relate to the vastness and richness of the universe as a means to offer compassionate services.

Tong len meditation is also called "sending and taking". The meditator willingly takes on the suffering of others as visualized during meditation and offers beneficence in its place. There are two aims of the practice: for the meditator to become free from suffering and confusion through direct experience of the nature of reality (ultimate bodhicitta) and to help others become free from suffering through the generation of compassion (relative bodhicitta).

Vajrayana practices and tantric healing rituals

The Vajrayana practices are called *tantric sadhanas*. They include the practice ritual to invoke the Medicine Buddha and other healing agents. There are various levels of tantra. The lower levels of Kriya, Carya and Yogatantra are primarily for purification and worldly aims, such as promoting longevity, avoiding catastrophe or attaining wealth. The practice of the Medicine Buddha falls within this category.

One tantric practice is a simulation of death by mimicking the changes in the subtle body that are associated with dying. Such practices help to remove fear and doubts about death, and there is a direct experience of the Clear Light of Mind. The practitioner then can identify the stages of transition in the *bardo*, or intermediate stages between death and rebirth. Death in Tibetan medicine is a process that is integrated with life, an event to be mastered through mindfullness training. Death can be imbued as a liberating, mindful experience of awareness – one that is a release from the samsaric bonds of suffering. Dying with awareness is one means to obtaining a good rebirth (Samuel, 1993).

Sadhana of Sangye Menla – the practice of the medicine Buddha

According to legend, the Medicine Buddha appeared and prophesied that Sakyamuni Buddha would manifest in the future as the Medicine Buddha to alleviate suffering throughout the world. The Medicine Buddha would manifest in India as Sangye Menla (Tibetan: *Sangs-rgyas sMan-bla*; Sanskrit: *Bhaishajya Guru*), also known as The Sovereign Healer (Figure 3).

According to historic texts, Sakyamuni Buddha entered into a meditative vision in which he created a palace built of five precious materials. From the center throne made of lapis lazuli, he manifested as Sangye Menla to convey teachings about compassion, medicine and the alleviation of suffering. The palm of his right hand faces out in the *mudra* (gesture) for bestowing protection. In his left hand he holds a bowl containing the myrobalan plant, with remarkable medicinal properties. He emanated the five *dhyani* (meditational) buddhas: Vairochana at the crown, Amitabha from the throat, Akshobhya at the heart, Ratnasambhava at the navel and Amoghasiddhi at the secret place (see Table 1) (Dhonden, 1986). Amitabha then requested the teachings on health and healing.

Figure 3 Sangye Menla. From Thrangu, Rimpoche, "The twelve great aspirations of the medicine Buddha." *Shenpen Ösel* 4(1), June 2000. Available at <http://www.shenpen-osel.org/>. Used with permission of the publisher.

Each of the other *dhyani* buddhas taught the various root texts in conversations with Amitabha. On another level of understanding, the Sage Rigpa Yeshe and other *bodhisattvas* taught the science of healing as a series of questions and answers, recorded in a text of gold leaves with ink made from lapis lazuli.

During the visualization of the Medicine Buddha, healing rays of light emanate in all directions from the visualized blue colored body. This light pacifies ailments arising from the three poisons – passion, aversion and ignorance. The breath of the practitioner when visualized as that of the Medicine Buddha can be used as a healing aid. The practitioner might gently blow upon the patient to bestow a blessing from Sangye Menla. Similar benefits arise from the well-motivated visualization practice with blessings of Sangye Menla directed to those in need, regardless of distance.

Anatomy of the subtle body

In the absence of enlightenment, our bodies, minds and ordinary phenomena appear to be composed of the five elements, when perceived by the five ordinary sense perceptions. From the perspective of Vajrayana, each element is associated with positive and negative qualities in a sacred view of the world. Since enlightenment is inexpressible, Tibetan teachings rely on symbols, iconography and metaphors to explain the Path to Enlightenment (see Table 1). Each purified emotion corresponds to the wisdom aspect of the five meditational buddhas, corresponding to the five buddha families, each with its own distinguishing qualities and characteristics. By analogy, the female consorts of each family correspond to the purification of related elements. The male and female deities represent *upaya* and *prajna*, method and wisdom. The sexual union of the deities visualized during meditation practice represents the bliss of enlightenment.

The emphasis on mind training within Tibetan Buddhism enables novel approaches to anatomy that are the focus of tantric meditations (Figure 4). Subtle anatomy plays a role in the tantric approach to ritual healing. The subtle body is composed of psychic centers (Sanskrit: *chakra*; Tibetan: *k'orlo*, wheel), channels (Sanskrit: *nadi*; Tibetan: *tsa*) and seeds of psychic energy (Sanskrit: *bindu*; Tibetan: *thigle*, drop) that flow along the channels (Figure 4). In this practice (Tibetan: *gyulū*; illusory body), all physical forms including the body are regarded as illusory in nature, while the subtle body is experienced with immense clarity.

Although the subtle body does not manifest itself in physical anatomy, an imbalance in relationships or functions of the subtle (psychic) anatomy is a cause of illness and suffering. Imbalances may be due to karma, inappropriate lifestyle, unethical conduct, transgression of vows, etc. The anatomy of the subtle body is "visualized", a term meaning that awareness is focused inwardly so that an image, feeling, sensation, and thought of the designated practices might occur within the mind of the meditator. Visualizations of subtle anatomy can deepen the stages of meditative awareness. The varieties of meditative experiences depend upon the openness, clarity and qualities of the channels.

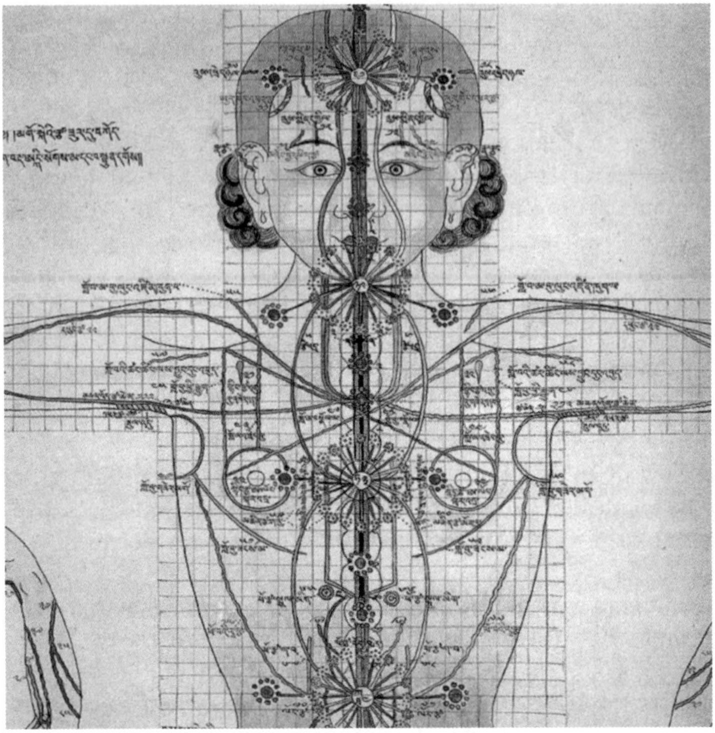

Figure 4 Tibetan subtle anatomy.

TIBETAN AND WESTERN MEDICINE: PHILOSOPHY, SCIENCE AND CULTURE

Tibetan medicine represents a challenge to the modern paradigms for scientific inquiry and discourse. Both traditional Tibetan medicine and the Western sciences are self-consistent in their theories and practices. Each shares the principle of causality, and they both utilize analytic methods to reach conclusions. Both provide a means to help those with adverse health conditions. Evaluation of clinical predictions is possible in both. Thus Tibetan medicine, like biomedicine, meets criteria as a science (Loizzo, 1998).

Tibetan medicine and Western biomedicine differ in scope and in dimensions for their respective applications. Suffering in Tibetan medicine is based on imbalances in the aggregates as a result of unhealthy lifestyles, karmic actions, obscured perceptions and self-centered attitudes. Each life experience is imbued with spiritual, mental and physical aspects. It is logically consistent that the interventions in Tibetan medicine to alleviate affliction while promoting health and happiness must also address all of these dimensions. Therefore, the spiritual nature of suffering is integrated in a manner that complements therapies for physical illness and disease. On the other hand, from the perspective of western science, the lack of objective measures to verify or refute key concepts, such as karma and klesas, are barriers to understanding the tenets of Tibetan medicine. This point has been raised as a key concern among leaders of biomedicine in the United States (Dalen, 1999). However, subjective reports also have a scientific basis in the western analytic sciences (Najman, 1981; Wallace, 2000).

Scientific methods and the evaluation of traditional Tibetan medicine

From a biomedical perspective, inquiries about a Tibetan medicine perspective can provide useful information about mechanisms of actions for herbal remedies. For example, Tibetan medicated baths were reported to modulate the responses of the immune system to patients with rheumatoid arthritis (Zhao, 1993). The traditional Tibetan herbal preparation known as Padma-28 shows anti-inflammatory activity involving functions of human neutrophils (Matzner, 1995). A newly discovered flavonoid contained in other traditional herbal preparations can suppress coughs (Peng, 1996). Clinical anecdotes about Tibetan treatments of HIV and AIDS, irritable bowel syndrome, cancer and other chronic diseases are found in popular texts (Dhonden, 1986, 2000; Tokar, 1998).

There are also reports about objective measures for traditional diagnosis. Research is emerging to support the objective validity of radial pulse diagnosis through statistical analysis of waveforms (Kosoburov, 1996; Shinohara, 1989). Transducers can detect changes in the form of pulse waves that correspond to qualities in traditional pulse diagnosis (Niu, 1994; Zhang, 1993; Fu, 1990). Changes in the characteristics of radial pulse waves are associated with experimental disturbances in circulation to other body parts (Dai, 1985). Computer software programs for cataloging Tibetan pulse diagnosis in association with symptoms are being developed (Boronoev, 1996). Research on tongue diagnosis and abdominal palpation in other traditions of medicine is also available (Shi, 1993; Wang, 1992).

If the goal of research about Tibetan medicine is to examine the utility of the traditional model, then a critical examination of research methods in the context of culture is needed (Nichter, 1992). Research methods that impose cultural assumptions onto a different tradition of medicine may yield hypotheses that lack validity from the native perspective. This is an error that arises from the application of the medico-centric bias to culturally different traditions of health (Steiner, 1992). The medico-centric bias may invalidate the results of clinical research efforts that do not consider traditional concepts embedded within the cross-cultural research hypothesis. In the biomedical model, disease is reified in accord with specific measurable criteria. However, the symptom complexes that are used to reify constitution and condition in the Tibetan medical model are different from the indicators of biomedical disease. The imbalances that are the basis for treatment in Tibetan medicine are often either not assessed or simply overlooked in research methods from the biomedical perspective. Thus, if biomedical case definitions are used to select and evaluate a traditional treatment, then the indications for the very use of that remedy may not be present from the perspective of Tibetan medicine.

The error of omitting culturally relevant variables from the research study may invalidate the cultural appropriateness of the intervention. Even though the research question may have been posed about evaluating an intervention from a biomedical perspective, it is a study question that was conceived without regard to the context of the native culture. That cultural context is the source of the construct for the traditional disease category and the corresponding treatment that are under investigation. It is quite possible that the disease of interest in biomedicine may not even exist within the framework of Tibetan medicine. Conversely, a simple biomedical disease may have many forms in Tibetan medicine, depending on the presence or absence of associated symptoms and traditional diagnostic findings in palpation of the pulse or inspection of the tongue or urine.

Therapy based solely on the presence of an isolated symptom may be insufficient to select optimal traditional treatments for evaluation in cross-cultural research. For example, the specificity of pharmacological action for a particular drug may be a key point of interest for Western medicine, but this approach may not be valid from the perspective of the Tibetan medical model. The goal of Tibetan medicine is to restore balance in the whole person based on redistribution of aggregates and elements rather than alleviating isolated symptoms. Re-establishing harmony reduces the symptoms. In addition, traditional remedies typically combine multiple herbs to reduce the side effects from any single active ingredient and to bolster any aggregate that may be adversely influenced during the dynamics of the healing process. Complex interactions among many ingredients in multi-herbal preparations reduce the statistical efficacy of modern methods of scientific investigation. In short, ethnopharmacologic evaluations of Tibetan medicines require special research methods (Anderson, 1992; Etkins, 1986).

Thus typical criterion-based methods for case-selection linked with interventions in biomedical research may not be valid to evaluate the outcomes of Tibetan medicine. If negative results are found from such research projects, they might be better attributed to poor study design with lack of insight into the cultural context, rather than interpreting the negative result as a lack of efficacy of the traditional treatment. The application of a traditional intervention without attention to the cultural context may miss the main objective of modern research methodologies – to provide valid results with minimal error.

There are simple research strategies that can overcome barriers to proper research designs about cross-cultural health issues. Observational studies using research designs which replicate measures over time (e.g., before and after traditional interventions), with the patient serving as his own control, are best suited for initial quantitative studies of culturally different health care traditions. This design strategy allows the Tibetan physician to prescribe complex treatments that are individualized to each patient. This approach is being utilized in current studies of Tibetan medicine for women with Stage IV breast cancer (Randal, 1999; Dhonden, 2000: 136–157). As psychometrically valid measures for the assessment of constitution and condition become available, then valid clinical trials may be designed to test the efficacy of complex treatments among groups of patients with similar standardized traditional clinical findings. Until such reliable measures are available, the optimal research design will include allowing master Tibetan physicians to assign traditional treatments to participating patients in research studies.

* * *

Tibetan medicine is a highly ordered, comprehensive health system. Although it is part of the culture of Tibet and based on Buddhist worldviews, it is a humanistic approach to medicine that can be applied to health care for patients in different societies.

Tibetan medicine addresses health in a positive sense (Seeman, 1989; Dash, 1991; Malathi, 2000); personal happiness and social harmony are explicit goals. The presence of dimensions of positive health and subjective wellbeing complement the indicators of disease in the biomedical model. Western models show that social support, self-control and spiritual beliefs can buffer the illness experience to improve quality of life (Cohen, 1985; Diener, 2000; Steiner, 1998).

Tibetan medicine offers a two-fold strategy for health: letting go of negativity and then attending to positive dimensions of health to assure a higher quality of life. The decisions to release attachments to samsara and to cultivate a more positive way of life are two separate courses of action. Both are necessary to realize the full potential of health. This model for positive and negative health contains the intriguing possibility that a person can be simultaneously healthy and ill. This dual approach with independent and interacting axes of health in Tibetan medicine may merit further investigation.

The presence of well-motivated and skillful physicians who are attentive to their patients and the community may carry as much potential for healing and the transformation of suffering as the development of modern medical technology. This statement is not to discount the importance of science and technology in medicine; indeed, the technologic advances in health care are manifestations of compassion and concern. Rather, Tibetan medicine emphasizes human relationships. The physician is necessary to assess the patients' qualities and to advise them properly about how to alleviate the root cause of suffering. It is the intention of the practice that defines the potential for benefit. This is akin to the position offered by the physician Sir William Osler (1849–1919), "It is more important to know the person with the disease than to know what sort of disease the patient has."

Modern scientific methods can be used to investigate Tibetan medicine. However, there are methodological barriers to be considered. For example, researchers make assumptions about the nature of disease and illness without being aware that there may be other perspectives that influence the validity of their research (Like, 1996). There are methods to overcome this bias. Hypothesis generation from the traditional perspective is one means; observational research designs in which the patients serve as their own controls and where treatments are provided in accord with traditional diagnostic methods may be another. The study of cultural traditions of health holds the potential to bridge the schism that exists today between the natural and the human sciences (Bernstein, 1983).

NOTES

[1] A terton is a revealer of hidden treasures, typically teachings concealed by previous master teachers for discovery at later dates.

[2] There are other versions of this story. See for example Tsarong, 1981.

[3] Ayurveda says that the body is made up of tissues (*dhatus*), waste products (*malas*) and *doshas* (loosely translated as Energetic Forces). It is the *tridoshas*' job to assist with the creation of all of the body's tissues and to remove any unnecessary waste products. The *tridoshas* influence all movements, all transformations, all sensory functions, and many of the other activities in the body and mind.

[4] *Samsara* is contrasted with *nirvana*. *Samsara* is the world of ordinary life; *nirvana* is a state of enlightenment in which all false ideas and conflicting emotions have been eliminated.

BIBLIOGRAPHY

Akong, Rimpoche. *Taming the Tiger: Tibetan Teachings on Right Conduct, Mindfulness and Universal Compassion.* Rochester, Vermont: Inner Traditions, 1996.

Anderson, R. 'The efficacy of ethnomedicine: research methods in trouble.' In *Anthropological Approaches to the Study of Ethnomedicine*, M. Nichter, ed. Philadelphia: Gordon and Breach Science Publishers, 1992, pp. 1–18.

Avelon, J.F., ed. *The Buddha's Art of Healing: Tibetan Paintings Rediscovered.* New York: Rizzoli International Publications, 1998.

Baker, Ian A. *The Tibetan Art of Healing.* San Francisco: Chronicle Books, 1997.

Bernstein, Richard J. *Beyond Objectivism and Relativism: Science, Hermeneutics and Praxis.* Philadelphia: University of Pennsylvania Press, 1983.

Birnbaum, R. *The Healing Buddha.* Boulder: Shambhala Publishers, 1979.

Bolsokhovea, Natalia, A. Kseniya and M. Gerasimova. 'The atlas of Tibetan medicine: treasures from the History Museum of Buryatia.' In *The Buddha's Art of Healing: Tibetan Paintings Rediscovered*, J.F. Avelon, ed. New York: Rizzoli International Publications, 1998, pp. 32–60.

Boronoev, V.V. and O.S. Rinchinoiv. 'Several aspects of software implementation of a computerized pulse diagnostics complex.' *Meditsinskaia Tekhnika* 5: 36–38, 1996.

Burang, Theodore. *Tibetan Art of Healing.* London: Robinson and Watkins, 1974.

Chopra, Deepak. *Perfect Health: The Complete Mind/Body Guide.* New York: Harmony Books, 1991.

Clark, Barry. *The Quintessence Tantras of Tibetan Medicine.* Ithaca, New York: Snow Lion Publications, 1995.

Clifford, Terry. *Tibetan Buddhist Medicine and Psychiatry: The Diamond Healing.* New York: Samuel Weiser, 1990.

Cohen, S. and T.A. Wills. 'Stress, social support, and the buffering hypothesis.' *Psychological Bulletin* 98(2): 310–357, 1985.

Dai, K., H. Xue, R. Dou and Y.C. Fung. 'On the detection of messages carried in arterial pulse waves.' *Journal of Biomechanical Engineering* 107(3): 268–273, 1985.

Dalai Lama XIV, Thupten Jinpa and Dominique Side, eds. *The Four Noble Truths: Fundamentals of the Buddhist Teachings.* London: Thorsons, 1997.

Dalai Lama XIV. 'The relevance of Tibetan medicine today.' *Alternative Therapies in Health and Medicine* 5(3): 67–69, May 1999.

Dalen, J.E. 'Is integrative medicine the medicine of the future? A debate between Arnold S. Relman, MD and Andrew Weil, MD.' *Archives of Internal Medicine* 159(18): 2122–2126, 1999.

Dash, Vaidya Bhagwan. *Tibetan Medicine: Special Reference to Yoga Sataka.* Dharamsala, India: Library of Tibetan Works and Archives, 1976.

Dash, Vaidya Bhagwan. *Positive Health in Tibetan Medicine.* Delhi, India: Sri Satguru Publications, 1991.

DeCharms, Christopher. *Two Views of Mind: Abhidharma and Brain Science.* Ithaca, New York: Snow Lion Publications, 1997.

Dhonden, Yeshe. *The Ambrosia Heart Tantra: The Secret Oral Teachings of the Eight Branches of the Science of Healing*, Volume 1, Jhampa Kelsang, trans. Dharamsala, India: Library of Tibetan Works and Archives, 1977.

Dhonden,Yeshe and Topgay Sonam. 'Pulse diagnosis in Tibetan medicine.' Transcribed and edited by R.P. Steiner. In *Tibetan Medicine.* Series 1. Dharamsala, India: Library of Tibetan Works and Archives, 1980.

Dhonden, Yeshe. *Healing Through Balance: An Introduction to Tibetan Medicine.* Ithaca, New York: Snow Lion Publications, 1986.

Dhonden, Yeshe. *Healing From the Source: The Science and Lore of Tibetan Medicine*, Alan Wallace, ed. and trans. Ithaca, New York: Snow Lion Publications, 2000.

Diener, Ed and E.M. Suh, eds. *Culture and Subjective Well-Being: Well-Being and Quality of Life.* Cambridge, Massachusetts: MIT Press, 2000.

Dummer, Tom. *Tibetan Medicine and Other Holistic Health Care Systems.* London: Routledge, 1988.

Etkins, Nina L., ed. *Plants in Indigenous Medicine and Diet: Biobehavioral Approaches.* Bedford Hills, New York: Redgrave Publishing, 1986.

Finckh, E. *Studies in Tibetan Medicine.* Ithaca, New York: Snow Lion Publications, 1988.

Finckh, E. *Foundations of Tibetan Medicine.* London: Robinson and Watkins, 1978.

Frey, J.J. and J. Frey. 'A literature analysis in family medicine and general practice.' *Family Medicine* 32(6): 398–402, 2000.

Fu, C.Y. 'An approach to objective indexes for differentiating floating, sunken, feeble and forceful pulses in traditional Chinese medicine.' *Chung Hsi i Chieh Ho Tsa Chih/Chinese Journal of Modern Developments in Traditional Medicine* 10(10): 603–605, 581, 1990.

Gadamer, Hans Georg. *Philosophical Hermeneutics.* Berkeley: University of California Press, 1977.

Gadamer, Hans Georg. *The Enigma of Health: The Art of Healing in a Scientific Age*, Jason Gaiger and Nicholas Walker, trans. Stanford, California: Stanford University Press, 1996.

Gyamtso, Tsultrim Rimpoche. *Progressive Stages on Meditation on Emptiness*, Shenpen Hookham, ed. Oxford: Longchen Foundation, 1988.

Hayward, Jeremy W. *Shifting Worlds, Changing Minds: Where the Sciences and Buddhism Meet.* Boston, Massachusetts: Shambhala Publishers, 1987.

Hayward, Jeremy W. and Francisco J. Varela, eds. *Gentle Bridges: Conversations with the Dalai Lama on the Sciences of Mind.* Boulder: Shambhala Publishers, 1992.

Janes, C.R. 'The transformations of Tibetan medicine.' *Medical Anthropology Quarterly* 9(1): 6–39, 1995.

Karmapa XVI Ranjung Rigpe Dorje. *Advice on Meditation*, Nudrop Tersing Buckhar, trans. Woodstock, New York: Karma Triyana Dharmachakra Monastery, audiotaped on-site August, 1980.

Karthar, Rimpoche. *Dharma Paths*, Laura Roth, ed. Ithaca, New York: Snow Lion Publications, 1992.

Kleinman, Arthur. *Patients and Healers in the Context of Culture.* Berkeley: University of California Press, 1981.

Kongtrul, Rimpoche. *The Path of Great Awakening: An Easily Accessible Introduction for Ordinary People: A Commentary on the Mahayana Teachings of the Seven Points of Mind Training*, Ken McLeod, trans. Boston: Shambhala Publishers, 1987.

Kosoburov, A.A. 'Transducer for pulse diagnostics.' *Meditsinskaia Tekhnika* 5: 35–36, 1996.

Kuzel, A.J. 'Naturalistic inquiry: an appropriate model for family medicine.' *Family Medicine* 18(6): 369–374, 1986.

Lad, Vasant. *Ayurveda: The Science of Self-Healing.* Santa Fe, New Mexico: Lotus Press, 1984.

Lauf, Detlef Ingo. *Tibetan Sacred Art: The Heritage of Tantra.* Berkeley, California: Shambhala Press, 1976.

Leslie, Charles M. and Allen Young, eds. *Paths to Asian Medical Knowledge: Comparative Studies of Health Systems and Medical Care.* Berkeley: University of California Press, 1992.

Like, R.C., R.P. Steiner and A.J. Rubel. 'STFM core curriculum guidelines. Recommended core curriculum guidelines on culturally sensitive and competent health care.' *Family Medicine* 28(4): 291–297, 1996.

Lock, Margaret and Nancy Scheper-Hughes. 'A critical interpretative approach in medical anthropology: rituals and routines of discipline and dissent.' In *Medical Anthropology: Contemporary*

Theory and Method, Thomas Johnson and Carolyn Sargent, eds. New York: Praeger, 1990, pp. 47–72.

Loizzo, J.J. and L.J. Blackhall. 'Traditional alternatives as complementary sciences: the case of Indo-Tibetan medicine.' *Journal of Alternative and Complementary Medicine* 4(3): 311–319, 1998.

Malathi, A., A. Damodaran, N. Shah, N. Patil and S. Maratha. 'Effect of yogic practices on subjective well being.' *Indian Journal of Physiology and Pharmacology* 44(2): 202–206, 2000.

Matzner, Y. and S. Sallon. 'The effect of Padma-28, a traditional Tibetan herbal preparation on human neutrophil function.' *Journal of Clinical and Laboratory Immunology* 46(1): 13–23, 1995.

Najman, J. and S. Levine. 'Evaluating the impact of medical care and technologies on the quality of life: a review and critique.' *Social Science and Medicine* 15: 107–115, 1981.

Nichter, M., ed. *Anthropological Approaches to the Study of Ethnomedicine*. Philadelphia: Gordon and Breach Science Publishers, 1992.

Niu, X., X.Z. Yang and C.Y. Fu. 'Three-dimensional motion of the radial artery and attributes of position, rhythm, shape and variance of pulse diagnosis.' *Chung-Kuo Chung Hsi i Chieh Ho Tsa Chih* 14(7): 435–437, 1994.

Peng, J.N., Y.C. Ge and X.H. Li. 'Studies on the chemical constituents of Rhodiola fastigita.' *Yao Hsueh Hsueh Pao/Acta Pharmaceutica Sinica* 31(10): 798–800, 1996.

Porkert, Manfred. *The Theoretical Foundations of Chinese Medicine: Systems of Correspondence*. Cambridge, Massachusetts: MIT Press, 1974.

Randal, J. 'Diagnosis, Tibetan style, underlies small herbal study of advanced breast cancer.' *Journal of the National Cancer Institute* 91(7): 587–588, 1999.

Rechung, Rimpoche. *Tibetan Medicine*, Jampal Kunzang, ed. Berkeley: University of California Press, 1976.

Sachs, R. and Lobsang Rapgay. *Health for Life: Secrets of Tibetan Ayurveda*. Santa Fe, New Mexico: Heartsfire Books, 1995.

Samuel, Geoffrey. *Civilized Shamans: Buddhism in Tibetan Society*. Washington, DC: Smithsonian Institution Press, 1993.

Seeman, J. 'Toward a model of positive health.' *American Psychologist* 44(8): 1099–1109, 1989.

Shi, Z.X., Z.M. Wu and J.L. Zhang. 'Doppler sphygmogram of uneven pulse and spectral changes of tongue.' *Chung-Kuo Chung Hsi i Chieh Ho Tsa Chih* 13(4): 205–207, 1993.

Shinohara, K. and M.F. Gilula. 'The arterial pulse analyzer as a potential replacement for manual pulse palpation in Oriental medicine.' *Acupuncture and Electro-Therapeutics Research* 14(3–4): 235–252, 1989.

Steiner, R.P. 'Culture, family medicine and society.' *American Family Physician* 46(5): 1398–1400, 1992.

Steiner, R.P. 'Tibetan medicine: Part I: Pulse diagnosis in Tibetan medicine: translated from the first chapter for the Fourth Tantra (rGyud-bzi).' *American Journal of Chinese Medicine*. Part I: 14(1–2): 83–88; Part II: 15(3–4): 165–170, 1987; Part III: 16(3–4): 173–178, 1988; Part IV: 17(1–2): 79–84, 1989.

Steiner, R.P. *Exploring the Association of Perceived Quality of Life and Depressive Symptoms among Medicare Patients Attending Primary Care Clinics*. Ph.D. Dissertation, Chapel Hill: University of North Carolina School of Public Health, 1998.

Tai Situpa Rimpoche and Lea Terhune, ed. *Relative World, Ultimate Mind*. Boston: Shambhala Publications, 1992.

Tamdin, S. Bradley. 'Commentaries: The tree of health and disease.' In *The Buddha's Art of Healing: Tibetan Paintings Rediscovered*, J.F. Avelon, ed. New York: Rizzoli, 1998, pp. 76–77.

Tarthang, Tulku. Kum Nye. *Relaxation*. Part 1: *Theory, Preparation, Massage*. Berkeley, California: Dharma Publishing, 1979.

Thakkur, Chandrashekhar. *Introduction to Ayurveda: The Indian Art and Science of Medicine*. New York: ASI Publishers, 1974.

Thondup, Tulku. *The Healing Power of Mind: Simple Meditation Exercises For Health, Well-Being and Enlightenment*. Boston: Shambhala Publishers, 1998.

Thrangu, Rimpoche. *Buddha Nature: Ten Teachings on the Uttara Tantra Shastra*. Hong Kong: Rangjung Yeshe Publications, 1993.

Thrangu, Rimpoche. *The Practice of Tranquility and Insight: A Guide To Tibetan Meditation.* Ithaca, New York: Snow Lion Publications, 1993b.

Thrangu, Rimpoche. 'Teachings on the Medicine Buddha Sadhana.' In Tibetan, with transliteration, word-for-word translation, and full literary translation. *Shenpen Ösel* 4(1), June 2000. (Available at <http://www.shenpen-osel.org/>, accessed on January 16, 2000).

Thrangu, Rimpoche. 'Commentary on the medicine Buddha Sutra.' *Shenpen Ösel* 4(2), June 2000. (Available at <http://www.shenpen-osel.org/>, accessed on January 16, 2000).

Thrangu, Rimpoche. 'The twelve great aspirations of the medicine Buddha.' *Shenpen Ösel* 4(1), June 2000. (Available at <http://www.shenpen-osel.org/>, accessed on January 16, 2000).

Tokar, E. 'Seeing to the distant mountain: diagnosis in Tibetan medicine.' *Alternative Therapies in Health and Medicine* 5(2): 50–58. 1999.

Tokar, E. and A. Vora. 'Building a means of discourse for integrative medicine, a Tibetan medical perspective on irritable bowel syndrome.' Originally published in *Alternative and Complementary Therapies* 4(5), October 1998, accessed on website <http://www.tibetanmedicine.com/html/article2.html#a> on January 20, 2001).

Tsarong, T.J., J.G. Drakton and L. Chomphel. *Fundamentals of Tibetan Medicine: According to the rGyud-bzhi.* Dharamsala, India: Tibetan Medical Center, 1981.

Wallace, B. Alan. *Choosing Reality: A Buddhist View of Physics and the Mind.* Ithaca, New York: Snow Lion Publications, 1996.

Wallace, B. Alan. *The Taboo of Subjectivity: Towards a New Science of Consciousness.* London: Oxford University Press, 2000.

Wang, J. and K.J. Chen. 'Abdominal diagnosis in blood-stasis syndrome.' *Chung-Kuo Chung Hsi i Chieh Ho Tsa Chih* 12(12): 726–729, 1992.

Zhao, S.L., P.L. Geng and J. Sang. 'Immune function of rheumatoid arthritis treated by edicated-bath therapy in Tibetan medicine.' *Chung-Kuo Chung Hsi i Chieh Ho Tsa Chih* 13(8): 467–470, 1993.

Zhang, Z.H., Z.H. Zhang and S.H. Wu. 'Assessment of early pregnancy detection with sphygmogram of smooth pulse.' *Chung-Kuo Chung Hsi i Chieh Ho Tsa Chih* 13(5): 276–277, 1993.

VIGGO BRUN

TRADITIONAL THAI MEDICINE

Western medicine was first introduced in Ayudhaya, the former Thai capital, during the reign of King Narai (1658–88), but it was only from around 1830 that missionaries began to practice cosmopolitan medicine in Thailand on a regular basis. Biomedicine is now widely accepted by the Thai public, as it has been Thailand's official medical system since the beginning of this century. At the same time there exists a Thai medical tradition based on the use of natural products and local disease concepts which has been practiced by local physicians for centuries. In present-day Thailand, the public makes use of both traditions – predominantly the official cosmopolitan medicine and, to a lesser extent, traditional medicine.

Traditional medicine has until recently been tolerated but not supported by the state. Traditional doctors had to apply for permission to practice, and since the 1950s, the state has also required that traditional doctors pass state-controlled exams in traditional pharmacy or medicine. Thus, the state has attempted to register traditional doctors and also to assure a certain amount of quality control, although these measures have not been strictly enforced. In spite of a restrictive state policy, there has been a continual demand for traditional medicine, which has survived quite well in the rural areas and which, since the late 1950s, also has been promulgated by some traditional medical schools in the major urban centers.

Over the last decade the official attitude towards traditional medicine has changed, and the policy is now to integrate traditional medicine into the public health system. Still, Western trained doctors in Thailand remain skeptical towards traditional medicine, although the Thai public is showing an increasing interest in its healing methods.

GENERAL CHARACTERISTICS

In villages all over Thailand one finds people referred to as traditional doctors (*moo phaen booraan*) who mainly use herbs (*samunphraj*) and other natural substances to cure diseases. The sum of knowledge and practices of these experts is labelled "traditional Thai medicine" or "folk medicine". There is no

H. Selin (ed.), Medicine Across Cultures: History and Practice of Medicine in Non-Western Cultures, 115–132.

traditional doctor who knows the entire body of traditional knowledge, and there are many local and individual variations of Thai traditional medicine within Thailand – in the way plants and diseases are labelled, in the way diseases are diagnosed, and in the way plants and other materia medica are used in the curing process.

Traditional Thai medicine includes knowledge about the identification of plants (and minerals and animal components) and their curing properties, about diagnosis, cause and development of diseases, about prescriptions and about relevant incantations and ceremonies. The prescriptions, which exhibit the concrete relationship between specific diseases and the plant world, are – together with the incantations – considered to be the essential and most valuable part of the tradition. The prescriptions contain any number of ingredients, normally from five to ten, but sometimes less and often more, even comprising as many as fifty different components. For each ingredient the prescription will ideally specify which part of the plant should be used – root, leaf, bark, fruit etc. – and normally in what quantity. While much of the medical knowledge is transmitted orally, the prescriptions are written down. There are a vast number of medical manuscripts in Thailand, both in temple libraries and in the care of the herbalists, and almost all of these manuscripts are merely collections of prescriptions without any further information. In order to access the information in the manuscripts as well as to copy them and write down the prescriptions, it was necessary to be able to read and write. In former times these skills were taught only in temples to boys, so the traditional doctors were exclusively male. Monks and former monks were central in the transmission of all written traditional knowledge, including medicine, and Buddhism has – as we shall see below – left a deep imprint on the way traditional medicine has been organised.

To a certain extent, the general public also possesses medical experience and knowledge which people use to talk about diseases, to chose between available curative alternatives, and for self-medication. There are a number of simple traditional remedies known to villagers, commonly referred to as "household remedies" (*jaaklaangbaan*).

The traditional doctors are often called to a career in medicine after they themselves have been sick. They will seek out a traditional doctor, get cured, and thereafter ask to be accepted as a student of that doctor and learn the prescription that cured them. After that they start treating people with this particular disease. If the treatment is successful, and if the new doctor feels he has a knack for traditional medicine, he seeks out more knowledge from other teachers and thus gradually expands his curing range. Another path is to learn the trade from a practicing relative. Few traditional doctors become true generalists, though, and few become full-timers. For most of them, curing provides them with supplementary income, although a successful traditional doctor will be respected and receive higher social status.

Traditional Thai doctors neither perform surgery, except for lancing abscesses, nor do they undertake dissection. Thus they do not have very precise or systematic knowledge of human interior organs and how they function. A

Figure 1 Preparation of herbal medicine. Various dried woods are grinded (*fon*) against a stone into a container with cold water (Northern Thailand, 1978. Photo: Viggo Brun).

certain amount of anatomical knowledge is gathered from partitioning animals, and Buddhist scriptures do enumerate the body components.

Some traditional doctors perform bone setting. The broken bones are pulled back into place and the area around the fracture stabilized with sticks of wood tied together, after which incantations are blown onto the fracture. Traditional bone setting may result in the bones growing together at unusual angles, so most people prefer to go to Western-style hospitals to be treated for fractures.

Massage and midwifery are separate branches of the local medical tradition. Traditional midwifery is fast disappearing, while Thai massage has become tremendously popular – even outside Thailand. A number of drawings of the human body in Wat Pho, the largest and oldest temple in Bangkok, pertain to Thai massage, indicating points on the exterior of the body on which applications of pressure have a therapeutic effect for specific ailments. Other drawings of the human body show lines (*sen*) connecting various points on the body. The meaning and practical application of these lines and how they connect up with the rest of traditional medical theory and practice are not clear.

Traditional saunas, where steam from herb-steeped water is directed into a small, closed room, are also used. These steam baths are regarded as generally healthful but are also used specifically to cure certain skin or respiratory diseases.

In the countryside people used to rinse their mouths with water after eating, and they used toothpicks and sometimes rubbed their gums with salt. Now most people use toothbrushes and toothpaste. Previously toothache was simply considered to be bad luck. When the pain became intolerable, the tooth was pulled out. Until recently there were peddlers appearing in the villages trying to convince people that their toothache was caused by small worms eating the teeth – they could miraculously produce such worms from the mouth of any bystander with toothache – and that their bottles of medicine would kill these worms. This medicine was very popular. The bottles they sold merely contained a simple painkiller.

TRANSMISSION

The medical tradition in Thailand is said to have been transmitted through a chain of teacher-pupil relationships which goes back to the tradition's original teacher, Jivaka Komarabhacca, the legendary physician of the Lord Buddha, and each practitioner pays respect to his teacher at annual rituals. On every occasion in which he uses the knowledge from the tradition, he should ceremonially invoke the whole chain of teachers – living and dead – to be present and thus infuse his actions with their sacred power. Medical knowledge is restricted to those who have been ritually initiated into the tradition, and thus every potential doctor has to find a teacher who will accept him as a pupil and share his knowledge with the student. The teacher is important because the tradition consists of both oral and written information. The written manuals contain the essence of the tradition – the prescriptions – but the manuals contain almost no information about each particular disease or plant, beyond their names. All the background knowledge and detailed information which the manuals presuppose are transmitted orally. The student needs a teacher to give him this oral information and from whom he can observe the finer points of medical practice.

According to Indian myths of genesis, a god is portrayed as possessing the totality of knowledge, which is then transmitted orally to another god, who in turn abbreviates it and passes it on to another god who again abbreviates it and transmits it and so on, until it is finally reduced to a size and form compatible with the human capacity for memory and understanding. At this point it is revealed to a human agent, who becomes the primordial *guru* of a chain of teachers and pupils through which the knowledge is passed on up to the present. The ideal would be if the complete primordial knowledge were passed on verbatim through faithful repetition and copying. Still, time involves change and decay, and it is recognized by everyone that losses, fragmentation, external additions and copying mistakes have occurred all along the transmission chain. Medical manuals are no longer complete and homogeneous. The actual transmission of knowledge is no longer only faithful copying but also a continual process of selection and reorganization, and new knowledge may even be added to the tradition. The standard way of justifying additions is to present them as revelations, often referred to as *tamraa phii book* (manual told by the spirits). In other words, additions and innovations are permitted as long as they are received directly from the "other world", that is, as long as they are disguised as restoration, rediscovery or reestablishment of something original.

Medical knowledge is ideally closed and complete, static and systematic, but reality shows it to be the opposite. The units of knowledge found in the practice of traditional medicine do not add up to a systematic whole.

TRADITIONAL PHARMACY

There are altogether probably several thousand plants (and ingredients from the mineral and animal kingdoms) which are found in traditional Thai medical prescriptions, and these should – alone or in combinations – have a curing property of one kind or another. A traditional doctor is not necessarily able to recognize these plants in nature; there are a number of herb stores in the larger towns where he can buy the ingredients he needs. Still it is convenient if the traditional doctor has the plants he uses most frequently ready at hand. Many herbalists therefore grow medicinal plants in their gardens and possess some botanical knowledge.

A number of monks also practice as traditional pharmacists or doctors. In Buddhist temples one can sometimes find extensive gardening of medicinal plants. There are also a number of specialists who produce traditional medicines for sale. These traditional pharmacists do not necessarily have any experience in treating diseases and sell their products at markets and fairs or distribute them locally through stores. Thus people have easy access to ready-made traditional medicines. They are cheaper than the available Western drugs, and although it is widely viewed that they may take a longer time to produce results than cosmopolitan medicines because they are weaker, for example, it is also claimed that they have fewer side effects than modern medicines.

The Ministry of Health stipulates that traditional drugs should be approved before public sale, but this law is not strictly enforced.

ETIOLOGY

Many of the disease causes recognized by traditional Thai doctors are also causative factors in modern medicine. Some diseases are, for example, regarded as contagious and some as inherited; some are caused by changes in the climate and others by contact with external agents such as metal chains or washing powder, resulting in skin eruptions. Some are caused by insect or animal bites and others by spoiled food or bad smells. There are also natural causes for fracture, dislocation, contusion, burns, or wounds.

The close relationship between food and disease can be seen in the monastic rules (*Vinaya pitaka*) where medicines are classified as "nonsubstantial food". Many of the most common medicinal plants are also ingredients in food. Food prohibitions are integral to the treatment of many diseases. Such prohibitions might be against meat, frogs and fish with skin, seafood, pickled food, certain

Figure 2 Man selling traditional herbal medicine in the market (Northern Thailand, 1978. Photo: Viggo Brun).

vegetables, eggs, liquor, or foods with certain tastes, all depending on the disease in question. There are also specific dietary rules for the post-natal period. Adding and withdrawing certain foods from the diet can also function as preventive measures.

A causative factor specific to traditional medicine is "imbalance in the elements (*thaat*)", i.e. imbalance between wind, water, earth and fire. These are the elements of which our bodies (and indeed all matter) are composed, according to Thai medical philosophy. In addition, traditional Thai medicine operates with other types of causes, which modern science summarily dismisses as superstition. These can be classified as spirits, black magic and *karma* (past actions). If the diagnosis is that a patient in some way has offended and angered a spirit, who in retaliation has caused the offender to fall sick, then the curing process may consist of ritual offerings only.

There are cases characterised by sudden violent and abusive behaviour towards one's nearest relatives (typically daughter-in-law towards mother-in-law living in the same house). These cases may be attributed to possession by an inferior, evil spirit. A specialist is then called upon who will talk to the spirit and ask it who it is, where it lives and what it wants. If the spirit does not want to leave the body after offerings have been given, it is mercilessly beaten out. In such cases the patient is not held responsible for what she did or said during possession.

Another type of mental disturbance which results in the patient's becoming introverted and neglecting her social responsibilities may, after all other treatments have proved fruitless, be attributed to a superior, good spirit who wilfully has inflicted this kind of suffering on the patient to force her to become its medium. When the person accepts mediumship, the spirit immediately withdraws the ailment, and the person recovers.

Piercing or lancing pains in the joints, abdomen, or chest may, if unresponsive to treatment, be attributed to black magic. This means that an object – like a nail or a lump of skin – has been inserted into the patient's body by a doctor using special incantations and acting on the instructions of an enemy of the patient. The only solution for the victim is to find a doctor who has even stronger incantations to counteract the original ones who can thus remove the object or even send it back to attack its owner.

The Buddhist concept of karma may also be used as an ultimate causative factor, but normally only after other explanations have failed. Using karma as the cause implies that the disease is a deserved and inescapable result of a previous action by the patient. People do not favour this explanation, because it means that a cure is impossible and that a person simply must endure his or her fate. However, people confronted with the inevitable deterioration and malfunctioning of their bodies, such as the elderly and those with incurable diseases, use karma to indicate that they have accepted the inevitable. This resignation can also be a relief for relatives, who no longer need to feel obliged to search for new remedies and pay even more bills.

The recitation of holy words plays an important role in many contexts in Thailand, and there is a strong belief that recitation of holy stanzas, if performed by the right persons, produces power and that this power is transferable. Incantations are an integral part of traditional medicine and are in many cases recited over or blown on to the medicines and the patients by the traditional doctor. Treatments against rabies, for example, consist of incantations only.

Magical practices are a central part of Thai culture and thus also of traditional Thai medicine. Still this must not overshadow the fact that traditional Thai medicine is basically a rational tradition based on experience. The traditional doctor will always first seek a natural and rational explanation to a disease. Only if this does not work will he resort to other types of explanations and treatments. And if one accepts the premises of magic, such as the existence of spirits and other levels of existence, it too possesses its own rational logic.

HETEROGENEITY

A central feature of the Thai medical tradition is its heterogeneity: it generates different answers to the same question. This is true at all levels of the tradition. The names for one and the same plant may vary from locality to locality, and one and the same name may be used for different plants in different places (which is not peculiar to Thailand, of course). The claimed medical properties of plants in isolation – the type of answer one gets when one asks, "What diseases does this plant cure?" – vary from informant to informant.

The content of the local disease names varies according to the informant. Furthermore, disease names known by one expert may not be known by another even if they live in the same area. And actual diagnosis varies a lot.

When we are confronted with the heterogeneity of medical practice, it is tempting to assume that the prescriptions, which link specific sets of plants to specific diseases or symptoms, would contain the order that is lacking when we ask for information in isolation. A closer scrutiny of the prescriptions may help to verify this assumption. One characteristic of the traditional medical prescriptions is that they normally contain many ingredients and that they are used against several diseases or symptoms. Thus there is not a one-to-one relationship between one plant and one disease in the prescriptions. On the contrary, one finds a much more diffuse many-to-many relationship. How similar are prescriptions against the same disease? To answer this, let us analyse prescriptions against *niw* (gallstones), because

1. the symptomatology of this disease is clear and not easily confused with other indigenous disease concepts;
2. there are a number of prescriptions which claim to cure only *niw*;
3. the traditional *niw* remedies have a reputation for being effective, and
4. gallstones are a common condition in Northern Thailand.

I collected 21 prescriptions from oral and written sources in Northern Thailand specifically against *niw*, which contained a total of 67 different ingredients. A frequency study of the ingredients reveals the following:

- 51 of the 67 ingredients occurred only once in the 21 prescriptions;
- 9 ingredients occurred twice;
- 6 ingredients occurred three times;
- 1 ingredient occurred four times.

Thus, 51 ingredients do not occur in more than one prescription, and the most frequent plant (pineapple) occurs in only 4 out of 21 prescriptions!

Below I have analysed the prescriptions in more detail by arranging them so that those with common components have been placed side by side. (The 51 ingredients occurring only once have not been named, but are indicated by a dash, while the names of the recurring ingredients have been written out in full. Each prescription has been given a number.)

As can be seen the prescriptions against *niw* fall into four groups, among which there are no common members (except for two instances).

If we look more closely at these four groups we observe that:

- Group 4 consists of six short prescriptions, of which four are single-ingredient prescriptions. One would have expected that the very short prescriptions would contain the "essence" of the tradition, that is, the specifically *niw*-

Table 1 Group 1–5 prescriptions

1	20	2	6	10
———	———	hin som	hin som	hin som
———	manad	manad	manad	manad
———	bafyang	bafyang	———	———
badoej	badoej	badoej	———	
———	huag	———	———	
———	———			
hoom	———			

Table 2 Group 2–6 prescriptions

11	21	9	22	8	17
hoomlaeng					
huag					
faang	faang	———	———		
	khacaaw	khacaaw	———	———	
	———	cumpuu	cumpuu	———	———
	———		bapin	bapin	———
	baduug		———	baduug	baduug
	———		———	khwaengkhaaw	khwaengkhaaw
				som hed	som hed

Table 3 Group 3–3 prescriptions

7	15	16
chakhaj	cakhaj	cakhaj
khawmin	khawmin	khawmin
puuloej	puuloej	puuloej
maliw	maliw	———
———	———	———
———	———	———
———	———	hoom

Table 4 Group 4–6 prescriptions

3	5	12	13	18	19
———	———	———	———	———	———
				———	———
					———

active plants. It is therefore surprising to find that these prescriptions have no ingredients in common with each other, nor do they have ingredients in common with any of the other *niw* prescriptions. There is thus no evidence at all to indicate that they are promising as *niw*-curing plants.

- Group 3 consists of three prescriptions with several recurring ingredients – *chakhaj* (lemon grass), *khawmin* (curcuma [turmeric]), *puuloej* (zingiber sp. [ginger-like root]), and *hoom* (onion) – all of which also occur in many other prescriptions against other diseases, and are widely used as ingredients in food. Thus none of them seems promising as having specifically *niw*-curing properties.

- Group 2 is centred around prescriptions 8 and 17 with three recurring ingredients, to which four other prescriptions are loosely linked.
- Group 1 contains five closely-knit prescriptions with four recurring ingredients, with prescription 2 as the statistically most promising.

The above result confirms the picture of heterogeneity, but if we interpret the result historically – and presume that trial-and-error is a guiding principle for the development of the local medical tradition – then I would be inclined to say that the tradition seems to be in the process of crystallizing its knowledge about *niw*-curing plants in two directions, centred around the plants in group 1 and group 2 respectively. But because of the lack of communication (i.e. competition) between the herbalists and the lack of records of the experiences of the individual herbalist, the crystallization process is slow and erratic.

There is another source of information one could add to that found in the prescriptions, namely information about the curing properties of plants found in nature. On botanical field trips we came across 15 plants which the botanical informant claimed had medical properties specifically against *niw*. Nine of these 15 plants did not occur in the above *niw* prescriptions! Six did, but two belonged to the non-recurring group, while only four were found among the recurring ingredients. These four were: *manad, bafyang, badoej* (Group 1) and *huag* (Group 1 and 2).

The heterogeneity of the prescriptions and the information about curing properties of isolated plants is remarkable. One can try to overcome this heterogeneity by using statistics, but even quite detailed studies yield only rather vague statistical indications of where to search for active ingredients, at least in the above case.

The lack of standardization of disease concepts, the heterogeneity of practice, and the lack of an explicit epistemological framework make the study of popular traditional medicine time-consuming and confusing to students of this tradition. The more people he consults, the more conflicting answers he will get. This does not deny the positive function that the local herbalists perform for their patients, nor does it deny that some traditional medicines do have curative properties. But if one were looking for traces of theoretical frameworks, then instead of studying folk medicine, one would do better to study the court medical tradition.

THE COURT MEDICAL TRADITION

During the reign of King Chulalongkorn (1867–1910), a committee of royal physicians was appointed to collect and edit old medical manuals. The result of their work became known as *tamra luang*, the Royal (Medical) Treatise, and was kept in the library of the palace, available to court physicians only. When Siriraj Hospital was established in 1887, the curriculum at the Siriraj Medical School included both cosmopolitan and traditional Thai medicine. To standardize the teaching of traditional medicine, a printed version of the *tamra luang* was introduced as the textbook. Thus the court medical tradition became the official medical tradition until the teaching of traditional medicine was discontinued around 1906. After a lapse of some 50 years, the Association of the School of Traditional Medicine was established at Wat Pho (Bangkok) in 1957, and both the teaching and the officially supervised examinations there were based on reprints of the *tamra luang*. The Wat Pho School later established schools at a number of temples in major towns. Since 1957, in order to practice traditional medicine legally one has had to pass examinations from these schools, where courses in traditional pharmacy, midwifery, and general medical practice are offered. As an innovation women have been allowed to participate on equal footing with men.

In the court medical tradition we find a theoretical framework which is based on the four elements (earth, water, wind, fire), the three humours (bile, wind, mucus), and the ten tastes (hot, cool, mild, astringent, sweet, poisonous, bitter, oily, salty and sour). The fact is, though, that the theoretical framework lacks internal coherence and consistency and that its relationship to actual medical practice is weak; it is not readily applicable to diagnosis and treatment of concrete cases of diseases. One could say that the theoretical framework we find in the court tradition is divorced from practice and functions only as a frame of reference, a model. This framework legitimates practice but does not dictate it. Thus the royal tradition has an explicit theoretical framework which the village tradition lacks, but the actual medical practice of the two are still very similar.

During the Aythaya period, the capital was a lively international trading port with many foreign communities from various Asian as well as European countries. A number of these foreigners were employed in the King's service. It is quite likely that these foreigners left an imprint on the court medical tradition, although it is not clear how and to what extent this actually happened. It is quite obvious, on the other hand, that there is an element of Sanskrit influence in the court tradition. A number of court medical manuscripts on which the *tamra luang* is based have Sanskrit names – although no one has yet been able to identify these names with titles of Sanskrit medical manuals in India – and many of the concepts and the disease names are borrowed from Sanskrit. Still, no comprehensive comparative studies on this matter have, to my knowledge, been undertaken, so it is difficult to ascertain the extent and exact nature of Ayurvedic influence on the court medical tradition.

Many Thais regard the court medical tradition as more advanced than the village tradition and thus accord it the status of *the* Thai medical tradition, considering the village tradition but a crude simplification with many local variations (see e.g. Somchintana, 1989: 280). This attitude is unfortunate because it glosses over the fact that there have not been any court physicians proper practicing traditional medicine for the last 100 years, so that the oral tradition that goes with the court medical tradition has been broken. Today there are therefore hardly any people who can explain the theories, concepts, organization, and the quite numerous unintelligible passages in the court medical manuals with authority. As for Thai-ness, it is obvious that the court medical tradition contains many more influences and borrowings from non-Thai sources than does the village medical tradition.

BUDDHISM AND TRADITIONAL MEDICINE

Buddhism is a crucial factor in Thai culture, but still there exists no study of the influence of Buddhism and Buddhist medical lore on the Thai medical tradition. To approach this question I shall base my ideas on Kenneth Zysk's seminal study of Buddhism and medicine, and point out some similarities with the Thai medical tradition.

In his study of the development of Buddhist medical knowledge within a broader Indian medical tradition, Zysk points out that Buddhist monks at the time of the Buddha already acted as healers for their fellow *bhikkhus* (monks), and that "from around the mid-third century BC monk-healers and the monasteries extended medical care to the population at large" (Zysk, 1991: 41). Portions of the medical knowledge and experience of the Buddhist *bhikkhus* were recorded in a chapter on medicines (*Bhesajjakhandhaka*) in the *Mahavagga* of the *Vinaya Pitaka*, the monastic rules, thereby giving rise to a Buddhist monastic medical tradition. The medical knowledge in the *Vinaya Pitaka* represents the earliest codification of medical knowledge in India, and "the Buddhist *sangha*, or monastic community, soon became the principal vehicle for the preservation, advancement, and transmission of Indian medical lore" (Zysk, 1991: 38).

Medicines were part of the five requisites a monk could possess, and originally the monks were allowed only a very limited number of basic medicines. As time went by the list of medicines allowed was increased and became in principle open-ended.

> The materia medica of the Buddhist monastery, representing the medicines requisite in sickness (*gilanapaccayabhesajja*), included initially the five basic medicines (ghee, or clarified butter; fresh butter; oil; honey; and molasses, or treacle), to which was added a more extensive pharmacopoeia of fats, roots, extracts, leaves, fruits, gums or resins, and salts. Medicines of the Buddhist monastic materia medica were considered to be foods but classified as nonsubstantial nourishment, allowing them to be consumed at any time. ... (Zysk, 1991: 73).

The medical knowledge in the *Vinaya* is organized in the form of case histories, where Buddha tells about the disease of a certain person and the remedies used to cure that person.

> The section on medicines in the *Mahavagga* (of the Vinaya Pitaka) ... is characterized by reference to actual cases, and functioned as a handbook and guide for the treatment of common afflictions ... The medical importance of these Buddhist records, however, is their recounting of patients' medical problems and the corresponding treatments. The academic medical treatises of classical *ayurveda* offered no such case-by-case medical instruction.
>
> ... Between those two traditions [i.e. the Buddhist Pali records and the classical Indian medical treatises] many similarities, but also numerous differences, exist – significantly, the Buddhist emphasis on practical application *devoid of the theoretical considerations of disease etiology* (italics added) that dominate the [non-Buddhist] medical books. This difference supports the view that codified Buddhist monastic medicine, with its emphasis on materia medica and case-based therapies, represents an early attempt to provide a manual of medical practice and in some sense legitimated the formalized collections of prescriptions ... (Zysk, 1991: 71–72).

A closer study of the chapter on medicines (*Bhesajjakhandhaka*) of the *Mahavagga* reveals that the description of the disease categories is empirical and descriptive. The categories which imply causation are "the nonhuman diseases" which imply possession by some spirit or demon as cause, various wind diseases, which imply "wind" as a cause (although "wind" here is not explicitly related to the element "wind") and the category "body filled with 'peccant' humors", which – Zysk notes – is "... the first time, the notion of 'peccant humor' (*dosha*) is implied in Buddhist monastic medicine, hinting at a more specialized meaning corresponding perhaps to that of dosha in the medical treatises" (Zysk, 1991: 108).

As for knowledge of anatomy we find that,

> In the *Mahasatipatthanasuttanta* of the Dighanikaya (in Sutta Pitaka), the four intents of contemplation (*cattaro satipatthana*) are detailed. The first of these was the human body (*kaya*) in all its parts, aspects, and impurities. The monk was to endeavor, through persistent contemplation, to realize the fundamental impermanence of his physical and mental constitution by meditating on the body:
>
> And in addition, O monks, a monk contemplates this very body, up from the soles of the feet (and) down from the crown of the head, bound by skin (and) full of manifold impurities. "There is in this body (the following): hair of the head, hair of the body, nails, teeth, skin, flesh, sinews, bones, bone marrow, kidney, heart, liver, pleura, spleen, lungs, bowels, intestines, stomach, excrement, bile, phlegm, pus, blood, sweat, fat, tears, grease, saliva, mucus, serous fluid, and urine ..." There is in this body "earth element, water element, fire element, and wind element" (Zysk, 1991: 34).

This list includes anatomical parts, but does not give any further descriptive details, neither how they function and relate to each other nor how they relate to diseases. The four elements are mentioned, but only to underline that the body, like all other matter, is ultimately composed of these elements.

Buddhist medical knowledge encompasses concepts such as the four elements, the three humours, and the 32 constituents of the body, and shows a very practical attitude to the curing process, but it does not attempt to construct specifically medical theoretical frameworks. Thus when Zysk concludes that, "The medical doctrines codified in the monastic rules probably provided the literary model for the subsequent enchiridions of medical practice" (Zysk, 1991: 118), it is not difficult to agree that this has indeed also been the case for traditional Thai medicine.

TRADITIONAL AND WESTERN MEDICINE

If we look at patients' choices of curing methods in the countryside nowadays, we find, according to some statistics, that 95% prefer Western medicine as a first choice, while only a small minority chooses traditional medicine. Furthermore, traditional medicine is in many cases resorted to only after treatment with Western medicine has failed. This means that patients who no longer can afford modern treatment or people with terminal, incurable, or chronic diseases will resort to traditional medicine. But there are also certain illnesses where traditional medicine has a reputation of being more efficient than modern medicine. These include: hemorrhoids, gallstones, and diseases with "wind" as a prominent symptom. In addition, the simple fact that traditional doctors are local people who know their patients' social situation, speak the local dialect, talk in a way the patients understand, and have more time to listen may give them an advantage over the more sterile and impersonal relationships the patients have with the Western-trained physicians and hospitals.

Still, there is from the patient's point of view not an either-or situation. Many patients will choose both traditions. People who are sick often go to several doctors and several pharmacies at the same time, both Western and traditional, and thus follow parallel treatments. This has the consequence that they stuff themselves with too many drugs, thus rendering it impossible to decide which medicine actually cured the disease.

While cosmopolitan trained doctors are quite skeptical of the efficacy of traditional medicine, the reverse is not true. Traditional doctors recognize quite readily the qualities of cosmopolitan medicine, such as the surgical skills and the value of vaccination and antibiotics. They also in some cases refer patients to modern doctors and they may (re)sell certain Western drugs they find useful. Ideologically they also feel akin to Western medicine in that, as one traditional doctor put it, "Diseases are a terribly complicated business, so traditional doctors must use the same method as the Western trained ones, namely the trial-and-error approach: If one prescription does not work, I will try another, and yet another, until I have found one that works, just like the pharmacists and the modern doctors".

But cosmopolitan medicine does indeed also take an interest in the traditional pharmacopoeia, trying to identify and extract medically active chemical components from traditional medical herbs in the hopes of being able to design new drugs. Transnational – and Thai – pharmaceutical firms are involved in this search, and the Government Pharmaceutical Organisation has in fact for years been producing herbal remedies for sale based on their own research.

Western medicine is also adopting traditional Thai disease terms for modern diseases, thus eroding the traditional concepts and causing some confusion. Modern medicine has, for example, adopted the traditional term *mareng* for cancer, while the term originally was a vague disease term covering various skin diseases with itching eruptions. So it seems rather far-fetched to claim – as some doctors do – that traditional prescriptions against certain skin diseases (*mareng*) also cure cancer (*mareng*) in general, simply because the names for the two diseases coincide. Also, some traditional doctors are claiming to have found or invented herbal medicines which can cure other new diseases, such as diabetes and AIDS, and herbal mixtures have also been used to cure heroin addicts. One could say that this shows that traditional medicine is capable of adapting to changes.

CURRENT DEVELOPMENTS

Over the last decade or so, the Thai government has radically changed its policy towards traditional medicine. The general context for this change has been the new WHO (World Health Organization) policy to encourage governments to incorporate traditional medicine into their official medical systems. Furthermore, public expenses for medical care have continued to increase, and so has the bill for imported medicines and medical supplies. There has also been a growing awareness of and pride in one's national heritage, and "nature", "ecology" and "holistic approaches" have become very popular concepts and part of the current discourse on health. Some even feel that there is an incongruity between modern technology and the Thai way of life and indigenous culture. All these factors have contributed to strengthening the hand of traditional medicine and also in changing the attitude of the bureaucracy. One of the first official steps in this direction was the Seventh National Economic and Social Development Plan for 1992–96, which stated that, "The promotion of people's health entails the efforts to develop traditional wisdom in health care, including Thai traditional medicine, herbal medicine and traditional massage, so as to integrate it into the modern health service system". The Thai government has followed up on this plan by adopting the goals of the plan in its official health policy and by establishing in 1993 the National Institute of Thai Traditional Medicine (NITTM) under the Ministry of Public Health, headed by the energetic and knowledgeable Dr. Phennapa Subcaroen. The basic aim of the NITTM is to integrate Thai Traditional Medicine (TTM) effectively into the national health service system. To achieve this aim, NITTM must, according to a brochure it has published, "systematize and standardize the body of TTM knowledge" and thus "gather knowledge, revise, verify, classify and explain TTM knowledge" as well as to "compare and explain the philosophies and basic theories of TTM and to produce textbooks on TTM".

Thus in order to revitalize traditional medicine and make it acceptable to the official medical system, the NITTM has deemed it necessary to create a new traditional medicine with a coherent medical theory, a consistent diagnosis and a consistent use of medical herbs, as well as a safe and efficient medical practice, or in other words, to replace heterogeneity with homogeneity. This recreation of tradition is a long process which has just started and which will go on for many years in cooperation with traditional doctors, NGOs, and other interested parties, but in the initial stages Dr. Phennapha's own understanding, based on many years of meticulous studies of existing medical theories and practices, seems to play a crucial role. One result is the change in nomenclature from "ancient traditional medicine" to "Thai traditional medicine".

Another important step taken by the NITTM has been to protect Thai traditional medicine from being appropriated and exploited by foreigners. Thus the NITTM drafted the Traditional Thai Medicinal Wisdom Protection Bill which, after several years of discussion, was approved and came into effect in May 2000. This bill makes Thailand one of the first third world countries to regulate foreigners' access to all aspects of traditional medical knowledge. The law drew quite irate comments from the United States embassy in Bangkok but was lauded by the local media. In a commentary in the *Bangkok Post* of June 18, 1997 for example, the editor Sanitsuda Ekachai praised this initiative to "defend our indigenous plants and age-old knowledge of their use from being hijacked by richer countries". She deplored the fact that rich countries with advanced biotechnology and sophisticated laboratories can use the genetic resources of developing countries free of charge, extract the effective chemicals, patent them and sell them back as expensive western medicines. She also referred to the unfair global trade laws, and concluded, "The Traditional Medicine Bill is a small effort by Thailand to defend itself against such injustice. No one can rob us of this right".

The increased awareness about the value and potentials of traditional medicine in Thailand and the attempts to revitalize this tradition will hopefully contribute to a better and more self-reliant public health system, but it has also made traditional medicine a highly political issue.

CHINESE MEDICINE

Chinese medicine is quite visible in Thailand, especially in the towns, where drugstores sell Chinese herbs and offer acupuncture treatment. Still, Chinese medicine remains quite a distinct tradition used mainly by the Sino-Thai population. Although the use of certain Chinese herbs and prescriptions may have entered the Thai medical tradition over the centuries, no systematic study has been made to verify the extent of such influence. The popularity among the Chinese of using the horns, bones, or bile of certain animals as aphrodisiacs has contributed to the endangering of mammals like the rhinoceros, tiger, wild boar, bears and certain snakes, among others.

BIBLIOGRAPHY

This bibliography includes works on Thai traditional medicine defined narrowly; thus works on botany, pharmacy, medical practice in general as well as contemporary health problems and policy issues have not been included. I have also left out numerous studies on rituals and religion, many of which have a bearing on traditional medicine. Furthermore, the bibliography confines itself to Thailand and does not include relevant works from neighbouring countries, and it contains only works in Western languages, thus omitting a number of studies in the Thai language. For further bibliographic references to works on traditional medicine in Thailand, including works in Thai, see Shigeharu Tanabe (ed.), *Religious Traditions among Tai Ethnic Groups: A Selected Bibliography*. Ayutthaya (Thailand): Ayutthaya Historical Study Centre, 1991, which contains special sections on Traditional Medicine.

The bibliography below reveals that most studies in Western languages on traditional Thai medicine are articles on specific aspects of medicine; there are only a few more comprehensive studies. Golomb (1985) gives a broad anthropological view of medical practices in Southern Thailand; Bamber (1987) studies disease classification from a linguistic point of view; Brun and Schumacher (1987) concentrate on medical practice in Northern Thailand using the traditional doctors as informants; Mulholland (1987) studies traditional paediatrics based on manuscripts from the court tradition, and Somchintana (1989) studies the concepts and system of traditional court medicine in Bangkok from a cognitive anthropological point of view.

There are but few works on the historical development of traditional medicine in Thailand, for the simple reason that there is very little material in Thai which has survived from before around 1800. Scattered evidence can be found, but thorough and critical studies are extremely time-consuming and would not necessarily yield substantial results. Furthermore, detailed and critical comparative studies, both historical and contemporary, between the Thai medical tradition(s) and similar traditions in India, Sri Lanka, Burma, Cambodia, Laos and China have yet to be undertaken. Thus how and to what degree these traditions have influenced each other is still an open question. What role Buddhism and the Buddhist *sangha* have played in the development of Thai traditional medicine awaits further studies. There is also a huge number of medical manuscripts which are in need of critical study which would also deepen our understanding of regional variations – just to mention some obvious areas of study which are needed to give us a better understanding of traditional Thai medicine.

Thais today conduct most of the research on traditional medicine, with basically two aims: to preserve tradition and to make it applicable to the health needs of contemporary Thailand.

Bamber, Scott. *Trope and Taxonomy: An Examination of the Classification and Treatment of Illness in Traditional Thai Medicine.* Ph.D. thesis. Canberra: Australian National University, 1987.

Beyer, C. 'About Siamese medicine.' *Journal of the Siam Society* 4: 1–9, 1907.

Bradley, D.B. 'Siamese theory and practice of medicine.' *Bangkok Calendar*, 1865. Reprinted in *Social Science Review* (Bangkok) 5(2): 103–119, 1967.

Brun, Viggo. 'Traditional manuals and the transmission of knowledge in Thailand.' In *The Master Said: To Study and ... To Søren Egerod on the Occasion of His Sixty-Seventh Birthday*, Birthe Arendrup *et al.*, eds. Copenhagen: University of Copenhagen. East Asian Institute Occasional Papers 6, 1990, pp. 43–65.

Brun, Viggo and Trond Schumacher. *Traditional Herbal Medicine in Northern Thailand.* Berkeley: University of California Press, 1987.

Chantana Banpasirichote. *The Indigenization of Development Process in Thailand: A Case Study of the Traditional Revivalist Movement (Thai Massage).* Ph.D. thesis. Ontario: University of Waterloo, 1989.

Cunningham, C.E. 'Thai "injection doctors": antibiotic mediators.' *Social Science and Medicine* 4: 1–24, 1970.

Damrong, Khun Ying. *Siamese Midwifery in the Old Days.* Bangkok, 1931.

Dhani Nivat. 'The inscriptions of Wat Phra Jetubon.' *Journal of the Siam Society* 26: 143–170, 1933.

Feltus, G.H. *Samuel Reynolds. House of Siam, Pioneer Medical Missionary 1847–1876.* New York: Fleming H. Revell, 1924.

Golomb, Louis. *An Anthropology of Curing in Multiethnic Thailand.* Urbana and Chicago: University of Illinois Press, 1985.

Griswold, A.B. 'The Rishis of Wat Pho.' In *Felicitation Volumes of Southeast-Asian Studies Presented to His Highness Prince Dhaninivat Kromamun Bidyalabh Bridhyakorn.* Bangkok: The Siam Society, vol. 2, 1965, pp. 319–328.

Hanks, J.R. *Maternity and its Rituals in Bang Chan.* Data Paper 51, Southeast Asia Program. Ithaca, New York: Cornell University, 1963.

Hofbauer, R. 'A medical retrospect of Thailand.' *Journal of the Siam Society* 34: 183–200, 1943.

Ionescu-Tongyonk, J. 'Explication of wind illness in Northern Thailand.' *Transcultural Psychiatric Research Review* 18: 42–45, 1981.

Konoshima, M., M. Tabata and D. Pecheraply. 'On a Thai native doctor and his folk medicine.' *Tonan Ajia Kenkyu* (Southeast Asian Studies) 11: 238–255, 1973 (In Japanese).

Kunstadter, P. 'Do cultural differences make any difference? Choice points in medical systems available in Northwestern Thailand.' In *Culture and Healing in Asian Societies*, A. Kleinman *et al.*, eds. Cambridge, Massachusetts: Schenkman, 1978, pp. 185–217.

Lafont, P.-B. 'Practiques médicales des Thai noirs du Laos de l'ouest.' *Anthropos* 54: 819–840, 1959.

Le Grand, Amanda. *Herbal Drugs in Primary Health Care. Thailand: The Impact of Promotional Activities on Drug Consumption, Drug Provision and Self-Reliance.* Bangkok: Mahidol University, 1989.

Lord, D.C. *Mo Bradley and Thailand.* Grand Rapids, Michigan: Wm.B. Ferdonans, 1969.

Mougne, Christine. 'An ethnography of reproduction: Changing patterns of fertility in a northern Thai village.' In *Nature and Man in South East Asia*, P.A. Stott, ed. London: School of Oriental and African Studies, 1978, pp. 68–106.

Muecke, M.A. 'An explication of "wind illness" in Northern Thailand.' *Culture, Medicine and Psychiatry* 3: 267–300, 1979.

Mulholland, Jean. *Medicine, Magic and Evil Spirits. Study of a Text on Thai Traditional Paediatrics.* Canberra: Australian National University, 1987.

Mulholland, J. 'Thai traditional medicine: Ancient thought and practice in a Thai context.' *Journal of the Siam Society* 67: 80–115, 1979.

Mulholland, J. 'Traditional medicine in Thailand.' *Hemisphere* 23: 224–229, 1979.

Ooms, Arno. *If You Really Want to Know. Individual Manipulation of Relationships and Self-Diagnosis of Skin-Diseases in Pai, Northern Thailand.* Amsterdam: University of Amsterdam, Anthropological Sociology Centre, n.d. [*ca.* 1997].

Riley, J.N. and S. Sermsri. *The Variegated Thai Medical System as a Context for Birth Control Services.* Working Paper no.6. Bangkok: Institute of Population and Social Research, Mahidol University, 1974.

Samruay Subchareon. *Thai Traditional Medicine System and Practice.* M.A. thesis. Bangkok: Mahidol University, 1989.

Scarpa, A. 'La medicina tradizionale del Siam secondo un maoscritto su scorza d'albero.' *Castalia* 20: 3–8, 1964.

Smith, M. *A Physician at the Court of Siam.* London: Country Life, 1947.

Somchintana Thongthew Ratarasarn. *The Principles and Concepts of Thai Classical Medicine.* Bangkok: Thammasat University, Thai Khadi Resarch Institute, 1989. (Originally Ph.D. thesis, University of Wisconsin-Madison, 1986.)

Tambiah, S.J. *The Buddhist Saints of the Forest and the Cult of Amulets. A Study in Charisma, Hagiography, Sectarianism, and Millennial Buddhism.* Cambridge: Cambridge University Press, 1984.

Vilaiporn Bharabhutanonda. 'The medical role of the "Doctor Bhikku" in Bangkok and rural communities.' *Journal of National Research Council of Thailand* 12: 11–60, 1980.

Wales, H.G.Q. 'Siamese theory and ritual connected with pregnancy, birth and infancy.' *Journal of the Royal Anthropological Institute of Great Britain and Ireland* 63: 441–451, 1933.

Weisberg, Daniel H. 'The practice of "Dr. Paep": Continuity and change in indigenous healing in Northern Thailand.' *Social Science and Medicine* 18(2): 117–128, 1984.

Yoddumnern, Benja. *The Role of Thai Traditional Doctors.* Bangkok: Mahidol University, Institute for Population and Social Research, 1974.

Zysk, Kenneth G. *Asceticism and Healing in Ancient India. Medicine in the Buddhist Monastery.* Oxford: Oxford University Press, 1991; reprinted Delhi: Motilal Banarasidass, 1998.

DON BAKER

ORIENTAL MEDICINE IN KOREA

Oriental medicine goes under several different names in Korea, varying according to the time and place. In North Korea, it has usually been called *Dongui* (Eastern medicine), a traditional term used to distinguish the Korean practice of Oriental medicine from the Chinese versions of the same basic medical system. Recently, North Koreans have started referring to Oriental medicine as *Goryeo uihak* (Korean medicine) to emphasize their view that Korean medicine is a medical system in its own right and should not be seen as the eastern form of a broader pan-national medical tradition.

In the Republic of Korea (South Korea), *Hanui* is the preferred name for Oriental medicine. Until the 1970s, Hanui was written with Sino-Korean characters which can be literally rendered as "Chinese medicine". The word Hanui is still used in South Korea, but the original Sino-Korean character for the first syllable, *han*,[1] has been replaced with a homophone which means Korea, so that today Hanui, just like Dongui, literally means "Korean medicine". However, when South Koreans discuss their traditional medicine in English, they generally refer to it as "Korean Oriental Medicine" or simply "Oriental Medicine".

This variety of names for Oriental medicine in Korea reveals how ambivalent Koreans have become about acknowledging the Chinese roots of their traditional medicine and how proud they are of the modifications they have made in the medical practices they adopted from China. However, when the theory and techniques of Oriental medicine first arrived on the Korean peninsula over 2,000 years ago, the Chinese origins of Oriental medicine were no cause for embarrassment. Quite the contrary. Oriental medicine came to Korea bundled together with a package containing mathematics, astronomy, architecture, sculpture, literature, law, government institutions, music and ritual, all seen as intertwined and inseparable components of the only advanced civilization Koreans knew, the Chinese.

Exactly when the concepts and techniques of Oriental medicine first began to be utilized in Korea is, however, still disputed. Some nationalists claim that needle-shaped bits of stone found in excavations of Neolithic villages on the Korean peninsula prove that Koreans were using acupuncture before the

H. Selin (ed.), Medicine Across Cultures: History and Practice of Medicine in Non-Western Cultures, 133–153.

Chinese were. Others claim that the ancient myth of Dan'gun, a legendary Korean ruler over 4,000 years ago, proves that much of what others call Chinese medicine actually existed in Korea over four millennia ago, since that myth refers to mugwort and garlic, two common ingredients in Oriental medicine prescriptions (Kim Dujong, 1981: 14–28).

It is more likely that Koreans first acquired elements of Chinese medical theory and practices a little over two millennia ago, when Han dynasty forces established four outposts in and around the peninsula. Three of those outposts quickly disappeared, but one, Nangnang (Chinese: Lo-lang), in the vicinity of the modern city of Pyongyang, survived for four centuries (108 BCE–313 CE). Since Nangnang was a Han dynasty city like all other Han dynasty cities (except that it was over a month's walk away from other centers of Chinese population), we can assume that its residents were familiar with the medical concepts and practices seen in the *Huangdi neijing* (*Yellow Emperor's Classic of Internal Medicine*), the *Shennong Bencaojing* (*Classic of Herbal Medicine*), *Shanghanlun* (On Cold-induced Maladies), and other medical classics from this period. Although Nangnang did not exercise direct political control over much of the Korean peninsula, the residents of Nangnang traded various commodities with the other inhabitants of the peninsula. They probably also traded ideas and practices. Moreover, some of the earliest Chinese pharmaceutical texts mention Korea as the source of some of the herbs and animal parts used in prescriptions at that time (Kim Dujong, 1981: 42–43). It is not clear that the Koreans who provided those materials knew what they were to be used for. Nevertheless, it is likely that Koreans first became acquainted with Oriental medicine through interaction with the Chinese residents of Nangnang.

THE THREE KINGDOMS

Koreans may have gained more detailed knowledge of Oriental medicine after the fall of Nangnang. Starting in the latter half of the fourth century, Buddhist monks from China and farther west began moving onto the peninsula. The first Korean kingdoms had appeared by this time and were vying for control. Believing that adopting the latest technology and techniques from China would give them an advantage over their rivals, those three kingdoms welcomed the Buddhist missionaries, since they brought with them not only their religion but also various secular components of Chinese civilization. Among those components was Oriental medicine.

Unfortunately, we have few written records from this period in Korean history. The oldest extant history of Korea dates back only as far back as the twelfth century, although that work, the *Samguk sagi* (*History of the Three Kingdoms*), draws on earlier materials, now lost. Combining information from that text with brief references to Korea in Chinese and Japanese materials from the same period provides some hints of what medicine and healing practices were like on the peninsula from the fifth century through the early tenth century.

For example, an eighth-century Japanese source, the *Nihon Shoki*, reports that a specialist in herbal medicine was sent to Japan from the southeastern

Korean kingdom of Silla in 414. That date appears to be too early, since the sinification of Silla did not begin until a century later. The first Korean kingdom to adopt Oriental medicine was probably Goguryeo, which was located much closer to China than the other two kingdoms were. The *Nihon Shoki* claims that 164 Chinese medical books reached Japan by way of Goguryeo in 561. If that date is correct, the people of Goguryeo had probably been reading Chinese medical manuals for several decades, if not a century or more, before that date. It is possible that Oriental medicine entered Goguryeo culture as early as 313, if there were any physicians among the inhabitants of Nangnang when Goguryeo forces captured it that year (Kim Dujong, 1981: 50–51, 56–58).

For the Three Kingdoms period (300–668), however, we have more evidence for Oriental medicine in Baekje than in either Goguryeo or Silla. That is because the *Nihon Shoki* mentions practitioners of Oriental medicine from Baekje more often than it mentions physicians, acupuncturists, or pharmacists from Goguryeo or Silla. Baekje lay in the southwestern quadrant of the Korean peninsula, giving it direct access across the Yellow Sea to coastal China. Baekje also had the closest relations with Japan of all the three kingdoms, so it played a much larger role in relaying Oriental medicine from China to Japan. However, those Japanese references to Baekje practitioners of Oriental medicine are not very detailed. We are told, for example, that Baekje dispatched physicians and pharmacists to Japan in the mid-sixth century. No other information about them is provided. For more information on Baekje medicine, we can turn to the *Ishimpo* (*Essentials of Medicine*), a tenth-century Japanese work, which mentions a couple of Baekje prescriptions. Those prescriptions, in the materia medica they use, the ways they are prepared, and the ways they are supposed to be ingested, closely resemble Han dynasty prescriptions, indicating that Baekje was familiar with the medical traditions of China (Heo, 1992: 31–37).

THE SILLA KINGDOM

Silla won the struggle among those three kingdoms for supremacy on the peninsula in 668 and maintained its dominant position for more than two centuries. Silla's victory ensured that we have more information about Silla, particularly after 668. For example, the *History of the Three Kingdoms* states that in 692 Silla established a medical office staffed with two physicians. Those physicians, we are told, were expected to base their medical practice on six basic Chinese sources: the *Yellow Emperor's Classic of Internal Medicine*, the *Classic of Herbal Medicine*, the *Jiayijing* (*Classic of Proper Ordering*), the *Maijing* (*Classic of the Pulse*), the *Mingtangjing* (*The Classic of the Hall of Light*), and the *Nanjing* (*Classic of Difficult Problems*). This is classic Oriental Medicine, focusing on acupuncture and correlative physiology. Baekje and Goguryeo may have established similar royal Oriental Medicine clinics earlier, but any documents which could have confirmed the existence of such clinics and provided details about them perished with those two kingdoms (Kim Dujong, 1981: 66).

We also have more information about healing practices in Silla than we do for Baekje or Guguryeo. But our primary source for information is a history of the three kingdoms written by a Buddhist monk in the thirteenth century, the *Samguk yusa* (*Memorabilia of the Three Kingdoms*). Since the author was a monk, he emphasized the efficacy of Buddhist chants and spells over herbs and needles. For example, the *Memorabilia* informs us that a monk named Hyetong cured a Silla king and a Silla princess by uttering certain esoteric incantations he had learned in China. It also tells us that another monk, named Milbon, once healed Queen Seondeok (632–647) of Silla by reading aloud a passage from the *Sutra of the Buddha of Medicine* and then killing the two evil spirits who were responsible for her illness (Baker, 1994: 60).

The Buddhist focus of Silla medical practice is confirmed by the Silla prescriptions recorded in the Japanese-authored *Essentials of Medicine*. Not only does that book give the sources of those prescriptions as texts named "Prescriptions of a Buddhist master from Silla" and "Secret Prescriptions of a Buddhist master from Silla", but also one of those prescriptions is a prayer to the Buddha of Medicine. Two more are described as secret Buddhist formulas for heightening sexual performance. Only the fourth, and last, citation is an actual prescription in the traditional medical sense (Kim Dujong, 1981: 74–77).

From the limited evidence we have available, it appears that Koreans knew the basics of Oriental medicine by the sixth century at the latest but that Oriental medicine was practiced alongside, and was seen as supplementary to, religious healing. This multi-faceted approach to health and healing characterizes medicine in Korea up through the modern period, although over the centuries secular Oriental medicine has grown increasingly more respectable than religious healing.

THE GORYEO DYNASTY

This growing respectability can be seen in the preference given to Oriental medicine in government institutions, as Koreans imitated the Chinese shift away from Buddhism toward a greater reliance on Confucianism at the official level. That shift became obvious after Silla fell and was replaced by the Goryeo dynasty (918–1392), which ruled over the Korean peninsula for four and a half centuries. The state religion of Goryeo was Buddhism, but as early as the second half of the tenth century, Goryeo began using Confucian-style civil service examinations to identify qualified candidates for government medical posts. By the middle of the dynasty, if not earlier, secular Oriental medicine rather than religious medicine was the primary subject tested on those examinations.

The first detailed information we have about the scope of those examinations comes from the first half of the twelfth century. We are told that applicants for government medical posts were tested on their knowledge of the basic principles of internal medicine, as found in the same classical medical classics studied in Silla. Yet there was a second section of the medical exam which tested their knowledge of external medicine (as defined in Oriental medicine), primarily acupuncture, moxibustion, and the treatment of wounds. That section included questions on material introduced in more recent medical texts, including some books from China's Sui (581–618) dynasty. The section of the test on internal medicine was called a medical exam. The second part, although it too tested secular medical concepts and practices, was called an "incantation" exam. The label "incantation" might represent a Goryeo memory of an earlier era when incantations were as much a part of the physician's repertoire of healing techniques as acupuncture and herbal concoctions were. Or it might simply represent Goryeo imitation of China's Tang dynasty (618–907), which included incantations among the four medical specialties in its Imperial Bureau of Medicine (Kim Dujong, 1981: 133–135).

Although China's Han, Sui, and Tang dynasties were the primary sources of medical concepts and practices, as well as pharmaceutical knowledge, in the first part of the Goryeo dynasty, there were other sources as well. Arab merchants traded in Chinese and Goryeo ports in the first part of the eleventh century, and they may have made medicinal material available to Korea from Southeast Asia and even farther away, which was not part of the standard Chinese materia medica.

A tumultuous international environment in northeast Asia brought further changes to Goryeo medicine. Starting in the tenth century, the same century in which Goryeo replaced Silla, tribal peoples living north of China and west of Korea began coalescing into a series of powerful states which blocked easy Korean access to the materia medica of Song dynasty (960–1270) China. At first medicine in Korea benefited from this challenge to China's domination of East Asia. In order to reward Goryeo loyalty and ensure that Goryeo did not switch its allegiance to the Liao kingdom (907–1119) on China's northern border, Song emperors presented gifts to the Goryeo court, including gifts of recent medical publications as well as rare pharmaceutical material. For example, Goryeo received a copy of the Song dynasty encyclopedia of prescriptions, the *Taiping Shenghuifang* (*Prescriptions Courtesy of the Wise Taizhong Emperor*), in 1017. Guguryeo also received copies of many of the titles produced by the Song dynasty's Bureau for the Re-editing of Medical Books, which was established in 1057 to produce authoritative editions of canonical medical texts. In 1079, the Song emperor sent one hundred prized pharmaceutical items to the Goryeo king when he learned that the king was ill (Kim Dujong, 1981: 117–124).

As a result of this Song attempt to keep Korea in the Sino-centric world order, Korea was introduced to many of the latest advances in medical thought and practice in Song China. However, when the Jin dynasty (1115–1234) replaced the Liao in the twelfth century and pushed farther south into China than the Liao had done, Korea found itself cut off from regular contact with China. This forced Goryeo to search for local replacements for many of the ingredients from China which it had used in prescriptions. This gave rise to *hyangyak*, the use of locally available herbs, animal parts and minerals in prescriptions adapted from the prescriptions in Chinese medical manuals. The first indication that Goryeo was becoming more self-reliant came in 1226, when a medical handbook, *Eo-ui chwaryobang* (*A Royal Collection of Essential*

Medical Prescriptions), was compiled. That work, no longer extant, apparently was the first medical text in Korea to include *hyangyak* prescriptions (Kim Dujong, 1981: 139–141).

The military crisis which Goryeo confronted in the thirteenth century gave further impetus to the indigenization of Oriental medicine in Korea. Armed incursions by Mongol forces from the north not only made it more difficult for Korea to obtain medicines from China but also created a more urgent need for readily available forms of medical treatment. In 1245 an attempt to meet that need was made by the publication of *Hyangyak gugeupbang* (*First Aid Prescriptions Using Native Ingredients*), the oldest extant Korean medical text. Not only did the prescriptions in this medical manual use ingredients available locally, but also some of them were drawn from local folk medicine and therefore called for only one or two ingredients, in contrast to the complex prescriptions that Koreans identified with prescriptions originating in China. This was just one of several medical manuals produced in the last part of the Goryeo era which emphasized locally available ingredients as well as prescriptions of local origin.

Hyangyak was not the only place Koreans turned for medical assistance during this turbulent century. They also looked to Buddha for help against the Mongol invasions and against the medical problems which came in their wake. In the thirteenth century, the besieged Goryeo government carved over 81,000 woodblocks for printing Chinese-character versions of sacred Buddhist texts, hoping that this display of devotion to the Buddha would convince that deity to intervene on their side in their struggle against the Mongols. Included among the over 1,500 titles in this collection were several which appear to be as much concerned with medicine as with religion. For example, one book in this collection is called *Foshuo Foyi jing* (*The Buddha Speaks on Buddhist Medicine*). There are volumes with titles such as *Foshuo zhou shiqibing jing* (*The Buddha Speaks on Prayers for Seasonal Disorders*), *Foshuo zhou chi jing* (*The Buddha Speaks on Prayers for the Teeth*), and *Fushuo zhou mu jing* (*The Buddha Speaks on Prayers for the Eyes*). One set of woobdlocks is called *Chu yiqie jibing tuoluoni jing* (*Spells for Eliminating Diseases*); another is called *Zhi chanbing miyao jing* (*Essential Secrets for Dealing with Medical Problems Arising from Meditation*). The Buddhist healing techniques in these books have both supernatural and secular components, drawing on both the Buddhist tradition of chanting sacred formulas and the Ayurvedic tradition of India (Kim Dujong, 1981: 127–129).

Further evidence that religious medicine remained viable in Goryeo despite the dominance of secular Oriental medicine at the official level is a twelfth-century report by a Song visitor to Goryeo. Xu Jing, in his *Gaoli Tujing* (*An Illustrated Account of Goryeo*), reported that the common people on the Korean peninsula knew little of formal medicine and instead turned to spirits for assistance when faced with medical problems (Miki, 1962: 55–58). Also, for much of the four and a half centuries of the Goryeo kingdom, the official names for the palace medical office alternated between Da-ui-gam (The Central Medical Bureau) and Da-ui-sa (The Central Medical Temple), suggesting that

Goryeo continued to perceive a close connection between Buddhism and healing (Kim Dujong, 1981: 149–150).

THE JOSEON DYNASTY

The Joseon dynasty (1392–1910), which followed the Goryeo dynasty, dethroned Buddhism as the official religion of the government, enshrining in its place the secular philosophy of Neo-Confucianism, newly imported from China. The rulers of Joseon Korea then embarked on a restructuring of Korean government and society along Neo-Confucian lines. That restructuring embraced both medical institutions and medical practice, further enhancing the official status of secular Oriental medicine at the expense of religious medicine.

The palace medical office for Joseon, for example, was quickly renamed the Cheonuigam (Directorate of Medicine). Never again would an official government medical center be called a "temple". The Directorate of Medicine, which provided medical treatment for members of the royal family as well as for top officials and their family members, was not the only central government medical facility. There was also a Naeuiweon (Royal Clinic), which provided both internal and external medicine for the king and his relatives. The government also established the Hyeminseo (Capital Medical Clinic) to provide medicine primarily for those who lived within the walls of the capital city to lessen the danger of disease spreading from that urban population to the court.

In an effort to keep possible carriers of disease from entering the capital, the government also opened Hwarinseo (Suburban Medical Clinics) just outside the main gates of the capital. Known under the Goryeo dynasty as Daejaweon (Institute for the Display of Buddhist Benevolence), these public health facilities distributed medicine at no cost, in contrast to the Capital Medical Clinic within the city walls which charged for the medicine it distributed. The Suburban Medical Clinics also served as quarantine centers, providing food, clothing, and sometimes housing to those suffering from possibly contagious diseases. Moreover, to make this facility even more attractive to those among the general population in need of medical attention, the government assigned both Buddhist monks and shaman healers to them, making them the only government medical facilities which employed practitioners of religious medicine (Kim Dujong, 1981: 196–199, 408–416).

The central government also established medical facilities in various regional centers, primarily to deal with outbreaks of epidemics among the general population as well as to collect materia medica from local areas. Three types of officials were assigned to the various government medical facilities in the capital area and in the countryside. A Confucian scholar normally led the higher-level medical offices. Under him would be physicians who had passed the civil service medical examination. There would also be medical technicians on the staff of a medical office.

Joseon Korea was a highly stratified society, in which both heredity and education played a role in determining both social status and access to government positions. The Confucian scholars were the elite of society. As in post-Song China, Joseon Korea appointed most of its top government officials on the basis of their performance on literati civil service examinations. To mitigate the social mobility associated with the Confucian examination system, Joseon normally permitted only those with high-status parents to take those examinations. However, although a good family background was usually a necessary condition for high office, it was not sufficient to guarantee an official appointment. In order to be considered qualified to lead a government agency or head a local government, a man usually had to prove through his performance on the literati civil service examinations that he had a solid command of Confucian philosophy and literature. As a result, the heads of central government medical agencies were usually Confucian scholars with more expertise in Chinese poetry and ancient Chinese history than in the details of medical practice.

Professional physicians ranked one step below Confucian scholars on the social ladder. They were members of a class of technical specialists which included translators, mathematicians, and law clerks as well as specialists in Oriental medicine. Each specialty had its own civil service examination, administered by the Board of Rites. Although there was no legal requirement that those sitting for a specialist examination be descended from someone who had passed the examination previously, the strong preference in the Joseon dynasty for hereditary status led to a few families' monopolizing those technical fields. Over the last four centuries of the Joseon dynasty (we lack reliable records on medical officials for the first century), slightly more than 1,500 people passed the medical civil service examination. Since those examinations were not held every year, and there were only 166 examinations between 1498 and 1894 (when the last examination was held), that averages out to slightly more than nine successful passes for each exam. Over half of those candidates who went on to serve in the most prestigious medical office, the Royal Clinic, came from just twenty families, living in or around the capital city of Seoul. Ten families alone accounted for almost 30% of all officially certified physicians between 1498 and 1894 (Yi Gyugeun, 1997; Kim Jin, 1999).

Not all physicians were required to pass the medical examination. That examination was for those who sought an appointment to a mid-level civil service position. Until 1900, medical practitioners among the general population did not need any official certification of their medical expertise, nor were they subject to government oversight. However, even among those with informal credentials, family background was still important. A popular saying during the Joseon dynasty warned villagers, "Do not accept medicine from a doctor who is not at least a third-generation physician."

In addition to Confucian scholars and certified physicians, government medical offices also employed medical technicians. Some of those technicians were entry-level civil servants, the equivalent of interns gaining the knowledge they would need to sit for an examination later through on-the-job training. Others were from an hereditary class of low-level government servants who passed on basic medical skills from father to son. A third group of medical specialists who did not share the high status of Confucian scholars, or even the relatively high status of certified physicians, were women.

Women were employed as medical technicians because of the constraints of Confucian ethics. Physicians often found it necessary to touch the bodies of other human beings in order to properly diagnose and treat them. That meant that a physician might have to come into physical contact with women in need of medical attention, violating the Confucian directive calling for a rigid separation of the sexes among upper-class adults. To ensure both the health of his consorts and respect for Confucian tradition, King Taejong (*r.* 1400–1418) ordered that a few women be selected for training in acupuncture, moxibustion, and pulse measurement (an important diagnostic tool). Male pharmaceutical specialists could prepare medicines for an upper-class woman who was ill, but they were not allowed to place the acupuncture needle or moxa on her skin nor could they place their male fingers on her veins to feel her pulse. After 1406, such tasks therefore were assigned to a small core of women medical practitioners at court. Despite the importance of their work, however, Korea's women medical technicians were accorded even less respect than were male medical specialists. By the second half of the dynasty, they began to be assigned social as well as medical duties and were sometimes referred to as "courtesan doctors" (Heo, 1992: 214–217).

Official medicine in Joseon Korea, whether practiced by Confucian scholars, certified physicians, or medical technicians, was based primarily on the canonical texts of Chinese medicine. Koreans supplemented the medical theory and clinical advice in those classical works with more recent Chinese publications as well as with Korean texts which provided local counterparts for the Chinese ingredients in Chinese prescriptions.

Song, Jin, and Yuan medical texts introduced Joseon Korea to the new emphasis on systematic correspondence in pharmacology, which had by the fourteenth century established itself at the center of medical theory and practice in China. Early in the Joseon dynasty Koreans learned that a variety of approaches to health and healing had appeared in China. They found the writings of Li Gao (1180–1251) and Zhu Zhenheng (1281–1358) particularly persuasive. Li focused on the stomach and the spleen and the benefits of strengthening those two vital organs. Zhu argued that nurturing internal yin energy to overcome the harmful side effects caused by excessive yang energy could treat many medical problems. [Editor's note: See the article on Chinese medicine for more details on *yin* and *yang*.] Koreans generally preferred the priority Li and Zhu gave to reinforcing the body's strengths over approaches which favored purging the body of harmful influences. Joseon physicians tended to see illness as due more to internal weakness than to some invading pathological agent. They therefore prescribed preventive medicine and the consumption of tonics such as ginseng more often than they prescribed emetics or other purgatives (Kim Ho, 2000: 126–133).

Early Joseon Korea not only absorbed the latest medical knowledge from the Song, Jin, and Yuan dynasties, it also continued the project begun by Goryeo to adapt Oriental medicine to the Korean environment. Both trends are particularly evident during the reign of Joseon's fourth and greatest King, King Sejong (*r.* 1418–1450). Among the many volumes compiled and published under King Sejong's direction, two are particularly important for the history of Oriental medicine in Korea.

The first is the 85-volume *Hyangyak jipseongbang* (*Great Collection of Native Korean Prescriptions*). Completed in 1433, this encyclopedic pharmaceutical guidebook identified 959 different diseases, arranged under 57 different categories. It described 703 different mineral, vegetable, or animal medicinal products available on the peninsula, told how to identify them, when to collect them, and what diseases they were effective against. Many of its prescriptions were extracted from approximately 160 different Chinese medical manuals, especially *Prescriptions Courtesy of the Wise Taizhong Emperor*, which Korea had first received from Song China in the early part of the Goryeo dynasty. However, it also drew on a few Goryeo texts which contained simple folk remedies which required only one or two readily available ingredients. Moreover, whenever possible it substituted native ingredients for expensive or rare Chinese ingredients in the Chinese prescriptions (Kim Tai-jin, 1976: 70–73).

This comprehensive reference work listed almost 1,500 ways acupuncture and moxibustion could be used. Moreover, appended to this massive work is a section called *Prescriptions of the Daoist Immortals*. Among the entries in this section are several which resemble the recipes of external alchemists. One such prescription promises to turn grey hair on the heads of old men black again and grow teeth in the empty spaces in the mouths of the elderly. Another promises a one hundred-fold increase in energy to those who properly ingest the potion it recommends.

Although Oriental medicine in the Joseon dynasty was secular medicine, it included elements which might appear to modern eyes more supernatural than natural, such as Daoist techniques for halting or even reversing the aging process. That is true not only of the *Great Collection of Native Korean Prescriptions*, but also of medical encyclopedias drawn entirely from Chinese sources, such as the *Uibang yuchwi* (*Classified Collection of Medical Prescriptions*). The *Classified Collection* was, at 365 volumes, even larger than the *Great Collection*. Compiled under King Sejong's direction in 1445, it consisted of a list of various symptoms, each followed by a list of the diseases associated with those symptoms and then a list in roughly chronological order of the various treatments Chinese have prescribed for those diseases over the centuries, adding up to over 50,000 separate prescriptions. It also included separate sections on treating childhood illnesses and mental disorders, on childbirth, on first-aid remedies, and on treating fractures (Kim Tai-jin, 1976: 80–83).

Six volumes of the *Classified Collection of Medical Prescriptions* are subtitled "promoting longevity". That section opens with the statement that, in the days of old, human beings regularly used to live to be a hundred years old. These days, it complains, when men and women reach only half that age, they are already weak and frail. That is because human beings no longer practice the techniques which can prolong life. The editors of this section then proceed to provide excerpts from a number of Daoist guides to achieving longevity, including guides to lower abdominal and embryonic breathing, Daoist calisthenics, and vitality-enhancing sexual practices. There are introductions to what are labeled Indian massage and Daoist massage, as well as instructions for avoiding the consumption of cereals. This section also includes the popular Daoist advice to practice swallowing your own spit, and it provides a list of taboos to avoid if you want a long life. There is also an excerpt from the scripture of the Jade Emperor and a section explaining Daoist meditation on the "three-in-one". Many of the texts excerpted in this Korean medical encyclopedia are taken from China's Daoist canon.

Both the *Great Collection of Native Korean Prescriptions* and the *Classified Collection of Medical Prescriptions* were compiled for the use of physicians at court. They were not made available to the general population. Nevertheless, they provide evidence of how much Oriental medicine had advanced in Korea from the thirteenth through the sixteenth centuries. One contemporary Korean historian claims that progress in medical knowledge, particularly pharmaceutical knowledge of locally available materia medica, was responsible, along with advances in agricultural technology, for a sharp increase in the Korean population over these centuries. Yi Taejin estimates that the number of people living on the Korean peninsula rose from a little more than two million in the beginning of the twelfth century to 5.5 million in 1391 and 14 million by 1591. He finds confirmation of his hypothesis that there were more Koreans on the peninsula because Koreans were taking better care of their health and living longer by comparing the average life spans of kings in the early part of the Goryeo dynasty to the life spans of kings in early Joseon. Early Goryeo kings lived to be less than forty years old, on the average. Kings in the second half of the Goryeo dynasty lived to be almost 48 years old. By fifteenth-century Joseon, if we exclude one king who was forced off the throne and then forced to commit suicide at an early age, kings did not die until they were at least 55 years old (Yi Taejin, 1997).

The rapid population growth of late Goguryeo and early Joseon ended with the Japanese invasions of the 1590s and the Manchu invasions of 1627 and 1636. After the sharp drop in population caused by foreign soldiers marching through Joseon's villages and fields, population figures rebounded but at a much slower rate. Two centuries of a global little ice age brought cooler temperatures to the Korean peninsula in the seventeenth and eighteenth centuries, which meant both a less favorable environment for agriculture and a more favorable environment for the spread of infectious diseases such as influenza, smallpox and measles. As a result, Korea endured more frequent famines and epidemics after 1600 than it had in Joseon's first two centuries. The consequent rising death rate was naturally of concern to the government, which relied on a healthy population for tax revenues (Baker, 1990).

An early sign of the government's concern was the publication in 1613 of *Dongui bogam* (*A Treasury of Eastern Medicine*), compiled by royal decree under the direction of Korea's most revered physician, Heo Jun (1539–1615). The text provided an encyclopedic overview of Oriental medicine from a

Figure 1 Heo Jun, the chief author of *A Treasury of Eastern Medicine*. Reprinted courtesy of the Association of Korean Oriental Medicine and ChiniKorea (Acupuncture Korea).

Korean perspective. At 25 volumes, it was more compact than either of King Sejong's two massive medical encyclopedias and therefore could be more easily distributed to government offices throughout the peninsula. Read by Confucian scholars as well as physicians, it shaped medical thought and practice in Korea for the rest of the Joseon dynasty and remains influential today.

A Treasury of Eastern Medicine opens with explanations of the basic concepts of Oriental medicine. Heo placed particular emphasis on *jeong* (Chinese: *jing*), *ki* (Chinese: *qi*), and *shin* (Chinese: *shen*). Those three Sino-Korean terms are used in such a wide variety of contexts with such a wide range of meanings that it is difficult to find English equivalents for them. As Heo uses them, they refer roughly to essential bodily fluids, bodily energy, and refined bodily energy respectively. All three terms refer to the fundamental cosmic energy which animates the universe. Within the human body that vital energy takes three different forms, differentiated according to differences in visibility, tangibility, and density. Jeong is visible vitalizing energy, such as semen or blood. Ki is invisible but palpable vitalizing energy, discerned when a doctor feels the pulse of a patient to determine the state of his or her internal energy. Shin is ethereal vitalizing energy, so rarefied that it provides the physiological basis for consciousness. Heo places fostering and reinforcing vital energy at the core of his approach to Oriental medicine.

After introducing jeong, ki, and shin, Heo discusses various external windows into the state of vital energy within, including blood, phlegm, saliva, dreams, and the sound of the patient's voice. Only in his third chapter does he turn to the coagulation of vital energy into the five major governing organs (liver, heart, spleen, lungs, and kidneys) and the six supportive organs (gallbladder, stomach, large intestine, small intestine, bladder, and triple burner). He follows that discussion of the inside of the human body with a discussion of the outside, from the head to the feet. He complements this analysis of the basic components and processes of the body with a detailed analysis of the various ways those components can become dysfunctional and as a result damage the body's other components, preventing them from functioning the way they should. He also includes a catalogue of the medicinal substances available in China and Korea, where they can be found, how to prepare them for human consumption, and when to prescribe them. His final chapters are a detailed description of when and how to use acupuncture and moxibustion.

Heo called his comprehensive medical manual a treasury of "Eastern Medicine" in order to distinguish Oriental medicine in Korea from what he called the "Northern" Chinese medicine of Li Gao and the "Southern" Chinese medicine of Zhu Zhenheng (Kim Ho, 2000: 132). However, the basic principles of both his medical theory and his practical suggestions for health and healing were derived from the Oriental medicine of China. His approach differed only in the way he combined elements of Oriental medicine and in his advocacy of the use of native Korean ingredients.

We can identify three significant components of *A Treasury of Eastern Medicine*. First, Heo introduced to Korea the attempts in Ming China to absorb and reconcile the various theoretical orientations of the Jin and Yuan schools of medicine. There are at least twenty titles from Ming China in his bibliography, more than from any other period in Chinese history (Kim Ho, 2000: 133). Second, he integrated the concepts and practices of Daoism, particularly the internal alchemy school of Daoism, into Oriental medical theory. Heo wrote that those who wished to remain healthy and enjoy a long life should focus their attention first on Daoist techniques for strengthening the production and harmonious circulation of vital energy within their bodies so that they would not need the healing techniques of medicine. The prescriptions and needles medicine prescribed became important only after those Daoist techniques had failed (Kim Ho, 2000: 136–174). Third, he continued the incorporation of native ingredients and folk prescriptions into Oriental medicine, rewriting Chinese prescriptions to replace expensive Chinese materia medica with Korean equivalents and also providing simple prescriptions for those who could not afford the multitude of ingredients many Chinese prescriptions demand (Kim Ho, 2000: 39–58, 228–234).

The book gained its persuasive power from the way Heo wove those three strands together. Previous medical encyclopedias were primarily compilations of medical texts which did not evaluate the relative effectiveness of the various prescriptions and other medical strategies those texts recommended. Heo provided an interpretive framework for understanding which healing and health-enhancing techniques were likely to work better than others. He also provided evaluations of which medicines or procedures were likely to be the most effective in specific situations.

Despite the advances in medical thought represented by the publication of *A Treasury of Eastern Medicine*, medical problems increased rather than decreased over the next three centuries. The first serious new threat to Korean health was syphilis. Introduced onto the peninsula early in the sixteenth century, it became a major medical problem during the Japanese invasions at the end of that century. That deadly sexually transmitted disease was joined soon afterwards by one even more deadly because it spread from victim to victim through more casual contact. Sometime in the seventeenth century, a new virulent strain of measles appeared in Korea, taking the lives of thousands in periodic epidemics from the late seventeenth through the nineteenth centuries. The nineteenth century brought one more new biological brake on population growth. Cholera reached Korean shores in 1822, returning again several more times that century, each time killing tens of thousands.

The impact on population growth was striking. Instead of growing rapidly, as it had in the first part of the Joseon dynasty, the Korean population first stagnated and then began shrinking. The Korean government counted almost half a million fewer Koreans in 1850 than it had found a half a century earlier. That 1800 figure itself was only a slight advance, of less than 100,000, over the census figures for 1750. Both epidemics and famines were taking their toll (Baker, 1990).

The response of Korea to this demographic decline was a popularization of Oriental medicine. *A Treasury of Eastern Medicine* was too large for distribution among the general population, so much smaller pamphlets dealing with prescriptions for specific diseases were printed and distributed to meet the growing popular demand for medical information. Heo Jun himself wrote a short pamphlet on scarlet fever and other rash-producing fevers the same year his magnum opus was published. Many more booklets on smallpox and measles by other writers followed, particularly in the second half of the eighteenth century when a major outbreak of measles in 1752 focused attention on that deadly disease. There were also a number of booklets published on childbirth and pediatrics in the seventeenth through the nineteenth centuries.

Comprehensive medical manuals for the general population also began to appear. One such work was *Sallim kyeongje* (*Farm Management*). Written by Hong Manseon (1643–1715) in the early eighteenth century, this multi-volume collection offered practical advice for dealing with whatever problems a farming household might encounter, from problems raising silkworms and storing harvested grain to medical problems. Since *Farm Management* was intended to be a reference work which farmers could turn to whenever they faced a particular problem, it did not discuss much medical theory. Instead, it provided simple prescriptions using locally available ingredients for a variety of medical situations, including not only diseases but also drowning, frostbite, choking, and dealing with troublesome spirits (Kim Daeweon, 1998: 192–199).

Another medical work worthy of note from this period is *Sumin myojeon* (*Some Excellent Advice on Prolonging Life*). That two-volume work is significant more for who wrote it than for its contents. It is the only medical manual from the entire five centuries of the Joseon dynasty compiled personally by a king.

It is based on the notes King Jeongjo (*r.* 1776–1800) took while taking care of his predecessor and grandfather, King Yeongjo (*r.* 1724–1776), over the last ten years of his grandfather's life. Since King Jeongjo was not a trained physician, his book does not contain many prescriptions or other concrete medical advice. Instead, it is primarily a reflection on medical theory. Recognizing that theory alone would not provide much help for people facing a medical emergency, King Jeongjo had his officials prepare a simplified and abridged but also updated version of *A Treasury of Eastern Medicine*, called *Jejung sinpyeon* (*A New Collection of Prescriptions for Saving the People*), for the use of the general population (Kim Ho, 1996).

More concise versions of medical manuals were not the only evidence of the popularization of Oriental medicine in the second half of the Joseon dynasty. The opening of Oriental medicine markets in four Korean cities from the mid-seventeenth century indicates that more and more Koreans were beginning to trust prescriptions more than the shamans and monks they had relied on in the past. By the middle of the eighteenth century there was enough demand for Oriental medicine that a new occupation appeared in Korea – pharmacists. Koreans began to frequent private shops run by men skilled in diagnosing disease according to the categories of Oriental medicine who would then prepare prescriptions in accordance with their diagnosis (Kim Daeweon, 1998: 190–192).

Privatized oriental medicine, although it was growing popular, was expensive. Starting in the seventeenth century, groups of families, to ensure that medicines would be available when they needed then, began to join together in pharmaceutical cooperatives. They would pool both their labor and their financial resources to support the local growing and collection of materia medica as well as the purchase of needed medicinal ingredients at a distant Oriental medicine market. In effect, in the middle of the Joseon dynasty, Koreans began creating their own indigenous form of health insurance (Kim Daeweon, 1998: 200–205).

This growing popularity of Oriental medicine among the general population stimulated not only the greater use of indigenous materia medica but also the indigenization of medical theory. Near the end of the nineteenth century, a military official turned amateur physician supplemented the medical theory imported from China with some additional reflections on the relationship between differences in individual physical constitutions and differences in responses to medical treatment. Lee Je-ma (1838–1900) suggested that every human being could be assigned to one of four separate and distinct physiological categories, depending on which of four emotions (sorrow, anger, joy, and pleasure) dominated that individual's personality. He recommended that physicians determine which of those four constitutional types best described a patient before they treated him or her, since he believed that the same outward symptoms were manifestations of different internal malfunctions in different physiological types. The physiological category applied to a patient was determined by the emotion that patient tended to display to excess. Someone prone to sudden outbursts of anger, for example, would damage his liver and would therefore have a physiology different from someone who was too quick to indulge in sensual pleasure and would therefore have damaged lungs.

Lee's unique "Four Constitutions" approach to diagnosis and treatment has become increasingly popular in Korea in recent decades, as Koreans have sought to create an Oriental medicine with roots in Korea. However, before Koreans could create a Koreanized Oriental medicine based in part on Lee Je-ma's hypotheses, they first had to overcome a drastic fall in status for Oriental medicine in Korea at the end of the nineteenth century and the first half of the twentieth.

In 1884, Christian missionaries from North America entered Korea carrying both Bibles and doctor's satchels. Within a decade, government legitimation of Oriental medicine through state civil service examinations had ended and a hospital practicing the new biomedicine from the West had opened in Seoul with support from the Korean government. Biomedicine began to replace Oriental medicine as Korea began to transform its institutions and practices in order to meet the demands of a radically transformed international environment, in particular the threat posed by the already modernizing society of Japan, by borrowed ideas, institutions, and technology from the more powerful and more advanced societies of the late nineteenth century.

Practitioners of Oriental medicine did not vanish from Seoul. There were so few physicians trained in biomedicine on the Korean peninsula that the government found that it had to assign Oriental medicine doctors as medical officers for the modern army it began trying to build in 1896. Moreover, Oriental medicine physicians remained on the staff on the Royal Clinic and the Directorate of Medicine, since the royal family was still more comfortable with traditional medicine than with the new medicine from abroad. Recognizing that most of the general population were also more comfortable with Oriental medicine, and also recognizing that there were very few Koreans trained in biomedicine, the Joseon government decided to include Oriental medicine in its plans for the modernization of its public health system. In 1899 it opened the Gwangjeweon (Seoul Public Hospital), which it staffed primarily with specialists in Oriental medicine. The next year, the government inaugurated a licensing system for all practitioners of Oriental medicine on the peninsula. This was the first time in Korean history that medical specialists who operated outside of the civil service had received any official certification of their expertise (Yeo, 1999: 293–301; Shin, 1999).

JAPANESE COLONIAL RULE

In 1905, however, the Joseon dynasty lost most of the power to make its own decisions when the rising imperial power of Japan imposed Japanese advisors on the Korean government. Five years later, in 1910, the last vestiges of Korean autonomy ended when Japan turned Korea into a colony under direct Japanese control. The Japanese immediately embarked on a campaign to change Korea into a society more closely resembling modernizing Japan. Oriental medicine, in their view, was one of many pre-modern features of Korean culture and society which impeded the economic and social modernization of Korea and therefore should be replaced with a "modern" medicine.

One of the first steps the Japanese took to modernize Korean medicine was to seize control of the Seoul Public Hospital in 1905 and replace all the Oriental medicine doctors there with physicians trained in biomedicine. Then, in 1913, the colonial government established new procedures for the certification of medical practitioners. It allowed those who were already operating as Oriental medicine doctors and pharmacists to continue doing so, but it demanded that any future Oriental medicine specialists pass an examination which tested knowledge of biomedical concepts and techniques (Yeo, 1999: 305–307).

There were only 164 trained Korean physicians of biomedicine in 1910, along with 425 Japanese and 22 foreigners. That made a total of 611 biomedical doctors for over 20 million Koreans, leaving the Japanese government no choice but to tolerate the continued existence of Oriental medicine specialists on the peninsula. However, the Japanese apparently intended that over time the older generation of Oriental medicine practitioners would retire and be replaced by a younger generation trained in biomedicine. If that was their plan, they made progress over the three and a half decades of colonial rule. The number of certified herbalists (that is the way the Japanese defined practitioners of Oriental medicine) declined over the decades they ruled Korea from 5,813 in 1913 to 4,699 in 1925 and 3,601 in 1940. At the same time, the number of doctors of biomedicine increased from less than 1,000 in 1915 to 3,660 in 1940 (Yi Jonghyeong, 1977: 284–312).

There was a slight improvement in the Japanese attitude toward Oriental medicine after full-scale war with China erupted in 1937 and Japan found itself in desperate need of medical specialists wherever it could find them. A school of Oriental medicine was allowed to open in 1937 near Seoul. It produced 300 graduates before wartime austerity forced it to close in 1944 (Yi Jonghyeong, 1977: 299). However, the colonial government continued to favor biomedicine. Oriental medicine had to wait until the Americans defeated Japan in 1945 and Korea recovered its autonomy before it could regain anything close to the respectability it had enjoyed prior to Japanese colonial rule.

MODERN KOREA

When two competing Korean governments appeared on the peninsula in 1948, they adopted different policies toward Oriental medicine. The Communist government of North Korea redefined Oriental medicine as the traditional medicine of the Korean people and therefore worthy of government support. An Institute for the Study of Eastern Medicine was established within North Korea's National Medical Science Center in the 1950s, decades before a similar research center received government funding in South Korea. North Korea also incorporated Oriental medicine into the national public health system long before South Korea assigned Oriental medicine doctors to its rural health centers. Moreover, the North Korean government supported research intended to place Oriental medicine on a modern scientific foundation. One such research project in the early 1960s claimed to have discovered anatomical structures supporting the claim in Oriental medicine that, besides blood vessels, there was a separate network of channels through which vital energy flowed.[2]

In pro-Western South Korea, Oriental medicine had to overcome the assumption among South Korea's westernized elite that Oriental medicine represented a past Korea needed to leave behind if it wanted to win respect as a modern nation. A shortage of medical practitioners during the Korean War forced the government in 1952 to recognize Oriental medicine doctors as physicians formally. This paved the way for official recognition of Oriental medicine clinics as well as a state licensing examination for new Oriental medicine doctors (Yi Jonghyeong, 1977: 316). Formal recognition, however, did not win Oriental medicine doctors the high status and high incomes biomedical physicians enjoyed. Given the traditional Korean respect for educational credentials, parity with biomedicine had to wait until there were respected medical schools for Oriental medicine in Korea.

The first college in independent South Korea dedicated to preparing doctors of Oriental medicine opened in 1953. Four years later it opened an attached Oriental Medicine Hospital. However, it ran into financial difficulties and was absorbed by the privately run Kyunghee University in 1965, which created an Oriental Medicine College within the Kyunghee University medical school. Kyunghee University quickly inaugurated a graduate program in Oriental medicine and in 1971 opened South Korea's first large Oriental medicine hospital as part of a newly opened Kyunghee University Medical Center. A year later Won'gwang University, also privately run, opened South Korea's second medical school for Oriental medicine. Won'gwang University added its own affiliated Oriental medicine hospital in 1975 (Yi Jonghyeong, 1977: 317–319).

There are now eleven medical schools for Oriental medicine in South Korea, all established by private universities. There are also over 100 Oriental medicine hospitals, all privately run, and over 7,000 Oriental medicine clinics. As yet, there are no schools of Oriental medicine at public universities, nor are there any government-run Oriental medicine hospitals. However, the growing respectability for Oriental medicine led the South Korean government to open the Korea Institute of Oriental Medicine in 1994 to coordinate research on the theory and practice of Oriental medicine.

That respectability has also been reflected in changes in the 1990s in South Korea's laws regulating providers of medical care. Graduates of colleges of Oriental medicine are now allowed to fulfill their obligatory term of military service as medical officers in the South Korean military, just as graduates of other medical colleges have done since the establishment of the Republic of Korea in 1948. They may also be assigned to responsible positions in public health centers, something not permitted before the 1990s. In addition, in 1996 the law regulating pharmacists was changed to recognize those who had graduated from a pharmacology department of a college of Oriental medicine as certified specialists in Oriental pharmacology. In 1992 South Korea's national medical insurance system recognized this increased respectability of

Oriental medicine by adding coverage for a few basic prescriptions to the coverage established in 1987 for acupuncture, moxibustion, and vacuum cupping.

Oriental medicine regained respectability in the last decade of the twentieth century for three reasons. First of all, South Korea's ascent from poverty, accented by Seoul's hosting of the 1988 Olympic games, gave South Koreans enough self-confidence to reexamine their traditional culture and re-evaluate its various components according to their own values and beliefs instead of relying on criteria imported from Japan and the West. Second, Koreans have discovered that biomedicine is not always able to cure chronic pain or persistent fatigue and that it is equally helpless in the face of the debilitation that accompanies growing older. Third, Korean Oriental medicine has become more compatible with modern urban life by adopting such markers of modernization as standardization and mechanization.

Traditionally, an Oriental medicine physician or pharmacist would modify prescriptions to meet the particular needs of a particular patient. It is still possible to obtain individually tailored prescriptions in South Korea today. It is also possible to walk into a drug store and purchase packages of Oriental medicine prescriptions in mass-produced pill, granular or liquid form, no different in appearance from many medicines produced in the factories of Western pharmaceutical companies.

Similarly, patients who visit an Oriental medicine hospital for a physical examination might find that their blood pressure is checked the same way it would be checked in a biomedical clinic, although they may also have their pulse checked by a machine which will produce a result framed in terms of the traditional 27 measures of the pulse in Oriental medicine. Besides an electrocardiogram evaluating how well their heart is functioning, they may encounter another mechanical device which will evaluate the flow of vital energy through their hands and feet.

Treatment, particularly acupuncture, has also been modernized. The traditional insertion of needles into specific points along the vital energy channels is still practiced, but now those needles sometimes are electrified to provide additional stimulus to vital energy. Lasers are also used to enhance the effectiveness of acupuncture needles. For the treatment of pain, sometimes bee venom is injected into an acupuncture point in a treatment called "medicinal acupuncture". Taping is another variant on traditional acupuncture. Tape is stretched over an acupuncture point, stretching the muscles and stimulating the vital energy flowing through that point.

South Koreans claim that they have added some unique variants to modern acupuncture therapy, ensuring that Oriental medicine is truly Korean medicine. One such variation focuses on those acupuncture points which fall below the elbow or the knee and the relationship between those points and the Five Phases of traditional correlative medical theory. For example, according to this Korean therapy, if the liver is underactive, an acupuncturist should stimulate a point on the lower arm or leg associated with water, since the liver is associated with wood, and water nourishes wood. South Koreans claim that this technique was discovered four centuries ago by a Korean monk but was forgotten until his lost manuscript was rediscovered in the mid-twentieth century. Japanese acupuncturists reject that Korean claim, insisting that Five Phases acupuncture actually originated in Japan.

No one, however, disputes the Korean claim to have originated hand acupuncture. Yu Tae-u, the founder of hand acupuncture, claims that he discovered in the 1970s that all the acupuncture points on the body have counterparts on the hand and that stimulation of those points on the hand is just as effective as stimulation of the corresponding points on the other parts of the body. Few mainstream doctors of Oriental medicine in Korea accept Yu's restriction of acupuncture to the hand, although hand acupuncture has become popular among the general population as a form of self-medication. It has also gained adherents in Japan and North America (Yu Tae-u, 1988).

Whether Oriental medicine in Korea is a unique Korean approach to health and healing or whether it is a Korean variant of traditional Chinese medicine, Koreans see it today as a legacy of their traditional culture worth preserving. Oriental medicine has not been supplanted by biomedicine in twentieth-century Korea. On the contrary, Oriental medicine flourishes alongside biomedicine, in a complementary rather than an antagonistic relationship. That complementary relationship is revealed in the presence of family practice biomedicine physicians staffing the emergency rooms of Oriental medicine hospitals. It also appears in the advice of a biomedicine physician to a patient recovering from a stroke to visit an acupuncturist, or in the advice of an Oriental medicine physician to a patient suffering from appendicitis to hurry to the emergency room of a biomedicine hospital. For most Koreans today, biomedicine remains the medicine of choice for acute medical problems or for problems requiring surgery. However, Oriental medicine remains the medicine of choice for chronic pain, fatigue, or the ravages of old age. If recent past history is any guide, it is likely that Oriental medicine will remain a viable and vital partner of biomedicine in Korea for decades, if not centuries, to come.

NOTES

1 The original *han* means China or Chinese.

2 North Korea abandoned that claim in the late 1960s when no researchers outside of North Korea were able to locate such structures (Gong, 1993: 71–94, Association for Korean Oriental Medicine, 1998: 49–51).

BIBLIOGRAPHY

Association of Korean Oriental Medicine. *Oriental Medicine of Korea*. Seoul: Association of Korean Oriental Medicine, 1998.

Baker, Donald Leslie. 'Sirhak medicine: measles, smallpox, and Jeong Tasan.' *Korean Studies* 14: 135–163, 1990.

Baker, Donald Leslie. 'Monks, medicine, and miracles: health and healing in the history of Korean Buddhism.' *Korean Studies* 18: 50–75, 1994.

Bowman, Newton. 'A history of Korean medicine.' *Transactions of the Korea Branch of the Royal Asiatic Society* 6(15): 1–34, 1915.

Gong Dongcheol. *Hanyageun jugeossda* [Oriental Medicine is Dead]. Seoul: Hakminsa, 1993.

Han, Gil Soo. *Health and Medicine under Capitalism: Korean Immigrants in Australia.* Madison, New Jersey: Fairleigh Dickinson University Press, 2000.

Heo Jeong. *Esei uiryo Han'guk sa* [Essays on the History of Korean Medicine]. Seoul: Hanul Publishing Co., 1992.

Kim Daeweon. '18 Segi ui min'gan uiryo ui seongjang' [The popularization of medicine during the eighteenth century]. *Han'guk saron* 39: 187–238, 1998.

Kim Du-chong (Kim Dujong). 'Middle Eastern and Western influence on the development of Korean medicine.' *Korea Journal* 2(12): 5–7, 1962.

Kim Dujong. *Han'guk uihaksa* [A History of Medicine in Korea]. Seoul: Tamgudang, 1981.

Kim Ho, 'Jeongjodae uiryo jeongchaek' [Medical policies during King Jeongjo's reign]. *Han'guk hakbo* 82: 231–257, 1996.

Kim Ho. *Heo Jun ui Dongui bogam yeon'gu* [A Study of Heo Jun's Treasury of Eastern Medicine]. Seoul: Iljisa, 2000.

Kim Jin. 'Joseon sidae uigwan seonbal' [The selection of medical officials during the Joseon dynasty]. *Dongbang hakji* 104: 1–93, 1999.

Kim, Nam Il. 'Ki Kyujun's study on *Huangdi neijing* in the late Choson era.' In *Current Perspectives in the History of Science in East Asia*, Yung Sik Kim and Francesca Bray, ed. Seoul: Seoul National University Press, 1999, pp. 371–375.

Kim Tai-jin. *A Bibliographical Guide to Traditional Korean Sources.* Seoul: Asiatic Research Center, Korea University, 1976.

Landis, E. Barr. 'The Korean Pharmacopae.' *Korean Repository* 5: 448–464, 1898.

Lee Je-ma (Yi Jema). *Longevity and Life Preservation in Oriental Medicine*, Choe Seung-hoon, trans. Seoul: Kyung Hee University Press, 1996.

Magner, Lois N. 'Diseases of the pre-modern period in Korea.' In *The Cambridge World History of Human Diseases*, Kenneth R. Kiple, ed. Cambridge: Cambridge University Press, 1993, pp. 392–408.

Magner, Lois N. 'Diseases of the modern period in Korea.' In *The Cambridge World History of Human Diseases*, Kenneth R. Kiple, ed. Cambridge: Cambridge University Press, 1993, pp. 392–408.

Miki Sakae. *Chosen igakushi oyobi shippeishi* [The History of Medicine and Diseases in Korea]. Sakai city, Japan: privately published, 1962.

No Cheong-u. 'Chinese medicine in Korea.' *Korea Journal* 11(2): 24–29, 35; 11(3): 15–18, 1971.

Shin, Dongwon. 'The license system for Korean herbal practitioners in 1900.' In *Current Perspectives in the History of Science in East Asia*, Yung Sik Kim and Francesca Bray, ed. Seoul: Seoul National University Press, 1999, pp. 478–483.

Yeo Inseok. 'Joseon gaehang ihu Hanui ui dongt'ae' [Oriental Medicine in Korea after the opening of Korea to the outside world]. *Tongbang hakji* 104: 291–324, 1999.

Yi Gyugeun. 'Joseon hugi naeuiweon uigwan yeon'gu' [A study of the medical staff of the Royal Clinic]. *Joseon sidae sahak* 3: 5–49, 1997.

Yi Jonghyeong. 'Han'guk Donguihak sa' [A history of Oriental medicine in Korea]. In *Han'guk Hyeondae munhwa-sa daegye*, vol. III: *Gwahak gisul sa* [A Survey of Modern Korean Culture: vol. 3, The History of Science and Technology]. Seoul: Goryeo daehakkyo minjok munhwa yeon'guweon, 1977, pp. 263–336.

Yi Pu-yeong. 'Illness and healing in the Three Kingdoms period: A symbolic interpretation.' *Korea Journal* 21(12): 4–12, 1981.

Yi Taejin. 'The influence of Neo-Confucianism on 14th-16th century Korean population growth.' *Korea Journal* 37(2): 5–23, 1997.

Yu Tae-u. *Koryo Hand Acupuncture: Koryo Sooji Chim.* Seoul, Korea: Eum Yang Mek Jin Pub. Co., 1988.

MARGARET LOCK

GLOBALIZATION AND CULTURES OF BIOMEDICINE: JAPAN AND NORTH AMERICA

The anthropologist Clifford Geertz has written that the task of "other-knowing" is a delicate business (1995). As part of the escalating process of globalization, borders dissolve, and boundaries drawn to differentiate self from other, whether they be political, economic, or cultural, become less and less meaningful. Under these circumstances it is questionable as to whether one can write with any confidence about culture as largely accounting for variation among contemporary medical knowledge, practice, and technologies in various geographical locations. Nevertheless, some striking differences are evident throughout the world; those between Japan and North America are all the more remarkable given that it is well over 100 years since European-style medicine was adopted as the official, state-backed, medical system of Japan.

In this chapter I will focus primarily on three topics: life cycle transitions and their medicalization, notably menopause; informed consent and the disclosure of diagnoses; and recognition of brain death as the end of human life and the management of brain-dead patients in Intensive Care Units (ICU). Attention will be directed primarily to Japan, but some observations from the North American situation will be introduced by way of comparison. These illustrative examples will make it clear how medical knowledge and practices in all societies are inevitably associated with moral judgments and with ideas about what is normal and abnormal. Such assessments are in turn intimately associated with widely shared values present in society at large that change over time in light of the introduction of new knowledge and technologies to do with the body. Despite these changes, and even with globalization, certain continuities are evident which ensure that culturally specific forms of medical practice can readily be detected everywhere.

Before turning to the main themes, two caveats are in order. First, biomedicine has taken root and been consolidated where indigenous medical knowledge and practices already exist. Furthermore, in no case, not even in Europe and North America, has biomedicine entirely usurped other forms of healing practices already present. Pluralism is the norm and, paradoxically, with globalization, diverse forms of medical practice have actually proliferated both in

H. Selin (ed.), Medicine Across Cultures: History and Practice of Medicine in Non-Western Cultures, 155–173.

countries that modernized early and in those where the process commenced much later. The existence of pluralism means that patients often have a choice with respect to different types of medical practitioners, a situation that has been particularly evident throughout most of Asia during the last century. In fact, the introduction and spread of medicine from Europe to Asia stimulated, more often than not, a lively revival of indigenous medicine, commonly associated with nationalism and a celebration of the traditional (Croizier, 1976; Leslie, 1976; Lock, 1980).

Interaction takes place across these different domains of healing to some extent, so that both indigenous medicine and biomedicine are modified as a result. Of more importance is that medicine of all kinds functions within an overarching social, political, and cultural milieu that has a profound influence on the way in which medical knowledge is interpreted and informs both incentives and constraints with respect to its practice. Although biomedical training and knowledge is reasonably uniform world wide, when put into practice and when associated policies are formulated, considerable variation can readily be detected that often has major implications for patient care and therapeutic outcomes.

The second caveat has to do with "culture". A concern exists among anthropologists today that culture is too often understood as a "totalizing", all or nothing concept, particularly amenable to appropriation by those with nationalistic interests. Culture can readily be turned into an "exclusionary teleology" (Daniel, 1991: 8), one that mobilizes the notion of an idealized shared past – a reinvented history. Further complications arise because, as noted above, the world is in a state of flux and the global economy is best interpreted as a "complex, overlapping, disjunctive order" (Appadurai, 1990: 5). Under these circumstances Dominquez argues that culture should be regarded as something invoked, not as something that simply "is" (1992: 23). I would add that this applies equally to those who lay claim to participation in a culture, *and* to those who make claims on behalf of others about their participation in or exclusion from a culture. Unequal relationships of power and of inclusion and exclusion exist within virtually all sociocultural complexes, and assertions about culture are more often than not political and moral claims.

In short, it is not appropriate to argue for essentialized cultural differences among groups of people, nor yet for essentialized similarities within any given culture. The culture concept can be understood as a *mise-en-abîme* – an endless series of self-reflecting regressions (Pollock, 1992: 2). Culture is a fluid, contested entity comprised of sets of practices, ideas, imagination, and discourse, continually subject to revision, although the motivations for many of these activities are barely available to consciousness.

FEMALE MID-LIFE AND ITS MEDICALIZATION

After the formation in 1868 of the Japanese State at the Meiji Restoration, the government undertook to standardize and "normalize" the lives of its citizens along several clear trajectories. In common with other technologically advanced societies, social and medical phenomena were enumerated and tabulated making use of the new method of statistics, and the way was opened up for what Ian Hacking has termed the "making up of people" as populations in which attributes are shared in common (1986). Such populations are thought of as being "at risk" for specific medical problems.

Maturation is simultaneously a social and a biological process. However, commencing at approximately the end of the 18th century in northern Europe and North America, life course transitions were gradually subject to increasing medicalization. Birth, infancy, adolescence, middle age and menopause came to be considered as events in which the biological changes that individuals were undergoing placed them at risk. The result was medicalization of these transitions and also of death itself. Attention was gradually confined ever more intently to the body physical, relegating the accompanying social changes to the background.

A French physician, Charles Pierre Luiz de Gardanne, invented the word "menopause" in 1821. From the middle of the 19th century this term gradually came into widespread use in Europe and North America, although relatively few physicians were much interested in this phenomenon until the latter part of the 20th century, by which time medication was readily available to counter the symptoms commonly associated with this stage of the life cycle.

In Japan, although the medical profession has shown an interest in the management of the life course since the early part of the 20th century, clinical gynecology (with the exception of surgery) was late in achieving the status and power accorded to it in Europe and North America. Medicalization of the female life course, especially its latter phases, has not taken root to anything like the same extent as in the West, although over the past decade this situation has changed considerably. However, practitioners of East Asian medicine have long associated certain characteristic symptoms with female middle age and the end of menstruation. It is argued in classical texts that "stale blood" remains in the body once the menstrual cycles have ceased – the cause in some women of numerous non-specific symptoms that often last a few years, including dizziness, palpitations, headaches, chilliness, stiff shoulders, a dry mouth, and so on (Nishimura, 1981).

Many other events also result in stale blood and similar symptoms, and no specific term was used by the *kanpo-i* (herbal doctors) of pre-modern times to gloss discomfort associated specifically with the end of menstruation. This changed towards the end of the 19th century, when numerous Japanese doctors went to Germany to study medicine in Europe and, under the influence of German medicine of the day, created the term *kōnenki* to convey the European concept of the climacterium. Nishimura (1981) has suggested that until the middle of the 20th century *kōnenki* was used to refer to all life cycle transitions regardless of the age at which they occurred, both male and female. This interpretation is very close to the meaning given to the term climacterium until the middle of the 19th century in Europe. Contemporary Japanese has no term that expresses in everyday language the event of the end of menstruation per se, although there is, of course, a medical term, much as menopause was a medical term in English, little used in daily parlance until as recently as forty years ago.

Turn of the century German discourse about the climacterium made use of the newly formulated "autonomic nervous system", a concept that made good intuitive sense to the Japanese medical world. First clearly articulated in 1898, this concept caused a stir in medical circles everywhere at the time (Sheehan, 1936), but perhaps particularly so in Japan where it fitted with the integrative physiological approach characteristic of Sino/Japanese medicine. Later, in the 1930s, when a close association was postulated between the endocrine system and the autonomic nervous system (Sheehan, 1936), Japanese physicians comfortably adopted this idea, and argued for a relationship between *kōnenki* and disturbances in the autonomic nervous system – an association that the majority of Japanese physicians and women recognize today (Lock *et al.*, 1988; Rosenberger, 1992). The dominant discourse about *kōnenki* is one in which distress is not linked directly to a decline in estrogen levels, as is the case in North America and most of Europe, but is said to be due to a destabilization of the autonomic nervous system. Both sexes are vulnerable, but women are thought to be more so than men because of the added impact of declining estrogen levels in their aging bodies.

Another factor that no doubt historically worked against the construction of a narrowly focused medical discourse about the aging ovary and declining estrogen levels was that Japanese doctors, unlike their Western counterparts, had practiced little surgery prior to the 20th century. The powerful, physiologically oriented herbalists of the East Asian medical system disparaged surgery. Furthermore, anatomy as it was conceived in Enlightenment medical discourse in Europe had relatively little impact in Japan until the twentieth century, and autopsies and dissection were not widely practiced (Kuriyama, 1992; Low, 1996). The result was that Japanese physicians did not have first hand experience of removing and dissecting many hundreds of ovaries, as was the case for the majority of late 19th century European and North American gynecologists (Laqueur, 1990), and they remained predominantly physiologically rather than anatomically oriented. This was in strong contrast to the approach that came to dominate in Europe.

Historical records strongly suggest that European attitudes towards aging have changed greatly over the years. Writing in 1813 Henry Halford made the following observation, "I should observe, that though this climacteric disease is sometimes equally remarkable in women as in men, yet most certainly I have not noticed it so frequently, nor so well characterized in females." Halford was writing about that "period of life at which the vital forces begin to decline," commencing from about 45 until 60 years of age. He described a general decay of strength, tiredness, loss of weight and appetite. "The patient sometimes suspects he has a fever and might also experience head and chest pains, vertigo, rheumatic pains, swollen legs and sluggish bowels but "above all, anxiety of mind and sorrow have laid the surest foundation for the malady." Halford concluded by wondering if it was the prospect of death that "inflicted the wound in the patient's peace of mind" (1813: 325).

It was not until the middle of the 19th century that the few physicians who specialized in diseases of women began to counter what was taken as common sense knowledge for the time. Barnes, writing in 1873, stated, "Physicians do, indeed, talk of the climacteric in man; but the analogy is more fanciful than real." He went on, "There is nothing to compare with the almost sudden decay of the organs of reproduction which marks the middle age of woman. Whilst these organs are in vigor, the whole economy of woman is subject to them. Ovulation and menstruation, gestation and lactation by turns absorb and govern almost all the energies of her system. The loss of these functions entails a complete revolution" (1873: 263–264).

The views of Barnes and his colleagues prevailed, and by the late 1930s one or two European and North American gynecologists had gone further and likened menopause to a deficiency disease – specifically to diabetes and thyroid deficiency disease (Shorr, 1940). However, in spite of the availability of medication to counter the supposed deficiency caused by declining estrogen levels, very few gynecologists participated actively in the medicalization of menopause until the 1970s or later. On the contrary, most physicians were vociferous in describing menopause as a normal event, and were actively opposed to the use of estrogen replacement therapy except under extenuating circumstances, largely because of a concern about iatrogenesis (Bell, 1987). In Japan, by contrast, *kōnenki* has never been conceptualized as a deficiency disease, and doctors made little effort to involve themselves with this part of the life cycle until the last decade of the 20th century. More so than was the case with menopause in Europe and North America, *kōnenki* has been understood as a normal transition, one that is not conceptualized as a medical problem or a disease-like state.

One advantage of thinking about menopause as a disease is that the psychological and moral discourse formerly associated with this stage of the life cycle has gradually disappeared in the West. Aside from a lingering association with depression, women going through this transition are no longer usually thought to be subject to psychosis, neurosis, or even "empty nest syndrome". In Japan, however, at least until the 1990s, it was assumed that most women would simply ride over *kōnenki*, and that those who complained of symptoms and suffered at this stage of their life were living lives of luxury with too much time on their hands. Self-indulgent middle class women were assumed to be particularly vulnerable, and their complaints were not taken very seriously. In this sense alone *kōnenki* is thought of as a disease – a luxury disease (*zeitakubyō*) – associated specifically with modernity.

The nuclear household in which approximately 60 per cent of Japanese live these days, lacking both enshrined ancestors and the elders, is thought by many commentators to be a fragile "pathological" conglomeration (Mochida, 1980; Eto, 1979). Members of such families – men, women and children – are believed to be particularly vulnerable to what has been termed "diseases of civilization" or "modernization", including a whole range of neuroses, behavioral disorders, and deviant behaviors. These diseases are made factual by means of catchy diagnostic labels such as "school refusal syndrome", "high rise apartment neurosis", "moving day depression", "death from overwork", and so on (Lock, 1988). A biological component is not denied, but of overwhelming importance is the social significance of these conditions, and women who complain of symptoms at *kōnenki* fall into this category. When asked if he thought that all women experience trouble at *kōnenki*, a Kobe gynecologist answered,

> No, I don't think so. Women who have no purpose in life [*ikigai*] have the most trouble. Housewives who are relatively well off, who have only one or two children and lots of free time come to see me most often. This is a sort of "luxury disease" [*zeitakubyō*], I'm sure women never used to come running to a doctor before the war with this kind of problem (Lock, 1993: 264).

Under these circumstances it is not surprising that relatively little attention has been paid to *kōnenki* or that very few women in Japan have gone to physicians complaining of symptoms, but this is only part of the story.

Extensive survey research carried out in the mid 1980s shows that, compared with statistically comparable Americans and Canadian samples, over 1300 Japanese women experience and report remarkably few symptoms at the end of menstruation, including hot flashes and night sweats (Lock, 1993), and this is the case whether they be blue collar workers, farmers, or middle class housewives. It is notable that there is no word that specifically signifies a "hot flash" in Japanese. On the basis of these findings I have argued for a recognition of "local biologies" (1993). In other words, sufficient biological variation exists among populations of women that the physical effects of lowered estrogen levels on the body characteristic of mid life are not the same in all geographical locations. There is evidence from other parts of Asia in addition to Japan, and from one or two other geographical locations, of considerable variation in symptom reporting at menopause (Beyene, 1989; Lock and Kaufert, 2001; Shea, 1998). This variation, to which genetics, diet, environment, and other cultural variables contribute, accounts for marked differences in the subjective experience and associated symptom reporting at this stage of the life cycle. The differences in experiences between North American and Japanese women are sufficient to produce an effect on (but *not* determine) the creation of knowledge, both popular and professional, about this life cycle transition (Lock, 1993).

The majority of Japanese women, when asked in interviews to describe their experience of *kōnenki*, responded along the following lines:

- I've had no problems at all, no headaches or anything like that ... I've heard from other women that their heads felt so heavy that they couldn't get up.
- The most common problems I've heard about are stiff shoulders, headaches, and aching joints.
- I get tired easily, that's *kōnenki* for sure, and I get stiff shoulders.
- My eyesight became worse, and sometimes I get ringing in the ears. I hear that some housewives get so depressed that they can't go out of the house.

A small number of women, 12 out of 105 interviewed, made statements that sound more familiar to North American ears: "The most noticeable thing was that I would suddenly feel hot; it happened every day, three times or so. I didn't go to the doctor or take any medication, I wasn't embarrassed and I didn't feel strange. I just thought it was my age."

Japanese gynecologists when interviewed in the late 1980s indicated that shoulder stiffness is the most common symptom reported by Japanese patients, followed by headaches and various non-specific symptoms; some physicians did not list hot flashes at all when asked to describe the symptoms of *kōnenki*. Recently Japanese gynecologists have become involved in international meetings on the subject of menopause, and some of them have published widely in women's magazines in Japan where they often describe hot flashes as the "typical" symptom of menopause. It will be interesting to see to what extent symptom reporting by women changes in the future, if at all, as a result of this new-found medical interest.

That the hot flash has not been the key signifier of distress at *kōnenki* has meant that there has been relatively little incentive over the years for Japanese doctors to prescribe hormone replacement therapy for symptom relief. Furthermore, the Pill has not been generally available in Japan for contraceptive purposes until very recently. Both doctors and women regularly report in survey research that they are concerned about possible iatrogenesis caused by long-term use of the Pill (Mainichi Daily News, 1987), and this fear extends to the use of hormone replacement therapy (HRT), the chemical composition of which is very similar to the Pill, a fact that the majority of Japanese women are well aware of.

THE AGING SOCIETY AND THE MEDICALIZATION OF *KŌNENKI*

In Japan obstetrics and gynecology have not usually been separated in primary care practice, and individual physicians own and run small hospitals where their income is derived largely from deliveries, abortions, and minor surgery. Over the past twenty years gynecologists have found their medical practices less lucrative than was formerly the case because the birth rate is low and in any case women have chosen to have their babies in tertiary care facilities. The abortion rate has also gone down during this time. Due in part to the economic pressures under which many gynecologists in private practice find themselves some have become interested in middle aged women and their health, so that *kōnenki* is now taken much more seriously as a medical matter. The formation of international menopause societies and the holding of regular conferences on this subject have also proved very influential in promoting an interest in menopause among Japanese physicians. But above all, an interest in *kōnenki* has taken off because of a concern about the "greying of society" as it is known in Japan, and the burden that elderly people, particularly elderly women because they live longer, are likely to put on the health care system.

Over the past two decades in virtually all of the so-called developed countries, a concern is evident about the aging society. Governments and the health care professions alike are concerned about escalating health care expenditures, and older women have become particular targets of attention. In part because menopause is conceptualized as a deficiency disease, it was relatively easy in

North America to argue that all women once past menopause are at an increased risk for various late onset diseases that have come to be associated with lowered estrogen levels, among which heart disease and osteoporosis are the most commonly cited. It is widely recommended that virtually all women should take hormone replacement therapy on a regular basis until the day they die, despite ceaseless debate as to whether this powerful medication is indeed effective in preventing heart disease and the dangers it poses for increasing the risk of breast cancer.

Japanese physicians, even though they are well aware of the aging population, live in a society actively dedicated to preventive medicine. They keep abreast of the medical literature published in the West, but even so have not considered it appropriate to promote HRT to the extent that is done in North America. Here again, local biology plays a part: mortality from coronary heart disease for Japanese women is about one quarter that of American women (WHO, 1990), and it is estimated that, although Japanese women become osteoporotic twice as often as do Japanese men, nevertheless, this rate is approximately half as often as in North America (Ross *et al.*, 1991). When Japanese physicians recommend HRT it is for osteoporosis prevention, to reduce the risk of Alzheimer's disease, and above all as protection against stroke. Japanese people have higher rates of stroke than is common in North America, and claims have been made in one or two studies that HRT is effective in reducing the risk of stroke. Even so, recommendations for HRT use remain more cautious than is the case in North America, and for patients who increasingly come to see physicians at this stage of the life cycle, herbal medicine may be recommended as commonly as HRT. It is of interest to note that several of the world's leading experts in menopause are currently revising their opinions about blanket recommendations for use of HRT, and that a disease-like approach to this stage of the life cycle is being subjected to second thoughts.[1]

Most Japanese gynecologists recognize that *kōnenki* is not the same concept as the English term menopause. Whether a medical term that simply means the end of menstruation, such as *heikei*, will take on popular usage in Japanese as this stage of the life cycle becomes increasingly medicalized is a matter for speculation. It is also possible, as dietary practices change in Japan, that symptom reporting at menopause may change. It is equally possible that some changes might occur in North America as a result of the recent attention that is being paid to diet. It would be patently unwise to assume that the Japanese case will necessarily become more like that of North America, or that Japanese medical practices will fall fully in line with those that are common in North America today, particularly as these practices are under dispute in their home territory. Ultimately, the way in which aging is understood and dealt with by society and the medical profession is a mix of politics, culture, and biology.

INFORMED CONSENT AND THE DISCLOSURE OF DIAGNOSIS

Throughout the 1960s in North America, research in which human subjects were left in the dark about what was being done to them came increasingly under fire, leading eventually to the institutionalization of informed consent not only for research subjects but for all patients (Rothman, 1991). This move was in part precipitated by concerns expressed within the medical profession itself, but the consumer movement of the 1970s, accompanied by an increasing demand from the public at large about the "right to know", was also influential in these changes.

It was only in the 1990s that *infōmudo konsento* (informed consent) was definitively introduced in Japan, after the Japan Medical Association put out a report on the subject *(Nihon Ishikai Seimei Rinri Kondankai,* 1990*)*. Even following the introduction of the concept, the actual practice of informed consent has taken a very different path in Japan than in North America. For one thing its meaning can range in Japan from the idea of a "courteous dialogue" between doctor and patient to offering the patient a choice of treatments (LeFlar, 1996).

Contemporary Japanese medicine, influenced at the turn of the century by German military medicine, has been decried both within Japan and from without as simultaneously feudalistic and paternalistic (Ikegami, 1988; LeFlar, 1996: 6). It has usually functioned until recently with the unspoken assumption that a frank discussion between doctors and patients is not necessary because a "tacit" understanding inevitably exists due to their shared language and cultural heritage. Associated with this assumption is the belief that vagueness on the part of the doctor is, in effect, a virtue – a buffering of patient anxiety in the face of unpleasant diagnoses. Moreover, because the word usually used to express paternalism in Japanese, *omoiyari*, conveys the idea of sympathetic understanding and support (Ikeda, 1993), the very notion of paternalism is softened through Japanese usage, avoiding an entirely negative association with such behavior. Furthermore, patients are steeped in the idea that one "puts oneself in the hands of the doctor" – he is the expert – all decisions should be up to him and are based on his superior knowledge. When drugs are prescribed, for example, patients are not customarily told of possible side effects. The idea of autonomous decision making by patients, or even decisions made by patients in partnership with their doctor, has simply not been part of the Japanese cultural or medical tradition until very recently.

Over the past five years or so there has been a proliferation of media presentations, books, and articles in Japan about human rights and in particular about the "right to know". People are now encouraged to take control of decision making with respect to their own diseases. Contentious issues such as the making of living wills, euthanasia, terminal care, and similar subjects, formerly not open to public debate, are now very much part of public consciousness in Japan. But despite these changes the extent to which the practice of informed consent and full disclosure of diagnoses is followed is debatable. The work of Susan Long in connection with disclosure of cancer diagnoses (1999, 2000a) and cultural decisions about life-support treatment (2000b) is highly informative in this respect.

Long points out that as recently as twenty years ago the practice of non-disclosure of diagnoses of terminal illnesses was not questioned at all in Japan

(1999, 2000a). A widespread belief existed among both the public and the health care profession that such disclosures were harmful to patients, possibly even driving them to suicide. During her research in the late 1970s Long was told that patients give up hope when presented with "a death sentence", and without hope they will die a rapid, unpleasant death (Long and Long, 1982). The idea of a "peaceful" death is highly valued, and disclosure was actually thought of as unethical.

Today things have changed, and the approach to informed consent that comprises "explanation and agreement" (*setsumei to dōi*) used thus far in Japan is being debated. By no means is there an uncontested push for applying the concept in a way similar to that which is customary in North America. A good number of people continue to argue for a cautious approach and claim that full disclosure goes against Japanese culture (Long, 1993: 32). For these individuals even the idea of a dialogue is going too far. It is often stressed that Japanese patients do not want to know about their diagnoses, and respecting their autonomy involves, therefore, protecting them from this information (Asai, 1995; Hayashi, 1996). Even the courts in Japan, although they recognize the right to informed consent, are supportive of the "customary" practice of non-disclosure (Long, 1993: 32).

Despite public discussion about informed consent, actual clinical practice has been slow to change. A 1992 survey revealed that in only 20% of cases where a patient had died did the family report that the doctor had informed the patient of the diagnosis. However in an additional 44% of cases families reported that they believed that the patient had nevertheless surmised the severity of his diagnosis (Ministry of Health and Welfare, 1994).

When in 1996 Long interviewed 41 physicians in Japan, most of them in internal medicine, the majority responded that their decisions to disclose diagnoses are made on a case-by-case basis. Long found that doctors were unaware of any departmental or hospital policies for or against disclosure of diagnoses, even though policies often existed. It is quite clear from Long's research that doctors agonize about disclosure and that their decisions are based on whether they judge the patient wants to know the diagnosis or not. It is very often assumed that if patients do not ask directly then they do not want to know. Judgments are also made as to whether patients can cope with the news based on assessment of their supposed personality, character, or mental state. Formerly it was very often assumed that women could not handle bad news of this kind, whereas men bore up better, but overtly at least, this gender-biased assessment has been dropped.

Doctors also worry about how they themselves would handle the situation if the patient or his family reacts badly (1999: 35). A few doctors have taken a strong stand either for or against always informing patients, but even so, these doctors, together with the majority, remain convinced that the family is in the best position to judge whether or not the patient can handle with composure disclosure about a terminal illness. Long reports that, according to the doctors she interviewed, most families, once approached, ask doctors not to disclose diagnoses with a poor prognosis. Doctors then "strategize" with the family, with the result that on occasion full disclosure follows, at other times gradual disclosure, or else no disclosure at all. Long also notes that although nearly 70% of people who participated in a national survey responded that they would want to hear about a diagnosis of a terminal illness in their own case, only 28.5% responded that they would want to inform family members of their diagnoses. One other survey of over 3000 adults found that only 19% thought that the doctor should tell the patient in all cases about a diagnosis of a terminal illness (1996).

It is clear from the research of Long and others that Japanese families, although outwardly very respectful of the medical profession, believe very strongly that they should be involved in the decision making process about all major medical matters, particularly those involving end of life decisions. As Long puts it,

> Most family members and most doctors agree that life is frail and is given meaning through fulfilling roles in human society; that lack of emotional control threatens life itself; that the patient is part of a social unit larger than the individual; and that this social unit has not so much the right as the obligation to serve as his or her surrogate (Long, 1999: 39).

This is the case even though it is recognized in Japan today that families come in a variety of forms and that not everyone has close living relatives, and the idea of individual interest is openly recognized as a worthy end under certain circumstances. Leaving someone to face on his or her own decisions as death approaches is not acceptable, even if it means that individual autonomy will be compromised in some way. On the whole, Japanese doctors strongly support this position.

Long is very cautious about accounting for her findings in terms of rather easily identifiable cultural differences. In her comparative work in connection with the use of life support treatment she notes that the clearest examples of medical paternalism are evident in the United States and not Japan. Long also found that families are involved in decision making in both settings, and that the ideal of patient autonomy upheld so strongly in the United States may be compromised in order to achieve valued goals aspired to by families and doctors (2000b: 247). Detailed ethnographic research is an invaluable tool for prizing apart assumptions and stereotypes about medical institutions and their practices, revealing the face-to-face negotiations that inform medical care.

THE DEATH OF BODIES AND OF PERSONS

For over thirty years in Japan the "brain-death problem", as it is known, has proved to be far and away the most contentious of bioethical debates, and during that time twenty-four members of transplant teams, in eight separate incidents, have been accused of homicide for procuring organs from brain-dead bodies. In the wake of enormous media coverage, lively demonstrations by opposition groups, and pressure by the Japan Medical Association, the Organ Transplant Law was finally passed in the autumn of 1997, making it legal to obtain organs from patients diagnosed as brain-dead. However, donor cards must be signed ahead of time not just by potential donors themselves, but also by family representatives, before procurement of organs can take place. Patients who become brain-dead and who have not signed donor cards are not counted as dead, and continue to be given life support until such time as the family is in agreement that further "treatment" is futile. In the four years since the law was passed only ten procurements have taken place that made use of brain-dead donors.

When commenting on this situation, many outsiders assume that religious beliefs or lingering superstition must be implicated – that something fundamentally pre-modern remains at the core of Japanese society. However, representatives of the major Buddhist sects in Japan have at no time been particularly vocal in opposition to the recognition of brain death, and Japan is, in any case, a much more markedly secular society than is the United States. Many Japanese, when polled about attitudes towards brain death, indicate that respect for their deceased relatives is of importance to them, but this Confucian-derived sentiment is not reckoned as a religious belief but rather as a basic value widely held in Japan. Some individuals indicate that they are uncomfortable with the idea of organs being taken from the recently deceased, because they believe that this may inflict yet more pain and suffering (Lock, 2002). However, research in Sweden and the United States suggests that numerous people in these countries feel exactly the same way, and Siminoff *et al.* (1999) have shown in a study of 400 families living in the United States who were approached about donation that more than half of them declined.

Given the ferocity with which the Japanese media has opposed recognition of brain death, questioned whether brain-dead patients are indeed dead, and exposed what are at times clearly unethical medical behaviors, the media has without doubt been more influential than religious belief in making the recognition of brain death a contentious issue. Also of great significance is what happens in intensive care units. It is widely recognized in Japan that death, even in ICUs, is above all a familial event, and it is considered to be inappropriate for physicians to intrude on grieving family members to make direct inquiries about organ donation. Second, the idea of the "person" is not located in the brain for large segments of the Japanese public, including some doctors, making it difficult to recognize brain death as the end of meaningful human life.

When persons linger in bodies

Abe Tomoko, a pediatrician employed for many years in a hospital which specializes in neurological disorders, has spent considerable energy during the past decade working with the grass roots movement in Japan against the legalization of brain death as the end of life. In discussing her objections with me at one of our several meetings, she emphasized that the concept of brain death was created primarily for the purpose of facilitating organ transplants. She is emphatic that when a dying person is understood as the focus of both a concerned family and a caring medical team, then it is difficult to interpret brain death as the demise of the individual. Her opinion is derived, Abe states, from reflection on her own subjective feelings as a pediatrician. "The point is not whether the patient is conscious or unconscious, but whether one *intuitively* understands that the patient is dead. Someone whose color is good, who is still warm, bleeds when cut, and urinates and defecates, is not dead as far as I am concerned. Of course I know that cardiac arrest will follow some hours later – but I think even more significant is the transformation of the warm body into something which is cold and hard – only then do the Japanese really accept death." When asked why this is so, Abe replied, "It's something to do with Buddhism, I suppose. I'm not really a Buddhist but it's part of our tradition in Japan." Abe is completely opposed to organ transplants that are dependent on brain dead donors, and also has strong reservations about living related organ donations (Lock, 2000).

Among the 19 Japanese physicians who work in ICUs whom I interviewed in 1996, the majority of whom were neurosurgeons, I did not find anyone who took a position as extreme as Abe's. However, her sentiments and those of the many others (including other physicians) who think as she does are well known among the Japanese public because of their numerous television appearances and publications. The physicians with whom I talked believed that brain death is an irreversible condition, provided that no errors have been made, but they also believed that a brain dead body is not dead. In contrast to Tomoko Abe, however, they are not adamantly opposed to organ transplants, although none of them has ever actually been involved with procurement of organs for donation.

Although I conducted my interviews in the year before the law was changed, I would be surprised if the practices of Japanese neurosurgeons have changed very much. I returned briefly in 1998 and 1999 and had this impression confirmed. The position of the majority of physicians working in intensive care in Japan has been that to declare brain death and then abruptly ask the family about donation is inappropriate. If the family does not raise the question independently, as they rarely do (although this is changing with the new law), then the matter will not be routinely raised. The fact that it has been illegal to procure organs from the brain-dead, and murder charges may be laid for doing so has, of course, ensured that physicians have been hesitant to approach families. But, above all, it is the intrusion into the privacy of the family that holds doctors back. Because organ procurement will not take place there has been no haste, no pressure, and no need for an accurate brain death diagnosis in Japanese ICUs. This situation continues even after the enactment of the law because, aside from the relatively few university hospitals designated as procurement centers, other hospitals are not permitted to do so.

The neurosurgeons interviewed all agreed that they "more or less" follow the Takeuchi Criteria, that is, the standards set out by the Ministry of Health and Welfare in 1985 for determining brain death. However, several of them added comments to the effect, "We do not always make the diagnosis, even when we suspect brain death. We often guess, which is much easier for the patients and their families." What is meant here is that in cases that appear hopeless the attending neurosurgeon will do one or more simple clinical tests on the basis of which he concludes that the patient is either brain dead or very close to it. He then informs the family that their relative is *hobo noshi no jotai* (almost brain dead), or alternatively that the situation looks hopeless. The ventilator is not turned off at this juncture but is kept in place until the family requests that it be turned off, often several days after the diagnosis. One physician commented, "Perhaps this is unique to Japan, but we believe that it is most appropriate to tell the family that we are continuing to do our best for their relative even though brain death is 'approaching', rather than to say as they do in America, 'the patient is brain dead, here are the test results, we are going to terminate all care'."

Commenting on the actions of his colleagues a neurosurgeon pointed out that, "brain death is a kind of 'end stage', in other words, there is nothing more that we can do for the patient, but we are ambivalent because brain death is not human death." He added,

> There was a case I had a while ago where a child stayed alive for six or seven days after a diagnosis of brain death, even after the ventilator had been turned right down. If the family had said early on that they wanted to donate organs it would have been different but there was no suggestion of this. As far as they were concerned, I would have been killing their child if I had turned off the ventilator – and in a way they were right. After, all we don't sign the death certificate until the heart stops beating (cited in Lock, 2002).

With the passing of the Organ Transplant Law, in those hospitals designated to procure organs there is now, of course, an enormous concern about precision in the determination of brain death, particularly so because the media expects to be fully informed about the most minute of medical procedures in an effort to make medical practices in Japan fully transparent and open to public scrutiny.

In complete contrast to the responses given in North America by medical professionals (see below), although an acute sensitivity exists about the ambiguous nature of a brain-dead patient, no one in Japan talked about the shell of a body remaining although the person or soul had departed. In my opinion there are at least two reasons for this in addition to the fact that doctors do not think it appropriate to persuade families that their relative is no longer alive. Although many physicians and surgeons stated clearly that, for them personally, once consciousness is permanently lost and a patient is unable to breathe independently a patient is as-good-as-dead, they do not believe that most families think as they do. Japanese indigenous medical knowledge holds that life is diffused throughout the body in the substance of *ki*, and it is assumed on this basis that most Japanese are not willing to equate a permanent loss of consciousness with death. Perhaps of even more importance is that "person" is not usually understood as an autonomous entity firmly encased inside a brain. Among fifty Japanese with whom I have talked on this matter, only one third clearly locate the center of their bodies in the brain; the others, of varying ages, selected *kokoro* as the center, a very old metaphorical concept that represents a region in the thorax where "true" feelings are located (Lock, 2002).

The idea of individual rights has gained a strong foothold in Japan, particularly over the latter part of the 20th century, but this concept has to battle against the powerful flow of indigenous knowledge in which an individual is conceptualized as residing at the center of a network of obligations, so that personhood is constructed out-of-mind, beyond body, in the space of ongoing human relationships. "Person" remains, perhaps for the majority in Japan, a dialogical creation, and what one does with and what is done to one's body are by no means limited to individual wishes, as the work of Susan Long also clearly shows. In this climate, in which doctors are themselves immersed, the majority are unlikely to impose their interests on families of dying patients.

Ironically, even though it is in Japan where the public has most clearly articulated fears that the lives of traumatized patients may be ended too soon in order that organs may be procured, in reality the chance of this happening is very slim as compared to North America or Europe. As Long's work also shows, physicians are less likely in many circumstances to behave paternalistically in Japan than they do in other parts of the world.

When bodies linger and persons are gone

Interviews with 32 intensivists – persons working in intensive care units – in Canada and the United States reveal that not one of them, similar to their Japanese counterparts, believes that brain death signals the end of biological life, although there was unanimity that the brain is irreversibly damaged and that this condition will lead, usually sooner rather than later, to complete biological death. It was agreed that the organs and cells of the body, including small portions of the brain, remain alive thanks to the artificial brain stem supplied by the ventilator that is necessary for the continued existence of a patient once brain-dead. Indeed, if organs are to be transplanted, then they must be kept alive and functioning and as close to normal as is possible; as Youngner *et al.* note, "maintaining organs for transplantation actually necessitates treating dead patients in many respects as if they were alive" (1985: 321).

For these intensivists a brain-dead patient is not simply on the path to biological death. Such patients have already completed a second irreversible change, in that the person and/or spirit is no longer present in the body. The patient has assumed a hybrid status – that of a dead-person-in-a-living-body. In order to convey their certainty about the death of the patient, in addition to explaining about tests and examinations to families, internists emphasize that the "person" is no longer present, even though the appearance of the individual lying in front of them usually does not give visual support to this argument.

The majority of interviewed intensivists inform assembled families that, "The things that make her her are not there any more," or "He's not going to recover. Death is inevitable." One intensivist pointed out that it is difficult to assess what is best to say to the family in part because in most cases one does not know if they have religious beliefs or not.

> I believe that a 'humanistic' death happens at the same time as brain death. If I didn't believe this, then I couldn't take care of these patients and permit them to become organ donors. For me the child has gone to heaven or wherever, and I'm dealing with an organism, respectfully, of course, but that child's soul, or whatever you want to call it, is no longer there. I don't know,

of course, whether the family believes in souls or not, although sometimes I can make a good guess. So I simply have to say that 'Johnny' is no longer here (cited in Lock, 2002).

Another physician stated that it is essential that the doctor take control "a bit" when discussing brain death. Families, she argued, often find it difficult to accept that there is no possibility of reversibility, and this is where the doctor cannot afford to appear diffident or equivocating. One doctor stated, "You can't go back to the family and say that their relative is brain dead, you've got to say that they are dead – you could be arrested for messing up on this." He recalled that during his training he had described a patient as "basically dead" to his supervisor, who had responded abruptly by insisting: "He's dead. That's what you mean, basically."

It is clear that the intensivists have no second thoughts about reversibility, but it is also evident that many of them nevertheless harbor some doubts about the condition of a recently declared brain dead patient, and it is often those with the longest experience who exhibit the most misgivings. A physician with over 15 years of experience said that he often lies in bed at night after sending a brain dead body for organ procurement and asks himself, "Was the patient *really* dead? It is irreversible – I know that, and the clinical tests are infallible. My rational mind is sure, but some nagging, irrational doubt seeps in." This doctor and the majority of other intensivists interviewed take some consolation from their belief that to remain in a severely vegetative state is much worse than to be dead.

One doctor professed to a personal belief in a soul that takes leave of the body when it dies. For her, this happens when the patient's brain is clearly irreversibly damaged, and it permits her to feel reassured about procurements. Another intensivist insisted at first, as did many of the people interviewed, that he had no difficulty with the idea of brain death. "It seems pretty straightforward to me. Do the tests, allow a certain amount of time; a flat EEG and you're dead." Then, ten minutes later this doctor said, "I guess I equate the death of a person with the death of the spirit because I don't really know about anything else, like a hereafter. I'm not sure anyway, if a hereafter makes a difference or not." When asked what he meant by the word "spirit", this doctor replied, "I guess one would have to take it as meaning that part of a person which is different, sort of not in the physical realm. Outside the physical realm. It's not just the brain, or the mind, but something more than that. I don't really know. But anyway, a brain dead patient, someone's loved one, won't ever be the person they used to know. Sure their nails can grow and their hair can grow, but that's not the essence."

Another senior doctor, struggling to express his feelings, imbued the physical body with a will. "The body wants to die, you can sense that when it becomes difficult to keep the blood pressure stable and so on." This intensivist, although he accepted that brain death is the end of meaningful life, revealed considerable confusion in going on to talk about the procurement of organs. "We don't want this patient to expire before we can harvest the organs, so it's important to keep them stable and alive, and that's why we keep up the same treatment after brain death." Yet another informant acknowledged that the "real" death happens when the heart stops. "The patient dies two deaths."

A utilitarian approach to the brain dead appears to be dominant in North America despite many lingering doubts about the condition of the brain-dead body. Widely shared cultural values have contributed, it seems, so that physicians can in good conscience send patients for organ procurement. In Japan, because death remains above all a family matter, and the commodification of human bodies is not widely supported, the transplant enterprise has been severely curtailed. In the end, it may be the genetic engineering of organs that gets Japan off the hook. Only then will it be possible to circumvent the narratives that we inevitably create about the death of persons, brains, and bodies.

Already another set of arguments is beginning to be heard internationally in connection with the cloning of life, in particular with the use of human stem cells taken from embryos. To date Japan has had relatively few disputes about the beginning of life, and it may well be that this technology will encounter much more resistance in the Western hemisphere than will be the case in Japan.

NOTES

[1] In July 2002, the United States government halted a large long-term study of hormone replacement therapy, noting that it seemed to cause more heart disease, breast cancer, and stroke than it prevented. Medical opinion on the use of HRT has now radically changed.

BIBLIOGRAPHY

Appadurai, Arjun. 'Disjuncture and difference in the global cultural economy.' *Public Culture* 2: 1–24, 1999.

Asai, A. 'Should physicians tell patients the truth?' *Western Journal of Medicine* 163: 36–39, 1995.

Barnes, Robert. *A Clinical History of the Medical and Surgical Diseases of Women.* London: J. and A. Churchill, 1873.

Bell, Susan. 'Changing ideas: the medicalization of menopause.' *Social Science and Medicine* 24: 535–543, 1987.

Beyene, Yewoubdar. *From Menarche to Menopause: Reproductive Lives of Peasant Women in Two Cultures.* Albany: State University of New York Press, 1989.

Crozier, Ralph C. 'The ideology of medical revivalism in modern China.' In *Asian Medical Systems: A Comparative Study*, Charles Leslie, ed. Berkeley: University of California Press, 1976, pp. 341–355.

Daniel, Valentine. *Is There a Counterpoint to Culture*? The Wertheim Lecture 1991. Amsterdam: Centre for Asian Studies, 1991.

Dominquez, Virginia R. 'Invoking culture: the messy side of "cultural politics".' *South Atlantic Quarterly* 91: 19–42, 1992.

Eto, Jun. 'The breakdown of motherhood is wrecking our children.' *Japan Echo* 6: 102–9, 1979.

Geertz, Clifford. 'Culture war.' *The New York Review of Books* 42: 4–6, 1995.

Hacking, Ian. 'Making up people.' In *Reconstructing Individualism*, Thomas Heller, Morton Sosna and David E. Wellberry, eds. Stanford: Stanford University Press, 1986, pp. 222–236.

Ikeda, Mitsuho. 'Kindai byōin no naka no dentōteki 'shi':Makki kanja to kōzō ka sareta patānarizumu.' In *Jirei o chūshin to shita tāminaru kea*, Kazuyo Yotsumoto and Reiko Kawaguchi, eds. Tokyo: Hirokawa Shuppan, 1993, pp. 11–26.

Hayashi, C. *Nihonjin no honshin o sagareba hyaku pasento gan kokuchi nado higenjitsuteki* (If we look at the true feelings of Japanese, one hundred percent cancer disclosure is unrealistic). Nihon no Ronten 1997. Tokyo: Bungeishunjū, 1996, p. 455.

Halford, Henry. 'On the climacteric disease.' In *Medical Transitions*, vol. 4. Royal College of Physicians of London. London: Longman, 1813.

Ikegami, Naoki. 'Health technology development in Japan.' *International Journal of Technology Assessment in Health Care* 4: 239–254, 1988.

Kuriyama, Shigehisa. 'Between mind and eye: Japanese anatomy in the eighteenth century.' In *Paths to Asian Medical Knowledge*, Charles Leslie and A. Young, eds. Berkeley: University of California Press, 1992, pp. 21–43.

Laqueur, Thomas. *Making Sex: Body and Gender from the Greeks to Freud*. Cambridge, Massachusetts: Harvard University Press, 1990.

Leflar, Robert B. 'Informed consent and patients' rights in Japan.' *Houston Law Review* 33(1): 1–112, 1996.

Leslie, Charles. 'The ambiguities of medical revivalism in modern India.' In *Asian Medical Systems: A Comparative Study*, Charles Leslie, ed. Berkeley: University of California Press, 1976, pp. 356–367.

Lock, Margaret. *East Asian Medicine in Urban Japan: Varieties of Medical Experience*. Berkeley: University of California Press. Paperback edition, 1980, with a new introduction, 1984.

Lock, Margaret. 'A nation at risk: interpretations of school refusal in Japan.' In *Biomedicine Examined*, M. Lock and D.R. Gordon, eds. Dordrecht: Kluwer Academic Publishers, 1988, pp. 391–414.

Lock, Margaret. *Encounters with Aging: Mythologies of Menopause in Japan and North America*. Berkeley: University of California Press, 1993.

Lock, Margaret. 'On dying twice: culture, technology and the determination of death.' In *Living and Working with the New Medical Technologies: Intersections of Injury*, M. Lock, A. Young and A. Cambrosio, eds. Cambridge: Cambridge University Press, 2000.

Lock, Margaret. *Twice Dead: Organ Transplant and the Reinvention of Death*. Berkeley: University of California Press, 2002.

Lock, Margaret and Patricia Kaufert. 'Local biologies, material difference, and enhancement genetics.' *Human Biology*, 2001.

Lock, Margaret *et al.* 'Cultural construction of the menopausal syndrome: the Japanese case.' *Maturitas* 10: 317–332, 1988.

Long, Susan Orpett. 'Family surrogacy and cancer disclosure: physician-family negotiation of an ethical dilemma in Japan.' *Journal of Palliative Care* 15(3): 31–42, 1999.

Long, Susan Orpett. 'Public passages, personal passages, and reluctant passages: notes on investigating cancer disclosure practices in Japan.' *Journal of Medical Humanities* 21(1): 3–13, 2000.

Long, Susan Orpett. 'Living poorly or dying well: cultural decisions about life-supporting treatment for American and Japanese patients.' *Journal of Clinical Ethics* 11(3): 236–250, 2000.

Long, Susan O. and Bruce D. Long. 'Curable cancers and fatal ulcers: attitudes toward cancer in Japan.' *Social Science and Medicine* 16: 2101–2108, 1982.

Low, Morris F. 'Medical representations of the body in Japan: gender, class and discourse in the eighteenth century.' *Annals of Science* 53: 345–359, 1996.

Mainichi Daily News. *Interview with Takuro Kobayashi, Pill Researcher*. 23 February, 1987.

Ministry of Health and Welfare (Kōseishō). Sorifu (Office of the Prime Minister). *Seron Chōsa Nenpō (Yearbook of Public Opinion Polls)*. Tokyo: Office of the Prime Minister, 1994, p. 466.

Mochida, Takeshi. 'Editorial comment: Focus on the family.' *Japan Echo* 3: 75–76, 1980.

Nihon Ishoku Gakkai Shakai Mondai Kentō Tokubetsu Iinkai, ed. *Zōki Ishōku e no apurōchi IV: Zōki Ishoku to Hōdō; Zōki Ishoku to rinri i inkai* (Approches to Organ Transplant IV: Organ Transplant and Media Report, Organ Transplant and Ethics Committee). Osaka: Medika Shuppan, 1991.

Nishiyama, Hideo. *Josei to kanpō*. Osaka: Sogensha, 1981.

Pollock. *Reading Against Culture: Ideology and Narrative in the Japanese Novel*. Ithaca: Cornell University Press, 1992.

Rothman, David J. *Strangers at the Bedside: A History of How Law and Bioethics Transformed Medical Decision Making*. New York: Basic Books, 1991.

Rosenberger, Nancy. 'The process of discourse: usages of a Japanese medical term.' *Social Science and Medicine* 34: 237–247, 1992.

Ross, P.D. *et al.* 'A comparison of hip fracture incidence among native Japanese, Japanese Americans, and American Caucasians.' *American Journal of Epidemiology* 133: 801–809, 1991.

Shea, J.L. *Revolutionary Women at Middle Age: An Ethnographic Survey of Menopause and Midlife Aging in Beijing, China.* Ph.D. Dissertation, Cambridge, Massachusetts: Harvard University, 1998.

Sheehan, Donald. 'Discovery of the autonomic nervous system.' *AMA Archives of Neurology and Psychiatry* 35: 1081–1115, 1936.

Shorr, Ephraim. 'The menopause.' *Bulletin of the New York Academy of Sciences* 16: 453–74, 1940.

Simionoff, Laura A. and Kata Chillag. 'The fallacy of the "gift of life".' *Hastings Center Report* 29: 34–41, 1999.

World Health Organization. *World Health Statistics: 1990.* Geneva: World Health Organization, 1990.

Youngner, Stuart *et al.* 'Psychological and ethical implications of organ retrieval.' *New England Journal of Medicine* 313: 321–324, 1985.

DAYALAN DEVANESEN AND PATRICK MAHER

TRADITIONAL ABORIGINAL HEALTH PRACTICE IN AUSTRALIA[1]

Australia is the only continent to have been occupied exclusively by nomadic hunters and gatherers until recent times. Carbon dating of skeletal remains proves that Australian Aboriginal history started some 40,000 years ago, long before Captain Cook landed on the eastern coast in 1770. This history is not completely lost. It is retained in the minds and memories of successive generations of Aboriginal people, passed on through a rich oral tradition of song, story, poetry and legend. According to Aboriginal belief all life forms – human, animal, plant and mineral – are part of one vast unchanging network of relationships which can be traced to the great spirit ancestors of the Dreamtime.

The Dreamtime continues as the "Dreaming" or *Jukurrpa* in the spiritual lives of Aboriginal people today. The events of the Dreamtime are enacted in ceremonies and dances and chanted incessantly to the accompaniment of didgeridoo or clapsticks (Isaacs, 1980). The Dreaming is the source of the rich artistry, creativity and ingenuity of the Aboriginal people.

In Australia, Western health services have been superimposed on traditional Aboriginal systems of health care. However, these traditional systems have survived despite the dramatic influence of cultural contact, and Aboriginal medicine is still widely practised in the Northern Territory today (Tynan, 1979; Nathan and Japanangka, 1983; Reid, 1983; Soong, 1983; Devanesen, 1985; Maher, 1999). The Northern Territory occupies one sixth of the Australian land mass. It has a population of just under 200,000 of which 28% are Aboriginal. The vast majority of the Aborigines live in small remote communities scattered across the Territory.

Traditional Aboriginal medicine is a complex system closely linked to the culture and beliefs of the people and to knowledge of their land and its flora and fauna. Its survival is explained by its "embeddedness" in the social fabric of Aboriginal culture. Reid (1983) has shown that, although Aborigines living at Yirrkala in the Northern Territory choose Western biomedicine to treat the majority of their sicknesses, they continue to explain the causes of these sicknesses through their traditional beliefs. This may be because the health beliefs

H. Selin (ed.), Medicine Across Cultures: History and Practice of Medicine in Non-Western Cultures, 175–189.

continue to play a role in providing meaning to events and thereby helping people to cope with serious illness and death (Reid, 1983).

The Aboriginal approach to health care is a holistic one. It recognises the social, physical and spiritual dimensions of health and life. Their concept of health in many ways is close to the World Health Organisation definition of health, "a state of complete physical, mental and social well being and not merely the absence of disease or infirmity". The Warlpiri Aboriginal tribe have described health as "life" or *Wankaru.* Their definition takes in a whole of life cycle. The front of their Health Centre at the Aboriginal settlement of Yuendumu is adorned with the painting in Figure 1. It shows family life, food, shelter, warmth, water and exercise, all of which are essential for health (Devanesen, 1983).

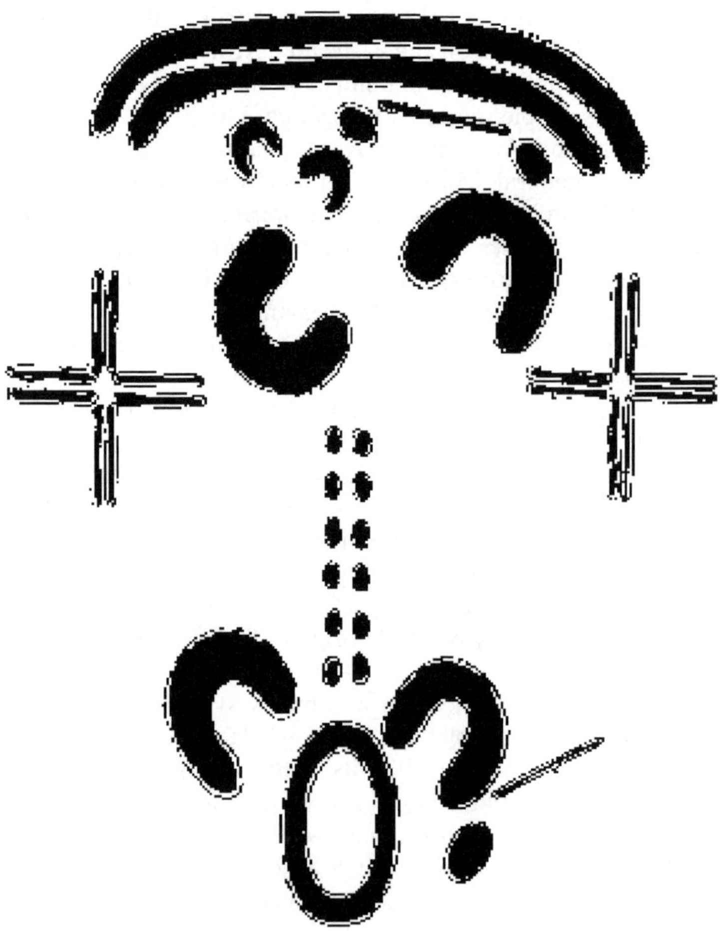

Figure 1 Health symbol for Yuendumu: *Napangardi* Women's Jukurrpa or Dreamtime.

TRADITIONAL INDIGENOUS HEALTH SYSTEMS

The traditional health beliefs of Aboriginal people are interconnected with many aspects of Aboriginal life such as the land, kinship obligations, and religion (Tynan, 1979). The sociomedical system of health beliefs held by Aboriginal people places emphasis on social and spiritual dysfunction causing illness. This approach emphasises that "individual well-being is always contingent upon the effective discharge of obligations to society and the land itself" (Morgan *et al.*, 1997: 598). People's social responsibilities and obligations may take precedence over their own health because of the priority given to social relationships in this model.

Sorcery and supernatural intervention are part of the perceived reality of Aboriginal life (Waldock, 1984), and in Aboriginal society explanations in terms of sorcery are often used. The deaths of infants or the very old or chronically ill are considered to be in the normal course of events, while deaths outside these groups may have a supernatural influence, especially if they are regarded as premature, unexpected and sudden (Reid and Mununggurr, 1977).

There are many beliefs associated with supernatural interventions and sorcery:

- Sorcery exists in many forms. Its effect is to manipulate and alter behaviour and cause morbidity and mortality.
- Sorcerers can be specialists or non specialists.
- Distant groups have the most virulent sorcery and are the most feared.
- Many diseases come from dangerous, secret sacred sites. They are manifestations of the forces or power emanating from those sites.
- Unskilled or uninitiated people may release forces from a dangerous site by disturbing the site.
- Sorcery is carried out in secrecy.
- Retribution sorcery is directed serially at members of a family or lineage. Therefore, the serious illness or death of one member is followed by the illness and death of others.
- A traditional healer can apply counter measures to identify the cause and source of illness and death, but the healer should not interfere if it is the result of legitimate punishment (Bienoff, 1982).

The ill effects of sorcery will not necessarily be felt only by the "offender", but may also be felt by his or her family and descendants. While the thought of sorcery is prominent in Aboriginal life, people do not live in constant fear of it. Sorcery is usually an explanation that is applied retrospectively to explain deaths, serious illness or injury (Reid and William, 1984).

The concept of supernatural intervention and sorcery plays an important function as "it explains why one person and not another died or became ill at a certain time and not at another" (Reid and Mununggurr, 1977: 39). It provides the explanations of "why me" and "why now" which cannot be answered in terms of Western medical theory. It provides the answer to the ultimate cause of the event.

Many have acknowledged the cultural diversity, particularly in regard to health beliefs, between Aboriginal groups and communities (Reid, 1983; Elkin, 1994). We have therefore chosen to describe the features of the traditional indigenous medical system by particular reference to one tribe, the Warlpiri.

TRADITIONAL HEALTH SYSTEM OF THE WARLPIRI ABORIGINES

The Warlpiri Aborigines comprise one of the largest tribes in the Northern Territory. They are scattered over many Aboriginal communities in the north-west of Central Australia. The main components of the Warlpiri health system are the *ngangkayikirili* or traditional healers, commonly referred to as *ngangkari* or *ngangkayi* (healing power), *Yawulyu* ceremonies; healing songs and herbal medicine. In addition there are laws governing behaviour that are aimed at preventing sickness.

Ngangkari

Traditional healers have a variety of roles, including providing strong spiritual and social support (Reid, 1983; Soong, 1983); determining the ultimate cause of a serious illness or injury (Nathan and Japanangka, 1983; Reid, 1983; Peile, 1997; Tynan, 1979); determining cause of deaths at an "inquest" (Nathan and Japanangka, 1983; Reid, 1983; Peile, 1997; Tynan, 1979); and employing counter sorcery to remove the evil influences causing illness (Reid, 1983; Tynan, 1979). They are believed to have many powers to undertake these roles and utilise numerous different healing techniques (see Table 1).

Professor Elkin (1994) referred to the traditional healers as Aboriginal men of high degree. The healers are kindred to Amerindian "men of power" and shamans. These healers are specially chosen and trained to remove the influence of sorcery and evil spirits and to restore the well being of the soul or spirit. Their role is extremely important because most serious illness is thought to be brought about by loss of a vital substance from the body (soul loss), introduction of a foreign and harmful substance into the body (spirit intrusion or possession), or violation of taboos and sorcery (singing). The traditional healers usually gain the power to heal through inheritance or through special spiritual experiences. They possess a spirit called *mapanpa* which is associated with healing power. This is different from the spirit that every Warlpiri person has "like a shadow" (Tynan, 1979).

The traditional healer carries out a healing ritual which often includes sucking the sick person. After sucking, the healer usually spits out a wooden object called *yarda* which is covered in blood. The *yarda* represents the evil influence. Sometimes, the traditional healer massages the patient, manipulates the body or sings during the ritual. The traditional healer may diagnose the state of the spirit, e.g. *kurrunpa yulangu* (the spirit is sad). The traditional healers do not use herbal medicine in their practice.

Yawulyu ceremonies and healing songs

Warlpiri women frequently perform *Yawulyu* ceremonies. These ceremonies improve the health of sick people but cannot remove the influence of sorcery.

Table 1 The powers and healing techniques of Aboriginal traditional healers (Sources: Nathan and Japanangka, 1983; Reid, 1983; Elkin, 1994; Peile, 1997; Cawte, 1974; Taylor, 1977; Tonkinson, 1982; Toussaint, 1989; Berndt, 1964; Soong, 1983; Brady, 1995; Hunter, 1993; Eastwell, 1973). From Maher, 1999. Used with the permission of Blackwell Science Asia.

Powers/capabilities	Healing techniques
– have assistance from the spirit world to assist healing	– provide positive emotional support
– telepathy	– physical contact and touching
– divination	– massage
– X-ray vision	– extraction of objects causing illness from a patient's body via sucking, rubbing or massage
– clairvoyance	– smoking used ritually (after initiation or death) or therapeutically for mothers and babies after childbirth
– telesthesia	– use of healing songs and chanting
– control the weather	– counseling
– mind reading	– using objects with healing powers
– walk on fire	– dreaming
– magically heal wounds including internal wounds	– inserting healing objects into patient's body
– travel at fast pace	– cleansing the patients' internal organ(s) while they sleep
– protect people against attack from spirits via counter sorcery	
– commune with spiritual beings including the dead	
– manipulate a patient's spirit	
– spirit travel	
– able to project their spirit into the body of patient	
– able to replace patient's blood with their own	
– able to communicate with the Spirit World	

The ceremony consists of singing songs and painting designs on the sick person. These designs are derived from the power of the Dreamtime (see Figure 2). Each ritual is carried out by the *kirda* (owners) or *kurdungurlu* (managers) of a particular "Dreaming". Sometimes the songs and designs appear to the people in their dreams and are thought to be revealed by spirit creatures called *yinawuru* (Munn, 1973). During the ceremony the sick person may be massaged with fat and red ochre. These materials derive special potency from the songs. In some cases senior men and women sing songs without the ceremony to strengthen sick people. Songs are sometimes sung to ensure safe childbirth. The *Yawulyu* ceremonies and songs assist in providing strong family support for the sick person.

Herbal medicine

Herbal medicine and knowledge of plants is not the domain of any particular group in the Warlpiri system. The whole family shares its knowledge and use. The Warlpiri have extensive knowledge of plants and have published their own book which lists several plants and their medicinal uses (Henshall *et al.*, 1980).

Medicinal plants are mainly used symptomatically for coughs and colds,

Figure 2 Yawulyu designs on women – yam dreaming.

pains and aches. Some are used as dressings for wounds and sores. Herbal medicine was the first component of the Warlpiri health system to be eroded by the introduction of Western medicine. However, the movement of the Warlpiri people back to their traditional land has led to a renewal of interest in the use of herbal medicine.

The Warlpiri health system

The Warlpiri health system can be represented as in Figure 3.

When someone falls sick one of the three main components of the health system is tried. If it does not work, another component is used or the same component tried again until there is a definite outcome.

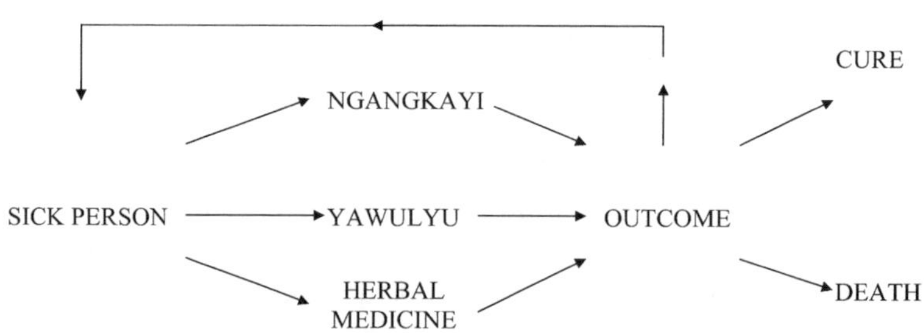

Figure 3 The Warlpiri health system (adapted from Tynan, 1979).

Preventing sickness

Reid (1982) gives a good description of prevention at the level of personal relationship and religious injunctions.

> Preventive measures can include avoiding foods prohibited during ceremonies or life crises, obeying ritual proscriptions, taking care not to abuse ones' [sic] land or trespass on territories of others, avoiding prohibited sacred sites or approaching them with ritual protection, observing debts and obligations to others, containing anger, violence or jealousy, exercising caution in interactions with strangers and taking steps to avoid sorcery or often conflict with others.

The methods of preventing illness link in directly with what are regarded as the ultimate causes of illness under the Aboriginal model of causation of illness. In summary, good health is associated with strict adherence to approved patterns of behaviour and avoidance of dangerous places, people and objects (Biernoff, 1982).

THE PLACE OF WESTERN MEDICINE

Western medicine has been incorporated into this system at the same level as herbal medicine. By doing this, the Warlpiri are able to retain their belief in spirit causation of illness while using Western medicine for the relief of symptoms.

Changes are taking place in the illness-related beliefs of Aborigines in the Northern Territory. Reid's study at Yirrkala shows that this change is characterised by the "gradual addition of causes and elaboration of the existing causes within the aetiological domain". Reid lists three categories of causes: (1) social and spiritual causes, e.g., sorcery or breaking the law; (2) causes other than social or spiritual, e.g., emotional state, old age, assault; (3) emergent causes, e.g., alcohol, sin, smoking (Reid, 1983).

Aboriginal people use the strategy of domain separation to divide illness into Aboriginal and Western causes. This involves thinking in terms of separated cultural or social domains and deciding when to apply the rules of each (Harris, 1988). This strategy occurs not only in the area of health beliefs but also in other areas to respond to cultural uncertainty, to reduce social complexity and stress, and to deal with social dilemmas (Harris, 1988).

There are a variety of behavioural patterns of seeking medical assistance that traditional people use during illness:

- sequential (use one practitioner then another kind, e.g., Western then traditional or vice versa);
- compartmental (using traditional medicine for conditions which have established traditional explanations); and
- concurrent (concurrent use of traditional and Western forms of health care (Armstrong and Fitzgerald, 1996).

Aboriginal people use all these patterns (Reid, 1983; Peile, 1997; Tynan, 1979; Elliot, 1984; Tonkinson, 1982; Gray, 1979; Berndt, 1964). This model is best expressed in Figure 4.

Generally combinations of traditional and Western medicines are used (Reid,

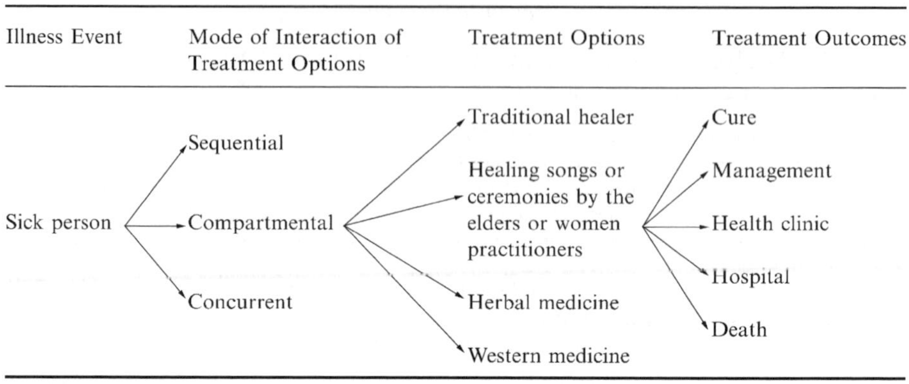

Figure 4 Model of Aboriginal behavioural patterns of seeking medical assistance. From Maher, 1999. Used with the permission of Blackwell Science Asia.

1983; Tynan, 1979). Western medicine may relieve symptoms and provide explanations to the mechanism of how something occurred while traditional explanations provide the reason it occurred and the ultimate cause. In cases of supernatural intervention Western medicine is used to treat the symptoms and to hasten the cure, provided it does not conflict with traditional beliefs, but it is not able to remove the cause of the illness (Tynan, 1979; Waldock, 1984). In contrast only Western medicine can affect emergent Western illnesses (Waldock, 1984; Scrimgeour *et al.*, 1997).

In spite of this attempt to incorporate Western medicine into the traditional system, there are areas of conflict. Western medicine is based upon particular Western explanatory models. Variation in the underlying beliefs, assumptions and general medical information has been implicated as the basis for the conscious rejection of Western health care by some Aborigines (Hamilton, 1974). It is well known that differences in underlying knowledge systems impede even willing compliance between culturally divergent groups. The lack of a common conceptual framework within which patient and practitioner can interact may result in decreased compliance and satisfaction (Maher, 1999).

DEVELOPMENT OF TRADITIONAL PRACTICE IN HEALTH SERVICES TO ABORIGINAL PEOPLE

There has been government support with recognition for Traditional Aboriginal Medicine since the 1970s. A report on Aboriginal health by the Australian Parliament recommended that,

> Aboriginal cultural beliefs and practices which affect their health and their use of health services such as their fear of hospitalisation, their attitudes to pain and surgery, the role of traditional healers and the differing needs and roles of Aboriginal men and women, be fully taken into account in the design and implementation of health care programs (Commonwealth of Australia, 1979).

Bicultural medicine

Two-way medicine is the term that has been coined by Aboriginal health workers to describe a bicultural approach to health care. It is based on the principle that "if you can use what is best in modern medicine together with what is best in traditional healing, the combination may be better than either one alone" (Werner, 1977). The Northern Territory Department of Health's first policy on Aboriginal health stated that "traditional medicine is a complementary and vital part of Aboriginal health care, and its value is recognised and supported" (Northern Territory Department of Health, 1982). The Northern Territory Department of Health over the years has established several programs that recognise the traditional health system, Aboriginal values and beliefs.

Support for traditional healers/Ngangkari

Traditional healers were employed by the Northern Territory Department of Health at various rural health centres in Central Australia in the early 1970s. A training course to teach traditional healers about Western medical practices was attempted in 1974. It was soon realised that it would be better to train a separate group as Aboriginal health workers and to leave the traditional healers to their vitally important roles (Devanesen and Briscoe, 1980). The employment of traditional healers was ceased, and a training program for Aboriginal health workers commenced. However, rural health centres continue to recognise and cooperate with traditional healers in the management of sick people. A meeting of *Ngangkaris* was held near Ayers Rock in April 2000. Over 40 *Ngangkaris* gathered together to discuss the production of an information manual about their work, history and traditions (Mullins, 2000).

Some remote Health Centres run by Aboriginal Community Controlled organisations have recommenced the employment of traditional healers. The healers often act also as consultants for determining culturally appropriate service delivery.

A recent proposal to establish an Aboriginal Healing Centre close to the Alice Springs Hospital is being examined. The centre would develop a place to promote spiritual health supported by a network of traditional Aboriginal healers.

Aboriginal health workers

There are over three hundred Aboriginal health workers in the Northern Territory today. The Aboriginal Health Worker Training Program is recognised as one of the main strategies for improving Aboriginal health. Aboriginal health workers are selected by their own communities and trained in various Western medical skills. They have proved highly successful in treating common health problems in Aboriginal communities, such as diarrhoea, chest infections, trachoma, and ear and skin infections. They also act as mediators between Western and traditional medical systems. The Northern Territory has registered

Aboriginal Health Workers through the Health Practitioners and Allied Professionals Registration Act 1985.

Aboriginal Health Workers bridge the "cultural chasm" separating the traditional and Western worldviews. They relate Western beliefs to an Aboriginal conceptual framework, making it possible for Aboriginal patients to understand what is being said and to assess the validity of the statements. They make it possible for the health centre teams to communicate with Aborigines in language and concepts that they understand. In most Aboriginal communities, the people's point of entry into the Western health system is through the Aboriginal health worker who may refer them to a nurse or doctor or back to the traditional system as shown in Figure 5.

The Bush Medicine or Herbal Medicine Program

Aboriginal expertise regarding plants has been acknowledged for many years. Webb (1969) has shown that many Aboriginal bush medicines contain biologically active compounds. Bitter Bark (*Alstonia constricta*), used to prepare a tonic, contains reserpine, a tranquilliser and antihypertensive. Plants used on sores and wounds contain proteolytic enzymes that help healing. Spilanthes, a native daisy used to treat toothache, has been shown to contain spilanthol – a local anaesthetic. Over half the world's supply of the drugs hyoscine and scopolamine come from an Australian native tree, Duboisia, which was used by Aborigines as an emu and fish poison (Pearn, 1981).

In 1973, the Northern Territory Department of Health started collecting information regarding the Aboriginal use of plants. The collection was systematised in 1979, and by 1982, over fifty different medicinal plants had been recorded. The Aboriginal health workers and tribal elders were the key personnel in this program. Doctors, nurses, pharmacists and botanists have also been involved. This program has helped non-Aboriginal staff appreciate the great knowledge and complexity of the Aboriginal health system (Devanesen and Henshall, 1982).

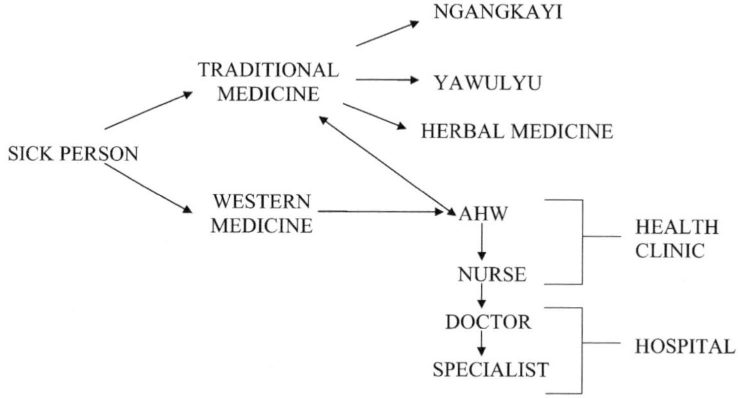

Figure 5 Aboriginal medicine and Western medicine.

The Aboriginal Pharmacopoeia

In 1988 the first Aboriginal Pharmacopoeia for the Northern Territory was published. This major work with beautiful photographs lists 70 plants and six other natural substances used by Aborigines. Forty Aboriginal communities contributed to the pharmacopoeia by describing and demonstrating the use of their herbal remedies. All the plants were carefully identified by botanists and then screened for essential oils, minerals, saponins, tannins, triterpenes and steroids (Aboriginal Communities of the Northern Territory, 1988).

The main conditions that herbal remedies were used for reflect the types of sickness that Aborigines had before contact with Western society. They were: joint and muscle pain, toothache and sore mouth, gastro-intestinal disorders, symptoms of colds and flu, e.g. fever, headache and ill-defined pain, congestion, cough, general malaise, sore throat, sores, boils, cuts, scabies, bites, stings, burns and major wounds, warts, allergy rash and itchy skin disorders, ringworm, other tinea form skin infections, eye disorders and fever.

Ongoing work in Aboriginal herbal medicine

In 1995 staff at a remote Health Centre in the Northern Territory carried out a study to compare the effectiveness of wound healing by the use of a traditional remedy, Bauhinia root (*Lysiphyllum cunninghamii*) and a Western preparation in the treatment of boils, sores and scabies. The study concluded that the herbal medicine was as effective as the Western preparation. In addition the Aboriginal people felt more comfortable using the traditional remedies and felt a sense of pride in their own traditional knowledge and culture (McLean *et al.*, 1996).

Herbal remedies continue to be studied in the Northern Territory. The Conservation Commission of the Northern Territory put out their own publication called *Traditional Aboriginal Medicines In the Northern Territory* that lists over 50 herbal remedies. There has been some special interest in plants used by Aborigines to control fertility and lactation.

The Bush Food Program

Aboriginal people often view food as their medicine. Many foods are known to strengthen the body against sickness or promote healing. Some prized foods such as the witchety grub (*Xyleutes*) are crushed and used for treatment of burns and wounds. The grubs are nutritious as well, with protein (15.1%), fat (19.2%), 100 mgs thiamine and 5mg vitamin C per 100 gms (see Figure 6).

In 1981, the Northern Territory Department of Health launched the Bush Food Program which sought to establish a durable record of traditional Aboriginal food practices and beliefs and develop a more relevant and acceptable style of nutrition education. This program, initiated by departmental dieticians and Aboriginal health workers, has come a long way, and an exciting nutrition education program has been developed.

A publication on the nutritional composition of forty-two bush foods collected through this program has some interesting results (Brand *et al.*, 1983).

Figure 6 Witchety grub (*Xyleutes*).

The green plum (*Terminiala ferdinandiana*) contains 3150 mg/100 g of Vitamin C, making it the richest source of Vitamin C in the world. The bush banana (*Leichardtia australis* and *L. leptophylla* [see Figure 7]) and the water lily root (*Nymphae macrosperm*) contain very high proportions of protein, and the list goes on.

Like the Bush Medicine Program, the Bush Foods Program has led to the stimulation of reciprocal learning processes between two cultures and the self-examination of attitudes and values.

* * *

The current health status of Aboriginal people is characterised by unacceptable levels of morbidity and mortality. Aboriginal life expectancy is twenty years less than other Australians'; Western medicine has not solved many of the Aboriginal health problems.

Traditional medicine is part of Aboriginal culture. Its recognition can bolster the self-confidence of Aboriginal people and improve the delivery of health services to their communities. Two-way medicine needs to be supported and developed with ongoing research to evaluate the therapeutic value of traditional medicine. The increasing worldwide popularity and use of complementary and alternative medicine may assist in the development and sustainability of Aboriginal traditional medicine and healing in Australia.

Figure 7 Bush banana (*Leichardtia australis*).

ACKNOWLEDGEMENT

All references to the Warlpiri tribe are with the kind permission and consent of the Yuendumu Council.

NOTE

[1]This paper contains photographs of people from the Warlpiri tribe of Central Australia. It may cause distress to people from this region to view the photograph if these people have subsequently died.

BIBLIOGRAPHY

Abbott, K. *Proposed Site of Development Ngalka Jana.* Unpublished, 1998.

Aboriginal Communities of the Northern Territory of Australia. *Traditional Bush Medicines. An Aboriginal Pharmacopoeia.* Greenhouse Publication, 1998.

Armstrong, M.J. and M.H. Fitzgerald. 'Culture and disability studies: an anthropological perspective.' *Rehabilitation Education* 10: 247–304, 1996.

Berndt, CH. 'The role of the native doctor in Aboriginal Australia.' In *Magic, Faith and Healing: Studies in Primitive Psychiatry Today,* Kiev A., ed. New York: The Free Press, 1964, pp. 264–282.

Biernoff, D. 'Psychiatric and anthropological interpretations of "aberrant" behaviour in an Aboriginal community.' In *Body, Land and Spirit. Health and Healing in Aboriginal Society,* Janice Reid, ed. Brisbane: University of Queensland Press, 1982, pp. 139–153.

Brady, M. 'Culture in treatment, culture as treatment. A critical appraisal of developments in addictions programs for Indigenous Northern Americans and Australians.' *Social Science and Medicine* 41: 1487–1498, 1995.

Brand, J.C. *et al.* 'The nutritional composition of Australian Aboriginal bushfoods.' *Food Technology in Australia* 6: 293–298, 1983.

Cawte, J. *Medicine is the Law: Studies in Psychiatric Anthropology of Australian Tribal Societies.* Honolulu: University Press of Hawaii, 1974.

Commonwealth of Australia. *Aboriginal Health.* House of Representatives Standing Committee on Aboriginal Affairs. Canberra: Australian Government Publishing Service, 1979.

Devanesen, D. 'Traditional art in the health worker training program.' *The Aboriginal Health Worker* 7(3): 4–8, 1983.

Devanesen, D and T.S. Henshall. 'A study of plant medicines in Central Australia.' *Transactions of the Menzies Foundation* 4: 161–166, 1982.

Devanesen, D. 'Traditional Aboriginal medicine and bicultural approach to health care in Australia's Northern Territory.' Proceedings of the 2nd National Drug Institute, Alcohol and Drug Foundation, Canberra, 1985.

Eastwell, H. 'The traditional healer in modern Arnhem Land.' *Medical Journal of Australia* 2: 1011–1017, 1973.

Elkin, A.P. *Aboriginal Men of High Degree: Initiation and Sorcery in the World's Oldest Tradition.* Brisbane: University of Queensland Press, 1994.

Elliot, D.M. *Aboriginal Perceptions of Disability and the Formulation of an Appropriate Method of Providing Rehabilitation Services to Clients on Remote Communities.* Darwin: Commonwealth Rehabilitation Service, 1984.

Gray, D. 'Traditional medicine on the Carnarvon Aboriginal Reserve.' In *Aborigines of the West: Their Past and Their Present*, R.M. Berndt and C.H. Berndt, eds. Nedlands: University of Western Australia Press, 1979, pp. 169–182.

Hamilton, A. 'The traditionally oriented community.' In *Better Health for Aborigines*, B.S. Hetzel *et al.*, eds. Brisbane: University of Queensland Press, 1974.

Harris, S. '"Coming up level" without "losing themselves": the dilemma of formal tertiary training for Aborigines.' In *Learning My Way: Papers from the National Conference on Adult Aboriginal Learning*, B. Harvey and S. McGinty, eds. Perth: Institute of Applied Aboriginal Studies, 1988, pp. 169–187.

Henshall, T. *et al. Ngurrju Maninja Kurlangu. Yapa Nyurnu Kurlangu Bush Medicine.* Yuendumu: Warlpiri Literature Production Centre, Inc., 1980.

Hunter, E. *Aboriginal Health and History: Power and Prejudice in Remote Australia.* Cambridge: Cambridge University Press, 1993.

Isaacs, J. *Australian Dreaming.* Sydney: Lansdown Press, 1980.

Maher, P. 'A review of "traditional" Aboriginal health beliefs.' *Australian Journal of Rural Health* 7: 229–236, 1999.

McLean, M., W. Dow, R. Bathern *et al. A Study of the Comparison Between the Traditional Aboriginal Medicines and Western Preparations in the Treatment and Healing of Boils, Sores and Scabies.* Unpublished, 1996.

Morgan, D.L., M.D. Slade and C.M.A. Morgan. 'Aboriginal philosophy and its impact on health care outcomes.' *Australian and New Zealand Journal of Public Health* 21: 597–601, 1997.

Mullins, M. *Personal Communication,*2000.

Munn, N.D. *Warlpiri Iconography.* Ithaca, New York: Cornell University Press, 1973.

Nathan, P. and D.L. Japanangka. *Health Business.* Victoria: Heinmann Educational, 1983.

Northern Territory Department of Health. *Annual Report 1981/1982.* Darwin: Government Printer of the Northern Territory, 1982.

Pearn, J. 'Corked up. Clinical hyoscine poisoning with alkaloids of the native corkwood, Duboisia.' *Medical Journal of Australia* 2: 422–423, 1981.

Peile, A.R. *Body and Soul: An Aboriginal View.* Carlisle: Hesperian Press, 1997.

Reid, J. *Body, Land and Spirit. Health and Healing in Aboriginal Society.* Brisbane: University of Queensland Press, 1982.

Reid, J.C. *Sorcerers and Healing Spirits.* Canberra: Australian National University Press, 1983.

Reid, J. and D. Mununggurr. 'We are losing our brothers: sorcery and alcohol in an Aboriginal community.' *Medical Journal of Australia* Supp. 2: 1–5, 1978.

Reid, J. and N. Williams. '"Voodoo death" in Arnhem Land. Whose reality?' *American Anthropologist* 86: 121–133, 1984.

Scrimgeour, D., T. Rowse and A. Lucas. *Too Much Sweet. The Social Relations of Diabetes in Central Australia.* Darwin: Menzies School of Health Research, 1997.

Soong, F.S. 'Role of the margidbu (traditional healer) in western Arnhem Land.' *Medical Journal of Australia* 1: 474–477, 1983.

Taylor, J.C. 'A pre-contact Aboriginal medical system on Cape York Peninsula.' *Journal of Human Evolution* 6: 419–432, 1977.

Tonkinson, M. 'The mabarn and the hospital: the selection of treatment in a remote Aboriginal community.' In *Body, Land and Spirit: Health and Healing in Aboriginal Society*, J. Reid, ed. Brisbane: University of Queensland Press, 1982, pp. 225–241.

Toussaint, S. 'Aboriginal and non Aboriginal healing, health and knowledge: sociocultural and environmental issues in the West Kimberley.' *Aboriginal Health Information Bulletin* 12: 30–35, 1989.

Tynan, B.J. *Medical Systems in Conflict. A Study of Power.* Darwin: Government Printer of the Northern Territory, 1979.

Waldock, D.J. 'A review of Aboriginal health beliefs and their incorporation into modern Aboriginal health delivery systems.' *Australian Health Surveyer* 16: 3–13, 1984.

Webb, L.J. 'Australian plants and chemical research.' Offprint from *The Last of the Lands.* Milton, Queensland: The Jacaranda Press, 1969.

Werner, D. *Where There Is No Doctor.* Palo Alto, California: Hesperian Foundation, 1977.

CLUNY MACPHERSON AND LA'AVASA MACPHERSON

WHEN HEALING CULTURES COLLIDE: A CASE FROM THE PACIFIC

Over a period of some two hundred years, the indigenous cultures and societies of the Pacific have been exposed to other cultures imported by missionaries, colonists, capitalists and administrators. Those Europeans who settled in the region did so with the express purpose of transforming the people they encountered, albeit by different means and for somewhat different reasons. The newcomers used force, political and religious indoctrination and involvement in commerce, and persuasion and intermarriage. Their motives varied from altruism to commercial and personal gain. A common perception beyond the region is that in these "collisions of culture", various Western cultures successfully dominated and displaced the indigenous ones. In the case of healing and medicine, it is frequently supposed that the superiority of a healing system founded on Western biomedicine rapidly displaced a healing system based in religion, superstition and local plants. This view oversimplifies what happened in the Pacific and the process by which it happened. In fact, while Western biomedicine is well established and widely used in Pacific society, it has not displaced commitment to indigenous medicine.

THE PACIFIC CONTEXT

Some 5500 years ago, a group of explorers left what is now southern China and moved down the Malay peninsula before setting off to explore and settle the more than ten thousand islands dotted across the Pacific Ocean, which covers one third of the earth's surface. Their journeys involved a series of longer and longer moves which became possible as their shipbuilding skills and navigation technologies evolved (Lewis, 1972). The earliest journeys were made on rafts and in dugout canoes made from a single log; they travelled over short distances during which both the departure point and destination were visible at any one time. Later journeys involved larger sail-powered outrigger canoes over longer distances. These used celestial navigation and required knowledge of sea currents and meteorology to make return journeys over larger areas of ocean. At its zenith, Pacific voyaging involved huge, double-

H. Selin (ed.), Medicine Across Cultures: History and Practice of Medicine in Non-Western Cultures, 191–207.

hulled canoes joined by a platform and powered by triangular woven sails which carried between 500–600 people and covered distances of over 3000 kilometres (Oliver, 1989).

THE DISCOVERY AND SETTLEMENT OF POLYNESIA

These ever-longer journeys were punctuated by periods of settlement and consolidation along the way in the Solomon Islands, Vanuatu, and Fiji before the voyagers arrived in Tonga and Samoa some 3000 years ago. There they remained for some 800 years. During that time, a distinctive culture and language evolved, becoming the basis of what is now known as Western Polynesian culture. This region became known as the "cradle of Polynesia" because of its pivotal role in the evolution of the culture and language found in variant forms in Polynesian societies throughout the Pacific (Oliver, 1989).

From Samoa, their descendants, for reasons that remain a matter of speculation, set out to explore, settle and people the far-flung islands of the region which has become known as the Polynesian Triangle (Lewis, 1972). Their first move involved a journey of some 4000 kilometres eastward to the Marquesas Islands in what is now French Polynesia. From there, they pushed northward to the Hawaiian Islands, south eastward to Rapanui, or Easter Island, and southward to the Cook Islands before finally moving Southwest to Aotearoa or New Zealand some 1000 years ago. This made Polynesia the last region on this planet to be settled by humans. The Polynesian culture and language came to vary over time and in different places, but all Polynesian societies are founded on a culture and use languages which share common characteristics.

THE REDISCOVERY AND SETTLEMENT OF POLYNESIA

Although Magellan sailed into the Pacific Ocean in 1520 in search of gold and souls, Europeans only rediscovered the islands of Polynesia in 1569, when the Spaniard Alvaro de Mendaña landed on an atoll in Tuvalu on the fringes of Polynesia. Only later, in 1595, did he land briefly in Polynesia proper at Fatuhiva in the Marquesas. Most Spanish voyages, however, took a northerly route across the Pacific to their colonies in Micronesia and the Philippines, and it was not until some 200 years later that Europeans set about systematic exploration of the islands of Polynesia. These visitors were mostly from northern Europe. Throughout the 17th century, the Dutch dominated exploration, and in the 18th century the English and French were the dominant visitors (Beaglehole, 1975).

The first Europeans in the Pacific were explorers looking for routes to the lands in Asia where undiscovered fortunes were supposed to exist. Their visits were fleeting and had little impact on the societies they contacted. Their intermittent stopovers, relatively small numbers, and their dependence on the islanders for food and water, meant that they were little more than social curiosities initially. They were followed by whalers, sealers and traders who settled semi-permanently in small numbers on the edges of the societies of Polynesia and were adopted by people who sought access to their knowledge and skills. Their impact too, appears to have been somewhat limited. They were valued, not primarily for their culture, but for the skills which allowed their patrons to trade with occasional European visitors and to gain advantage over their traditional rivals. While Polynesian societies were transformed by the material culture and technology to which they gained access through these earliest visitors, their impact on the societies and populations with whom they lived was quite minimal. Nor is it clear that they wished to transform these societies. Many were ship deserters, escapees and refugees who sought to draw as little attention to themselves as possible lest their pasts, and the authorities, caught up with them.

Until the end of the 18th century then, Polynesian societies remained relatively unchanged. The major clashes of cultures occurred when traders, missionaries and the agents of colonial powers arrived and settled in Polynesia in larger numbers in the early nineteenth century. They brought with them clearly defined social, spiritual, political and economic missions. These encounters transformed both the indigenous peoples and the visitors to varying degrees and in different ways. The forms and consequences of the clashes differed widely throughout Polynesia. They were shaped by the arrival sequences of the various agents, the value of what they found in places, the nationalities of the agents and the degree of resistance which they encountered from the indigenous peoples. In these circumstances, it is impossible to generalise about *the* consequences of contact in Polynesia.

It is, however, possible to examine a particular sequence in a specific site, the Samoan archipelago, to show how the healing cultures of the indigenous people and those of the visitors collided and to review the consequences of 170 years of contact on the healing practices of the Samoans. It is possible to look at the process and to identify how and why various elements of two more or less discreet healing systems were combined in various ways at different times in a model of healing and healing practice.

THE SAMOAN CASE

The Samoan archipelago lies in the Pacific Ocean between 169 and 172 degrees west longitude and 13 and 15 degrees south latitude and comprises the independent nation of Samoa and the unincorporated territory of American Samoa. As the cradle of Polynesian culture, Samoan society had features which formed the substrate of later settled Polynesian societies.

It is possible to speak with some confidence about the collision of cultures, and specifically healing cultures, in Samoa, because of the quality of the historical record. The Samoan historical record is detailed because of the number of contributors and is interesting because it is not dominated by accounts from a single source. The early record comprises accounts from such varied sources as English Protestant and French Catholic missions, German jurists and a naval physician, a U.S. exploring expedition, English, German and U.S. consular officials, a few botanists and assorted visitors. The later record draws on the reports of colonial powers and the international organisations[1] which oversaw their endeavours. This reconstruction draws on all of these accounts to describe and analyse the collision of healing cultures in Samoa and to show how this process produced the contemporary health system.

The pre-contact paradigm

Samoans knew that their world was dominated by the many gods who had emerged from the void, created the natural world and, much later, its human inhabitants (Stair, 1983: 210–211). All humans were the descendants of the various gods, known as *atua* and *aitu*, whose activities had produced humans and conferred their reproductive power upon them. The hierarchy of gods assumed an immediate presence in people's daily lives. Individuals, families and villages were all protected by a number of gods, which appeared as natural phenomena such as rainbows and shooting stars and animals such as sharks, eels, turtles, owls, terns and so on.[2]

Connections between people and their protective spirits (*aitu*) were established at birth. During a woman's confinement, the names of several spirits were invoked in succession and the one who was being addressed at the time of birth became the child's god for life and an "object of veneration" (Turner, 1983: 17). A person might eat anything except the animal in which his or her god took its earthly form. Families were also protected by what Turner called "household gods" (*aitu fale*). The family gods were asked, during the family's evening prayers, to protect the family from "fines, sickness, war and death" (Turner, 1983). Families periodically made feasts for and offered libations of *'ava* to the gods to ensure their continued protection. *'Ava* is a drink made from water and the powdered, dried roots of the *Piper methisticum* plant which is consumed, usually by chiefs, on ceremonial occasions. Villages were protected by a separate category of gods, and each was also protected by its village god (Turner, 1983). Children born in a village were offered to that village's god, also at birth, and were protected by that god throughout life.

These gods were omnipresent in the Samoans' world. They were present symbolically in the spirits' houses (*fale aitu)* and temples of the gods (*O le mālumalu ole aitu)* which were "built in the usual shape and style" on raised stone platforms on the central common *(malae)* in most villages. These apparently inconspicuous houses were "rendered sacred by their being set apart as the dwelling-place of the god and...were regarded with much veneration by the Samoans". They housed "whatever emblems of the deity were in the possession of the village" and were guarded by the keepers of the war gods (Stair, 1983: 225–226).

The gods themselves made appearances among their followers in their animal forms. Their conduct on these occasions was watched carefully by those who lived under their protection for signs which were interpreted by certain family heads and village chiefs who were held to be their earthly incarnations or "anchors of the gods" (*taulāaitu*). Stair (1983: 220–225) distinguished four subclasses of *taulāitu*, which he glossed as priests of the war gods, keepers of the war gods, family priests, and prophets or sorcerers. They interpreted the messages embodied in the conduct of the gods and explained these to the families and villages which lived under their protection. They also explained how the requirements of the gods might be met, usually in the form of gifts of food and such things as canoes and houses, accepted the gifts from the people and thanked the givers on their behalf. Samoans acknowledged the omnipresence of these gods and their power over their individual and collective lives and readily gave what they were directed to give, as Turner noted, "to remove a present evil or to avert a threatened one" (1983: 18).

Illness and its management

These gods' influence was so pervasive and their powers so complete that they were held to be the principal authors of all human illness and their assistance was invoked early in cases of sickness. One of the earliest authors, the missionary J. B. Stair (1983: 234), noted that "at one period, all bodily pain was supposed to be occasioned by the *aitu* and strange things sometimes occurred in connexion with such notions". In some cases, the illnesses were the consequences of the gods' anger at a lack of respect or meanness shown them by the families and individuals under their protection. In others, the illness or misfortune was visited on people by a group of gods known as the *Taulā-aitu-vavalo-ma-fai-tu'i,*[3] or the anchors of the gods who prophesied and cursed and whose powers to curse could be invoked by people to make others ill. Since the cause of an illness was not immediately clear to the victim and the family,

> The sick were also taken to the Taulā-aitu, and they were consulted as to the occasion of the sickness and probable issue, at the same time they were besought to invoke the aid of the gods in the removal of the disease. In return for these services they received large presents of food and valuable property (Stair, 1983: 225).

Once the cause of the illness was determined, the *taulāitu* was asked to invoke the god's aid in removing it. This invocation took various forms according to those who documented it.[4] In the case of a family deity known as Taisumalie,

> ... this god was incarnate in an old man who acted as the doctor of the family. ... His principal remedy was to rub the affected part with oil, and then shout at the top of his voice five times the word Taisumalie, and five times also call to him to come and heal. This being done, the patient was dismissed to wait a recovery. On recovery, the family had a feast over it, poured out on the ground a cup of kava 'ava to the god, thanked for healing and health, and prayed that he might continue to turn his *back* toward them for protection, and set his *face* against all the enemies of the family (Turner, 1985: 59).

Samoan medicine consisted of more than prayers and incantations (Macpherson and Macpherson, 1990: 38–53). A limited amount of plant medicine was available, there were a number of forms of massage in which scented oils were used and some forms of surgery were in routine medicinal use (1983: 141). Of these, Stair concluded that the Samoans' use of massage was more successful than their use of "herbs and simples" and surgery.

Plant-based medicines were not used widely. Stair noted that, "Although they had much sickness their remedies were few and for the most part unreliable, notwithstanding the fact that the flora of the group included many medicinal plants and herbs of much value" (Stair, 1983: 164).[5] The missionary botanist, Thomas Powell, confirmed this observation; his botanical fieldwork identified some 336 plants in Samoa but mentioned only nine medicinal uses amongst all listed uses for them (1868). Heath listed only seven plant-based remedies and noted that while many trees were in use,

> ... of the most useful services which science has to render to Polynesia is, for some medical botanist to point out the medical properties of the plants and herbs in order that they may be applied to the cure of the people's diseases (1840: 15).

A small number of trees and plants were used in the treatment of certain supernatural illnesses such as those caused by the god Nifoloa, or long tooth. In these cases, when the "tooth" that was left behind in an abscess after an encounter with Nifoloa became visible under the skin, the wound was opened, the pus and blood drained and masticated leaves and bark of some trees applied for four days by the physician. The missionary John Williams reported the use of the "best of food" and a "native cough mixture" in healing. George Turner reported the use of an emetic (1884: 139), the use of "external applications for skin diseases" (1884: 219), and "shampooing and anointing the affected part of the body with scented oil" (1884: 223).

In surgery, Samoans were generally considered less expert than Tongan and Fijian surgeons, who were in Samoa at the time and who had had more opportunities to practice in intertribal wars in Fiji and in Tongans' colonial ventures. Turner's summary of Samoan surgical practice is succinct, although we should not rely completely on a foreigner's interpretation of indigenous practice.

> In surgery, they lanced ulcers with a shell or shark's tooth, and, in a similar way, bled from the arm. For inflammatory swellings they sometimes tried local bleeding; but shampooing and rubbing with oil were the more common remedies in such cases. Cuts they washed in the sea and bound up with a leaf. Into wounds in the scalp they blew the smoke of burnt chestnut wood. To take a barbed spear from the arm or leg, they cut into the limb from the opposite side and pushed it right through. Amputation they never attempted (Turner, 1983: 141).

For many years, Samoa's relative geographical isolation limited its range of endemic illnesses (Kramer, 1944: 128–129), the number of visits and therefore the number of illnesses which were introduced. With the repertoire outlined above, Samoans were able to explain the causes and patterns of illness which afflicted them, appropriate forms of intervention and management and the causes of their success or lack of success.

Introduced illness and medicine

This relative medical isolation ended when Europeans arrived in Samoa in the early nineteenth century, bringing with them new illnesses, ideas about illness and management practices. The epidemiology of these new illnesses was, in most cases, unlike those of pre-contact illnesses. Ironically, the isolation which had produced the relative epidemiological stability which existed before contact (Kramer, 1944), also prevented Samoans from developing immunity to the new illnesses. This promised that each introduced illness would have dramatic consequences for Samoan society.

The pre-contact healing paradigm might have survived a single, catastrophic epidemic. Such an event could have been explained as either a visitation by the gods on the Samoan people who seemed intent on abandoning them in favour of a new god, or as the act of an all powerful new god on a people determined to hold to elements of their pre-Christian worldview. Instead, Samoans faced a series of epidemics in the decades immediately after contact.

These began when the first missionary force, led by John Williams, landed crew members and teachers who were suffering from influenza at several places around Samoa in 1830 (Macpherson and Macpherson, 1990). They spread the disease among the Samoans. The influenza pandemic that arrived from Europe in 1918 killed some 25% of the Samoan population in a period of months (Field, 1984). In between these events a series of others illnesses were introduced, including new and severe forms of influenza in 1837, 1846 and 1891, whooping cough in 1849, syphilis in 1850, mumps in 1851 and measles in 1893 (Macpherson and Macpherson, 1990).

The effects of these illnesses were unlike those of other illnesses to which some immunity existed within the Samoan population. These hit entire sections that had no immunity and resulted in large losses of life within particular parts of the population. Influenza, for instance, killed significant numbers of older people who were unable to withstand it. The missionary physician George Turner, speaking of influenza in 1846, said,

> Ever since [1830] there have been returns of the illness almost annually ... It lasts for about a month, and passes off as fine weather and steady tradewinds set in. In many cases it is fatal to old people and those who have been previously weakened by pulmonary diseases. There was an attack in May 1837 and another in November 1846, both of which were unusually severe and fatal (1983: 138–139).

Each of these illnesses had distinct social and cognitive effects. Influenza, for example, killed many of the elderly and had a profound effect in a gerontocratic society where older people were guardians of much traditional knowledge and wielded significant social and political power. Their loss robbed their society of certain crucial religious, secular and genealogical knowledge essential to their social viability. The loss also created leadership vacuums because younger people who survived were thought to lack the knowledge needed for effective leadership.

In a society in which rivals quickly exploited vacuums in families' and villages' leadership structures, political instability resulted. The leadership vacuums increased levels of social instability as forces emerged to compete for the vacant chieftaincies.

These new illnesses and associated treatments revealed some of the limitations of pre-contact medicine and generated pressure for its modification. It was clear that their paradigm alone could not explain or protect them from these devastating illnesses. This realisation in turn produced dilemmas for the Samoans. Were they to abandon their belief system in favour of a foreign one associated with the missions and their new and powerful god? Or were they to graft elements of the new system on to their existing one in some hybrid form? As it happened, a hybrid healing culture emerged in a long and complex process which has been dealt with elsewhere (Macpherson and Macpherson, 1990) and can only be summarised here.

WHEN HEALING CULTURES COLLIDED

The Samoans' pre-contact healing paradigm[6] might have been rapidly and completely displaced by another one if at least four conditions had been met. First, the pre-contact and new systems would have had to be so clearly distinct as to have been mutually exclusive. In fact, the opposite occurred. In an attempt to help the Samoan people master key concepts of the Christian gospel, missionaries stressed the similarities between the orders which were, in many cases, more apparent than real. By emphasising similarities, the real differences between the cultures may have been blurred; the cultures, and especially the healing systems, looked remarkably similar.

At the centre of both the Samoan and Christian worlds were omniscient and omnipresent supernatural agencies which dominated the lives of mortals. These agencies observed, judged and intervened in human affairs and had the power to produce and remove illness in human populations as signs of their pleasure or displeasure. These supernatural agencies had, in each case, delegated some of their power to mediums which took human form and lived among people. In both cases, these mediums also enjoyed the power to cause and to relieve illness in human populations. Those who neglected their gods or violated standards of conduct which their gods set were routinely punished for their failure. In this respect, the powers of the Samoan gods and the Christian God, and the ways in which they were exercised, were remarkably similar. In both the Samoan and Christian worlds, the power of faith in, and obedience to, gods was more significant in healing than was human intervention. This was demonstrated by certain scriptural "facts": Jesus had succeeded where physicians had failed, Jesus healed those who, or whose friends and families, showed faith in and obedience to God without physical intervention.

The doctrinal similarities, which followed from increasing exposure to the scriptures, were confirmed in the conduct of the missionaries, whose responses to acute illness were rather similar to those of the Samoans. Missionaries often had only a limited medical knowledge and healing repertoire. As illnesses in the missionary community progressed, and where professional help was unavailable, they turned increasingly to prayer for healing and relief from illness. "Certain appeals common in such prayers...were not unlike those which, according to mission accounts, Samoans made to their deities in similar circumstances" (Macpherson and Macpherson, 1990: 62). Missionaries' records show that, without medical resources, they gathered around their brethren as they neared death, often in acute pain, unable to do anything but say prayers and sing hymns. While their passive acceptance of their God's will may have persuaded Samoans of the intensity of their commitment to their faith, it is not likely to have convinced them of the fundamental distinctions between the two healing systems or of the superiority of the European healing paradigm.

There were other similarities between the two. Missionaries experimented with plants which resembled the "herbs and simples" with which they were familiar, and Samoans also used some plants in healing. But both paradigms stressed the central importance of relationships between gods and humans in determining the outcomes of illness. The Samoans may have seen the plant experiments of the missionaries as less significant than they were to the missionaries.

Second, to have suppressed the indigenous healing paradigm, the missions and other Europeans would have had to act decisively and early to ban its use and to outlaw the activities of its practitioners. They would also have had to make clear the disjunction between each belief system. Indeed, the missionaries might have been expected to attempt to suppress the Samoan healing culture because it was closely associated with pre-Christian gods whose authority was associated with, and derived from, a "pagan" social order. Missionaries from the London Mission Society (LMS) had deliberately and systematically suppressed these gods and the practices associated with them in Eastern Polynesia only a little earlier. Had they done the same thing in Samoa the outcome for healing belief and practice might have been quite different. In the event, they chose not to, which begs the question of why they acted as they did in the Samoan case.

The Samoan chief who led the first LMS party from Tonga to Samoa, Fauea, warned its leader, the Reverend John Williams, against wholesale condemnation of Samoan cultural practices. He pointed out that such proscription would be resisted by the Samoans and would be counter-productive for the missionary endeavour. John Williams, conscious of the mission's dependence on chiefs for support and protection, accepted Fauea's counsel, and indeed thanked the Christian God for it. His decision to follow the advice contributed to the success of the LMS mission in Samoa (Moyle, 1984: 68, 80). It also had a profound effect on the healing paradigm which developed.

The missionaries did not immediately ban all non-Christian gods and force the Samoans to renounce them as they had elsewhere in Polynesia. The missionaries may have agreed with Stair who seemed satisfied that,

> The Samoans seemed conscious of thus suffering from the spiritual thraldom that surrounded them at the time the Gospel leavened and spread among them. ... The real extent of this suffering, however, both mentally and bodily, it is difficult to speak of or fully understand. Their whole lives were enshrouded and enslaved by it, and the time that they suffered from it was well termed by them "the days of darkness" (Stair, 1983: 260).

Yet, at the same time, the missionaries systematically investigated and documented the existence and relations of these pre-Christian gods. They were interested in the origins and forms of these gods, their activities and their powers. While they may have done this to understand better the forces ranged against them, one can only imagine how the Samoans who supplied the information that the missionaries sought construed their interest. But the missionary interest in these gods went beyond documentation, and other aspects of their conduct may also have given Samoans further reason to believe that their gods could live alongside the new one.

When Stair encountered events in which supernatural agency was suspected he routinely discussed possibilities with both missionary colleagues and Samoans and shared each group's thoughts and suspicions with the other in the search for a solution. These were not insignificant events. When, in 1840, a new and apparently substantial mission house was built for Stair, it was occupied by the Stair family and a number of other invisible supernatural agencies. Although these spirits were invisible, they rolled balls down the hallway at night and pounded at the door making life increasingly taxing for the Stair family and its visitors. With the help of Samoans in the area, Stair eventually identified these as spirits or *aitu* who had been inconvenienced by the new house. They were believed to have taken exception to the felling of an old tree on the house site, which may have been their temple, and to encroachment on old burial ground in the area.

Stair, whose wife's health was by this time visibly deteriorating, subsequently left the house on the advice of the Samoans and "at much loss and inconvenience". He reported that after their removal they "were happily freed from any further ghostly visitations" (1983: 264–265). Stair's acceptance of the Samoan analysis of the cause of the situation encouraged the Samoans to offer further assistance. Stair was offered the services of a man who was reputed to be willing and able to confront these spirits but "declined, not wishing to have any closer contact with our tormentor" (265–266). In the circumstances, the actions of a prominent missionary whose activities were visible to and scrutinised by large numbers of parishioners among whom they lived were significant. Their actions in such cases may have given Samoans clear reason to believe missionaries also acknowledged the existence and power of Samoan gods.

On another occasion, a missionary and his parishioners were gathered at a chestnut tree when a large bell suspended within the tree began to ring, "violently and without any cause. No hand was pulling it, but it kept on wildly clanging in full view of the whole party who looked on in amazement" (264–265). Searches failed to locate the cause and it was again agreed by all involved that it was a case of local *aitu*. The actions of the missionaries and their families in these cases could well have confirmed the Samoans' perception that missionaries too accepted and acknowledged the power of the Samoan gods. In the circumstances, it may have seemed to the Samoans that the existence and power of their pre-contact gods continued alongside the new God. Furthermore, since the *aitu* were apparently alive and well and existed alongside the new God, their power over the lives and health of their families presumably also continued. Thus, the second requirement – a decisive break between pre-contact and introduced healing paradigms – was not fully met.

Third, the abandonment of the pre-contact healing paradigm might have been more likely if there had been a "contest" between the two paradigms in which the introduced one was conclusively shown to be more powerful than the indigenous one. In the Samoan case, no such contest ever developed. While missionaries practiced some form of medicine among their Samoan followers, most did so with limited knowledge of medicine,[7] few formal diagnostic skills and a limited range of medicines and treatments (Macpherson and Macpherson,

1990). Those that were available to them were in many cases of limited value and, in some cases, simply dangerous. An American consular agent, J. B. Steinberger, who observed missionary practice, noted with some alarm that,

> They adhere to the old school of practice, and ignorantly dispense blue-mass, gray powders, calomel, and other preparations of mercury, while Dover's powders, podophyllum, preparations of arsenic, &c, are freely given. I foresee in this reckless issuance of drugs no little mischief in the future, as mercurial diseases must certainly develop themselves unless it is abandoned (Steinberger, 1874: 22).

Most practised in the belief that they could heal, even when their medicines failed, if their faith in God and their commitment to their people won divine intervention.[8] Their accounts reflect their belief that the gift of healing, bestowed on the disciples by Jesus, might also be given to them if they demonstrated similar faith and commitment to their mission.

The missionaries' lack of success was not confined to their Samoan patients. Their success with their own people was also limited, and this was very obvious to the Samoans among whom they lived. Missionaries and their families suffered greatly from a range of illnesses and from lack of professional assistance, and the missionary record contains numerous accounts of the protracted suffering and painful deaths of many of their number (Macpherson and Macpherson, 1990). In many of these cases the missionaries abandoned their medicines and turned to God in prayer and song. While missionary medical practice may again have demonstrated the extent of their religious commitment and their faith, it may also have done little to convince the Samoans that the Europeans possessed a more effective healing paradigm.

The final pre-condition for giving up the Samoan pre-contact healing paradigm was the abandonment of a worldview within which it was embedded. This did not occur. The missionaries proceeded slowly and avoided sweeping proscriptions of Samoan beliefs and practices. Their caution reflected their hope that the growing familiarity with the gospel message would persuade Samoans to abandon their old ways. In some respects they succeeded.

Samoan Christians were called on to acknowledge the superiority of the Christian God and to follow his teachings. They were not, however, called upon to deny the existence and power of other earlier gods. They had instead to acknowledge that these were like the evil spirits referred to in scripture. But they could, and would, continue to interfere in human affairs and would tempt humans to violate Christian laws. When this occurred, Samoan Christians could call on the greater power of their new God to overcome and cast out these evil spirits as he had at other times and places. In providing this re-classification of status of their earlier gods, the missions may have re-confirmed their existence for the Samoans. As long as these gods continued to inhabit the Samoan world and to intervene in their lives, it was necessary to maintain a way of thinking about and reacting to them.

For all of these reasons then, the Samoans were not persuaded by early contact with Europeans to abandon their pre-contact healing paradigm and to embrace a new one. They were well aware that neither system alone could successfully explain their new and expanded reality or provide responses to the new patterns of illness which they were experiencing. It was also clear that the pre-contact healing paradigm had serious shortcomings and that a new and more comprehensive one was needed.

TOWARD A NEW HEALING PARADIGM

Samoans opted for an augmented healing paradigm which incorporated significant elements of their pre-Christian one and others from the European, Melanesian and Chinese ones. They combined elements of various available forms of health knowledge and practice in a new healing system which responded to new patterns of illness.

The Samoans determined that a series of illnesses had existed before contact. These were known as Samoan illnesses (*ma'i samoa*) because of their association with pre-contact social order. The agencies which had produced these illnesses remained present in the Samoan world albeit re-classified as evil spirits. It stood to reason that these *aitu* would continue to inflict suffering and illness on those who had denied their power and had given their allegiance and resources to another God. Illnesses caused by these pre-contact supernatural agencies became known as illnesses of the ghosts or spirits (*ma'i aitu*). It also stood to reason that certain earlier forms of intervention might be relevant in dealing with these supernatural agencies and a number of specialists known as the anchors of the gods (*taulāitu*) continued to specialise in the management of these illnesses. Their practice included various prayers and incantations and a few plant-based medicines, known as *vaiao*, and resembled that of the *taulāitu* of earlier times.

Samoans were, however, increasingly aware of the role of plant-based medicines in other healing systems. The earliest experimentation and use of these medicines by the missionaries had awakened Samoan interest in these plant compounds' possibilities. Increasing numbers of visitors and residents from Hawai'i, Tonga and Fiji, throughout the later 19th century, extended the range of plants which Samoans saw in use in healing (Kramer, 1994: 134). From the mid-1800s, Samoans served as missionaries and teachers in the larger, high islands to the west where plant medicines were widely used and became familiar with local plants and their medicinal uses. Samoan missionaries were often dependent on their parishioners for food, shelter and medicine and learned about local plants and their medicinal uses throughout Western Polynesia and Melanesia. Those missionaries who returned to Samoa took some medicinal plants, which they planted and propagated. Even today one comes across pockets of exotic tropical plants which were grown by missionaries who served in Melanesia and particularly in Papua New Guinea over 60 years ago in some cases.

The greatest impetus for the adoption of and experimentation with plant medicines, however, undoubtedly came from the introduction of significant numbers of Chinese and Melanesian labourers into commercial plantations in Samoa late in the 19th and early 20th centuries. Both of these groups brought their own plant-based medicines and adapted and expanded them to include plants which they found in Samoa. While colonial powers made deliberate attempts to prevent their co-habitation with Samoans, their knowledge was readily shared with the Samoans with whom they worked. These people's knowledge of the uses of plant medicines, and their apparent success with them, provided new knowledge and generated growing Samoan interest in experimentation and specialisation in these medicines. This is reflected in the fact that, as ethnobotanists have observed, a significant proportion of those plants currently in use in "Samoan" medicine are in fact exotic (Whistler, 1984, 1987).

These new ideas, medicines and practices were frequently grafted on to the knowledge bases of those who already treated *ma'i aitu*. This trend was reflected in the emergence of groups of healers whose practices included the treatment of some spirit illnesses but were increasingly focused on plant-based and musculo-skeletal medicine respectively. These were the *fōma'i*, whose practice dealt principally with internal illnesses and skin complaints and who used increasing amounts of plant-based medicine, and the *fōgau*, whose practice dealt principally with fractures, dislocations, sprains and contusions and who employed various forms of massage, manipulation and immobilisation. Later a playing card reader, or *faipēlē*, was added to the ranks of practitioners who were able to assist with diagnosis of the illnesses caused by spirits.

These healers could, between them, diagnose and manage these earlier illnesses caused by the pre-Christian gods, and, with an expanded pharmacopeia, other illnesses which were caused by natural agencies and trauma which were no longer immediately assumed to be the consequences of vengeful gods.

But the Samoans were also acutely aware that new illnesses, associated with the arrival and settlement of Europeans or *papālagi*, were a fact of life in post-contact Samoa. They were aware that Europeans had brought these diseases and that they claimed knowledge of their aetiology and management. These diseases became known generically as the illnesses of the Europeans (*ma'i papālagi*). Specific illnesses were named: *ma'i suka* (diabetes), *toto maualuga* (high blood pressure), *toto maualalo* (low blood pressure), *kanesa* (cancer), *misela* (measles), *mami* (mumps) and so on (Macpherson and Macpherson, 1990).

Samoans knew that Europeans' greater familiarity with and knowledge of these illnesses allowed them to treat them with greater success. These illnesses were most effectively treated by practitioners trained in the *papālagi* healing tradition, and known as *foma'i papālagi*, using medicines which became known as *vai lā'au* or *vai papālagi* and practices, including surgery, which became known as *togafiti*. Samoan healers had to recognise the symptoms of these introduced illnesses so they could re-direct patients with *ma'i papālagi* to those trained in the *papālagi* healing system.[9] Samoans readily took their ill to the hospitals established by Europeans and, from early in the 20th century, began to construct dispensaries and small hospitals in villages and districts to ensure that they had access to this European healing. Increasing numbers of Samoans were sent abroad for training as doctors and nurses at the Fiji School of Medicine. More recently, Samoan physicians have been trained in New Zealand and the United States.

THE PRESENT

Over time the two subsets of illness and their associated treatments came to sit easily and logically alongside one another. Given this situation, diagnosis involved first identifying the likely origin of a given set of symptoms. This diagnosis allowed a patient to determine whether to seek Samoan medicine or *papālagi* medicine. In many cases, these diagnoses, often made with the assistance of a healer, were relatively straightforward. Treatments were administered and patients recovered and confirmed the efficacy of the healing system which had provided the diagnosis and treatment. Where an initial treatment failed, a further diagnosis was attempted and the patient was re-directed to another healer where necessary. In some cases patients were persuaded to transfer their search for a cure from Samoan medicine to European medicine and, in others, from the European to the Samoan medicine.[10] Where symptoms were ambiguous and clear diagnoses were not available, patients sought treatments in European and Samoan medicine simultaneously. When people became well again in these circumstances, it was often unclear which of the treatments had produced their cure and people simply, and pragmatically, resolved to take the two cures again in similar circumstances.

THE FOUNDATIONS OF CO-EXISTENCE

A very similar situation exists today in Samoa. Two healing systems co-exist alongside one another and are routinely used by many Samoans. This may seem surprising given the steadily increasing power and reach of Western biomedicine and raises the issue of why these two healing systems can co-exist in the early 21st century in much the same way as they did in the early 20th century. There seem to be a number of reasons for this situation.

Sociologists of health may view these dual healing systems as discrete entities with distinctive epistemologies which differ in certain fundamental respects. Logic dictates that one of these systems can explain illness more comprehensively and treat it more effectively, and that ultimately most people will opt for that system. Samoans also acknowledge these crucial differences. They acknowledge that Western biomedicine can deliver powerful cures for a range of ailments and routinely use this medicine. They remain equally certain that Western biomedicine cannot provide comprehensive explanation and treatment for a range of other illnesses which have their origins in the Samoan world. They see the dual systems not as mutually exclusive alternatives, but as necessary and complementary adjuncts. But this alone is not an adequate explanation of why this situation exists.

The two systems have never gone head-to-head in a contest which could prove conclusively that one was superior to another. Nor is this likely to occur. Each is known to be superior in certain respects to the other. Samoans acknowledge the value of antibiotics and use them widely and recommend them to others. But they are equally certain that antibiotics can have little effect in an illness which is caused by the deterioration of relations between the living and the dead.

Samoans hold a more holistic view of the nature and causes of their illness. They acknowledge the agency of the supernatural, natural and social in illness. This makes the diagnostic process both longer and more complex. Samoans then accept that the search for an explanation may require extensive and protracted discussions of parts of an individual and his or her family's recent history if only to eliminate supernatural and social causation as possibilities. It is scarcely surprising that many are highly sceptical of the diagnostic value of short consultations of the overworked physicians in the public health system. They ask how a physician can be certain of a particular diagnosis if their questioning could not possibly have uncovered and eliminated other possible causes. While such cursory, and primarily physiological, examination may reveal the causes of non-Samoan illnesses, it will clearly not be able to reveal the causes of Samoan illnesses. As long as these two subsets of sickness exist in the Samoan world it will remain necessary to maintain parallel subsets of diagnostic and treatment practices.

This view is confirmed for Samoans in the conduct of key figures in their society. Throughout our research we encountered narrative after narrative which was invoked to demonstrate the truth of this argument. The father of the current head of state, for instance, was a highly respected and powerful person in both the Samoan and European worlds. He was also a respected medium for the nationally known spirit named Telesā. Those whose lives were disrupted by Telesā routinely took their concerns to his residence in the political capital of Samoa and sought his intervention, which was always freely given. In the same way, Samoans saw a senior physician, whose life and education straddled the Samoan and European worlds, take his elderly, much-loved and acutely ill mother to a Samoan healer where she was apparently healed. The same physician returned later and publicly acknowledged the healer's talents and made her valuable traditional gifts. These sorts of events seemed, to those who observed them, clear confirmation of the complementary status of the healing systems.

The Samoan healing system has also incorporated a number of ideas and treatments from other systems so completely that their exotic origins are now forgotten. In the process, Samoan medicine may have experimented with and adopted, and discarded, ideas, practices and plants on a purely pragmatic basis so that over a period of time the range of treatments and their efficacy has increased. Thus, one finds in Samoan medicine such elements as ice, aspirin, kava 'ava, and sterilisation of equipment which can be tracked to priests and physicians who provided information to make Samoan treatments more effective. In some cases they did this to afford Samoans better treatments in the period before they had widespread access to public health and hospitals and, in others, because they sought to make more effective a set of practices which they realised they could not proscribe. Other Samoan practices of more recent origin including a dosage calculated on the basis of bodyweight which was incorporated into tradition after a healer had seen the practice used in a highly successful public health intervention into filariasis in the mid-1960s. Similarly with plants, one finds now in Samoan medicine, flowers, leaves and bark from species which were introduced as part of a village beautification program some 15 years ago, but which have become so widespread as to leave people believing that they are in fact indigenous.

* * *

The Samoan case shows that the collision of healing cultures in the Pacific led not to displacement of an indigenous healing paradigm and its replacement with another introduced one as is widely supposed. Many argue that missionaries rapidly and completely undermined the indigenous religions and those parts of worldviews and lifestyles, including health and illness, which were associated with them. But this sort of abrupt, total, disjunctive transformation was rare, if indeed it ever occurred, in the Pacific. As the Samoan case shows, the process which followed the collisions of healing cultures was one of borrowing, experimentation, incorporation and incremental expansion of the indigenous healing system. Alongside this was an enthusiastically embraced and widely used Western biomedicine which also continued to expand and to become more influential over time. These two look likely to sit alongside one another for considerable time yet, for the Samoans see no reason to believe that either has all of the answers to all of their health problems. As one elderly Samoan noted, as long as you can have the best of both worlds, why would you settle for only one?

NOTES

[1] The League of Nations supervised the New Zealand administration's mandate between 1918 and 1945, and the United Nations supervised the New Zealand administration's trusteeship between 1945 and 1962 when Western Samoa became independent.

[2] While there is agreement on these basic elements of the nature and role of the godly hierarchy, both early authors, such as Stair and Turner, and later writers, such as Meleisea *et al.*, acknowledge the existence of multiple accounts of both the creation and activities of gods.

[3] This term, *Taulā-aitu*, means the same as *taulāitu* above. Modern Samoan orthography has collapsed the two separate words into one.

[4] One problem with these accounts is that they were provided by missionaries who resented the control that these agencies had over the minds and lives of their adherents, and their accounts reflect this.

[5] The same observation could readily be made of the missionary medicine of the day which consisted of the use of such substances as mercury and the widespread use of "herbs and simples"

[6] Healing paradigm and healing system are used interchangeably in this discussion. Each refers to an integrated set of beliefs, knowledge and treatments which people use to explain, diagnose and treat their illnesses.

[7] Only one, Dr George Turner, was medically qualified and most in the Samoan mission field were chosen from the ranks of committed artisans on the basis of their religious commitment.

[8] Ironically, in the one case in which missionary medicine did succeed conclusively the success was invisible. They vaccinated Samoans against smallpox ahead of an approaching epidemic and may have saved significant numbers of lives. But, because few if any Samoans suffered from the disease it was difficult to demonstrate conclusively how the vaccination and the absence of illness were connected to their adherents.

[9] It is unlikely that European physicians were willing to familiarise themselves with Samoan illnesses so that they could direct patients to Samoan healers in earlier times. It is therefore ironic, that as Western biomedicine has become increasingly powerful, its younger practitioners have

become more willing to acknowledge the value of certain Samoan treatments and to consider partnership with Samoans healers.

[10] In the general hospital we were advised by a nursing sister that our son's fracture, which had not healed as quickly as was expected, should probably also be seen by a Samoan healer since some supernatural agency might be involved. She provided the names and locations of two whom she knew had been able to help others in similar circumstances.

BIBLIOGRAPHY

Beaglehole, J.C. *The Exploration of the Pacific.* London: Adam and Charles Black, 1975.

Field, M.J. *Mau. Samoa's Struggle Against New Zealand Oppression.* Wellington: A.H. and A.W. Reed, 1984.

Heath, T. 'The Navigator's or Samoa Islands.' *The Polynesian* 1, 1840.

Kramer, A. *The Samoa Islands.* Auckland: The Polynesian Press, 1994.

Lewis, D. *We, The Navigators. The Ancient Art of Landfinding in the Pacific.* Honolulu: University of Hawaii Press, 1972.

Macpherson, C. and L. Macpherson. *Samoan Medical Belief and Practice.* Auckland: Auckland University Press, 1990.

Macpherson, C. 'The persistence of chiefly authority in Western Samoa.' In *Chiefs Today. Traditional Pacific Leadership and the Postcolonial State,* G. White and L. Lindstrom, eds. Palo Alto, California: Stanford University Press, 1997, pp. 19–48.

Meleisea, M. *Lagaga: A Short History of Western Samoa.* Suva: Institute of Pacific Studies, University of the South Pacific, 1987.

Moyle, R. *The Samoan Journals of John Williams 1830 and 1832.* Canberra: Australian National University Press, 1984.

Oliver, D.H. *Native Cultures of the Pacific Islands.* Honolulu: University of Hawaii Press, 1989.

Powell, T. 'On various Samoan plants and their vernacular names.' *Journal of Botany* 6: 278–285, 342–345, 355–370, 1868.

Stair, J.B. *Old Samoa or Flotsam and Jetsam from the Pacific Ocean.* London: Religious Tract Society, 1897, reprinted Papakura: R. McMillan, 1983.

Steinberger, A.B. *Report on Samoa or Navigators' Islands.* Washington DC: 43rd Congress of the US, 1st Session (Executive Document #45), 1874.

Turner, G. *Samoa a Hundred Years Ago and Long Before.* Papakura: R. McMillan, 1983.

Whistler, W.A. 'Annotated list of Samoan plant names.' *Economic Botany* 38(4): 464–489, 1984.

LEWIS MEHL-MADRONA

NATIVE AMERICAN MEDICINE: HERBAL PHARMACOLOGY, THERAPIES, AND ELDER CARE

Many people who grew up in North America, especially Native Americans, had biases and romantic prejudices about what it was like in the past to be old in Native America. We wanted our culture to be somehow better for the aged than was that of the European settlers. We grew up with a variety of apocryphal stories about elders in pre-European times – about how we valued the elderly, even revered and respected them for their wisdom and teaching. Research shows, however, that there were changing patterns of age relations even before the European invasion.

Older Native Americans were generally a healthy lot. They walked extensively, even from village to village. At the turn of the century, Eastern Cherokee elders still ate a traditional diet of possum stew and squirrel stew, also called Brunswick stew. Older women picked greens for the evening meal from fields near their homes. These included poke greens, mustard greens, collard greens, dandelion greens, and whatever else grew in the vicinity. They ate fresh berries. They canned extensively from their large gardens, leading to a supply of squash, tomatoes, pumpkin, corn, okra, green beans, and many other foods. Meat was mostly wild or locally raised. Corn bread was the staple carbohydrate, along with corn grits and corn on the cob. Old people had little access to medical doctors, and even less trust of them. They relied upon local healers to minister to their woes.

Today's Native American elders are much different. Over 50% have diabetes. Heart disease is rampant. Diet consists of government surplus commodities, full of lard and sugar. White bread is the staple carbohydrate, along with soda pop and fried bread. Sugar consumption is massive, and meat quality is poor. Exercise is minimal, while smoking is preponderant. People's sense of dignity and meaning in life is compromised. Reservation life today tells us unfortunately little about Native elders of yesterday.

Nevertheless, I have made an attempt to describe what is known, recorded, remembered, and discovered about Native American ancestors and their way of life.

H. Selin (ed.), Medicine Across Cultures: History and Practice of Medicine in Non-Western Cultures, 209–224.

AGE ROLES: SCRIPT NEGOTIATION

Every society tends to think about its elders in categorical ways, naturalizing and universalizing the concept, creating fixed images that prevent us from appreciating age as a variable social script, changing with the conditions of the moment.

From the Hollywood image of Eskimos sealing their old people inside the igloo to die because they could no longer keep up with the tribes to the historical practice among the Lakota people in which the oldest woman has absolute veto power over the war council, we have constructed a series of perceptions of age in Native America.

The process by which societies conceive age and translate this concept into social roles is a fundamental dynamic of history and social life, much as Bruhns and Stothert (1999) have described for gender.

We use age to assign roles and status, open and close opportunities, empower and constrain. The negotiation of age relations can be viewed as script writing by the various involved generations. The parties seek to gain advantage or minimize disadvantage as they invent technologies, alter economies, change social relations, develop different politics, and generate ideology. These creative activities are what humans do as they solve the problems of organization and survival. All societies create cultural scripts that bind people and their activities together in ways that work to solve the basic problems of subsistence, mental and physical survival, and personal satisfaction. What remains hard for each culture to appreciate is the extent to which roles maintained for the elderly are just that, and not biologically requisite activities.

What is different about elders in ancient America when compared to elders in modern America? Since there was no written language, elders were vital for interpreting events and for constructing survival strategies, since the elderly had the longest memory of how the people had previously responded to changing conditions of the environment and to other challenges. Elders were historians, repositories for stories and ceremonies, interpreters of natural events, childcare and home care providers and family laborers.

The evidence indicates less stratification of age (and gender) roles in pre-agricultural societies than in Western Society today (Bruhns and Stothert, 1999).

CULTURE AND THE ELDERLY

One current representative of ancient hunter-gatherer societies is the Inglalik people of Alaska, who represent a material culture in which the elderly were integral, honored members. The Inglalik, also called Deg Hit'an, an Athabascan group, strongly engender and promote cooperation and mutual dependence among men and women, young and old. The primary foods and sources of material for artifacts are wild species. Manufacturing involves hand labor and individual skills applied to locally available raw material. This material culture carries non-material ideological elements. Artifacts are not just aids to material living, but also means to achieving social cooperation and interdependence.

"Differentiation of tasks and artifacts associated with these tasks was accompanied by a model of integration. This model represents how most hunter-gatherer societies of North and South America probably lived" (Osgood, 1940).

Hunter-gatherer societies have less differentiation of age and sex roles than emerging agricultural societies (Bruhns and Stothert, 1999). Among hunter-gatherers, age-related restrictions eventually limit a man's ability to hunt, propelling him into other activities, but women's roles remain relatively constant. Less scripting of roles is observed with people contributing to the common survival and community life as they are able.

With some notable exceptions, such as the Hopi, pre-European Native communities were mobile. Small settlements or villages of up to a few hundred people formed and disbanded as the seasons and food availability changed. During spring, which was the berry-picking, root-digging, and fish-spawning season, many kinship networks would gather and form a dense settlement for a brief period of time. Later in the fall, the same people would gather into smaller bands and hunt throughout a wide geographical area. Even tribes involved in agriculture were fairly mobile. After the crops were planted in the spring and after the first or second weeding, groups moved to other planting and gathering sites. In general, women were responsible for planting and caring for the fields and for gathering wild nuts, berries, and other plants. Late summer and fall would bring the groups back together to harvest corn and other crops, and to gather nuts and other wild plants (Jackson, 1994).

Elders had to be mobile to participate in this life. Skeletal remains suggest general healthiness and mobility, even though there is also some evidence of degenerative conditions.

RESPECT FOR THE AGED AND THE DEAD

The bones of the dead had special meaning for ancient people. The archaic Las Vegas people of coastal Ecuador are a well-studied example (Ubelaker, 1980). An individual found buried under the threshold of an early Las Vegas shelter was a woman well over age 45. Archaeological evidence has suggested that she was accorded an important role as lineage head and authority, which would accrue to the oldest members of a group.

Veneration of ancestors improves the status of the elderly, as is evidenced by the Zapotec and Mixtec religions, who believed that the ancestors propitiated the forces of nature (Marcus, 1983; Marcus and Flannery, 1996).

HEALTH IN NATIVE AMERICA

Traditional Native American cultures perceive health as a state of balance of spirit, mind, and body; illness is the result of disharmony or imbalance. Illness requires treatment at the many levels, including personal, family, community, and spiritual. Traditional medicine has included herbalists, shamans, purification ceremonies, healing rituals, emotional therapies, manipulative medicine, teas, herbs, and special foods (Jackson, 1994).

Aztec curing, for example, consisted of medicines or mechanical manipulations, along with incantations, prayers, and rituals. Women healers were prominent at the time of Conquest, but their roles were reduced by Christian pressure, which defined traditional medicine systems as witchcraft (Ortiz de Montellano, 1990). Nevertheless, Aztec healers used advanced methods in the treatment of injuries, broken bones, pregnancy, and childbirth. Much of this knowledge was lost due to the conquest and the suppression of traditional medical practice by the settlers. [Editor's note: See Carlos Viesca's chapter on Mesoamerican medicine in this volume.]

Native American surgeons repaired lacerations with bone needles and human hair. They set bones in plasters made of downy feathers, gum, resin, and rubber. They lanced boils, removed tumors, amputated legs, made artificial legs, removed teeth, castrated men and animals, sucked out venom to treat snake bites, and used tourniquets and cauterization. They concocted emetics, purgatives, febrifuges, skin ointments, deodorants, toothpaste, and breath fresheners. They gave enemas with rubber hoses. North Americans used animal bladders for the purpose of irrigating wounds and suctioning mucous.

Native people took baths in rivers, lakes, streams, or ponds, or had elaborate medicinal baths. Sweathouses and steam rooms abounded in North America, called *timescalli* by the Aztec, who built beehive-shaped structures of stone and brick. Stones were heated outside and brought into the structure where the patient rested. Herbs were burned for smoke or added to steam. Leaves and ointments aided body massage. Every village had one or more *timescale*, used to treat everything from fever and boils to insect allergies, snakebites, exhaustion, and aching muscles. Quechua speakers and others used hot springs as sacred *huaca* for ceremonial and medicinal bathing.

Sweathouses or sweat lodges were ubiquitous, from California to Delaware, from the Arctic to the tip of Chile. Both Californians and Delaware peoples built semi-subterranean earthen structures entered by a tunnel. The Alaskans built similar baths covered by logs. The Creek people used hides and mats. People of the southeastern United States slept all night in sweathouses in the winter months and then jumped into cold creeks, ponds, or rivers in the morning. Plains people covered their lodges with branches and leaves, and later blankets. Some tribes included elaborate ceremonies as part of the sweat bath (Lakota, Cree), while others simply sat together in fellowship (Coeur d'Alene, Spokane).

Europeans of that time thought that frequent bathing was debilitating. Colonial officials outlawed bathing as harmful to the peoples they had conquered. Bathing decreased disease, and may have accounted for some of the general freedom from epidemic diseases among Native peoples before European conquest. The destruction of sweathouses and the forbidding of bathing may have contributed to the rapid spread of Old World epidemics in the New World (Weatherford, 1988).

FOOD AND HEALTH IN NATIVE AMERICA

The hunter-gatherer lifestyle decreases the incidence of free radical diseases, which increase with aging, increasing morbidity and mortality and decreasing longevity (Harman, 2001). Inherent in the lifestyle is food restriction and a high proportion of antioxidant consumption, while minimizing grain and dairy intake. Hunter-gatherer diets consumed by North and South American natives had relatively high levels of antioxidants compared to modern diets, as established by studies from the South Pacific on traditional hunter-gatherer diets of native populations previously unexposed to contemporary dietary practices (Lindeberg and Lundh, 1993; Lindeberg *et al.*, 1994, 1996, 1999; Srikumer *et al.*, 1994).

Following Columbus' voyage to the Americas, Native American spices, condiments, and foods had a very strong impact on the European diet. Yellow, orange, green, and red tomatoes from cherry to almost melon size and in round and oblong shapes found their way into the kitchen to be pickled, sliced, chopped, diced, dried, pureed, and made into hundreds of sauces. There was also a diverse set of American sweet peppers, varying in more sizes and shapes than the tomatoes and named bell, banana, and cherry peppers because their shapes reminded the cooks of something already familiar to them (Weatherford, 1988).

North Americans were not exposed to casein (the main protein of milk). Some studies have shown decreased life expectancy when casein is the only source of protein (Harman, 1978). Replacing casein with soybean protein might increase life expectancy. Controversy continues over whether Native Americans are genetically lactose intolerant; at any rate, pre-European elders did not eat dairy products.

Periods of caloric restriction were also part of the hunter-gatherer life style. Decreases in caloric intake are associated with proportionate decreases in oxygen utilization, decreasing mitochondrial superoxide radical formation, and ATP production (Harman, 2001). The interaction of these adaptations with the sudden appearance of the modern European diet may explain some of the sudden collapse in health found among colonized and assimilated Native populations.

Inevitably the nutrient in shortest supply was fat, since wild animals were lean. The major sources of fat were nuts, sunflower seeds, and, for some tribes, cotton seeds (Jackson, 1994). Cooking methods included boiling, roasting, steaming, and baking. Boiling was the preferred method for cooking most foods. Meat, fish, corn, wild rice, and other foods were boiled in clay pots, birch bark pots, or any container that would hold water. Both stone boiling (putting heated stones into a container of liquid) and boiling by directly heating a container of liquid were used. Some foods, such as clams, were steamed in a pit lined with hot rocks and wetted with plant materials. Root vegetables, such as wild turnips, were roasted over hot coals or baked in hot ashes. Meat and fowl were barbecued on sticks near a fire. Foods were preserved for later use by drying.

The writings of immigrants to North America tell of abundant food supplies. Early explorers encountered numerous wild plants, roots, berries, nuts, and lichens. When de Soto's expedition landed in Florida in 1539, the Spaniards found the area "... cultivated with fields of Indian Corn, beans, pumpkins and other vegetables, sufficient for the supply of a large army" (Spellman, 1948: 37, 41–42).

The early Spanish explorers who reached the Southwest in the 1550s witnessed extensive farming of the staple crops of corns, beans, and squash. In 1601, while visiting the Wichita communities in what is now south central Kansas, don Juan de Onate found the villages surrounded by fields of corn, beans, squash, sunflowers, and tobacco (Carlson, 1992). The Wichita people also gathered seeds, berries, roots, greens, piñon nuts, and acorns in season and hunted deer and small animals such as rabbits and prairie dogs.

Sharing food was basic to traditional Indian cultures. The 18th century naturalist, W. Bartram (1955: 401), conveyed the communal focus of the post-harvest activities among the Choctaws, Creeks, Chickasaws, and Cherokees:

> There is a large crib or granary, erected in the plantation ... and to this each family carries and deposits a certain quantity, according to his ability or inclination, or none at all if he chooses. [This produces] a public treasury, supplied by a few and voluntary contributions, and to which every citizen has the right of free and equal access, when his own private stores are consumed; to serve as a surplus to fly to for succour; to assist neighboring towns, whose crops may have failed; accommodating strangers, or travelers, afford provisions or supplies, when they go forth on hostile expeditions; and for all other exigencies of the state.

Native peoples usually ate one midday meal. Morgan (1891: 99) described the home life of native people of the late 1880s. It should be noted that Native American society was already in crisis by this time.

> The meal was prepared and served usually before the noon-day hour – 10–11:00. After its division at the kettle, it was served warm to each person in earthen or wooden bowls. They had neither table or chairs, nor plates, nor any room in the nature of a kitchen or a dining room, but ate, each by himself, sitting or standing or where most convenient for the person. Food which remained, was reserved for any member of the household when hungry. Toward evening the women cooked hominy, and put aside to be used cold for lunch in the morning or evening or for the entertainment of visitors. They had neither formal breakfasts or suppers. Each person when hungry ate whatever food the house contained. They were moderate eaters. This is a fair picture of Indian life in general in America when discovered.

What were the traditional foods of North America? As mentioned, still most popular among the world's cuisines is the pepper – from mild to spicy. Tomatoes are a close second. Prior to Columbus, the only pepper known to Europeans was the black powder made by grinding the dried berry fruit of the plant *Piper nigrum.* If the outer shell of the plant was removed before grinding, the powder was white. Native Americans used a completely unrelated plant, *Capsicum frutescens,* its fruits ranging from dark greens through bright oranges, purples, and yellows (Weatherford, 1988: 101). Hot peppers (including cayenne) were an important part of the diet along with potatoes, sweet potatoes, yams, and many fruits.

Squashes, including zucchini, were important, along with green beans, kidney, and other beans. Corn and beans were staples, along with venison, poultry, and other game. Maple sugar and syrup provided sweetening. Seafood was plentiful along the coast, replaced by fresh water fish inland. Massachusetts natives cooked clams in earthen ovens with seaweed. The Narragansetts boiled together whole corn kernels with lima beans in a mixture called succotash.

Cranberries accompanied wild turkey. Corn meal was fried into thick cakes much like Mexican tortillas, called pone by the Algonquians, or Shawnee-cake by European settlers. Spoonfuls of cornmeal were cooked in pots of hot bear fat, later called hush puppies. Jerusalem artichokes were cultivated in the south, along with tapioca from the cassava plant and native American pecans. Southeastern natives, such as the Taino of what is now called Puerto Rico, basted meats and fish with special sauces and cooked them over outdoor fires. The traditional North American diet did not include wheat or breads (except from nuts or corn – after about 1100 AD).

The Indian red bean was used by the Choctaw in what is now called Louisiana, along with gumbo filé, a sassafras flavoring, made from the leaves of the Sassafras tree, mixed with shrimp, crayfish, and other fish. Catfish stew involved potatoes and tomatoes, alongside this unusual American fish with a skin instead of scales. Squirrel stew was popular in what is now called Virginia and North Carolina, the meat being combined with corn, tomatoes, and beans. Jerky and dried meat sticks were common, as were chocolate and vanilla. Native fruits included pineapples and papayas, persimmons, papaws (*Asimina triloba*), and the maypop fruit (*Passiflora incarnata*).

The Muskogee and Creeks used 11 varieties of hickory nut and several of black walnut (from the tree *Juglans nigra*). Acorns were the staple of the California Native diet, along with pine nuts. Forty-seven types of berries have been found, further subdivided into 20 varieties of blueberries, 12 varieties of gooseberries, choke cherries (*Prunus serotina*), wild currants (*Ribes inebrians*), 4 varieties of elderberry (*Sambucus melanocarpa, S. mexicana, S. neomexicana,* and *S. coerulea*), wild grapes (*Vitis arizonica* and *V. californica*), ground cherries (*Purpalis pubescens* and *P. fendleri*), hockberries (*Celtis pallida, C. reticula, C. douglasa*), manzanita (*Arctostaphylos pringlii* (*A. pringens, A. patuli*), and squaw berry (*Rhus trilobata*).

HEALTH, HUNTER-GATHERERS, AND THE TRANSITION TO AGRICULTURE

Beginning about 10,000 years ago in the post-glacial period, foragers began to harvest some wild plants and animals more intensively than others and to engage in experiments designed to encourage the growth of their favorite species. This period of development, called the Formative Age in the New World (Neolithic in the Old World), was associated with increased cultivation, animal husbandry, and the crafts of pottery and weaving. Permanent settlements eventually developed, along with larger trading networks and complex societies. Population density increased with the creation of communities, as did diarrheal and other infectious diseases with their various vectors and parasites. Tapeworms and pinworms increased, leading to debilitation, poor absorption of food, malnutrition, anemia, and sometimes death.

This transition to agriculture was not necessarily healthy or desirable. Sonia Guillen and Marvin Allison studied the Chinchorro, who were foragers of the desert coast of northern Chile and southern Peru 9000 to 2000 years ago, and found them to be generally healthier than later agricultural people of the same area. Their skeletal remains showed that women and men had a good life expectancy for that time period. Some lived well over fifty years. Guillen's 1992 study of the Chinchorro cemeteries found little evidence of pathology in their skeletons (the normal pattern for foragers living in small groups and eating well).

Women were affected by spinal arthritis, found in one-third of all their skeletons. One-fifth of women suffered from compression fractures in their spines due to osteoporosis, perhaps aggravated by multiple pregnancies. Forty percent suffered from leg infections severe enough to damage their bones. Among 51 Chinchorro skeletons from around 2000 BC, 86.4% of adults had Harris lines, indicating some periods of starvation (4.8 per person) and/or nutritional stress. 86% of children had these lines (6.1 per individual).

An analysis of 200 skeletons of the La Paloma site on the desert coast of Peru (7000 to 4500 years ago) showed that ancient people were healthy by pre-ceramic standards, though they did suffer from tuberculosis, cancer, frequently broken foot bones, osteoarthritis of the spine, and back problems (Quilter, 1989). As these people moved toward agriculture, longevity decreased.

Worldwide, women who contribute to subsistence agriculture by cultivating plants introduce solid foods to infants earlier than hunter-gatherers. Grinding cultivated seeds and boiling them in water in ceramic vessels permits early weaning. Giving an infant gruel for lunch allows a woman greater mobility and relieves the stress of producing milk. But this practice also shortens the birth interval and spurs population growth, resulting in increased elderly mortality and decreased longevity (Crown and Wills, 1995). The average life span, for example, at Grasshopper Pueblo, from 1275 to 1400 AD, in Arizona, was 38 years (Ezzo, 1993), dramatically lower than what had been found for earlier hunter-gatherer peoples. Among the early Formative Valdivia people of coastal Ecuador (*ca.* 3000–1500 BC), comparative early adoption of horticulture led to a life expectancy at age 15 that was the lowest of all studied Ecuadorian populations – that of 38.7 years for men and 32.3 years for women. Increased mortality was related to increased infection, trauma, and interpersonal violence. Increased dental hypoplasia, caries, and abscesses were also noted in skeletal remains (Ubelaker, 2001). This pattern of health and disease mirrors the patterns seen with increased sedentary life styles, increased population densities, changing diet to include more grains, and increased warfare.

Changes in activities, scheduling, and sex and age roles all evolved concomitantly with changes in food-producing technology and the development of crafts. Gathering populations had been more respectful of the contributions provided by the elderly, whose memories included where to search for food and how to survive (Cohen and Armeledge, 1984; Cohen *et al.*, 1993).

Belief in the importance of the dead as continuing participants in community life influenced thoughts about elders. Women and men, living and dead, participated in important social activities (Guillen, 1992; Allison, 1984; Rivera, 1995). Deceased family members were expected at ceremonies and meals and were buried in the family compound, presumably because they continued to participate in family life.

Early classic Mexico showed similar changes and relationships at Teotihuacan, long in ruins when the Aztecs entered the Valley of Mexico. This city covered 29 square miles of densely packed buildings. Its citizens lived in apartment compounds. Adult women were malnourished and carried a high parasite load (Thomsen, 1969; Spence, 1974). Old people participated in spinning and weaving of textiles, among the most labor intensive and time consuming tasks performed in the home. Investigators at the site, Tlajinga 33, found virtually everyone living in compounds by the late years of the city, with 30–40% infant mortality, and 50% of family members dead by age 15. Deaths were highest at ages 3 to 5. Half of the inhabitants of Tlajinga were dead by age 40. Most of the remainder died within the next 15 years. Women's work was shared by the handful of elderly women available. The city collapsed by the 7th century AD.

Anemia was frequently seen, manifesting as porotic hyperostosis and cribra orbitalia, names of diseases found in bone, because of nutritional stress. It was common among maize-dependent populations. These signs are also seen among people with diets low in animal products.

Around 500 AD, endemic syphilis appeared to affect both sexes, indicating a 50% prevalence, because of both long-distance travel and trade. By the time of first European contact, women did all the agricultural labor. Men hunted, made war, and traded.

Maize began to become important around 800 AD. After 1100 AD, maize was the staff of life for Indians all over the East. It was the basis for the development of complex chiefdoms along the Mississippi and the Ohio River Valleys (Cook, 1984). Increasing agriculture and the cultivation of corn were associated with more arthritis of the left arm and the spine among skeletons analyzed from the post-1100 AD time frame.

The disease load increased, along with crowding, competition, and war. By 1200 AD, 25,000 people or more lived in the city of Cahokia, on the Mississippi River, suffering under crowded and unsanitary conditions, all of which fostered disease. These included no clean drinking water and no controlled sewage removal. Urban concentration was responsible for a dramatic increase in the prevalence of tuberculosis in North America. Dental disease increased with agriculture because of the presence of a sticky carbohydrate staple (Buikstra, 1984; Buikstra *et al.*, 1986).

Food production expanded through technological innovation and increased labor investment. Consequently, people worked much harder. Complex communities resulted, along with payment of tributes and taxation. Life expectancy decreased. By first contact, for example, the average life span at the Averbach site in Tennessee was 15 years for women and 17 years for men.

Eighteenth and nineteenth century Plains Indians' rank and status came from participation in aggressive activities. The Plains people had been river bottom farmers and food hunters, which quickly changed when horses appeared. War had been common, but horses and rifles resulted in an escalation of earlier patterns. Tribal ideology came to be associated with male military, political, and religious roles, with relative de-emphasis of metaphors involving creation, nurturance, motherhood, fertility, and sex (Medicine, 1983).

NATIVE AMERICAN HERBAL PHARMACOLOGY

No review of Native American medicine is complete without a discussion of the herbs that were part of medical practice.

Native herbal pharmacology was well developed when Europeans arrived in the Americas. From the Amazon came the cure for malaria – cinchona, the Peruvian bark, which contained quinine, previously used for many ailments, including cramps, chills, and heart rhythm disorders (Taylor, 1965: 78). Its cultivators, the Quechua-speaking Incas, who lived at an altitude of 1000 to 3000 meters, called this plant quina-quina.

Amazon natives used ipecac, a Quechua term for *Cephalaelis ipecacuanha* or *acuminata*, for expelling poisons and toxins and for ritual purifications. Ipecac became a cure for amebic dysentery. The Amazonians also used curare, a member of the genus *Chondodendron*, to treat lockjaw. High-altitude Incas used chilca (*Baccharis pentlandii*), a shrub of high, cold altitudes (1000 to 3000 meters), to treat inflammation, rheumatism, and bone and joint injuries. Coastal Incas used Pacific kelp, *Macrocystis*, to cure and prevent goiter.

Pine bark and needles were used to prevent and to cure scurvy, which came to European attention in 1535, when Jacques Cartier (1491–1557) became stuck in the ice with his three ships at the Huron city of Hochelaga (now Montreal). The Hurons treated scurvy with their pine concoction called *annedda*. It contained a massive dose of vitamin C, anti-oxidants, bioflavinoids, and pycnogenol.

Northern California and Oregonian Indians used *Rhamnus purshiana*, a shrub, to cure constipation. Called *cascara sagrada* by the Spanish, for sacred bark, the herb is still in widespread use.

Northeastern Americans used *Spigelia marilandica*, a vermifuge pinkroot, to treat intestinal worms. North American dogwood (genus *Cornus*) was used to reduce fevers. Bloodroot (*Sanguinaria Canadensis*, or *puccoon*) was used as an emetic, along with lobelia. Wild geranium, also called alumroot, (*Heuchera americana*) was used as an astringent. Boneset (*Eupatorium petrolatum*) was used as a stimulant (Driver, 1969).

The bark of the poplar tree or the willow tree was used for headaches and other aches and pains. This bark is now known to contain salicin, a relative of salicylate and a powerful anti-inflammatory agent. Witchhazel bark and leaves (*Hamamelis virginiana*) were used as an astringent and as a balm to soothe strained or tired muscles. Dried flowers of plants in the genus *Arnica* were applied to sprains and bruises to relieve pain and swelling.

Plants like Oregon grape root and pine bark were widely used and are now known to contain pycnogenol – a unique mixture of phenols and polyphenols, broadly divided into monomers such as catechin, epicatechin, and taxifolin, and condensed flavinoids, such as procyanidin B1, B3, B7. Pycnogenol also contains phenolic acids, such as caffeic, ferulic, and para-hydroxybenzoic acid as minor constituents (Packer *et al.*, 1999). Pycnogenol prolongs the lifetime of the ascorbyl radical (Cossins *et al.*, 1998) and protects endogenous vitamin E (Virgili *et al.*, 1998) and glutathione (Rimbach *et al.*, 1999) in human endothelial cells from oxidative stress. It modulates nitric oxide metabolism in stimulated macrophages by inhibiting both NOS mRNA expression and its activity (Virgili *et al.*, 1998). This means that it provides powerful antioxidant protection, thereby reducing the severity of or preventing the onset of chronic degenerative diseases. Antioxidants are known to delay the onset or to reduce the severity of diabetes, independent of other risk factors. Antioxidants like pycnogenol reduce the severity or slow the development of arteriosclerotic cardiovascular disease and are effective protectors against cancer promoters and mutagens.

NATIVE AMERICANS AND SUBSTANCE USE

Substance abuse accelerates aging. Contemporary Native American communities suffer from tobacco and alcohol abuse, while the youth have high rates of drug abuse. Potential substances of abuse existed in the New World, although actual abuse was rare. Woodland natives of North America smoked dried tobacco in pipes, but only for ceremonial purposes. Mexican natives and Southwestern United States natives rolled tobacco into cornhusk cigarettes and cigars for ceremonial use. The cigar was the ceremonial object for Caribbean natives and those of the Southeastern United States. Northern Pacific Coast peoples chewed tobacco with lime; Aztecs ate tobacco leaves. Creek Natives mixed tobacco with *Ilex cassine* leaves and other herbs to make the "black drink" used in rituals. Cherokee laws were particularly harsh on public intoxication, as were those of the Aztecs.

Other examples of spiritual and highly controlled use of hallucinogens include the use of *Lophophora williamsii* in Mexico and peyote or mescal cacti in Texas. The Native American Church currently uses peyote in its ceremonies, just as ancient Aztec priests did. The milder cactus, Dona Ana, or *Coryphantha macromeris*, Aztec *pipintzintli* (the leaves of *Saloia divinorum*), *ololiuqui* (seeds of *Rivea corymbsa* vine), and mescal beans of the Texas mountain laurel (*Sophora secundiflora*) were also used as spiritual hallucinogens. The Mayas and the Aztecs used *Psilocybe mexicana* and *Psilocybe cubensis* in ceremonies. The Aztecs also used *Paneolus campanulatus*, or *teonanacatl*, which translates as "food of the gods", in ceremonies. Canadian Plains peoples used a drink from the root of the marsh plant *Acorus clamus*, also called *cakanies*. Jimsonweed was used in the same way in North America. It was a hallucinogen from the genus *Datura*, and was named for its use in Jamestown, Virginia.

Native peoples in the Americas made wines and beers, but rarely with more than 3 to 4 percent alcohol. Ancient Mexicans fermented *Agave* and *Dasyliron* plants to make *pulque*, a vitamin-rich, fermented beverage. The Tahono O'odham and the Papago peoples of southern Arizona and northern Mexico made cactus wine and beer from mesquite, screwberries, maize, and even corn stalks. Atlantic Coast peoples made persimmon wine, popular with the

European colonists. North America boasted over 40 different types of alcoholic drinks at the time of Columbus' landing, all made from fruits and plants, including palm, plum, pineapple, mamey, and sarsaparilla wines (Driver: 110). Mayan mead was called *balche* and came from the fermented honey of a stingless bee.

MIND-BODY THERAPIES

Therapies that included meditation, guided imagery, and visualization were common in North America. An Arikara word for hypnosis translates as "putting them to sleep so that they dream like they're asleep, but they're really awake."

Native American healing was counseling intensive, with long discussions of the patients' emotional and family life, problems within the community, and behavior. Healers worked with spirits who advised them about the patient and helped the healer to make behavioral prescriptions for how the patient must change his lifestyle, relationships, and behavior in order to heal. I will cite one case in which an incest taboo was broken. The involved uncle confessed and begged for forgiveness. The community witnessed the healer drive the evil out of the uncle's body, and then agreed to his penance, which consisted of a series of selfless acts for the community to restore balance and therefore relieve an illness in the niece. This was an interesting example of the community's dealing with a serious problem in a healing manner so that the perpetrator's position in the community was restored along with the honor and health of the victim (Mehl-Madrona, 1997.

SPIRITUAL THERAPIES

Spiritual therapies, including ceremonies, were, and continue to be, among the most powerful examples of Native American healing. Many studies support the importance of religious commitment to longevity and health (Matthews *et al.*, 1993, 1995; Matthews, 1997). Other studies are beginning to appear on the effect of prayer on health. Byrd (1994), for example, studied 393 patients in the coronary care unit of the San Francisco General Hospital of the University of California at San Francisco Medical Center. All received standard medical care but were randomized to two groups. The group receiving distant prayer (even though they did not know that were receiving such prayer) fared better than those that did not. [Editor's note: For a detailed discussion of intercessory prayer, see Jim Tucker's article on Religion and Medicine.]

MANIPULATIVE AND ENERGY THERAPIES

Manipulative therapies were the hallmark of Native American medicine. Unfortunately, many of these therapies have disappeared, although a few devoted practitioners try to maintain them. Cherokee, for example, practiced an intricate form of manual therapy. Cherokee healing also included energy medicine with oils, crystals, and hands-on and hands-above the body techniques.

PLACE, HEALTH, MODERNIZATION, AND AGING

Despite migratory patterns, the concept of place was important to Native peoples, who had extensive traditions about how to interact with the land upon which they roamed. Migration, of course, was limited by the absence of any mode of transportation except feet and small boats.

In Native American society, the individual was subservient to place and existed within the environment in a complex web of inter-relationships. The young and the old were both important, with those of middle age intent on providing service to them.

Modern Western society is not necessarily good for the well being of elders. Tribal life, with its close-knit relationships, provides an environment for healthy aging that other cultures do not. Inadequate social programs cannot compete with life-long tribal participation for creating the sense of meaning and purpose so crucial to mental, physical, and spiritual health.

Within modern society, elders are more excluded from active participation in the lives of their children and grandchildren.

TODAY'S SITUATION

Elder care is dismal in North America today. Diabetes and other degenerative diseases are rampant. Nutrition is poor. Diets are high in refined carbohydrates, fat, and sodium, and low in meats, eggs, cheese, milk, vegetables, and fruits (Kumanyika and Hellitzed, 1985: 27).

A study of food intake patterns of members of the Standing Rock Reservation in the Dakotas in 1970 showed a high consumption of coffee, bologna, potato chips, and carbonated beverages. Frying and boiling were the most common methods of food preparation. Traditional foods were reserved for special occasions (Bass and Wakefield, 1974).

A dietary survey of 420 Hopi women and children conducted in 1974–1975 revealed that the contemporary Hopi diet consisted primarily of meat, mutton, eggs, potatoes, some canned vegetables, fruits and fruit juices, lard and other fats, coffee, tea, milk, and several commercial pastries and sweets (Kuhlein, 1981). Only one-fourth of those studied had eaten a traditional food during the 24 hours prior to the study.

Similar dietary intakes were found for Navajo women in 1979–1980, with infrequent use of traditional Navajo foods. Coffee, tortillas or fry bread, potatoes, eggs, and sugar were the foods most commonly consumed, followed by sweetened drink mixes or soft drinks, store-bought bread, mutton, beef, milk, butter or margarine, bacon, luncheon meat, chicken, and tea (Wolfe and Sanjur, 1988).

In contrast to the Hopi and the Navajo, 98.6% of Eastern Cherokee households questioned in a food practice survey conducted in 1982 stated that they obtained food from traditional sources such as hunting, fishing, and gathering wild foods. The most frequently consumed Native foods were corn, squash, beans, trout, pumpkin, and wild greens. Non-native foods frequently consumed included coffee, milk, eggs, wheat bread, soft drinks, potatoes, bacon, fruits, and fruit juices (Terry and Bass, 1984).

As with the Eastern Cherokee, Taos Pueblo residents continue to eat a variety of traditional foods (Aspenland and Pelican, 1992). Four varieties of corn are cultivated – sweet corn and yellow, blue, and white corn – and are used in a variety of dishes. Pumpkins, wild greens, wild plums and chokecherries, pinto beans, and other traditional foods are widely consumed.

Nevertheless, for the United States as a whole, Native American elders are eating poorly and are partaking of a diet conducive to diabetes, heart disease, and other degenerative disorders.

* * *

The traditional Native American lifestyle might have elements that promote successful aging more than contemporary lifestyles. These include a diet high in anti-oxidants and low in grains and dairy, an emphasis on exercise, and the importance of personal spiritual practice. Aging had meaning in Native America, and this sense of personal destiny and importance may be crucial.

Much can be learned from a study of ancient Native America that might be used to help create more graceful aging in modern society. Future studies may focus on the hunter-gatherer diet, the importance of dignity and meaning for the elderly, and the role of agriculture and cities in denigrating the role of the elderly.

BIBLIOGRAPHY

Allison, M.J. 'Paleopathology in Peruvian and Chilean populations.' In *Paleopathology at the Origin of Agriculture*, Mark N. Cohen and George J. Armelagos, ed. Orlando, Florida: Academic Press, 1984, pp. 515–530.

Anderson, M.M. *Kafka's Clothes: Ornament and Aestheticism in the Habsburg Fin de Siècle*. New York: Oxford University Press, 1992.

Aspenland, S. and S. Pelican. 'Traditional food practices of contemporary Taos Pueblo.' *Nutrition Today* 27: 6–12, 1992.

Bass, M.A. and L.M. Wakefield. 'Nutrient patterns and food patterns of Indians on Standing Rock Reservation.' *Journal of the American Dietetic Association* 64: 36–41, 1974.

Bridges, P.S. 'Changes in activities with the shift to agriculture in the Southeastern United States.' *Current Anthropology* 30: 385–394, 1989.

Bridges, P.S. 'Skeletal evidence of changes in subsistence activities between the Archaic and Mississippian time periods in Northwestern Alabama.' In *What Mean These Bones? Studies in Southeastern Bioarchaeology*, Mary Lucas Powell, Patricia S. Bridges and Ana Maria Wagner Mires, ed. Tuscaloosa: University of Alabama Press, 1991, pp. 89–101.

Bruhns, Karen Olsen and Karen E. Stothert. *Women in Native America*. Norman: University of Oklahoma Press, 1999.

Buikstra, J.E. 'The Lower Illinois River region. A prehistoric context for the study of ancient diet and health.' In *Paleopathology at the Origins of Agriculture*, Mark N. Cohen and George R. Armelagos, ed. Orlando, Florida: Academic Press, 1984, pp. 307–345.

Buikstra, J.E., L.W. Koningsberg and J. Bullington. 'Fertility and the development of agriculture in the Prehistoric Midwest.' *American Antiquity* 51(3): 528–546, 1986.

Byrd, R.B. 'Positive therapeutic benefit of intercessory prayer in a coronary unit population.' *Southern Medical Journal* 81: 826–829, 1988.

Carlson, P.H. 'Indian agriculture, changing subsistence patterns and the environment on the Southern Great Plains.' *Agricultural History* 68: 52–60, 1992.

Cohen, Mark N. and George J. Armelagos, ed. 'Summation.' In *Paleopathology at the Origin of Agriculture*, Mark N. Cohen and George J. Armelagos, ed. Orlando, Florida: Academic Press, 1984, pp. 393–424.

Cohen, M.N. and S. Bennett. 'Skeletal evidence for sex roles and gender hierarchies in prehistory.' In *Sex and Gender Hierarchies*, Barbara Diane Miller, ed. Cambridge and New York: Cambridge University Press, 1993, pp. 273–296.

Cook, D.C. 'Subsistence and health in the Lower Illinois Valley: osteological evidence.' In *Paleopathology at the Origin of Agriculture*, Mark N. Cohen and George J. Armelagos, ed. Orlando, Florida: Academic Press, 1984, pp. 235–269.

Crown, P.L. and W.H. Wills. 'The origins of Southwestern ceramic containers: women's time allocation and economic intensification.' *Journal of Anthropological Research* 51: 173–186, 1995.

Driver, H.E. *Indians of North America*, 2nd ed. Chicago: University of Chicago Press, 1969.

Ezzo, J.A. *Human Adaptation at Grasshopper Pueblo, Arizona: Social and Ecological Perspectives.* Archaeological Series 4. Ann Arbor, Michigan: International Monographs in Prehistory, 1993.

Guillen, S.E. *The Chinchorro Culture: Mummies and Crania in the Reconstruction of Pre-Ceramic Coastal Adaptation in the South Central Andes.* Ph.D. Dissertation, Dumbarton Oaks. Ann Arbor, Michigan: University Microfilms, 1992.

Harman, D. 'Free radical theory of aging: nutritional implications.' *Age* 1: 145–152, 1978.

Harman, D. 'Free radical involvement in aging: pathophysiology and therapeutic implications.' *Drugs and Aging* 3: 60–80, 1993.

Harman, D. 'Free radical theory of aging: role of free radicals in the origination and evolution of life, aging and disease processes.' In *Free Radicals, Aging and Degenerative Diseases*, J.E. Johnson Jr, R. Walford, D. Harman and J. Miquel, eds. New York: Alan R. Liss, 1995, pp. 3–49.

Harman, D. 'Aging: prospects for further increases in the functional life span.' *Age* 17: 119–146, 1994.

Harman, D. 'Aging: overview.' *In Healthy Aging for Functional Longevity: Molecular and Cellular Interactions in Senescence*, S.C. Park *et al.*, ed. Annals of the New York Academy of Sciences 928: 1–21, 2001.

Kuhnlein, H.V. 'Dietary mineral ecology of the Hopi.' *Journal of Ethnobiology* 1: 84–94, 1981.

Kumanyika, S. and D. Hellitzed. *Nutritional Status and Dietary Patterns of Racial Minorities in the United States.* Background paper for the U.S. DHHS Task Force on Black and Minority Health, 1985.

Matthews, D.A., D.B. Larson and C.P. Barry. *The Faith Factor: An Annotated Bibliography of Clinical Research on Spiritual Subjects*, vol I. Rockville, Maryland: National Institute for Healthcare Research, 1993.

Matthews, D.A., D.B. Larson and C.P. Barry. *The Faith Factor: An Annotated Bibliography of Clinical Research on Spiritual Subjects*, vol II. Rockville, Maryland: National Institute for Healthcare Research, 1995.

Matthews, D.A. 'Religion and spirituality in primary care.' *Mind/Body Medicine* 2(1): 9–19, 1997.

Lindeberg. S. and B. Lundh. 'Apparent absence of stroke and ischaemic heart disease in a traditional Melanesian island: a clinical study in Kitava.' *Journal of Internal Medicine* 233(3): 269–275, 1993.

Lindeberg, S. *et al.* 'Cardiovascular risk factors in a Melanesian population apparently free from stroke and ischaemic heart disease: the Kitava study.' *Journal of Internal Medicine* 236(3): 331–340, 1994.

Lindeberg, S., P. Nilsson-Ehle and B. Vessby. 'Lipoprotein composition and serum cholesterol ester fatty acids in nonwesternized Melanesians.' *Lipids* 31(2): 153–158, 1996.

Lindeberg, S., M. Eliasso, B. Lindah and B. Ahren. 'Low serum insulin in traditional Pacific Islanders – the Kitava Study.' *Metabolism* 48(10): 1216–1219, 1999.

Marcus, J. 'Rethinking the Zapotec urn.' In *The Cloud People*, Kent V. Flannery and Joyce Marcus, eds. Orlando, Florida: Academic Press, 1983, pp. 144–148.

Marcus, J. and K.V. Flannery. *Zapotec Civilization: How Urban Society Evolved in Mexico's Oaxaca Valley.* London: Thames and Hudson, 1996.

Medicine, B. 'Warrior women: sex role alternatives for Plains Indian women.' In *The Hidden Half: Studies of Plains Indian Women*, Patricia C. Albers and Beatrice Medicine, ed. Lanham, Maryland: University Press of America, 1983, pp. 267–280.

Mehl-Madrona, L. *Coyote Medicine: Lessons for Healing from Native America*. New York: Scribner, 1997.

Morgan, L.H. 'Houses and house life of American aborigines.' In *Contributions to North American Ethnology, Rocky Mountain Region*, vol. 4. Washington, DC: U.S. Government Printing Office, 1891.

Ortiz de Montellano, B. *Aztec Medicine, Health and Nutrition*. New Brunswick, New Jersey: Rutgers University Press, 1990.

Osgood, C. *Ingalik Material Culture*. Yale University Publications in Anthropology, No. 22. New Haven, Connecticut: Yale University Press, 1940.

Packer, L., G. Rimbach and F. Virgili. 'Antioxidant activity and biological properties of a procyanidin-rich extract from pine (Pinus maritime) bark, pycnogenol.' *Free Radicals in Biology and Medicine* 27: 704–724, 1999.

Quilter, J. *Life and Death at La Paloma*. Iowa City: University of Iowa Press, 1989.

Rimbach, G., F. Virgini, Y.C. Park and L. Packer. 'Effect of procyanidins from Pinus maritime on glutathione levels in endothelial cells challenged by 3-morpholinosyndonime or activated macrophages.' *Redox Report* 4: 171–177, 1999.

Rivera, M.A. 'The pre-ceramic Chinchorro mummy complex of Northern Chile: context, style and purpose.' In *Tombs for the Living: Andean Mortuary Practices*, T.D. Dillehay, ed. Washington, DC: Dumbarton Oaks, 1995, pp. 43–78.

Spellman, C.W. 'The agriculture of the early north Florida Indians.' *Florida Anthropology* 1: 37 and 41–42, 1950.

Spence, M.W. 'Residential practices and the distribution of skeletal traits in Teotihuacan, Mexico.' *Man* 9: 262–273, 1974.

Taylor, N. *Plant Drugs That Changed the World*. New York: Dodd Mead, 1965.

Terry, R,D. and M.A. Bass, 'Food practices of families in an Eastern Cherokee township.' *Ecology and Food Nutrition* 14: 63–70, 1984.

Thomsen, M. *Living Poor: A Peace Corps Chronicle*. Seattle: University of Washington Press, 1969.

Ubelaker, D.H. 'Health issues in the Early Formative of Ecuador: skeletal biology of Real Alto.' In *Formative Ecuador*, Richard Burger and Scott Raymond, ed. Washington, DC: Dumbarton Oaks, in press.

Ubelaker, D.H. 'Human skeletal remains from Site OGSE-80, a pre-ceramic site on the Santa Elena Peninsula, coastal Ecuador.' *Journal of the Washington Academy of Sciences* 70(1): 3–24, 1980.

Virgili, F., D. Kim and L. Parker. 'Procyanidins extracted from pine bark protect alpha tocopherol in ECV 304 endothelial cells challenged by activated RAW 264.7 macrophages: role of nitric oxide and peroxynitrate.' *FEBS Letters* 431: 315–318, 1998.

Virgili, F., H. Kobuchi and L. Packer. 'Procyanidins extracted from Pinus maritime (Pycnogenol): scavengers of free radical species and modulators of nitrogen monoxide metabolism in activated murine RAW 264.7 macrophages.' *Free Radicals in Bioogy and Medicine* 24: 1120–1129, 1998.

Weatherford, J. *Indian Givers: How the Indians of the Americas Transformed the World*. New York: Fawcett Columbine, 1988.

Wolfe, W.S. and D. Sanjur. 'Contemporary diet and body weight of Navajo women receiving food assistance: An ethnographic and nutritional investigation.' *Journal of the American Dietetic Association* 88: 822–827, 1988.

RUBEN G. MENDOZA

LORDS OF THE MEDICINE BAG: MEDICAL SCIENCE AND TRADITIONAL PRACTICE IN ANCIENT PERU AND SOUTH AMERICA[1]

> I believe it to be correct to state that no primitive or ancient race of people anywhere in the world had developed such a field of surgical knowledge as the pre-Columbian Peruvians. Their surgical attempts are truly amazing and include amputations, excisions, trephining, bandaging, bone transplants [?], cauterizations and other less evident procedures (R.L. Moodie, 1929. Ackerknecht, 1949: 637).[2]

Despite the wealth of pre-contact and colonial era documentary and archaeological sources attesting to the diversity and sophistication of medical practices in pre-colonial (pre-AD 1532) South America, much study remains to be done. This is particularly true in the case of ancient Peru and Bolivia, because some of what has been written relied on what we now see as ethnocentric interpretations (e.g., Bandelier, 1904; Moodie, 1929; Ackerknecht, 1949; Froeschner, 1992) and a radically divergent set of understandings (Hemming, 1970: 490;[3] Bernand, 1994: 88–91[4]).

Ancient America provides a compelling case study for examining the development of medical practices (Ortiz de Montellano, 1990; Schultes, 1994; Mendoza, 1997b; Mendoza and Wolter, 1997; Verano, 1998). Initial European contact with New World cultures of the early 16th century found aboriginal medical systems that embodied principles of healing that included mental, somatic, spiritual, and supernatural approaches. While the great centers of New World civilization provide our most complete record of medical practices, many regional indigenous traditions have contributed technical knowledge and expertise associated with herbal, chemical, surgical, extra-somatic, ritual, and institutionalized approaches to healing and public hygiene (Vogel 1970; Bastien and Donahue, 1981; Bastien, 1981, 1987; Sullivan, 1988).

NEW WORLD MEDICAL CONTRIBUTIONS

The corpus of ancient South American medical practices is as extensive as it is impressive. It includes gold and other metal-based dental fillings (Stewart, 1963a: 47; Stewart, 1963b, Plate 10), cranial trephination (MacCurdy, 1923;

H. Selin (ed.), Medicine Across Cultures: History and Practice of Medicine in Non-Western Cultures, 225–257.

Stewart, 1958; Hjortsjö, 1972), post-cranial surgery (Moodie, 1929; Urteaga-Ballon, 1991), coca-based and related anesthetics (Von Hagen, 1961; Martin, 1970; Bastien, 1987), intra-medullar nails (Ortiz de Montellano, 1990), "bone" transplant or replacement surgery (Moodie, 1929; Ackerknecht, 1949: 637), surgical amputation (Urteaga-Ballon, 1991; Anderson and Verano, 1996), foot surgery (Cushing, 1897), bulbed enema syringes and straight clyster tubes (Ackerknecht, 1949; Wassén, 1972), plaster (or *chicle*) casts and splints (Ackerknecht, 1949: 631; Bastien, 1987; Ortiz de Montellano, 1990), fine gauze and cotton bandaging (Ackerknecht, 1949: 637, plate 56; Verano, 1998), anesthetic snuffs (Holmstedt and Lindgren, 1972), surgical sutures and cauterization (Moodie, 1929; Cockburn, Cockburn, and Reyman, 1998), hypnosis (Cobo, 1990: 167), caesarean sections (Ortiz de Montellano, 1990), sinus surgery (Ackerknecht, 1949: 638, plate 56; Canalis *et al.*, 1981), emetics, topical anesthetics, poultices, laxatives, diuretics, and herbal (*coca*) fumigants (Chávez Velásquez, 1977; Cobo, 1990: 220), surgeons, midwives, masseuses, and psychotherapists (Guerra, 1971; Cobo, 1990: 164), birth control (Bastien, 1987), and abortion (Cobo, 1990: 164).

Because of the great diversity of American Indian medical practices, this chapter will review a small sampling specific to ancient South America, particularly from those areas that now constitute Ecuador, Peru, Bolivia, and Chile. I will further restrict my analysis by temporal considerations, as many of the great innovations addressed here came to fruition during the period extending from 650 BC through AD 1536. The information pertaining to mummies (Cockburn, Cockburn, and Reyman, 1998) extends back some 9,000 years with the Chinchorro peoples of southern coastal Peru and northern Chile (Arriaza, 1995: 68–89), while medicinal herbs have been documented as early as 13,000 years ago at the site of Monte Verde, Chile (Dillehay, 1989). Indigenous forms of neurosurgery, trephination, or craniotomy have persisted until quite recently (Bastien, 1987: 17).

The following discussion will consider (a) American Indian concepts concerning the causes and consequences of disease; (b) the pharmacological repertoire of ancient South American shamans, or medicine men and women, both prior to and since contact (*ca.* AD 1532); (c) evidence for a formally constituted and state-sponsored corps of pre-colonial medical specialists; (d) evidence for disease and trauma and medical practices intended to address such conditions, as derived from skeletal remains and mummified cadavers; and (e) cranial trephination and post-cranial and related surgical procedures. The concepts, evidence, and perspectives in question are informed by contact era and 16th through 20th century accounts, particularly from contact with the peoples and cultures of the Inca Empire. Some evidence is drawn from paleopathology and archaeology and from studies of primary data and secondary resources. In this regard, the peoples and traditions of highland Peru and Bolivia, coastal Ecuador and Peru, and the Amazon Basin and beyond are of particular relevance. These cultures lived in varied microenvironments that now correspond with the Sierra, the Coastal plain, Montaña, and the forest or selva. Their pharmacological

BUREAU OF AMERICAN ETHNOLOGY SIXTEENTH ANNUAL REPORT PL. III

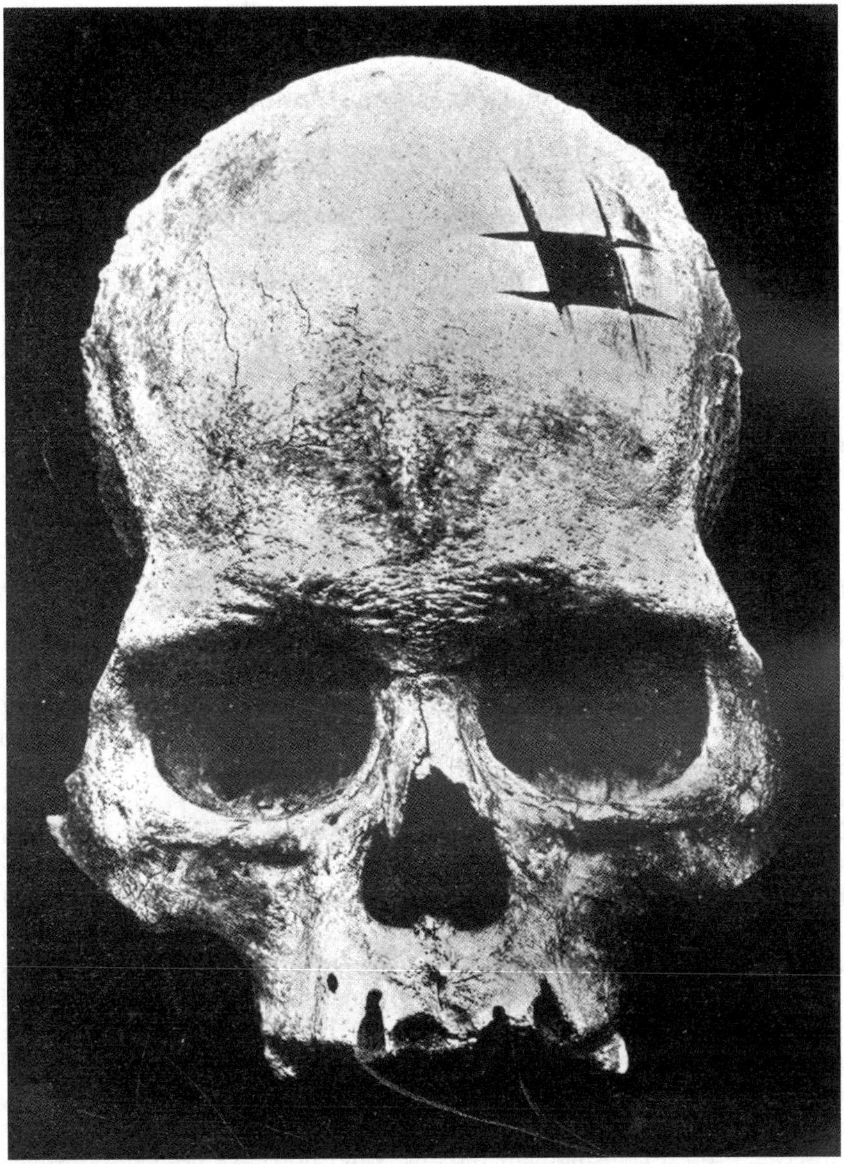

Figure 1 Anterior aspect of trephined cranium with unhealed rectilinear incision pattern (Plate III; From Muñiz and McGee, 1897).

repertoire and other medical practices are the byproduct of a much broader regional and temporal cultural continuum (see Figure 2).

HEGEMONIC DISCOURSE AND THE COLONIALIST PARADIGM

Any assessment of pre-colonial medicine must bear in mind that the European colonials condemned the exercise of American Indian religious and medical practice (Arriaga, 1621, 1968; Bastien, 1987). In fact, throughout the colonial era (*ca.* AD 1521/32–1822) and beyond, indigenous practices were considered the work of sorcerers, heretics, magicians, and charlatans in league with the devil (Guerra, 1971; Bastien, 1987; Cobo, 1990). Europeans sought to destroy ancient medical works, along with pagan practices and their practitioners. Regrettably, the practice of undermining indigenous medicine has been perpetuated in the modern era with the modernization of medical practice (Bastien, 1982: 795–796).[5] What little knowledge that remains of pre-colonial American Indian science and medicine is derived primarily from colonial era documentary and archaeological sources and recent paleopathology studies (Bastien, 1987).

THE "PRIMITIVIST" PERSPECTIVE IN THE NEW WORLD CULTURE HISTORY

While some early European chroniclers attempted to document American Indian medical practices, these same chroniclers disparaged such practices and traditions in their writings (Guerra, 1971; Cobo, 1990: 220–222). For example, Fray Bernabé Cobo's 17th-century accounts of the Inca Empire speak of the "primitive", "barbaric", and uncultured character of Inca beliefs and practices over a century after the collapse of the Inca Empire. Dr. Francisco Guerra's (1971) survey of the "pre-Columbian mind" provides a retrospective of primary documents illustrating the European predilection for writing about the more nefarious, barbarous, amoral, and brutal customs encountered in the New World. Yet only cursory treatment was given to the more mundane aspects of New World civilizations; this has left large gaps in our understanding. In those instances where a systematic scholarly effort was made to collect information on the medical practices of such groups as the Mexica Aztec, for instance (Sahagun, 1932), the distribution of such works was prohibited for centuries. And, where the sophistication of those New World contributions was seemingly incontrovertible, scholars argued that the advancements and innovations were themselves proof that American Indians could not have been responsible for their creation (Willey and Sabloff, 1974).

By contrast with earlier research, recent studies have tried to understand the epistemology of science in the non-Western world (Horton, 1967; Young, 1977; Rashed, 1980; Browman, 1981; Bastien, 1987; Harding, 1993; Schmidt and Patterson, 1995; Selin, 1997; Mendoza, 1997a, 1997b, 1997c; Mendoza and Jordan, 1997; Mendoza and Wolter, 1997; Mendoza, 2001a, 2001b). Paleopathologic examination of skeletal remains has now established a basis for documenting the existence of Andean scientific methodologies.

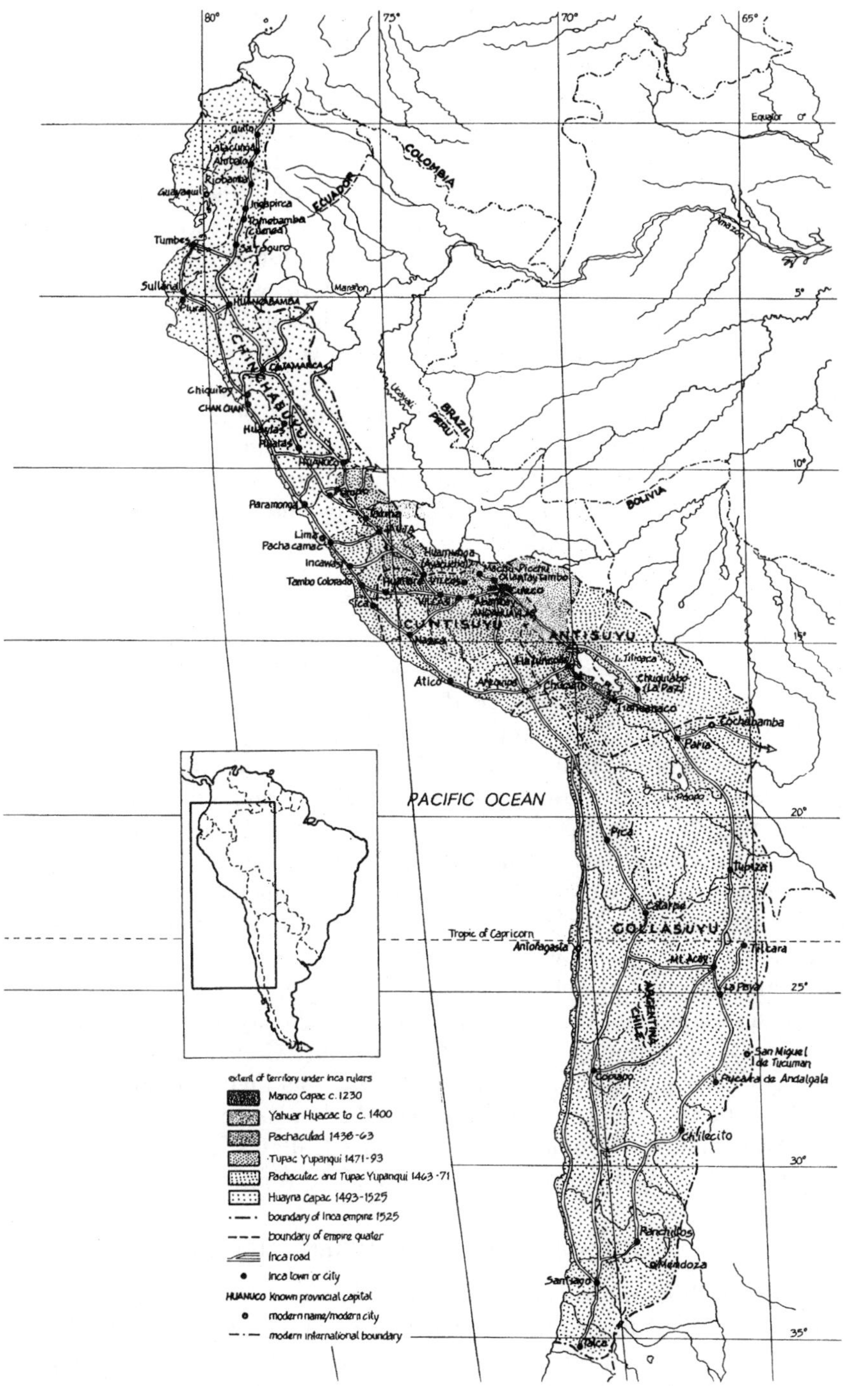

Figure 2 Map of the Inca Empire of South America. Redrawn by Juan Carlos Arias after Michael Coe, Dean Snow, and Elizabeth Benson, *Atlas of Ancient America*. New York: Facts on File Publications, 1986, p. 196.

ANDEAN COSMOLOGY AND THE HUMAN BODY

Joseph Bastien (1982, 1987) and Constance Classen (1993) have detailed the complex relationships between Andean cosmology and the human body. In addition to breaking down the human body into its constituent physical components, Andean peoples, and Inca physicians in particular, subdivided the human body into physical, cosmological (Classen, 1993) and metaphysical (Bastien, 1982) parts. Bastien (1987) has discussed Kallawaya or Qollahuaya body concepts that perceive the body as a finely tuned hydraulic system that circulates fluids and "semifluids" in the form of air, water, blood, semen, food, and related substances (Classen, 1993: 15). The human body was seen to mirror the physical cosmos, much as the ancient Inca capital city of Cuzco was seen to mirror the Inca Empire and cosmos.

Andean peoples conceived of the soul as integral both to the human body and to the cosmos that enveloped it and its parts. Stark (1969: 6) has studied the lexical structure of Quechua names for human body parts and concluded that the human body "has two members: *kirpu* (body) and *alma* (soul)", and "the soul is synonymous with the conscience and acts as the body's mentor" (see Table 1). The mentor was to be maintained in equilibrium with the body as mirror to the cosmos. Disruptions of the soul/mentor were tantamount to disruptions in the cosmos. This illustrates a way to understand the connection between the human body, moral transgressions, cosmic perturbations, and physical or spiritual illness and death.

Andean peoples believed that harmony was necessary for the balance between the spiritual and physical dimensions of human existence. Moral transgressions destabilized this balance; atonement for moral transgressions was thus critical to the healing process. According to George Urioste's (1981: 11–13) reading of a late 16th century Inca document known as the *Runa indio ñiscap machoncuna*, "among the Quechua the transgression – hucha in Quechua – is an objective state and therefore independent from both awareness and intention." The afflicted patient achieved atonement by confessing to the medicine man or woman (Acosta, 1589) who conducted the healing rite by "revealing" the transgression, applying the appropriate herbal remedy, conducting the necessary surgery, or initiating therapy (Acosta, 1589; Guerra, 1971: 145). Today, Kallawaya herbalists no longer practice surgery as their ancestors did, opting instead for herbal and extra-somatic remedies (Wassén, 1972; Bastien, 1982, 1987). According to Bastien (1987: 17), some Kallawaya herbalists no longer see surgery *per se* as an appropriate approach to treating illness and disease.

PALEOPATHOLOGY AND ANCIENT MEDICINE

Three primary bodies of evidence help us to comprehend the medical practices in pre-colonial South America. They are (a) contact period chronicles, (b) ethnographic or anthropological studies that report contemporary indigenous medical customs and beliefs, and, (c) the archaeological evidence that provides physical remains for the assessment of disease vectors and their treatment. These last examples also provide insight into mortuary and material

Table 1 Quechua somatic cosmology and geographical topology. After Stark, 1969, pp. 8–9; with revisions based on Bastien, 1987, p. 75. *Note:* Classen's 1993 study includes extensive appendices with hundreds of Quechua somatic terms of pre-Hispanic origin.

Simple lexemes	*Somatic meaning*	*Geographical association*
kunka	neck	small chasm in mountain
ñawi	eye	spring
rixra	shoulder	shoulder of mountain
sinqa	nose	mountain ridge
uma	head	mountain peak
wasa	back	other side of mountain
wixsa	stomach	small hill
siki	buttocks	foot of a hill
muqu	knee	hill
Compound lexemes	*Somatic meaning*	*Literal meaning*
caki muqu	ankle	foot hill
caki puxyu	arch of the foot	foot well
caki pampa	sole of foot	foot field
q'asqu puxyu	furrow between the chest and diaphragm	chest well
q'asqu wayq'u	area between breasts	chest ravine
q'asqu ñuñu	breast	chest hill
siki pata	buttocks	buttock ledge
siki wak'a	buttocks	buttocks furrow
wasa wayq'u	area between shoulder blades	back ravine
wasa muqu	shoulder blade	back hill
ñawih t'uqu	eye socket	eye hole
simi pata	area between nose and mouth	mouth ledge
simi wirp'a	lips	rounded mouth
simi uxu	mouth opening	mouth cavity
ninri t'uqu	ear opening	ear hole
ninri kapachu	outer ear	ear cape
sinqa t'uqu	interior of the nostril	nose hole
sinqa rapra	outer nose	nose

practices and the socioeconomic status of both patients and practitioners. Mummified remains are one of the best resources for understanding cultural responses to disease and its treatment.

Artificial mummification as cultural legacy

Artificial and natural mummies have been recovered from the many Inca era burial chambers (*chullpas*), caves, and other mortuary structures found throughout *Tawantinsuyu* (Land of the Four Quarters) – the region that once constituted the Inca Empire.

Because of the precocious[6] development of the art of mummification in coastal Peru, Paracas, and highland Bolivia, paleopathology holds great promise for a substantive body of evidence concerning the etiology of disease and its treatment (Cockburn, Cockburn, and Reyman, 1998). Several recent papers have revolutionized our understanding of how ancient South Americans treated disease. The recent recovery and examination of some 2200 mummy bundles from a 20-acre Inca period (*ca.* AD 1438–1532) cemetery in Tupac Amaru, Peru, promises to provide much new data for studies in this field.

Disease etiology and post-mortem diagnosis

Diseases and traumas identified with the peoples and cultures of ancient South America include:

(a) primary and secondary hyperparathyroidism, parathyroid hyperplasia, bilateral nephrocalcinosis, idiopathic hypercalciuria, renal lithiasis, and renal failure (Blackman *et al.*, 1991);
(b) *uta* or mucocutaneous (nasal) leishmaniasis (Urteaga-Ballon, 1991; Figures 5 through 7, and 9);
(c) tuberculosis (Buikstra and Williams, 1991), pneumonia, bronchitis, and the common cold (Von Hagen, 1961);
(d) yaws or endemic syphilis or syphilitic necrosis (MacCurdy, 1923: 289);
(e) arthritis and spondylitis deformans and arthritic spurs;
(f) tibial and Poirier's facets and Harris' lines;
(g) herniated discs, paraplegia, alteration of the lumbar column, and periostitis and osteoperiostitis;
(h) cholesteatoma;
(i) dental pathologies including dental caries, pyorrhea alveolaris, alveolar abscesses, dental fractures, antemortem dental absence, dental hypoplasias, periapical abscesses, palatal abscesses, antral fistula, tumors or cysts, and related dental diseases (MacCurdy, 1923: 277; Munizaga, 1991: 147);
(j) skeletal pathologies, including osteoporosis, osteomalacia, deformities, meningioma, and osteosarcoma with osteophytic excrescence (MacCurdy, 1923: 289; Verano, 1998; Figures 10.10 and 10.11);
(k) cranial and post-cranial tuberosities and lesions, orbitalia, parietalia, obelionic and auditory osteomas or external auditory exostosis (Arriaza *et al.*, 1998: 190), and other non-disease related pathologies resulting from
(l) malnutrition, anemia, or arsenic poisoning; and
(m) blunt trauma and bone fractures – cranial and post-cranial, including epidural hematomas, single and multiple fractures, and other injuries or lesions (Arriaza *et al.*, 1998; Verano, 1998) resulting in bone deterioration or disease.

Those diseases of the soft anatomy identified from mummified and skeletal remains or ethnohistorical accounts include:

(a) leprosy (Urteaga-Ballon, 1991: 98, Figure 10);
(b) epilepsy;
(c) mental disorder (Elferink, 1999);
(d) malignant and benign tumors;
(e) skin cancers (Verano, 1998);
(f) boils;
(g) gangrene;
(h) *sarumpi* or snow blindness (Von Hagen, 1961: 484);

(i) *sarochi* or altitude sickness (Von Hagen, 1961: 484), and many other skin diseases that included
(j) *verruga*; and
(k) *mal de pott* (Von Hagen, 1961: 484).

Medical treatments detected from the evidence of hard anatomy or mummy bundles include:

(a) surgical amputation (Urteaga-Ballon, 1991; Anderson and Verano, 1996);
(b) trephination (Aufderheide, 1985; Verano, 1998: 227, Figure 10.12);
(c) nasal surgeries (Moodie, 1929; Canalis *et al.*, 1981);
(d) bindings (Wassén, 1972) and tourniquets (Von Hagen, 1961: 486);
(e) the reduction or resetting of fractured bones (MacCurdy, 1923; Urteaga-Ballon, 1991); and
(f) bandages and cotton swabs (Ackerknecht, 1949; Von Hagen, 1961: 486).

Other illnesses or conditions for which ethno-historical information is available include hypoxia or high-altitude stress or altitude sickness (Fuchs, 1978), lethargy, and a host of conditions identified with the soft anatomy and skin. As with the case of endemic syphilis or yaws, which became a colonial era pandemic (Ortner and Aufderheide, 1991), no reported pre-colonial cases of malaria were identified in the literature reviewed for this study or reported by way of secondary accounts and surveys. It is assumed that malaria was a condition that arrived with the Europeans. Effective treatment appeared among 18th-century herbalists and healers soon after the identification of malaria in Peru (Bastien, 1987).

How Andean medicine men and women dealt with such a broad variety of disease types and trauma has been clarified in recent years through ethnobotanical and ethnomedical studies.

SOUTH AMERICAN MEDICAL PRACTICE

Much of our knowledge of contact era therapeutic practices derives from chronicles of the cultural history of both the Mexica Aztec and Inca civilizations. For example, based on such sources, medical anthropologist Bernard Ortiz de Montellano (1990) has subdivided Mexica Aztec concepts pertaining to the causes and treatment of disease into three categories: (a) supernatural or religious, (b) magical, and (c) natural or physical. The Mexica Aztec case closely parallels the interpretation and diagnosis of disease among the Inca and other Andean peoples. In fact, the following quotation might be modified to include references to the Inca solar deity, Viracocha, or the Inca royal house itself, and remain applicable to medical beliefs in Peru or the ancient American Southwest (e.g., Cushing, 1897).

> The origin of illness is complex, including and often intertwining two types of causes: those that we would call natural – excesses, accidents, deficiencies, exposure to sudden temperature changes, contagions and the like – and those caused by the intervention of non-human beings or of human beings with more than normal powers. For example, a native could think that his rheumatic problems came from the supreme will of *Titlacahuan*, from the punishment sent by

> the *tlaloque* for not having performed a certain rite, from direct attack by a being who inhabited a certain spring, and from prolonged chilling in cold water; the native would not consider it all as a confluence of diverse causes but as a complex (Lopez Austin, 1974: 216–217).

This "complex" approach to treatment required that the physician use a variety of dimensions in the course of diagnosis and treatment. Mexica Aztec, Inca or Quechua, Aymara or Kallawaya, and other American Indian doctors and physicians were required to balance herbal and surgical treatments with interpretive models of causation ranging from the supernatural and magical to the natural and physical. The supernatural and magical, encompassing astrological interpretations such as those prevalent in 16th-century Europe, were of the greatest interest to early contact-era European chroniclers; much of their scholarship consists of narratives seemingly fixated on religion and the supernatural.

Ethnobotanical knowledge

The breadth of the pharmacopoeias collected by indigenous curers, and the pharmacological knowledge developed by American Indian medicine men and women, necessitates a much broader treatment than can be provided in this essay (see Towle, 1961; Vogel, 1970; Chávez Velásquez, 1977; Lira, 1985; Bastien, 1987). The complexity of their pharmacological contributions is striking: "several species of medicinal plants" were recovered from the early Cordilleran settlement of Monte Verde, which is located in the subartic pine forests of south-central Chile (Dillehay, 1989; Moseley, 1992: 83) and is 13,000 years old (Moseley, 1992: 83).

Sources for the study of indigenous remedies

Scholarly studies provide some of the most comprehensive listings of their medical botanical specimens (Towle, 1961; Vogel, 1970; Chávez Velásquez, 1977; Lira, 1985; Bastien, 1987; Schultes, 1994). Towle (1961), Chávez Velásquez (1977), Lira (1985), Bastien (1987), and Schultes (1994) completed surveys of the ethnobotany and pharmacology of the peoples of pre-colonial, colonial, and post-colonial South America. Schultes (1994: 26) alone documented "more than 1,500 species valued by the Indians for their biological activity as medicines, narcotics or poisons."

However broad the diversity of the ethnobotanical specimens at our disposal, efforts to interpret ethno-historical and contact era chronicles is complicated by the fact that early chroniclers often modified indigenous wisdom in order to make it palatable to European audiences.

Ethnobotanical contributions

Studies of the "hidden chemical wealth of plants" used by the American Indian tribes of the Amazon rain forest provide other examples of their range and extent (Schultes, 1994). In his seminal study of the pharmacology of the Kofan and Witoto tribes of the Amazon Basin, ethnobotanist Richard Evans Schultes observed that, "the forest peoples' acquaintance with plants is subtle as well as extensive. The Indians often distinguish 'kinds' of a plant that appear indistinguishable, even to the experienced taxonomic botanist" (Schultes, 1994: 24). This taxonomic acuteness extends to the level of distinguishing between chemovars, or the basic chemical constituents of a specific sub-variety, by visual inspection alone. However, of the estimated 80,000 species of higher plants in the Amazon, fewer than 10 percent have been "subjected to even superficial chemical analysis."

There is evidence of use of such medicinal and hallucinogenic substances as coca (*Erythrox coca*), mescaline (*Lophophora williamsii*), curare or tubocurarine (*Chondrodendron tomentosum*), nicotine (*Nicotiana tabacum*), quinine (*Quina cinchona*), psilocyben (*Psilocybe mexicana*), belladonna (*Datura ferox*), Ayahuasca (vine of the souls, *Banisteria caapi*), dopamine (*Carnegiea gigantea*), anodyne analgesics (*Solandra guerrerensis*), the ergot alkoloid d-lysergic acid (*Ipomoea violacea*), and genipen-based antibacterial agents (*Chlorophora tinctoria*). Coca or *cuca* (*Erythrox coca*) and tobacco (*Nicotiana tabacum*) were among the most used (Bondeson, 1972; Wassén, 1972; Sullivan, 1988). Coca was known as the "divine herb", and state-owned plantations controlled and "vastly extended under Inca rule" tightly regulated its development (Von Hagen, 1961: 488). Its many ancient uses, and its application as a local anesthetic, are widely known, and according to the chronicler Garcilaso de la Vega (Von Hagen, 1961: 487), coca "preserves the body from many infirmities and our doctors ... use it in powder form to arrest and placate the swelling of wounds." The chewing of coca leaves, and the drinking of coca tea, is still used today to treat hypoxia in the high Andes.

To this list may be added hundreds of medications and related chemicals and supplements ranging from N-dimethylhistamine to atropine, seratonin, tryptamine, kaempferol, prosopine, pectin, and camphor to name but a few (see Table 2).

The South American medicine cabinet included diuretics, laxatives, sedatives, muscle relaxants, soporifics, purgatives, astringents, emetics, hemostats, hallucinogens, anesthetics, oxytocics, diaphoretics, and anthelmintics (Bastien, 1987; Schultes, 1994). Furthermore, evidence from both Mexico and Peru shows that there was a wide variety of antibiotic and antiseptic treatments for treating wounds, medicating infections and fractures, and performing surgery (Ackerknecht, 1949; Bastien and Donahue, 1981; Bastien, 1987; Chávez Velásquez, 1977; Cobo, 1990; Cushing, 1897; Fuchs, 1978; Holmstedt and Lindgren, 1972; Lopez Austin, 1974; Martin, 1970; Schultes, 1972, 1994; Towle, 1961; Vogel, 1970; Wassén, 1972). These include the herbal vasoconstrictor *comelina pallida*, *maguey* or *agave* sap, for its hemolytic, osmotic, and detergent effects, hot urine in lieu of other available sources of sterile water, and mixtures of salt and honey as antiseptics (Ortiz de Montellano, 1990).

Medical specialists and personnel

Some scholars have argued that while American Indians had significant medical innovations, the absence of formally sanctioned or professionally organized

Table 2 Selected ethnobotanical contributions of the Kallawaya. After Bastien, 1987, pp. 165–170. *Note:* This list represents a very small selection of Bastien's list.

Name	*Botanical taxonomy*	*Properties*	*Medicinal uses*
Ajenjo	*Artemesia absinthium*	Vermifuge	mate: intestinal parasites
Aji	*Capsicum pendulum*	Stimulant	topically: facial paralysis; colds, swollen gums
Alhucema	*Lavandula angustifolia*	Analgesic	massage: arthritis, rheumatism
Amaqari	*Bocconia integrifolia*	Expectorant	mate: colds
Amapola	*Papaver somniferum*	Tranquilizer	mate: insomnia, nervousness, coughing
Anís del Campo	*Tagetes pusilla*	Sudorific	mate: colds
Azucena	*Lilium candidum*	Emolient	cast or dressing: fractures, luxations, and blows
Bilyea	*Psoralea pubescens*	Febrifuge	mate: dysentery, typhoid, malaria, diarrhea
Borraja	*Borago officinalis*	Sudorific	mate: colds, respiratory congestion
Chilca	*Baccharis pentlandii*	Disinflamatory	wrap: arthritis, rheumatism
Chinchi Chinchi	*Solanum nitidum*	Vermifuge	wash: chiggers, fleas, lice
Chinchircuma	*Mutisia acuminata*	Cardiotonic	mate: heart palpitations, muscle/nerve spasms
Ch'ujriquella	*Eupatorium* sp.	Glutinous	plaster: strains, sprains, fractures
Ch'ulco Ch'ulco	*Oxalis* sp.	Disinflamatory	topically: burns
Coca	*Erythroxylon coca*	Analgesic	topically: wounds, headaches
"	"	Stomachic	mate: gastritis, dyspepsia
"	"	Stimulant	masticate: altitude sickness
"	"	Etiology	casting leaves: divination
Floripondio	*Datura sanguinea*	Linament	massage: rheumatism, arthritis, paralysis
Genciana	*Gentiana* spp.	Control sugar	mate: diabetes
Hediondilla	*Cestrum matthewsii*	Emmenagogue	mate: delay of menses
Llantén Silvestre	*Plantago tomentosa*	Stomachic	mate: diarrhea, indigestion
Macha Macha	*Gaultheria* sp.	Stimulant	massage: muscular debility
Manzanilla	*Matricaria chamomilla*	Biliary regulant	mate: bilious, biliary colic
Nacha Nacha	*Erodium moschatum*	Dissolvent	mate: tumors
Nuez moscada	*Myristica fragrans*	Stimulant	masticate: Bell's Palsy, paralysis
"	"	Absorbent	mate: intestinal debility
Ortega Colorada	*Cajophora* sp.	Nutritive	mate: vitamins
Ortega Negra	*Urtica echinata* L.	Analgesic	mate: maternity
Paico Lombrio	*Chenopodium*	Vermifuge	suppository: parasites
Pampa Marancela	*Astragalus uniflorus*		mate: heart disease, nerves
Pampa Yareta	*Azorella biloba*	Glutinous	cast or plaster: luxations, fractures, swelling
Quina Cascarilla	*Cinchona calisaya*	Febrifuge	mate: malaria, typhoid
Quina Quina	*Myroxylon balsamum*	Analgesic	topically: earache
Romero	*Rosmarinus officinalis*	Analgesic	gargle: toothaches, mouth infections
Sábila	Aloe vera	Emolient	topically: skin, eyelids
Salvia de la Puna	*Lepechinia* cf. *meyennii*	Analgesic	wrap: rheumatism
Salvia Grande	*Salvia haenkii*	Thermic	mate: hypothermia
Santa Maria	*Chrysanthemum parthenium*	Abortifacient	mate: abortions
Saúco	*Sambucus peruviana*	Pectoral	mate: pneumonia, Tuberculosis, asthma
Sayre	*Nicotiana rustica*	Linament	massage: Bell's Palsy, pimples, paralysis
Solimán	*Plumbago coerulea*	Insecticide	topically: insect, vermin
Tamarindo	*Tamarindus indica*	Stomachic	pulp: intestinal disorder
T'ula	*Lepidophyllum guadrangular*	Vulnerary	topically: wounds
Verbena	*Verbena hispida*	Biliary	mate: biliary colic
Waji	*Aristoloquia fragantissima*	Antidote	direct: snake bite
		Analgesic	direct: headaches, dizziness, sinus
Warankayso	*Ranunculus praemorsus*	Caustic	topically: warts
Yawar Chunka	*Oenothera multicaulis*	Hemostatic	poultice: black/blue injuries, swelling, hemorrhage

scientific guilds makes these contributions irrelevant to the broader tradition of Western science (Bastien, 1982: 795). Moreover, since early Europeans were intent on "the extirpation of idolatry" (Arriaga, 1621; 1968), they reduced the role of medical specialists from American Indian communities (Bastien, 1982, 1987). However, surviving documents both indicate the sophistication of medical practices and document the existence of medical confraternities and specialists.

The Yauyo medical community

Archaeologist C.A. Burland (1967: 68) informs us that, in the coastal mountain region above Lima, Peru, there existed an ethnic enclave known as the Yauyos with whom the Inca collaborated to train specialists in the art of cranial trephination or skull surgery. The Inca state paid for their services or allowed them to forego taxation in exchange for their expertise and medical work. During surgery the patient was drugged, and pain was alleviated by applying direct pressure to nerve endings in the affected area (Burland, 1967: 68). In the case of skull fractures, trephination was used only as a last resort, when the diseased or smashed bone was cut away from the skull and the affected area was then sutured (see Figure 3).

Other forms of surgery included the "removal of a torn spleen, cleaning out

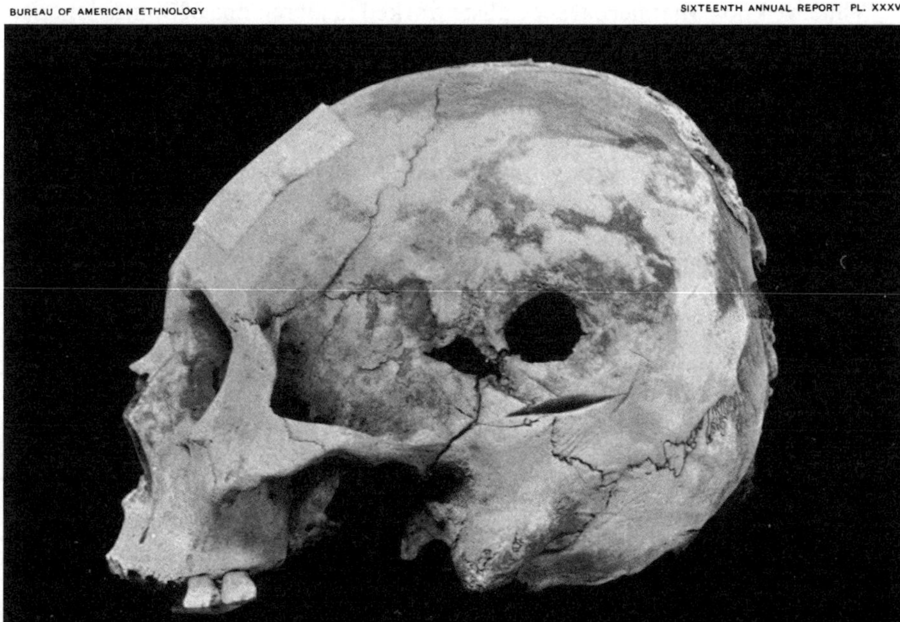

Figure 3 Left aspect of cranium 17, from Tarma, Peru, with evidence of both healed semicircular trephining and blunt force trauma sustained from axe blows. (Plate XXXV; From Muñiz and McGee, 1897.)

of ulcers, and the cleaning and after-care of wounds" (Burland, 1967: 68). Professional alliances between the Inca and Yauyo allowed for the exchange of technical knowledge and technicians. Julio Tello (1913; MacCurdy, 1923: 239–240) has suggested that the numbers of trephined crania from the Peruvian archaeological site of Huarochiri may provide indications of its importance as a center for cranial trephination and related surgical practices. Huarochiri has been attributed to the Yauyo peoples, and the evidence provides another point of corroboration for historical accounts regarding trephination in Yauyo society (MacCurdy, 1923: 239; Burland, 1967: 68). In Tello's day, "some 10,000 crania and mummies" were recovered from Huarochiri.

Similarly, we are informed that herbal specialists of highland Peru were apparently organized into local confraternities, while medical specialists and physicians (*camasca* or *soncoyoc*) were sponsored by the Inca state and "highly trained within their own kind of collegiate discipline" (Burland, 1967: 68; see also Tello, 1913). This latter point helps establish the existence of a specialized corps of medical experts and the makings of a formally constituted and state-sponsored scientific tradition specializing in the trephination of the human cranium (Tello, 1913: 75–83). Inca sponsorship of technical communities is also seen in the relationship of the Inca royal house with the Callajuayas or Kallawaya herbalists and medical practitioners of the province of Bautista Saavedra in the Department of La Paz, Bolivia (Stark, 1972; Saignes, 1984; Bastien, 1987).

While we know that herbal specialists worked in all regions of South America, the existence of a broader corps of trained medical specialists was less evenly documented. According to Burland (1967: 68), "They had a wide knowledge of herbal medicine, using herbs in a strictly scientific way. Details of cases were recorded, and the treatment and results discussed by the healers, who were loosely organized into local confraternities." Bastien's 1987 studies of the Kallawaya herbalists and surgeons support these observations, as do related contact era accounts and archaeological evidence (Chávez Velásquez, 1977; Bastien, 1987; Cockburn, Cockburn, and Reyman, 1998).

An ancient medicine man's implements

The ancient state of Tiwanaku or Tiahuanaco in the Bolivian altiplano has been compared to the ancient metropolis of Teotihuacán, Mexico (Millon, 1967). In its present state, the ruins of Tiwanaku are an impressive constellation of massive monolithic monuments, including stelae and gateways, monumental structures, sunken courtyards, elite residential sectors, and residential foundations and artifacts extending some 8 to 10 kilometers beyond the urban core (Moseley, 1992: 203). Tiwanaku's origins extend back to 400 BC, but the height of monolithic construction and imperial development encompass the period extending from approximately AD 100 through 700, with the decline and collapse of the metropolis taking place by AD 1000. In Classic Tiwanaku Phase 4 (*ca.* AD 375–600/700) "the metropolis attained truly imperial standing, establishing a hierarchy of administrative centers, as well as far-flung colonies and caravanserai throughout the altiplano, the dry puna, and the lowlands of southern Peru and northern Chile" (Moseley, 1992: 203). It was during this phase – *ca.* AD 400/650–850 – when the medicine man reported by Wassén (1972: 29) was buried in a tomb not far north of Tiwanaku at Calliicho (Bastien, 1987: 14). The burial content attributed to the medicine man identifies both medical paraphernalia and pharmacologically significant botanical remains.

Wassén (1972) initiated the study of the "specialized equipment" recovered from the tomb at Calliicho or Niño Korin (Wassén, 1972: 11). By inspection and comparative analysis, Schultes (1972) then identified a significant body of the preserved plant material as that of *Ilex guayusa* – a caffeine-rich plant used as a stimulant, purgative, and snuff. Holmstedt and Lindgren (1972) conducted an alkaloid analysis that identified *Ilex guayusa.* Hjortsjö (1972) investigated the cranium from the tomb at Calliicho that bore evidence of having been artificially deformed and trephined on three separate occasions. Bondeson (1972) identified the presence of tobacco (*Nicotiana rustica L.*) residues, stems, and leaf fragments contained in leather pouches. According to Wassén, the hoard contained "the equipment of a medicine man with containers, spatulas, enema syringes", as well as "interesting snuff trays and a corresponding tube, bag-like pouches with leaves of Ilex Guayusa, and, finally, an artificially deformed and trepanned cranium" (see Figure 4).

Much of the material recovered originated in highland, puna, and lowland forest, in much the same way that the so-called "health-tables" of the Kallawaya herbalists of today are mandated to incorporate elements from the mountains, montaña, forest, and coast (Bastien, 1981: 28, Figure 2.2). They also found a wooden probe and wood and bone pestles (Wassén, 1972: 84, Figure 30), as well as other spoon-like spatulas, portable mortars for macerating botanical materials, gourd containers for powdered herbs and minerals, and other reed or cane tubes and straws, bamboo containers, fur (howler monkey, alpaca, llama, and deer) pouches, baskets, textiles, an elongated shawl pin, and a staff or scepter with the image of a parrot (Wassén, 1972: 27, and 62, Figure 1).

The five elaborately carved wooden snuff trays and reed tubes from the tomb bore iconography directly identified with the site of Tiwanaku at AD 600–1000 and the region of Lake Titicaca, or perhaps much earlier (Wassén, 1972: 30). These snuff trays were very likely employed in the use of hallucinogenic snuffs, such as *Anadenanthera columbrina*, used at the ancient site of Tiwanaku during the Middle Horizon period. It has also been suggested that the enema bags and clysters tubes recovered from the tomb may well have facilitated the rectal ingestion of the botanical specimens that were also recovered (Wassén, 1972: 24).

Evidence from the region of Lake Titicaca makes clear how much the peoples of that area used medicinal plants. According to David Browman (1981: 115–118), "one surprising aspect of our Lake Titicaca archeobotanical materials was the amount of plants with only medical uses" at 1000 BC. Three thousand years later, Kallawaya healers of this same region of Bolivia supply much of the medicinal repertoire still used by Andean peoples.

Kallawaya curers can, for example, calm mental illness by *floripondio* (datura), set bones with

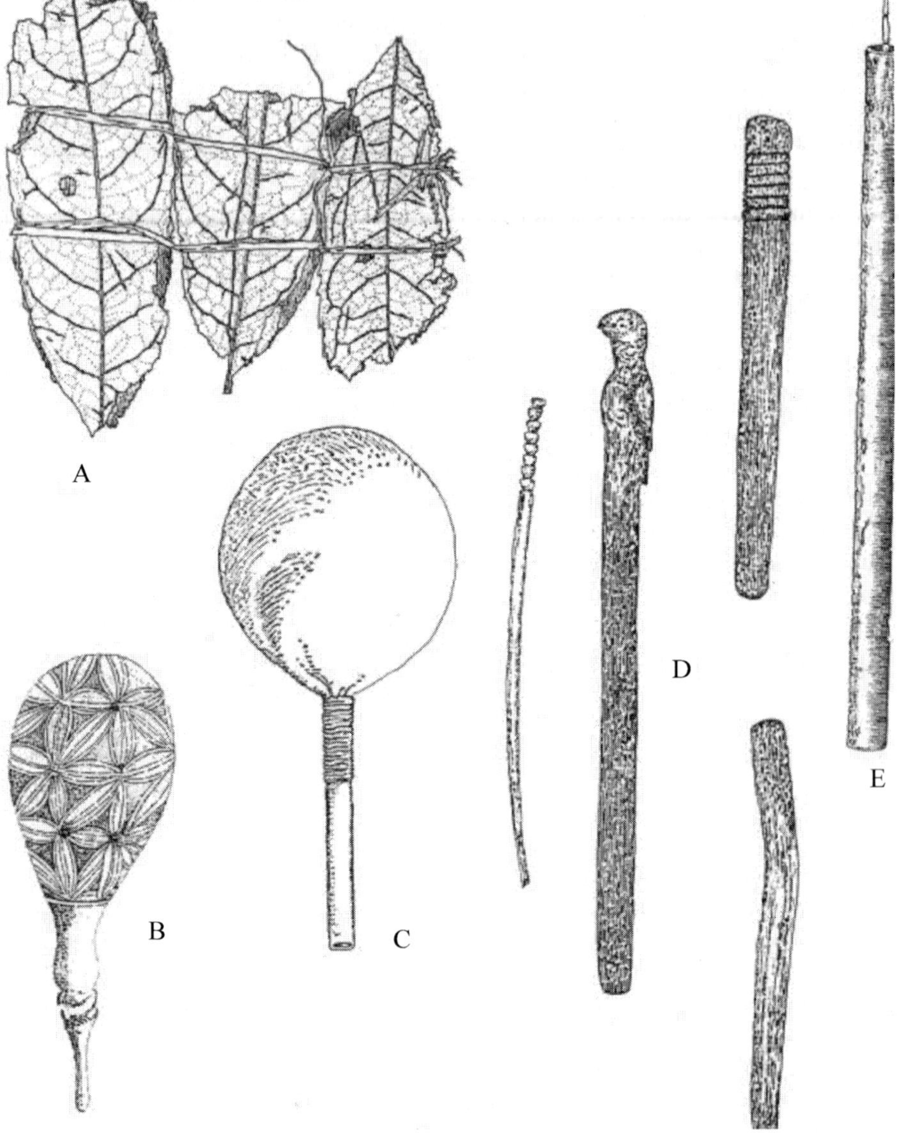

Figure 4 Examples of South American medical paraphernalia (not to scale) redrawn by Juan Carlos Arias. Legend: (A) *Ilex Guayusa* leaves from Bolivia. Original drawing by E.W. Smith, after Schultes, 1972, p. 119; (B) Enema syringe from the Guiana and Amazon region, after Ackerknecht, 1949, p. 630; (C) Enema syringe from a grave at Ollachea, Peru, consisting of a reed tube and leather bulb; after Ackerknecht, 1949, p. 630; (D) "Specialized equipment" from Niño Korin, pictured in clockwise order from left to right: a wooden probe, parrot pestle or ceremonial staff of wood, a second wooden pestle, and a pestle of bone, after Wassén, 1972, p. 84; and (E) Enema tube of the Chocó Indians of South America; after Ackerknecht, 1949, p. 631.

plaster of Paris casts, and stretch muscles with frog skins. Their pharmacy includes remedies that are nature's equivalent of aspirin, penicillin, and quinine and others that have yet to be discovered by modern medicine (Bastien, 1987: 5).

Browman (1981: 112–117) also documented anti-toxin agents from archaeological evidence, such as *Malvaceae sp.* (antiphlogistic, poultice, cataplasm, expectorant, calmative emollient, purgative enema, and mucilaginous) and plantago (diuretic, gargle, poultice, cataplasm for sores and ulcers, medicinal wash, enemas, anti-tumor agent, snake anti-venom).

"Land of the Medicine": The Kallawaya medical community

The Kallawaya community of highland Bolivia has maintained a tradition of medical practice for over one thousand years (Bastien, 1981: 20). The healers of that community were so highly regarded by the Inca royal house that they were called upon to serve as the honored litter bearers to a succession of Inca emperors (Guamán Poma, 1936: 331; Wassén, 1972: 16). The Aymara Indians classify healers into two groups: (a) herbalists or *curanderos* like those of the communities of Chajaya and Curva, who "for centuries ... have passed along from father to son a vast pharmacology of roots, plants, minerals, and exotica" and a host of medical applications that they gather "from the jungle to the mountain top," and (b) diviners who "specialize in curing, predicting, and other functions using ritual as their primary method" (Bastien, 1981: 20). Fray Bernabé Cobo (1990: 164) noted this same distinction for therapeutic specialists available to contact-era Inca or Quechua peoples; his narrative distinguishes between the *sortilegos* (diviners) and the *camasca* or *soncoyoc* (herbalists). In addition, the Inca referred to such healers as *hampi-camayoc*, literally, "remedy keepers" (Von Hagen, 1961: 20).

The Cashinahua peoples of southeastern Peru differentiate between *huni bata dauya* (sweet medicine), and *huni mukaya* (bitter medicine) herbalists based on the type of remedies employed (Kensinger, 1985: 283–288). The *huni dauya* is charged with treating two types of disease: "(1) those which the patient 'has' *haya*, including fungi, swellings, sores, dysentery, and boils; and (2) those which the patient 'feels' *tenei*, or 'is,' such as pain, nausea, laziness, sleeplessness, failure in hunting, anger, selfishness, and feelings of personal inadequacy." Both of these types have physical origins, whereas those diseases treated by the *huni mukaya* have supernatural origins. It is this latter class that has so often drawn the attention of European chroniclers and anthropologists. The use of divination, prophecy, hallucinogens, obligatory fasting, trance-like states, sucking, and the removal of foreign objects from the body of the afflicted are most commonly noted.[7] Even so, the shaman will still prescribe herbs that must be obtained from the "sweet medicine" herbalists (Kensinger, 1985: 286).

Bastien (1981) noted that the majority of Qollahuaya diviners were women. In the Aymara community and *ayllu*[8] of Ayllu Kaata, there were 46 male to 61 female healers. The herbalists, on the other hand, were predominantly men. Because of their facility and expertise, Qollahuaya diviners are as indispensable to curing as herbalists (Bastien, 1981: 21–23).

Because of their links with diverse microenvironments – ranging from the forests of the eastern slope of the Cordillera to the highest elevations of the Andes – Qollahuaya herbalists have been integral to the health system in the Andes for the past 1000 years. The "vertical archipelago" is the concept that John Murra (1972) formulated to identify and explain this system of indigenous adaptation to the extremes of altitudinal and seasonal variation. Similarly, because of their multilingual proficiency in the Aymara and Quechua languages and the Machaj-Juyai dialect, the "secret language of the Qollahuayas" (Stark, 1972), the Qollahuaya providc a critical bridge for medicine across the cultures of the Andes.

Other New World medical communities

According to medical historian Gordon Schendel (1968), in Mesoamerica technical distinctions were made between "old" and "new" school physicians. This same sort of distinction was made in Peru regarding the competence or experience of the physician (de la Vega, 1961; Cobo, 1990). At European contact, old school physicians were thought of as more traditional and more adept at conveying therapeutic and spiritual beliefs, whereas the physicians of the new school appeared to use specialized methods and procedures.

CRANIOTOMY AND ANCIENT SURGERY

Pre-colonial surgery was quite sophisticated. Paleopathologist John Verano (1998: 225) has noted that while osteological specimens that provide evidence of surgery for amputations are limited in the archaeological literature (Tyson and Alcauskas, 1980; Merbs, 1989; Urteaga-Ballon, 1991; Anderson and Verano, 1996), "evidence of trephination ... is extensive and widespread, both geographically and temporally." Verano alone has documented over 1000 individually trephined specimens from Peru. This number far exceeds the total number of trephined specimens recovered from archaeological sites worldwide. This is all the more significant when one considers that virtually all of the individual Andean specimens constituting the Muñiz collection (Muñiz and McGee, 1897) – comprised of some 1000 Andean trephined skulls collected by Muñiz himself for exhibition at the World's Fair – were destroyed in a fire resulting from an early 20th century Latin American insurrection.

The enigmatic silence of an "undocumented" tradition

Contact-period European and Inca chronicles are oddly silent with regard to Andean medical and surgical practices. Bastien (1987) and others have shown that Inca era surgeons and medical practitioners were secretive about their talents (Cobo, 1990: 222). This may well have roots in the relative prestige associated with the medical arts and in an Inca worldview that saw the human body and the cosmos as homologous entities (Bastien, 1987; Classen, 1993).

Cross-cultural analogs in New World surgical practice

Given the paucity of Inca era accounts, we must search other areas of ancient America for verification of surgical practices in other parts of the New World. The Mexica Aztec civilization of central highland Mexico or Mesoamerica is the most socially and culturally consonant with the Inca Empire.[9] Medical historian Miguel Guzman Peredo cites Fray Bernardino de Sahagun (1932):

> Cuts and wounds on the nose after an accident had to be treated by suturing with hair from the head and by applying to the stitches and the wound white honey and salt. After this, if the nose fell off or if the treatment was a failure, an artificial nose took the place of the real one. Wounds on the lips had to be sutured with hair from the head, and afterwards melted juice from the maguey plant, called *meulli*, was poured on the wound; if, however, after the cure, an ugly blemish remained, an incision had to be made and the wound had to be burned and sutured again with hair and treated with melted *meulli* (Guzman Peredo, 1985).

This illustrates the use of prosthetic or cosmetic devices (Urteaga-Ballon, 1991) and also describes the use of sutures and corollary therapeutic procedures. Moodie documented instances of successful Andean bone transplantation, replacement and regeneration (1929: Fig. 17, Skull 55, Cuyungo, Peru). The application of *maguey* or *agave* sap (cf. Towle, 1961: 2) as an antibiotic ointment has also been documented in Mesoamerica. The *Agave americana* species identified in Peruvian archaeological sites could only propagate beyond its indigenous Mesoamerican hearth by way of human transport; it cannot cover such distances by the natural dispersal of its fronds or shoots alone (Mendoza, 1990, 1992). This provides evidence for the long-distance transport of the fronds of such plants, by way of direct trade or exchange, from Mesoamerica into highland Peru. Similarly, the use of the sap or *meulli* for a variety of medical needs in both Mesoamerica and Peru is widely documented (Mendoza, 1990, 1992; Mendoza and Wolter, 1997).

Craniotomy, trepanning or trephination

One of the least understood surgical practices performed in ancient America is cranial trephination (also spelled "trepanation" after the Greek term *trypanon* for "borer"; cf. Margetts, 1967: 673), or the surgical modification or penetration and "removal of a portion of the skull vault" (Verano, 1998: 225–226). Trephination was practiced widely in the region of modern Peru and Bolivia for over 2000 years, from *ca.* 500 BC until the sixteenth century AD (Verano and Williams, 1992; Verano, 1996; Verano, 1998: 226). Other evidence extends the timelines for the antiquity of trephination from 650 BC at Paracas on the southern Pacific coast through AD 1980 in highland Bolivia and Peru (Bastien, 1987: 17).

Much of the history pertaining to trephination in Peru and South America is largely dated or is dominated by an older generation of primitivist viewpoints. These generally ascribe the practice of trephination to spirit release and possession (Bandelier, 1904; Margetts, 1967; Hemming, 1970; Froeschner, 1992; Bernand, 1994). However these analyses ignore the substantive body of evidence available from trephined specimens and a large body of surgical instruments and procedures documented in archaeological and historical studies (Daland, 1935; Hjortsjö, 1972; Tyson and Alcauskas, 1980; Aufderheide, 1985; Mann, 1991; Verano and Williams, 1992; Verano, 1996, 1998).

Medical historian Guido Majno (1975) notes that trephination was used as a cure for a number of ailments including skull fractures and related traumas, epileptic seizures, and insanity. Mann (1991) suggests that chronic ear infection may have been another cause for the operations. Canalis, Hemenway, Cabieses, and Aragon (1981) have documented sinus surgery that entailed trephination of the frontal sinus for the purposes of reducing nasal or mucocutancous leishmaniasis – a disease that results in the spread of a necrotic lesion and the ulceration of the sinus or nasal cavity. Many of the trephined specimens examined to date bear evidence of blunt trauma or other pathologies that may provide a more direct indicator of why the practice was carried out in the first place (MacCurdy, 1923; Daland, 1935; Canalis, Hemenway, Cabieses, and Aragon, 1981; Mann, 1991). Because Bolivian medicine men and women were still practicing trephination well into the 20th century (i.e., 1895 and 1976; Bandelier, 1904, and Bastien, 1987: 17, respectively), we have a better assessment of the variables underlying this medical procedure.[10]

Conflict and battlefield medicine

"As a result of chronic warfare, many people suffered serious lesions which demanded different types of surgical operations" (Urteaga-Ballon, 1991: 95). Moche or Mochica (Donnan, 1976) pottery provides some of the "most dramatic representations" of trephination on record anywhere in the ancient world. Moche ceramics depict Andean rulers, rituals, and daily life, and include images of warlords, weapons, warfare, and the casualties of battlefield injuries. Urteaga-Ballon's review of Moche pottery – replete with depictions of amputees, some of whom bear massive facial wounds – documents the use of pre-colonial prosthetic devices, such as pegs, "suction sockets", caps or plugs, and crutches. The pots portray the tragic consequences of war and violence, as well as those therapeutic measures used to heal the injured.

There was great variation in craniotomy procedures. In one analysis of a trephined skull collection consisting of 112 specimens (the Tello collection housed at the Peabody Museum of Harvard University), 53.6 percent of the skulls, or 60 specimens, bore trephined areas in the frontal area of the skull (26 of which were situated in the frontal bone, 12 in the region of the bregma, and another 12 crossing over the left coronal suture); 33 percent, or 37 specimens, were trephined in the parietal area (18 of which bore surgical openings over the left parietal, while 15 crossed the sagital suture), and 13.4 percent, or 15 specimens, were trephined over the occipital area – with 7 bearing trephined perforations in the region of the lambda and 4 in the occipital bone (Tello, 1913; Stewart, 1963a: 46). Most trephined specimens from the Harvard collection bear surgical excisions within the frontal areas of the cranium, and trephined openings are most highly correlated with blunt trauma and other pathologies (Muñiz and McGee, 1897; MacCurdy, 1923: 288). In these instances, the trephination was undertaken to relieve the effects of depressed fractures created by stone and metal shock weapons such as mace heads, clubs, and bola stones employed in combat (Pizzaro, 1921; MacCurdy, 1923: 288; Daland, 1935; Bram, 1941; Verano, 1996, 1998). Significantly, of some 273 crania examined by MacCurdy (1923: 288), 47 or 17 percent had been trephined at least once, and of these, fully 28 percent were apparently trephined for the purpose of "relieving depressed fractures."

Therapeutic objectives

Cranial trephination in Peru was primarily a therapeutic measure for treating (a) antecedent fracture, (b) simple trauma of the cranium that denuded the periosteum, (c) circumscribed periostitis, possibly of traumatic origin; or (d) lesions, of a probable syphilitic nature (MacCurdy, 1923: 240; Julio Tello, 1913: 75–83). While Tello's contemporaries largely agreed with these findings, although both McGee and Hrdlicka disagreed on the point about syphilitic lesions, MacCurdy did encounter evidence to support the idea of trephination for the purposes of removing diseased portions of the cranium, although they were not syphilitic. Both European chroniclers and Inca scribes documented the incidence of epileptic seizures, paralysis, and tumors, or osteophytic excrescence (or bony protrusions) among Inca-era peoples. As noted above, Bastien told of a modern trephination conducted in 1976, in which the medicine man used ancient methods in the excision of a malignant tumor.

MacCurdy (1923: 288) noted that of the 273 crania that he examined, several had undergone more than one, and some as many as five, trephining operations. MacCurdy concluded that "in combat, the left side of the head would be exposed to danger more often than the right, and it was precisely the left side that suffered most in the series," and "a study of the position of the fractures and of the trephining operations, leads to the conclusion that the high percentage of both marks a period of warfare among the Incas of the region in question" (MacCurdy, 1923: 288; see also Pizarro, 1921; Bram, 1941; Conrad and Demarest, 1984). MacCurdy's argument assumes that mace and club wielding right-handed Andean warriors meted out such a pattern of injuries.

While ancient Peru provides the largest body of trephined specimens available for study, other important specimens have been reported from throughout ancient America. Documented examples with evidence of osteitis (post-surgical bone regeneration) have been reported from the Mexican states of Chihuahua and Oaxaca, and from archaeological sites in Lamy, New Mexico and Accokeek, Maryland, in the United States, where trephined surgical openings are often no more than two centimeters in diameter. Specimens from Columbia County, Georgia, archaeological zones along the Skeena River, and in the sites of Ebume and Lytton, British Columbia and Kodiak Island, Alaska, bear openings that range between 3.8 and 6.0 centimeters in diameter. If all the putative cases of trephination reported in these examples are a reliable indication of areas in which the practice was known, then cranial trephination was quite widespread (Mendoza, 1997c).

Techniques employed in Andean craniotomy

Majno (1975) has identified five primary techniques of trephination. These include (a) scraping [construed as the most ancient method on record, this entails the scraping away of the cranial bone lying just beneath the scalp], (b) grooving, (c) boring, (d) cutting, (e) sawing, and (f) rectangular intersection incision [the most predominant method employed in early highland Peru, in which cut, sawed, or abraded grooves were used to perforate the skull in a rectilinear or curvilinear pattern] (Muñiz and McGee, 1897: 55; MacCurdy, 1923; Verano, 1998). "The linear cutting technique is most characteristic of the central highlands of Peru ... Circular grooving appears later, and is characteristic of the southern highlands of Peru during the Inca Empire ... A trend can be seen towards a reduction in the size of trephination openings through time, although there is substantial variability in the samples" (Verano, 1998: 226).

In linear cutting, once the linear incisions intersected, a rectangular or octagonal shaped plate of bone remaining within the intersection was removed with a spatula-like device. Schendel (1968) notes, "The trepanning technique used in Mexico was to punch a series of small holes in the skull outlining the fracture, or the area to be removed, then to cut between those holes and lift off the depressed section of cranial bone." After this initial procedure, the exposed brain was protected with a thin plate of hardwood and cotton pads and the scalp was sutured back together with human hair. Moodie (1929) disputed McGee's assertion that protective caps, suction sockets, or plugs were used (Muñiz and McGee, 1897). Subsequent finds, however, reveal that trephined specimens had retained such devices (Stewart, 1963a). Bastien documented the use of gourd plates as prosthetic devices. Specimens from Peru include trephined areas plugged with gold plates and shell fixtures or related prostheses (Tello, 1913; Stewart, 1963a: 46; Stewart, 1963b, Plate 10).

In Mesoamerica, when trephination was not deemed suitable because of the extent of skull trauma, the damaged portion of skull was encased in a protective plaster cast. In Peru, as in other areas of South America, they used *chicle* (hardened gum) or plaster for similar wraps or "casts" for limb fractures (Bastien, 1987). According to Ortiz de Montellano (1990), there were two types of plaster casts used in Mesoamerica. The first consisted of a mixture of feathers, egg whites, and resin, and the second was prepared from a mixture of animal ash, blood, and egg whites. One need only compare Frank Hamilton Cushing's (1897: 979) detailed observations of an 1890's Zuni Indian foot surgery – including the irrigation of the wound with cleansing agents and the subsequent application of a medicine-coated wrap or bandage – in order to see therapeutic parallels in post-operative procedures from other areas of the Americas. At the same time, it is clear that Peruvian surgeons often worked against great odds in their attempts to rescue victims of traumatic head injuries.

Specialized trephining tools and surgical kits employed by Inca physicians (Wassén, 1972) included flintknapped or flaked obsidian, volcanic glass, blades or lancets with micron-thin cutting edges (Whittaker, 1994: 219–242),[11] and *tumi* knives crafted from alloyed and annealed, or metallurgically hardened, copper and bronze metals (Mendoza, 1997b) (see Figure 5). The tumi knife consisted of a crescent-shaped blade, at the midpoint of which was attached a cylindrical or flat metal handle or other appendage. Tumi knives used in trephination often took the form of razor sharp, serrated, or other saw-toothed scalpels or knives. Additional instruments included bronze perforators, drills, tweezers, copper thumb forceps, needles, and chisels (e.g., Ackerknecht, 1949: 638). The basic surgical instrumentation utilized in cranial trephination long

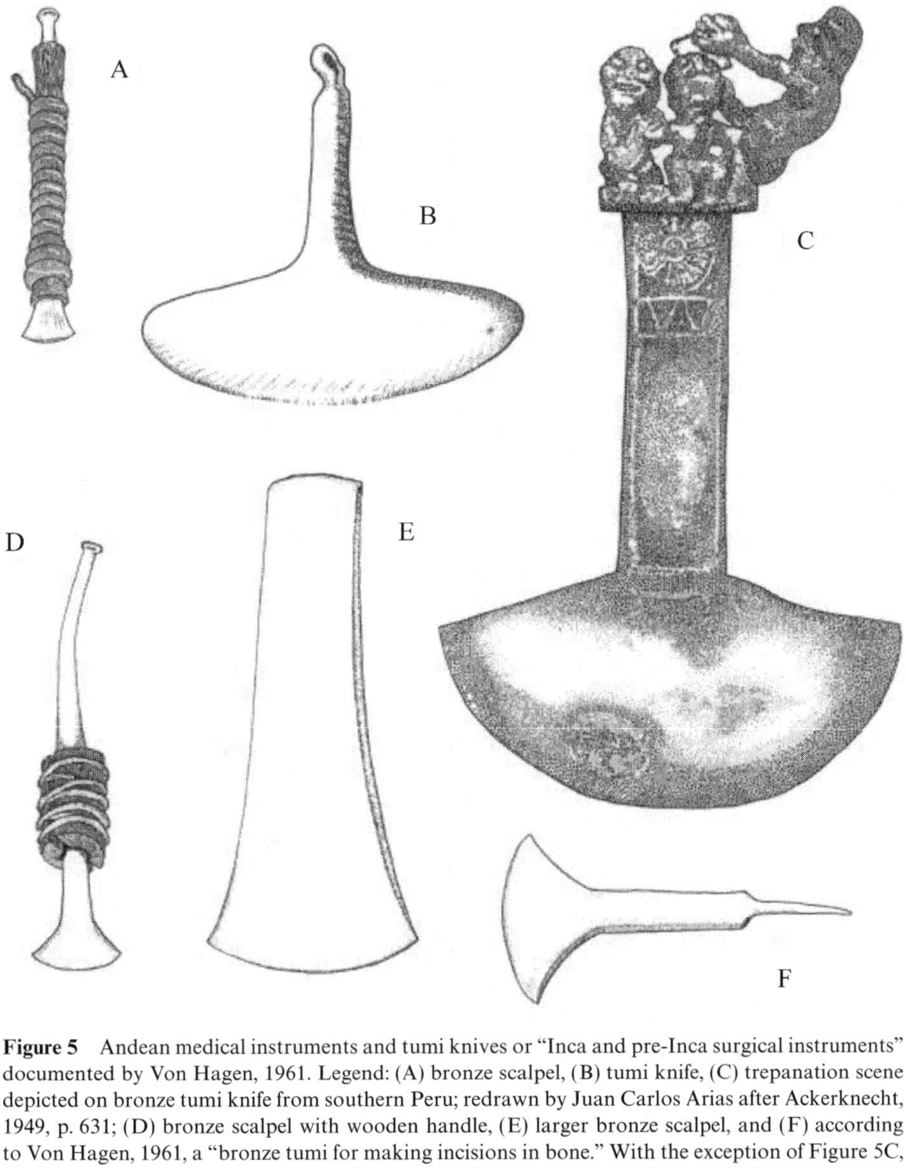

Figure 5 Andean medical instruments and tumi knives or "Inca and pre-Inca surgical instruments" documented by Von Hagen, 1961. Legend: (A) bronze scalpel, (B) tumi knife, (C) trepanation scene depicted on bronze tumi knife from southern Peru; redrawn by Juan Carlos Arias after Ackerknecht, 1949, p. 631; (D) bronze scalpel with wooden handle, (E) larger bronze scalpel, and (F) according to Von Hagen, 1961, a "bronze tumi for making incisions in bone." With the exception of Figure 5C, all figures adapted from and redrawn by Juan Carlos Arias after von Hagen, 1961, pp. 484–485.

predated the rise of the Inca state (Burland, 1967); cranial trephination and tumi knives, and other bronze and related metal perforators, are sometimes depicted on the carved and painted surfaces of Mochica vessels and metalwork from the north coast of Peru (*ca.* AD 1–750; Moseley, 1992: 22). In one example (Ackerknecht, 1949, Plate 55c), a bronze tumi knife was crafted with a hilt depicting a surgeon trephining the cranium of a patient who is accompanied or "restrained" by what appears to be an assistant or companion (see Figure 5c).

MODERN TECHNIQUES OF ANDEAN CRANIOTOMY

Bastien's account of a 1976 craniotomy provides an opportunity to scrutinize the ancient procedure. A Bolivian doctor, Oscar Velasco, witnessed a successful trephination and reported his observations to Bastien in 1984. The 42-year old patient, Rosendo Mollo, originally sought treatment from physicians in La Paz, but was unable to produce the 20,000 pesos – the equivalent of two to three years' salary at that time – required for medical treatment before the cranial tumor resulted in paralysis of the left side of the patient's body. The patient then consulted an herbalist who agreed to operate.

There were seventeen distinct steps:

1. The patient was directed to fast and purge the body for two days preceding the operation.
2. Medicinal plants were provided to decrease the risk of infection.
3. At the time of the operation, the patient was directed to chew coca (*Erythroxylon coca*) and consume alcohol until he was unconscious.
4. The herbalist washed the patient's head, shaved his hair over the area to be surgically opened, and proceeded to cut away the skin immediately over the tumor in the form of a "partial circle".
5. The herbalist cut and lifted the skin with a knife "of the type that Andeans use to castrate pigs".
6. The herbalist "raised the skin" and permitted the incision to "bleed profusely".
7. An "herbal powder" was administered in order to stop the bleeding.
8. Using the knife described in step 5, the herbalist "scraped away the aponeuroses [tendon]" until the bone of the cranial vault was visible.
9. He trephined a 4 to 5 centimeter diameter circle into the bone of the cranial vault.
10. After an hour of scraping and cutting the bone of the cranial vault, the herbalist removed the bone plug.
11. Upon removing the bone plug the herbalist inserted his fingers into the cranial vault, located the tumor between the *dura mater* and bone, and proceeded to loosen the tumor with his fingers.
12. He then removed the tumor, leaving a "bleeding hole" where the tumor had been lodged.
13. Once the tumor was dislodged and removed, salt water was poured into the "bleeding hole".
14. The trephined opening was then "cleansed" with a piece of cloth.

15. Herbs were sprinkled into the trephined incision.
16. The trephined perforation was then covered with the "peel of a gourd" and the aponeuroses was replaced.
17. The herbalist then proceeded to suture the skin over the opening with a "thorn needle and thread."

As a consequence of this surgical procedure, the patient recovered.

Projected survival rates

To account for the success of craniotomy, paleopathologists have examined trephined skulls for evidence of healing. John Verano (1998: 228) has observed that, "trephinations by the scraping and by the circular grooving technique seem to show the highest success rate, whereas trephinations by straight cutting or by drilling and cutting show the lowest." The latter two techniques might, it appears, penetrate or perforate the dura mater [tough protective layer] of the brain.

> Healing can be classified into three general categories, based on the degree of bony reaction to the trephination: (1) none, where there is no evidence of bony reaction, suggesting that death occurred during the procedure or shortly thereafter; (2) short-term survival, where evidence of osteoclastic activity, bone necrosis, or hypervascularity is evident around the margins of the trephination opening, indicating survival perhaps for several weeks, and (3) long-term survival, where there is evidence of extensive remodeling of the margins of the defect (Verano, 1998: 228).

According to Majno (1975), the skulls studied thus far suggest that the survival rate was near one hundred percent. However, a few cases of osteomyelitis, or bone deterioration, were documented in the cranial collection he examined. The deterioration was associated primarily with the size of the original injury and whether or not the trauma was located at the base of the skull. Patients suffering blunt trauma over the cerebral cortex or brain stem were least likely to benefit from treatment by trephination. In other studies, projected survival rates, based on the presence of osteitis, range from 62.5 percent – in one study that found healing or evidence of osteitis or post-operative bone regeneration in 250 of 400 skulls examined – to between 23.4 percent and 55.3 percent in other studies (Froeschner, 1992: 552). Direct osteological analysis of a broad array of trephined collections by Verano and Williams (1992) and Verano (1996) have documented a "long-term survival rate of over 70 percent for Inca trephinations, in contrast to a roughly 40 percent survival rate for the earliest trephinations from southern coastal Peru" (Verano, 1998: 228).

Majno concluded that the Inca were the "masters of the art of trepanning". He also noted that the survival rate for patients in Inca times was better than for more recent times. Peter James and Nick Thorpe (1994: 26) have documented "modern" survival rates for trephined patients at Saint George's and Guy's Hospitals in London for the period from 1870 through 1877. Of the thirty-two patients that underwent the operation during that period, only twenty-five percent survived the operation (e.g., Fischer, 1864). By contrast, in one sample of 214 trephined Inca skulls cited by James and Thorpe (1994: 26), 55 percent provide indications of complete post-operative healing, while only

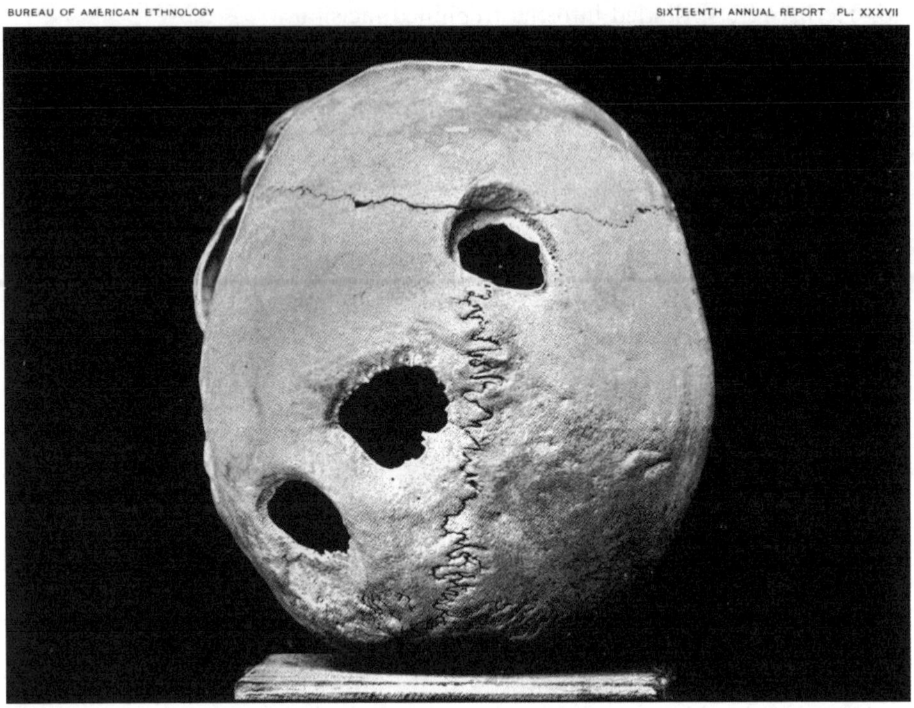

Figure 6 Superior aspect of cranium 18 from Cuzco, Peru, with multiple trephinations and osteitis or healing (Plate XXXVII; From Muñiz and McGee, 1897.)

28 percent did not recover. James and Thorpe caution that "we should remember that many of the 'failures' may not have actually died 'under the knife' but simply as a result of the injury or illness that prompted the trepanation in the first place ... [and] ... an extraordinary tribute to the skill of the ancient surgeons is provided by a Peruvian skull with no less than seven healed trepanations, evidence of as many successful operations." The authors of virtually every Peruvian paleopathologic study of trephined crania have echoed these observations (Muñiz and McGee, 1897; MacCurdy, 1923; Verano, 1998).

* * *

Our review of pre-colonial medical practices raises more questions than can be addressed in this essay. Significant parts of the system included (a) holistic concepts of health, (b) state-sponsored public health programs, (c) an extensive body of anatomical and pharmacological terminology, (d) the existence of confraternal medical associations and state-sponsored medical corps, and specific practices centered on medical innovations such as those pertaining to (e) cranial trephination, (f) post-cranial surgery and cauterization, (g) prosthetic and cosmetic devices, (h) antibiotic and antiseptic ointments and medications, (i) intramedullar nails and bone transplant procedures, (j) diagnostic and therapeutic procedures pertaining to dental health and hygiene, and (k) a large pharmacological repertoire of herbal remedies. Scholars have only scratched the surface in their assessments of pre-colonial American Indian science, medicine, and belief.

ACKNOWLEDGEMENTS

I would like first to thank Helaine Selin for the professionalism, collegiality, diligence, and patience that she has demonstrated in every interaction regarding this invited contribution. Dr. Vine Deloria, formerly of the University of Arizona, Tucson, first inspired my interests in and studies of American Indian science, technology, and medicine. The paper that was most influential was S. Linné's (1957) "Technical Secrets of the American Indians: The Huxley Memorial Lecture, 1957." The work of Roshdi Rashed (1980; revised 1997) encouraged my research into new ways of looking at the history of science in non-Western cultures. I would also like to thank the faculty and staff of the Social and Behavioral Sciences Center of the California State University, Monterey Bay, for their many contributions toward the development of a premier teaching, learning, and research environment. I would particularly like to thank the Center's Dean, Dr. Armando Arias, for his gracious support for faculty teaching and research. Both Ms. Lilly Martinez and Ms. Myisha Washington have been an invaluable source of clerical support. The Center's student assistants, particularly Mrinal "Ranu" Sinha, copied and collated interlibrary loan materials. Drs. Eddy Hogan and Steve Watkins and the CSU Monterey Bay library staff facilitated many of my requests for interlibrary loan materials. I extend my great appreciation to Juan Carlos Arias for the superb maps and illustrations that he drew on short notice. Finally, I would like to thank my wife, Linda Marie Mendoza, and my daughters Natalie Dawn Marie and Maya Nicole for their continuing support, encouragement, and patience.

NOTES

[1] Both Aymara and Quechua indigenous communities refer to Kallawaya medicine men and women as the *Qolla Kapachayuh*, or "Lords of the Medicine Bag" of the "Land of Medicine" (Bastien, 1981: 20).

[2] Although the sentiment of Ackerknecht's (1949) "quotation" of Moodie is echoed in Moodie's (1929) findings, the language does not correspond – word for word – to any statement specifically made in the version cited here.

[3] John Hemming (1970: 490) argued, incorrectly, that "Trepanning is generally thought to have been for magical or medical purposes, and is not associated with battle wounds of warriors. The absence of trepanned skulls in Machu Picchu therefore does not help identify the ruin as a religious center." Compare the osteological work and paleopathology studies of Verano (1998: 226–227), who notes that where trephination is concerned, "many South American trephinations are clearly associated with skull fracture," specifically, depressed fractures; and "the majority of these injuries were probably produced by blows from clubs and sling stones."

[4] Despite archaeological, medical, and paleopathological evidence to the contrary, Carmen Bernand (1994: 88–91) notes, "It seems that the anatomical knowledge among the Andean peoples was very limited. This did not prevent them from practicing trepanation, however, which involved making small holes in the skull using very fine copper knives. Despite the dangerous nature of this

operation, at least some patients must have survived, because skulls have been found in which the orifice had partially closed up again."

5 Bastien (1982) has documented the impact of capitalism and urbanization on medicine in Peru and its catastrophic consequences for rural and indigenous medicine. In its efforts to centralize and urbanize medicine, the modern medical profession played a key role in reducing and undermining the indigenous medical traditions of the Andes. As a consequence, hundreds of thousands of American Indians have been excluded from access to both modern and traditional health care.

6 Arriaza (1995) and other scholars, including Bittmann and Munizaga (1976), Allison *et al.* (1994), have documented the antiquity of the preceramic Chinchorro mummies of southern Peru and northern Chile and have concluded that they provide the oldest documented evidence for artificial mummification in the world. Radiocarbon evidence sets the age of the oldest Chinchorro mummies at 9,000 years of age.

7. See Cushing, 1897 for a reassessment of such practices in terms of their therapeutic value as rites of transference or displacement.

8 The *ayllu* is a clan or social formation that has served as the basic socioeconomic unit of Andean society since the time of the Inca. The *ayllus* also serve as a form of risk management in that the clans that constitute each *ayllu* generally consist of peoples from diverse ecological and altitudinal zones. The ability to draw on resources from different microenvironments serves to guard against starvation in the environmentally challenging conditions of the Andes, where most communities are to be found at elevations ranging from 6,000 to 17,000 feet above sea level (Bastien, 1981: 29).

9 In other words, each society was constituted by comparable political structures characterized in the literature as expansionist empires and conquest states (Conrad and Demarest, 1984). Similarly, each society lasted little more than a century and each saw battle with contact era Europeans. Each was ultimately defeated as the result of a combination of factors, not the least of which was the rapid proliferation and spread of deadly European diseases.

10 Bastien's (1987: 17) account of a 1976 surgical procedure provides one of the most complete descriptions of an Andean trephination on record.

11 The flintknapped or flaked prismatic obsidian blade remains among the sharpest cutting edge yet produced by humans. The archaeological record of the Americas shows it extending back to the earliest American Indian settlers in Alaska and the Aleutian Islands at 15,000 years ago. See Whittaker (1994) or Mendoza (2001a, 2001b) for a detailed overview of the flintknapping reduction process necessary for the production of these tools.

BIBLIOGRAPHY

Ackerknecht, Erwin H. 'Medical practices.' In *Handbook of South American Indians.* Vol. 5, J.H. Steward, ed. Bureau of American Ethnology Bulletin 143. Washington, DC: U.S. Government Printing Office, 1949, pp. 621–643.

Acosta, José de. *De natura Novi Orbis libri duo et de Promulgatione Evangelii, apud Barbaros, sive de Procuranda Indorum salute libri sex.* Salamanca: Guillermo Foquel, 1589.

Anderson, L.S. and J.W. Verano. 'A possible case of amputation of the feet, with evidence of healing, in a Moche burial from the north coast of Peru.' Paper presented at the 23rd Annual Meeting of the Paleopathology Association, Durham, North Carolina, April, 1996. Abstract in *Paleopathology Newsletter* 94(Supplement): 16, 1996.

Anton, Ferdinand. *Woman in Pre-Columbian America.* New York: Abner Schram, 1973.

Arriaga, P.J. *The Extirpation of Idolatry in Peru,* C. Keating, trans. and ed. Lexington: University of Kentucky Press, [1621], 1968.

Arriaza, Bernardo T. 'Chile's Chinchorro mummies.' *National Geographic* 187(3): 68–89, 1995.

Arriaza, Bernardo T., Felipe Cárdenas-Arroyo, Ekkehard Kleiss and John W. Verano. 'South American mummies: culture and disease.' In *Mummies, Disease and Ancient Cultures,* Aidan Cockburn, Eve Cockburn and Theodore A. Reyman, ed. New York: Cambridge University Press, 1998, pp. 190–234.

Aufderheide, A.C. 'The enigma of ancient cranial trepanation.' *Minnesota Medicine* 68(2): 119–122, 1985.

Bandelier, Adolf. 'Aboriginal trephining in Bolivia.' *American Anthropologist* 6: 40–46, 1904.

Bastien, Joseph W. 'Exchange between Andean and Western medicine.' *Social Science and Medicine* 16: 795–803, 1982.

Bastien, Joseph William. *Healers of the Andes: Kallawaya Herbalists and their Medicinal Plants.* Salt Lake City: University of Utah Press, 1987.

Bastien, Joseph W. and John M. Donahue, eds. *Health in the Andes.* Washington, DC: American Anthropological Association, 1981.

Bernand, Carmen. *The Incas: People of the Sun.* New York: Harry N. Abrams, 1994.

Blackman, James, Marvin J. Allison, Arthur C. Aufderheide, Norman Oldroyd and R. Ted Steinbock. 'Secondary hyperparathyroidism in an Andean mummy.' In *Human Paleopathology: Current Synthesis and Future Options*, Donald J. Ortner and Arthur C. Aufderheide, eds. Washington, DC: Smithsonian Institution, 1991, pp. 291–296.

Bondeson, Wolmar E. 'Tobacco from a Tiahuanacoid culture period.' In *A Medicine-Man's Implements and Plants in a Tiahuanacoid Tomb in Highland Bolivia*, Henry Wassén, ed. Goteborg: Goteborgs Etnografiska Museum, 1972, pp. 177–184.

Bram, Joseph. *An Analysis of Inca Militarism.* New York: J.J. Augustin, 1941.

Browman, David L. 'Prehistoric nutrition and medicine in the Lake Titicaca Basin.' In *Health in the Andes*, Joseph W. Bastien and John M. Donahue, eds. Washington, DC: American Anthropological Association, 1981, pp. 103–118.

Buikstra, Jane E. and Sloan Williams. 'Tuberculosis in the Americas: current perspectives.' In *Human Paleopathology: Current Synthesis and Future Options*, Donald J. Ortner and Arthur C. Aufderheide, eds. Washington, DC: Smithsonian Institution, 1991, pp. 161–172.

Burland, C.A. *Peru Under the Incas.* New York: Putnam, 1967.

Cambridge, Charles. 'Historical contributions of American Indians.' Workshop for Governmental American Indian Heritage Month, Region VIII, U.S. Department of Health and Human Services, Denver, Colorado. June, 1992.

Cambridge, Charles. 'On the question of American Indian prehistory.' Personal communication. University of Colorado, Denver, 1994.

Canalis, R.F., W.G. Hemenway, F. Cabieses and R. Aragon. 'Prehistoric trephination of the frontal sinus.' *Annals of Otology, Rhinology and Laryngology* 90: 186–189, 1981.

Chávez Velásquez, Nancy A. *La materia medica en el Incanato.* Lima, Peru: Editorial Mejia Baca, 1977.

Classen, Constance. *Inca Cosmology and the Human Body.* Salt Lake City: University of Utah Press, 1993.

Cobo, Bernabé. *Inca Religion and Customs*, Roland Hamilton, trans. and ed. Austin: University of Texas Press, 1990.

Cock, Guillermo A. 'Inca rescue.' *National Geographic* 201(5): 78–91, May 2002.

Cockburn, Aidan, Eve Cockburn and Theodore A. Reyman, eds. *Mummies, Disease and Ancient Cultures.* New York: Cambridge University Press, 1998.

Conrad, Geoffrey W. and Arthur A. Demarest. *Religion and Empire: The Dynamics of Aztec and Inca Expansionism.* Cambridge: Cambridge University Press, 1984.

Crosby, Alfred W. *Ecological Imperialism: The Biological Expansion of Europe, 900–1900.* Cambridge: Cambridge University Press, 1986.

Cushing, Frank H. 'A case of primitive surgery.' *Science*, n.s. V(130): 977–981, 1897.

Daland, J. 'Depressed fracture and trephining of the skull by the Incas of Peru.' *Annals of Medical History* 7: 550–558, 1935.

de la Vega, Garcilaso. *Historia General del Peru* (segunda parte de Los Comentarios reales de los Incas), Angel Rosenblat, ed. Buenos Aires: Emecé Editores, 1944.

de la Vega, Garcilaso. *The Incas: The Royal Commentaries of the Inca*, Maria Jolas, trans. and Alain Gheerbrant, ed. New York: Avon Books, 1961.

Dillehay, T.D. *Monte Verde, a Late Pleistocene Settlement in Chile: A Palaeoenvironment and Site Context.* Washington, DC: Smithsonian Institution Press, 1989.

Elferink, Jan G.R. 'Mental disorder among the Incas in ancient Peru.' *History of Psychiatry* 10(3): 303–318, 1999.

Fischer, H. 'Klinisches und Experimentelles zur Lehre der Trepanation.' *Archiv für Klinische Chirurgie* 6: 595–647, 1864.

Froeschner, Elsie H. 'Historical vignette: two examples of ancient skull surgery.' *Journal of Neurosurgery* 76(3): 550–552, 1992.

Fuchs, Andrew. 'Coca chewing and high-altitude stress: possible effects of coca alkaloids on erythropoiesis.' *Current Anthropology* 19: 277–291, 1978.

Guerra, Francisco. *The Pre-Columbian Mind: A Study into the Aberrant Nature of Sexual Drives, Drugs Affecting Behaviour and the Attitude towards Life and Death, With a Survey of Psychotherapy in Pre-Columbian America.* New York: Seminar Press, 1971.

Guzman Peredo, Miguel. *Medical Practices in Ancient America.* Mexico: Ediciones Euroamericanas, 1985.

Harding, Sandra. 'Introduction: Eurocentric scientific illiteracy – a challenge for the world community.' In *The Racial Economy of Science: Toward a Democratic Future*, Sandra Harding, ed. Bloomington and Indianapolis: Indiana University Press, 1993, pp. 1–22.

Hemming, John. *The Conquest of the Incas.* San Diego: Harcourt, Brace and Company, 1970.

Hjortsjö, Carl-Herman. 'Anthropological investigation of an artificially deformed and trepanned cranium from Niño Korin, La Paz, Bolivia.' In *A Medicine-Man's Implements and Plants in a Tiahuanacoid Tomb in Highland Bolivia*, S. Henry Wassén, ed. Goteborg: Goteborgs Etnografiska Museum, 1972, pp. 145–158.

Holmstedt, Bo and Jan-Erik Lindgren. 'Alkaloid analyses of botanical material more than a thousand years old.' In *A Medicine-Man's Implements and Plants in a Tiahuanacoid Tomb in Highland Bolivia*, S. Henry Wassén, ed. Goteborg: Goteborgs Etnografiska Museum, 1972, pp. 139–144.

Horton, Robin. 'African traditional thought and Western science.' *Africa* 37: 50–71, 1967.

Hultkrantz, Åke. 'Health, religion and medicine in Native North American traditions.' In *Healing and Restoring: Health and Medicine in the World's Religious Traditions*, L.E. Sullivan, ed. New York: Macmillan, 1989, pp. 327–358.

Hultkrantz, Åke. *Shamanic Healing and Ritual Drama: Health and Medicine in Native North American Religious Traditions.* New York: Crossroad, 1992.

James, Peter and Nick Thorpe. *Ancient Inventions.* New York: Ballantine Books, 1994.

Kensinger, Kenneth M. 'Cashinahua medicine and medicine men.' In *Native South Americans: Ethnology of the Least Known Continent*, Patricia J. Lyon, ed. Boston: Little, Brown and Company, 1985, pp. 283–288.

Lanning, Edward P. *Peru Before the Incas.* Englewood Cliffs, New Jersey: Prentice-Hall, 1967.

Leslie, Charles. 'Scientific racism: reflections on peer review, science and ideology.' *Social Science and Medicine* 31(8): 891–912, 1990.

Linné, S. 'Technical secrets of the American Indians: The Huxley Memorial Lecture, 1957.' *Journal of the Royal Anthropological Institute of Great Britain and Ireland* 37(2): 149–164, 1957.

Lira, Jorge A. *Medicina Andina: Farmacopea y Rituales.* Cusco, Perú: Centro de estudios rurales andinos 'bartolomé de las casas', 1985.

Lopez Austin. 'Sahagun's work on the medicine of the ancient Nahuas: possibilities for study.' In *Sixteenth Century Mexico: The Work of Sahagun*, M.S. Edmondson, ed. Albuquerque: University of New Mexico Press, 1974, pp. 205–224.

MacCurdy, G.G. 'Human skeletal remains from the highlands of Peru.' *American Journal of Physical Anthropology* 6: 217–329, 1923.

Majno, Guido. *The Healing Hand: Man and Wound in the Ancient World.* Cambridge: Harvard University Press, 1975, reprinted 1991.

Mann, G. 'Chronic ear disease as a possible reason for trephination.' *International Journal of Osteoarchaeology* I: 165–168, 1991.

Margetts, E.L. 'Trepanation of the skull by the medicine-men of primitive cultures with particular reference to the present-day native East African practice.' In *Diseases in Antiquity*, D. Brothwell and A. Sandison, ed. Springfield, Illinois: C.C. Thomas, 1967, pp. 673–701.

Martin, Richard. 'The role of coca in the history, religion and medicine of South American Indians.' *Economic Botany* 24: 422–437, 1970.

Mendoza, Ruben G. 'The Ome Tochtli Complex: Agave in Mesoamerican Ritual Context.' Paper presented for the symposium 'Agave, Maguey and Pulque,' American Anthropological Association, New Orleans, Louisiana, 1990.

Mendoza, Ruben G. *Conquest Polities of the Mesoamerican Epiclassic: Circum-Basin Regionalism,*

A.D. 550–850. Unpublished Ph.D. Thesis. Tucson: Department of Anthropology, University of Arizona, 1992.

Mendoza, Ruben G. 'City planning: Maya city planning.' In *Encyclopaedia of the History of Science, Technology and Medicine in Non-Western Cultures*, Helaine Selin, ed. Dordrecht: Kluwer Academic Publishers, 1997a, pp. 205–208.

Mendoza, Ruben G. 'Metallurgy in Meso and North America.' In *Encyclopaedia of the History of Science, Technology and Medicine in Non-Western Cultures*, Helaine Selin, ed. Dordrecht: Kluwer Academic Publishers, 1997b, pp. 730–733.

Mendoza, Ruben G. 'Trephination.' In *Encyclopaedia of the History of Science, Technology and Medicine in Non-Western Cultures*, Helaine Selin, ed. Dordrecht: Kluwer Academic Publishers, 1997c, pp. 985–986.

Mendoza, Ruben G. and Gretchen Jordan. 'Road networks in ancient Native America.' In *Encyclopaedia of the History of Science, Technology and Medicine in Non-Western Cultures*, Helaine Selin, ed. Dordrecht: Kluwer Academic Publishers, 1997, pp. 868–872.

Mendoza, Ruben G. and Jay R. Wolter. 'Medicine in Meso and South America.' In *Encyclopaedia of the History of Science, Technology and Medicine in Non-Western Cultures*, Helaine Selin, ed. Dordrecht: Kluwer Academic Publishers, 1997, pp. 702–706.

Mendoza, Ruben G. 'An archaeological approach to the teaching of American cultural diversity.' In *Cultural Diversity in the United States: A Critical Reader*, Ida Susser and Thomas C. Patterson, eds. Oxford: Blackwell Publishers, 2001a, pp. 414–433.

Mendoza, Ruben G. 'The Virtual Learning Lab: facilitated teaching and learning in the online environment.' In *Emerging Technologies in Teaching Languages and Cultures – Effective Use of Technology for Language Learning: What Works and What We've Learned*, Yoshiko Saito-Abbott, Richard Donovan, Thomas F. Abbott and Phillip Kennedy, eds. San Diego, California: LARC Press, Language Arts Resource Center, California State University, San Diego, 2001b, pp. 167–195.

Merbs, C.F. 'Trauma.' In *Reconstruction of Life from the Skeleton*, M.Y. Iscan and K.A.R. Kennedy, eds. New York: Alan R. Liss, 1989, pp. 161–189.

Metraux, Alfred. 'Religion and shamanism.' In *Handbook of South American Indians*. Vol. 5, J.H. Steward, ed. Bureau of American Ethnology Bulletin 143. Washington: U.S. Government Printing Office, 1949, pp. 559–599.

Millon, René. 'Teotihuacan.' *Scientific American* 216: 38–49.

Moodie, Roy L. 'Studies in paleopathology XXIII: surgery in pre-Columbian Peru.' *Annals of Medical History* 1: 698–728, 1929.

Moseley, Michael E. *The Incas and Their Ancestors: The Archaeology of Peru*. London: Thames and Hudson, 1992.

Muñiz, Manuel A. and W.J. McGee. *Primitive Trephining in Peru*. Washington, DC: Government Printing Office, 1897.

Munizaga, Juan R. 'Human skeletal pathology in pre-Columbian populations of northern Chile.' In *Human Paleopathology: Current Synthesis and Future Options*, Donald J. Ortner and Arthur C. Aufderheide, eds. Washington, DC: Smithsonian Institution, 1991, pp. 145–150.

Murra, John V. 'El "control vertical" de un maximo de pesos ecologicos en la economia de las sociedades andinas.' In *Visita de la Provincia de León de Huánuco [1562], Iñigo Ortiz de Zúñiga*, Vol. 1, Huanaco, Peru: Universidad Nacional Hermilio Valdizán, Facultad de Letras y Educación, 1972, pp. 429–476.

Ortiz de Montellano, Bernard R. 'Mesoamerican religious tradition and medicine.' In *Healing and Restoring: Health and Medicine in the World's Religious Traditions*, L.E. Sullivan, ed. New York: Macmillan, 1989, pp. 359–394.

Ortiz de Montellano, Bernard R. *Aztec Medicine, Health and Nutrition*. New Brunswick, New Jersey: Rutgers University Press, 1990.

Ortner, Donald J. and Arthur C. Aufderheide. *Human Paleopathology: Current Synthesis and Future Options*. Washington, DC: Smithsonian Institution Press, 1991.

Pizarro, Pedro. *Relation of the Discovery and Conquest of the Kingdoms of Peru*. New York: The Cortes Society, 1921.

Polanyi, Karl. *Primitive, Archaic and Modern Economies: Essays of Karl Polanyi*, George Dalton, ed. Garden City, New York: Anchor Books, 1968.

Poma de Ayala, Felipe Guamán. *Nueva Coronica y Buen Gobierno*. Paris: Travaux et Mémoires de l'Institut d'Ethnologie 23, 1936.

Rashed, Roshdi. 'Science as a Western phenomenon.' *Fundamenta Scientiae* 1: 7–21, 1980.

Rashed, Roshdi. 'Science as a Western phenomenon.' In *Encyclopaedia of the History of Science, Technology and Medicine in Non-Western Cultures*, Helaine Selin, ed. Dordrecht, The Netherlands: Kluwer Academic Publishers, 1997, pp. 884–890. [Revision of Rashed, 1980].

Sahagun, Fray Bernardino de Sahagun. *A History of Ancient Mexico*, Fanny R. Bandelier, trans. Nashville, Tennessee: Fisk University Press, 1932.

Saignes, Thierry. '?Quienes son los Callajuayas? Nota sobre un enigma historico.' In *Espacio y Tiempo en el Mundo Callahuaya*. La Paz, Bolivia: Universidad Mayor de San Andres, 1984, pp. 111–129.

Schendel, Gordon. *Medicine in Mexico: From Aztec Herbs to Betatrons*. Austin: University of Texas Press, 1968.

Schmidt, Peter R. and Thomas C. Patterson. *Making Alternative Histories: The Practice of Archaeology and History in Non-Western Settings*. Santa Fe, New Mexico: School of American Research Press, 1995.

Schultes, Richard Evans. 'Ilex Guayusa from 500 A.D. to the present.' In *A Medicine-Man's Implements and Plants in a Tiahuanacoid Tomb in Highland Bolivia*, S. Henry Wassén, ed. Goteborg: Goteborgs Etnografiska Museum, 1972, pp. 115–138.

Schultes, Richard Evans. 'Burning the library of Amazonia.' *The Sciences* 34(2): 24–31, 1994.

Seymour-Smith, Charlotte. *Dictionary of Anthropology*. Boston, Massachusetts: G.K. Hall, 1986.

Stark, Louisa. 'The lexical structure of Quechua body parts.' *Anthropological Linguistics* 11(1): 1–15, 1969.

Stark, Louisa. 'Machaj-juyai: secret language of the Callahuayas.' *Papers in Andean Linguistics* 1: 199–227, 1972.

Stewart, T. Dale. 'Stone Age skull surgery: a general review with emphasis on the New World.' *Annual Reports of the Smithsonian Institution, 1957*. Washington, DC: Smithsonian Institution, 1958, pp. 469–591.

Stewart, T.D. 'Deformity, trephining and mutilation in South American Indian skeletal remains.' In *Handbook of South American Indians*. Vol. 6, J.H. Steward, ed. Washington, DC: Smithsonian Institution, 1963a, pp. 43–48.

Stewart, T.D. 'Pathological changes in South American Indian skeletal remains.' In *Handbook of South American Indians*. Vol. 6, J.H. Steward, ed. Washington, DC: Smithsonian Institution, 1963b, pp. 49–52.

Sullivan, Lawrence E. *Icanchu 's Drum: An Orientation to Meaning in South American Religions*. New York: Macmillan, 1988.

Sullivan, Lawrence E. 'Religious foundations of health and medical power in South America.' In *Healing and Restoring: Health and Medicine in the World's Religious Traditions*, L.E. Sullivan, ed. New York: Macmillan, 1989, pp. 395–448.

Tello, Julio C. 'Prehistoric trephining among the Yauyos of Peru.' *Proceedings of the International Congress of Americanists*, Part 1. London: Harrison and Sons, 1913, pp. 75–83.

Thomas, David Hurst. *Skull Wars: Kennewick Man, Archaeology and the Battle for Native American Identity*. New York: Basic Books, 2000.

Towle, Margaret A. *The Ethnobotany of Pre-Columbian Peru*. Chicago: Aldine Publishing Company, 1961.

Tyson, R.A. and E.S.D. Alcauskas. *Catalogue of the Hrdlicka Collection*. San Diego: San Diego Museum of Man, 1980.

Urioste, George L. 'Sickness and death in preconquest Andean cosmology: the Huarochiri oral tradition.' In *Health in the Andes*, Joseph W. Bastien and John M. Donahue, eds. Washington, DC: American Anthropological Association, 1981, pp. 9–18.

Urteaga-Ballon, Oscar. 'Medical ceramic representation of nasal leishmaniasis and surgical amputation in ancient Peruvian civilization.' In *Human Paleopathology: Current Synthesis and Future*

Options, Donald J. Ortner and Arthur C. Aufderheide, eds. Washington, DC: Smithsonian Institution, 1991, pp. 95–101.

Verano, John W. 'La trepanación como tratamiento terapéutico para fracturas craneales en el antiguo Peru.' Paper presented at the VIII Coloquio Internacional de Antropología Física 'Juan Comas', Mexico City, November, 1996.

Verano, John W. 'Disease in South American mummies.' In *Mummies, Disease and Ancient Cultures*, 2nd ed., Aidan Cockburn, Eve Cockburn and Theodore A. Reyman, eds. London: Cambridge University Press, 1998, pp. 215–228 and 231–234.

Verano, J.W. and J.M. Williams. 'Head injury and surgical intervention in pre-Columbian Peru.' Abstract in *American Journal of Physical Anthropology*, Supplement 14: 167–168, 1992.

Vogel, Virgil J. *American Indian Medicine*. Norman: University of Oklahoma Press, 1970.

Von Hagen, Victor Wolfgang. *The Ancient Sun Kingdoms of the Americas: Aztec, Maya, Inca*. Cleveland: World Publishing Company, 1961.

Wassén, H., ed. *A Medicine-Man's Implements and Plants in a Tiahuanacoid Tomb in Highland Bolivia*. Goteborg: Goteborgs Etnografiska Museum, 1972.

Willey, G.R. and J.A. Sabloff. *A History of American Archaeology*. London: Thames and Hudson, 1974.

Wolf, Eric R. *Europe and the People Without History*. Berkeley: University of California Press, 1982.

Young, Allan. 'Analogy and efficacy in Ethiopian medical divination.' *Culture, Medicine and Psychiatry* 1: 183–199, 1977.

CARLOS VIESCA

MEDICINE IN ANCIENT MESOAMERICA

Mesoamerica is a wide territory contained between the 20th parallel north, just north of modern Mexico City, and the area situated immediately north of Darien Isthmus, in Panama. In this area a cultural continuum developed from the second millennium BC, resulting in complex and sophisticated civilizations, from the Olmecs to the Aztec and Maya. These cultures shared comparable worldviews, including a view of the universe as floors at different levels of a vertical axis. There were also similarities in language, including a linguistic plurality with some major groups, such as the Maya, Utoaztec and Zapotec, and many related idioms and dialects. There were, for example, at least twenty Mayan languages and a hundred dialects. The existence of similarities among all these cultures makes it possible for us to study Mesoamerican ancient cultures in an integrative way, but we must also be conscious of particularities and give each culture a distinctive treatment to obtain a better understanding of its reality (Lombardo and Nalda, 1996; Manzanilla and López Luján, 1994).

WORLDVIEW

The universe was conceived of as a series of floors disposed along a central axis, which, in Aztec and Nahuatl thought was compared to *malinalli*, a kind of herb characterized by its tortuous form. The Maya perceived the universe as a large tree, a *ceiba*, while their ancestors, the Olmec, conceived of a tree growing over a crocodile's body and provided with four sprouts, one in each corner angle, which aided it in holding up the heavens (Freidel, Schele and Parker, 1993: 59–90; Reilly, 1990 and 1996; Schele, 1996; Wagner, 2000).

There are many narratives about cosmic creation and organization, which do not coincide exactly, but it is still possible to formulate a general view. All of them concur in the existence of three different parts: the heavens, the central part and the underworld. Nahua tradition speaks about thirteen celestial floors, of which four are related to the center, with nine more in the underworld. There were two different visions. The first was composed of nine true heavens and nine underworld regions, making reference to a natural cosmic balance. The second, with thirteen heavens and nine underworld floors, illustrated the

H. Selin (ed.), Medicine Across Cultures: History and Practice of Medicine in Non-Western Cultures, 259–283.

differences between feminine and masculine natures and related the first with the womb of the universe, that is the interior of earth and the underworld, and the last with erection, arrows and with a hot substance. In terms of the creation order, the universe was disposed from the highest heaven below, and at some moment, at the center, it was necessary to configure that particular zone, giving form to the earth's surface and elevating the celestial vault that corresponded to the four central heavenly floors. The universe was described as a rectangular prism, with a central pole, herb or tree, put on over one of its short sides and with the flat surfaces or floors disposed horizontally with the pole in their middle (López Austin, 1994; Montoliú, 1989: 40–54).

Some descriptions, coming from Maya sources, apparently make reference to the central part of the universe, considering the heavenly floors as the sun, the moon, the stars and a repository for the souls for future humans, animals and plants. This central region was especially important, because it was inhabited by mankind, which was considered a central creature, neither celestial nor from the underworld, neither god nor pertaining to the death realm. In Nahuatl, the classical idiom in northern Mesoamerica, man is called *tlacatl*, the middle one.

HUMAN ANATOMY AND THE MICROCOSMOS

The extent of anatomical knowledge is different in pre-Hispanic and western medical systems. From ancient times knowledge of external and internal anatomical structures existed, but there is no evidence of any systematic study. Some scholars believe that human sacrifice was the source of anatomical knowledge, but the priests in charge of sacrifice had no connection to medicine or to the medical experts. We have long lists of anatomical terms making some fine distinctions, as well as a series of symbolic items related to anatomical structures.

In pre-Hispanic Mexico, the body was compared to the universe and was considered a microcosmos, with an axis in the vertebral column and organs and structures disposed at different levels. The diaphragm symbolized the earth; above it was the celestial realm, beginning with the heart, *yóllotl*, situated in the same way the sun is observed in our latitudes, at the left of the central line of heaven. This astronomical observation of the south solar deviation explains the fact that all solar deities, such as Huitzilopochtli or Camaxtle, for example, were left handed. The liver, *elli*, situated under the diaphragm and at the right side, was also a vital organ and was provided with a soul known as *ihíyotl*. The thoracic cavity with the lungs represented the airy heavenly regions, and this vertical disposition culminated with an imaginary organ composed of the top of the cranium, including hair, the posterior fontanel in children, the lamboid suture in adults, and the vascular and cerebral structures. This highest part of the skull was called *ilhuícatl*, signifying heaven, and functionally worked connecting the body with the cosmic heavenly floors.

The bodily underworld was imagined from the diaphragm down, including the abdominal cavity, the pelvic regions with reproductive organs, and the pelvic limbs. This lower part of the body established contact with the corresponding cosmic structures through the perineal holes, anus, vagina, and the soles of the feet. At these places the body was in touch with the earth's cavities and the underworld. In that way, the human body and the universe not only had a parallel disposition, but also had functional similarities (López Austin, 1980: 395–401; Viesca, 1997: 133).

The navel was also an interesting structure. Apparently it represented the configuration of another energetically charged organ, with a functional mission to maintain connections with other cosmic structures. Reminiscent of the relationship between the mother and her offspring and bringing to mind its continuity after birth and physical separation, that organ, called *tip té* by Quintana Roo Maya, became an important structure destined to coordinate body relations with cosmic dynamics and provide the body with balance and equilibrium (Villa Rojas, 1945: 377). No specific mention of this "organ" exists in Nahuatl writings, but rituals and indirect references permit us to infer this concept.

THE HUMAN BODY AND THE DYNAMIC COSMOS

Mesoamerican pre-Hispanic cultures had a dynamic worldview. Every part in the universe was charged with energetic properties and developed influences over its neighbors that produced short and long-term consequences. Every body was animated by a spirit and had its own energy; it also had what was necessary to maintain and increase it. Every body, and every part of the body, was related to other beings, spirits, and regions situated at different levels in the universe or in diverse places over the earth's surface. From every part in the cosmos, energy converged to the central area, making contact with creatures and influencing them in diverse ways. One was represented by numerical interpretations of cosmic influences, developed in forces capable of changing destiny and modifying humans and other living beings, even the gods (López Austin, 1994: 30).

Sometimes the influences were healthy. For example, at the very moment of conception the entity called *tonalli* descended from heaven, coming into the new creature's body, bringing it heat and life. This entity was also considered the person's destiny, being defined by the characteristics of the planets' relationships, the gods related to each of them, and the time and disposition of every cosmic body at the precise moment of his/her birth. All these possibilities provided an individual condition, defined by the sum of influences and their possibilities for interaction (López Austin, 1980, I: 330 and 399; Sahagún, 1993: f.282v; Viesca, 1997: 128 and 149; Viesca *et al.*, 1997: 143–144).

THE ASTRAL BODY

The heavenly vault was divided into twenty quadrants, similar to zodiacal divisions, each one representing a particular concentration of cosmic forces. The celestial bodies' positions at the moment of conception and then at birth defined the destiny of terrestrial beings. Reading celestial maps made it possible to know their future activities and their somatic and psychic characteristics, which could be modified by conducting purifying ceremonies. A calendric sign defined the twenty entities resulting from the heaven's division into quadrants. In this way, the birth date, intended as a result of astrological combinations not determined by their structure but mainly by balance and force relations, defined the person's destiny. But, when an ominous prediction was made, it was possible to conjure or change it, delaying the purifying ceremonies until an auspicious sign materialized.

Here are some illustrative examples of sign nature. Those born on a 1-Jaguar day (Ce-Océlotl) would be brave warriors, in a positive manifestation, or die fighting in useless street quarrels. When the birth day was 2-rabbit (Ome Tochtli), people would die from drunkenness. Longevity was the main characteristic of those born on 1-Vulture (Ce-Cozcacuauhtli). Those born on 1 Reed (Ce-Acatl), a day consecrated to Quetzalcoatl, the wind god, would lead lives as worthless as the wind (*Florentine Codex*, 1979, 4: 29).

The day when illnesses began or presented significant changes, or when wounds and other accidents occurred, was an important datum to be considered in prognosis. The *Codex Tudela* (f. 125r) explicitly stated that representations of astrological signs over a deer's skin, when related to organs or parts of the body, had this function. Their consideration was also important to treatment, as was proposed some years ago by Mexican historians of medicine such as Fernández del Castillo (1964), Aguirre Beltrán (1963: 49–51) and Martínez Cortés (1965). Unfortunately no document survives today explaining how it was done.

Human organs and parts of the body were also related to calendric signs. In an ancient codex, eight representations of these types of relations were shown, some on god bodies, some over deer skins, but only one, that in *Vatican A Codex* (Lam. 73), on a human body (Figure 1).

The signs and body parts formed a network recalling those related to a cardinal zone, or representing magical properties, as the relationship between the left arm and magic, or a myth, like the left foot and Tezcatlipoca history (Viesca *et al*., 1997). In addition, indigenous sources such as the *Florentine Codex* expressed a direct relationship between certain days, divine actions and illnesses, as was the case of the days designated as 1-Monkey (Ce Ozomatli), when they believed the *cihuateteo* descended to earth, producing twisted lips, shrunken mouths, crossed eyes or weak vision (1979, 4: 81). These goddesses were women who died in their first childbirth and also came down and produced those same illnesses on 1-Rain (Ce Quiahuitl) days (*Florentine Codex*, 1979, 4: 41; Sahagún, 1959, I: 49).

EMPIRICAL ANATOMY

The only texts providing us with human anatomical terms are those transmitted to Bernardino de Sahagún, the 16th century missionary and archaeologist, by his indigenous informants. He selected old people and *titici*, medical doctors, to inquire about human anatomy, illnesses and their treatments, with special

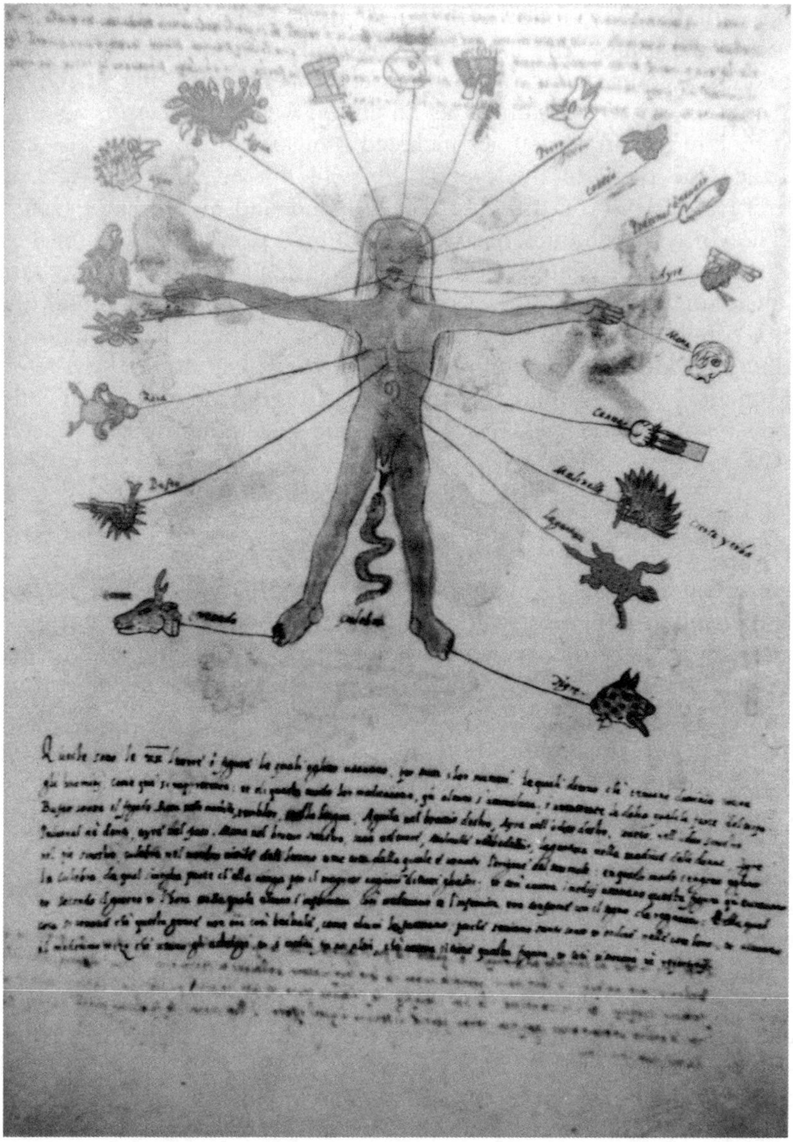

Figure 1 This human figure shows the correspondence between calendric signs and parts of the body. *Vatican Codex*, plate 73.

attention to medicinal plants and other items employed in therapeutics, such as animal parts, stones and minerals.

The anatomical texts were actually oriented to teach the Nahuatl language and not precisely to offer instruction in either medicine or anatomy (*Florentine Codex*, 1979, XI: 95–138). Scientific knowledge was intended only as a supplement to a better understanding of culture. Anatomical terms were explained starting with the skin, followed by the names of the internal and external organs and the joints. The first paragraph only provided the terms for skin, flesh and fat, accompanying each with a series of characterizing adjectives. The bodily containers were organized as an entity. Anatomical terms also characterized functions, and sometimes included a metaphorical sense. For example they called the pupil of the eye *tixtotouh* (a mirror), the eye *auic iauh* (it wanders) (*Ibid.*: 103), or the lung *técauh* (wind and also breath) (*Ibid.*: 120, 130). The voice was characterized as a functional part (*Ibid.*: 108) in the same way as the vision was related to the eye and was described with many characteristics, including a knowledge capacity. When talking about the blood, the text does not contain any allusion to blood components, but refers to its color and its freshness related to the relationship among tired and hot and the refreshing blood function, with its ability to strengthen. In the same way, when collecting the terms related to the liver, *elli*, most of them speak about bile characteristics (*Ibid.*: 131).

THE FUNCTIONAL BODY

The eye was considered a mirror, and it was transparent (*teuilotic*) and never clouded (*popoiutl*) under normal conditions. But the real instrument of seeing, *tlachialoni*, was the pupil, which was described as "our animated organ [to see]" (*toioiolcauh*) evoking movement like that of the heart, as the real mirror, putting in it the principal vision predicates (*Florentine Codex*, X, 28: 101–103). Some terms denote the appreciation of pulses, called *tlahuatl* and the perception of the heart stroke in the precordial zone, calling it *tetecuicaliztli.* Another term, *tlatemoniliztli*, suggests some ideas about digestion. The similarity between semen and seeds was expressed with the word *tlacaxinachtli*, which signifies the seed of man (*Florentine Codex*, X, 28: 130).

The functions of the body were closely related to the microcosmic structure concepts. The equivalence between the heart and the sun exactly corresponded with bringing life, giving heat and shining, all of which were considered heart functions, as were thinking, feeling and all other mental functions. In contrast, the liver was considered a nighttime sun which traveled inside the earth; it was also a blood rich organ absolutely necessary to life and also another site of thinking, this time inspired by Tezcatlipoca and magic.

Mental functions, knowledge, sensation and feelings were considered in a double and in some sense contradictory way. Indigenous sources ascribed them to the head, never mentioning the brain but instead the top of the head with all its adjacent structures, and also to the heart and the liver. This makes it possible to consider the existence of three souls, *tonalli, teyolía and ihíyotl*, situated respectively in the top of the head, in the heart and in the liver, each representing one of the cosmic divisions including the center, the human realm. López Austin (1980, I: 221) has described not only the properties and characteristics of these animic entities and their role in life and health maintenance, but also their bad influences, especially those originating from *ihíyotl* alterations and causing vigor diminution and energy loss manifested in very different ways and particular illnesses. I recently proposed (1997: 141–142) a bipartite conception, understanding the center as the confluence point of influences from the heavens and the underworlds and also from all the earth surface directions and represented by the central part of the body, the diaphragm and the heart above it and the liver below. This explanation permits us to visualize the possibility of an alternating heart and liver rule in the successive five suns characterizing the previous periods of world creation. The heart was perceived as the most important organ in humans and in all the fifth sun creatures, being not only the source but the center of life, mental functions, and centralizing impulses. Energy and influences were transmitted to it by the other animic centers, by the senses and by each contact with the surrounding world.

HEALTH AND ILLNESS

Health was defined as the equilibrate function of the human body and its souls and mental and spiritual functions, all of them accorded to a cosmic harmony. In this way every corporeal part or organ was maintained in proper limits. Illness appeared when any bodily part failed to maintain this balance and was more grave or fatal proportionally to a middle human range separation. Humans, living at the universe's center, participated in all the properties and characteristics owned by every creature in any of the celestial and underworld floors. The Nahuatl name for human is *tlácatl*, and it signifies etymologically "the [being] of the middle", remembering man's place of origin and the divine participation in his creation that is his mixed nature.

The first consequence of this was the coexistence of celestial and underworld characteristics in this strange being placed by himself at the cosmic center. Practically it was manifested as a definite and determined hot and cold presence in humans and, inside them, in every part of their bodies and in each one of their souls. The head was hotter than the foot and the abdominal cavity colder than the thoracic; the heart was hotter than the liver and the uterus colder than the penis. *Tonalli*, coming from heaven, was hotter than *ihíyotl*, which is earth related, and *teyolía*, representing the middle and working from the heart, was moderately hot and moderately cold. Changes in the quantity and distribution of hot and cold among the organs produced distemper and disease.

This hot-cold axis, or syndrome, as it was called by López Austin (1971: 21–23), has been identified as a central classificatory illness concept in ancient Nahuatl medicine and later was recognized as a central principle in Maya, Zapotec and other pre-Hispanic and contemporary indigenous groups. Cold and hot illnesses appeared in early sources; cold, aquatic fever or other kinds of fevers called *tlanatonahuiztli* were mentioned in the *Primeros Memoriales* collected by Sahagún at Tepeapulco in the mid-16th century (fo. 69v). Hot and cold fevers were distinguished because of the presence or absence of chills. Heat in the eyes, the throat, and the heart, or cold in the abdomen are mentioned in the *de la Cruz-Badianus Codex* (ff. 14, 29, 48 and 55).

Other illnesses were classified as hot or cold after considering their characteristics, as with *tlanatonahuiztli, tlallatonahuiztli* or *yoallatonahuiztli*, cold illnesses related to teeth, earth and "agua nocturna" (a kind of water associated with night gods) respectively, and, sometimes, only after comparing and establishing parallels between their treatments. This was the case with epilepsy, whose treatment included substances characterized by their refreshing properties, or that of gout, with heating qualities. This concept is crucial to understanding Mesoamerican medicine's transition from magical medicine. Some authors maintain that this conceptual organization arrived in Mexico with the Spanish, without considering that the Mexican system does not make use of wet and dry as physical primary characteristics and constructs its entire system only using hot and cold (Foster, 1987; López Austin, 1980: 303–317; Ortíz de Montellano, 1990: 213–221; Viesca, 1986: 101 and ff; Viesca, 1997: 180).

The question is not limited to labeling illnesses as hot or cold. In its practical applications it was necessary to determine the relative weight of multiple factors contributing to the cause and/or development of the illness and, finally, to define its hot or cold characteristics and their dynamics. For example, a group of illnesses was conceived as proceeding from or at least related to *tonalli* loss or weakening. All of them involved a heat loss and were defined as cold illnesses, but the cold degrees are as innumerable as the *tonalli* variability. López Austin (1980, I: 247–251) identified possibilities for *tonalli* direct damage beginning at the interior of the body, by means of divine punishment effects, possession by different spiritual beings, witchcraft, and physical repercussions from sin. When this entity was outside the body it might be affected by sorcerers, malignant spirits or some beings characterized as "energy greedy", such as the earth inhabitants of previous suns who, in the shape of stones emerging partially from the ground, waited for the opportunity to bite the leg of anyone stumbling against them, especially at night. Other energy greedy people were pregnant women, provided with a double *tonalli*, their own and also that of the foetus; albino people, marked by Tlaloc and expressing heat absence through their colour absence; and twins, who, obliged to share the maternal warmth with their counterparts, suffer from chronic heat avidity. All these people endured very general symptoms such as weakness, headaches, and anxiety, being simultaneously sick and dangerous to others.

A very particular illness was that called *netonalcahualiztli*. This corresponded to *susto* or magical fright, which is its equivalent in colonial and modern Mexico. Some authors, among them George Foster (1953), think that this illness came with the Spaniards in the 16th century, but the provenance of *netonalcahuiliztli* is clearly pre-Hispanic and constitutes one important correspondence between ancient Mexican and Spanish medical traditions which facilitated medical interchanges between conquerors and conquered. Segre (2000) has recently proposed a conceptual revision noting the differences between the pre-Hispanic *netonalcihuiztli*, a mixed physical and psychological ailment, and the postconquest *susto*, with its socially derived components.

Another *tonalli* illness was the "fatigue of those administering the government and holding public affairs" as it was called in the *de la Cruz-Badianus Codex* (fo. 30v). Contrary to lassitude, whose treatment was described in a folio before it in which it was referred to as physical tiredness and related to the legs and feet, this peculiar fatigue came from *tonalli* debility and the consequent impossibility to maintain effective communication with the gods and spirits (Viesca, 1977). This illness involved a relative coldness and was in clear opposition to common weariness, which was considered hot because of the warming up following physical activity.

Another taxonomical axis applied to illness classification was that organized around specific organs. The heart and the liver were charged with a great significance as regulator centers for very different functions, and their illnesses represented the confluence of multiple factors. Starting with magical causation we had a conceptually mixed, cardiac and mental, illness consequent to a definite kind of sorcerer action, in which he exerted his power by overturning the heart of some person who consequently became mad. Another sorcerer type is *teyollocuani*, that is "the heart devourer", who acted by eating the *teyolía*, the cardiac entity, which diminished every vital function, including mental life, ending habitually in death. Contrasting with these magical causes was an illness described as oppression of the chest (*de la Cruz*, fo. 27), removing anguish as a primary symptom, which is easily recognizable as coronary disease. The treatment sequence and some statements about the cure permit us to identify certain types of mucous substances, called *alláhuac*, capable of severely burdening the heart and, if spilled around and over this organ, provoking death (*de la Cruz*, fo. 62).

Alláhuac also constitutes an important concept regarding pathology and illness etiology. Initially it was identified with phlegm, but this term, evoking European humoral medicine, is too narrow. *Alláhuac* included a variety of pathological substances differentiated by their color and their hot or cold characteristics. *Ezállahuac*, for example, was used to signify composite secretions with blood and mucus. This concept permits one to establish another classificatory axis around pathological substances secreted by or accumulated in the body.

Dropsy was called *ytixihuiztli*, but we do not have evidence to affirm if it was related to the liver. A big, round belly, with umbilical hernia and emaciated limbs and chest, was represented skillfully in terracotta figurines, but this is not sufficient to diagnose hepatic cirrhosis. Jaundice was considered, but the same term designating it, *tetzauhcocoliztli*, was also used for pthisis, denoting in the same way the yellowish skin and the accentuated pallor.

There are hundreds of illnesses and diseases listed. The Nahuatl data, i.e., those coming from Central Mexico in areas with an old tradition stretching from Teotihuacan to the Aztec Empire, have been extensively studied through the scarce existing clinical commentaries and also the linguistic items rendered by ancient vocabularies and quotations and compared with those surviving in indigenous communities. All this knowledge reveals a highly sophisticated taxonomy system, correspondent to a similarly complex worldview and provided also with complicated medical subsystems such as therapeutics and materia medica.

Among the Maya the studies are mainly descriptive, recording a reasonable number of diseases. In the *Dresden Codex*, the Moon Goddess introduced hot illnesses to the earth related to the glyph *kak* (*Dresden Codex*, lam.17a). The codex also distinguished *tancás*, a category including all kinds of aggression, which represented a transitional group of illnesses such as spasms, convulsions, fainting, madness, weakness, speech difficulties, disordered sexual appetite, drunkenness, and lassitude in the legs. The last two were associated with chronic states after drinking alcoholic beverages and after spirit possession when traveling. Convulsions were a special type of *motancás* or macaw attack with high fever. Today we have only very fragmentary and incomplete records about how Maya culture conceived illnesses; Mayan illness classification needs more and deeper studies, with attention to their worldview and to semantic items.

In general terms, the illness concepts and classifications in the most important Mesoamerican cultures were part of a congruent system with its own rationale and logic, capable of explaining the most important questions about health and illness.

GODS, GODDESSES AND MEDICINE

Divine intervention was the rule in ancient Mesoamerica, pervading all facets of life. Health, illness and medicine were not exceptions. In the Nahuatl world, the main medical goddess was *Tzapotlatenan* (Figure 2). She was originally a very prestigious curer in a village called Tzapotla and discovered the use of and the way to prepare *óxitl*, a pine resin still used in the treatment of wounds, scars, itch and other skin diseases; after her death she was venerated as a goddess (Sahagún, 1959, I: 49). Other deities were also related to medicine in very different ways: Quetzalcóatl was considered the Fifth Sun creator and then god of medicine, astronomy, mathematics and all cultural and artistic activities; he was also the protector deity of the *calmecac*, the school for noble children, and was reputed to be a healer god. In his main temple at Cholula, deaf and blind people, and those suffering from palsy, lameness, or gout prayed to the god for his help (*Códice Ramírez*, 161). But, as Ehécatl, the god sent rheumatism, gout and neck stiffness to people exposed to night or cold winds, and his auxiliary spirits, the *ehecame*, produced muscular pains, stiffness and facial paralysis (Sahagún, 1959, I: 45).

Another important god, Tlaloc, Lord of Rain, had an important role in health (Figure 3). He protected the crops and food, but he also sent some illnesses to selected people, who ended by going to the *Tlalocan*, the special heaven where Tlaloc and the *tlaloques*, his collaborators, took care of all the embryos of future generations, of nutriments, seeds and every kind of life germs (Sahagún, 1959, I: 257).

A very interesting case was that of Ixtlilton, a god represented as a little child and involved in caring for children. If a sick child looked into a big earthenware jar containing water, and if his reflected features were fuzzy, Ixtlilton could diagnose *tonalli* loss, claiming that one of the parents was an adulterer. A similar situation occurred when, during the god's festival, priests found straw or dust inside the water jars of the house where the celebration was taking place (Sahagún, 1959, I: 52–53).

Figure 2 Tzapotlatenan, the medicine goddess as represented in Sahagún's *First Memorials*.

Nahuatl mythology also had some goddesses related to different aspects of procreation. They constituted a truly divine complex, as multiple in its functions as in the nature and conditions of each one of the divinities in it (Ortíz de Montellano, 1990: 46). The moon goddesses were considered fertility deities. The old grandmother, Toci, reigned over the *temazcal*, the sweat house where pregnant women, those with difficult deliveries and women after childbirth were cared for and treated; she consequently ruled over all prenatal care. Quato y Caxoch took care of delivery.

Women who died in childbirth, especially in their first delivery, acquired a semi-divine quality; their *teyolía* went down with the sun on its way from the zenith to the earth's surface and then caused serious illnesses in little children, puerperal women, convalescents or debilitated people (Figure 4).

In Maya religion the main gods involved with health and illness were Itzam Ná, a solar deity also related to rain, fecundity and medicine, and Ix-chel, the moon goddess (Landa, 60–70; Montoliú, 1984: 287). The nine underworld gods were reported in *Popolo Vuh* to send particular illnesses: Xiriquipat and Ahalganá, bleeding and haemorrhages; Ahalpuh, swollen legs with suppuration and *chuganal*, an icteric (jaundice) disease; Chamiabac and Chamiabolom, gradual weakening and death; Ahalmez and Ahaltogob, injures and accidents;

Figure 3 Tlaloc, the rain god, was related to water and some water related illnesses. *Magliabecchi Codex*, fo. 22.

and Xic and Patan, vomiting blood, perhaps due to yellow fever (*Popol Vuh*, 124).

In ancient Michoacán, illness was generically considered a god's punishment. Curicáveri, the sun and war god, struck his enemies with illness. Cuerauáperi, goddess of the crops, fertility and abundance, also had a contrary role as Auicamine, misery. Xarátanga, the moon goddess, sent gastric illnesses and provoked vomiting in people who offended her (Recinos, 1956: 10; Sepúlveda, 1984: 355–356; 1988: 59–61).

In late Mesoamerican cultures, there was a tendency to recognize universal deities, such as the Netzahualcóyotl's Unknown god or the Ipalnemohuani of Aztec poets and wise men. They represented absolute gods assuming all the attributes of the gods existing in the overpopulated pantheon. Sometimes a tendency to abstraction and the image of a distant and powerful god was evident (León Portilla, 1960). In other contexts, figures like Quetzalcóatl in Central Mexico or Kukulkán in Mayan Yucatan centralized religious thinking around cultural and humanizing heroes who took care of humankind (Sejourné, 139 y ss).

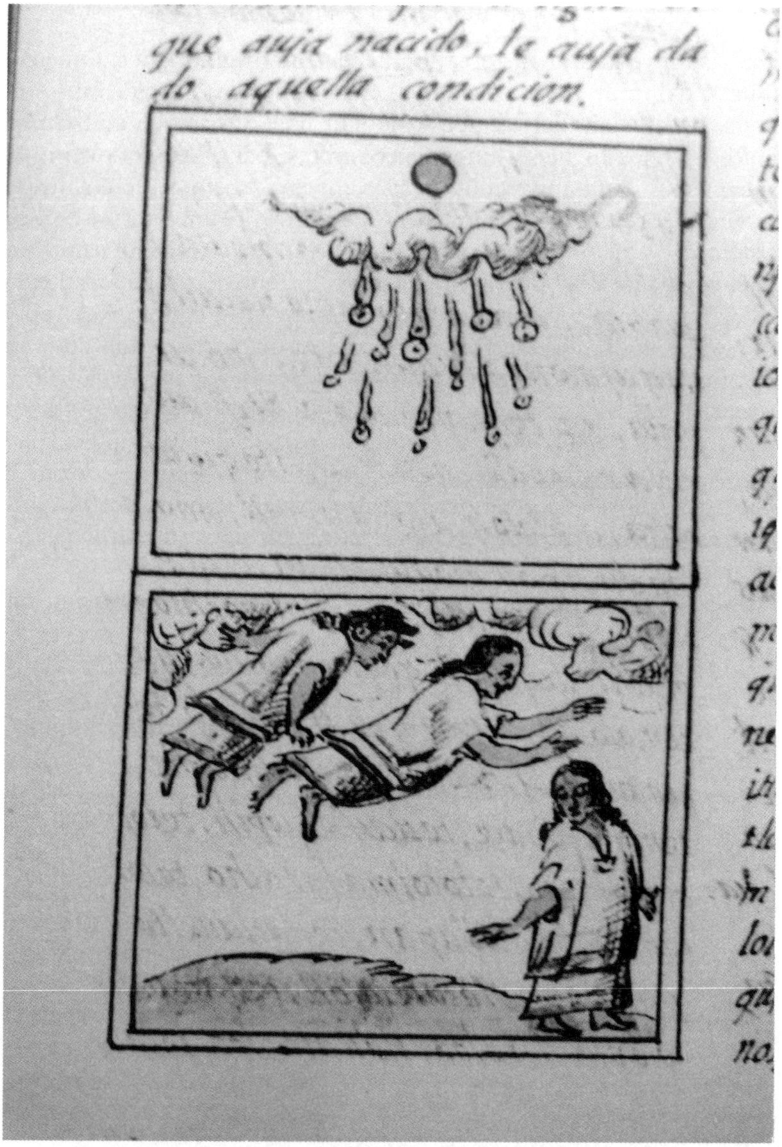

Figure 4 *Cihuateteo*, the souls of women who died in their first childbirth, descending to the earth's surface on One Rain days, provoking fright, convulsions and other grave ailments in children, women and weak people (*Florentine Codex*, IV).

Some gods were direct influences in causing diseases; others acted by means of the energy represented by astrological influences. Sometimes, gods were subjected to "natural", i.e. cosmic, forces, becoming active only because of his or her dynamic characteristics.

THE INVENTION OF MEDICINE

Medicine was invented at the creation, when the original gods, Ometéotl and Omecíhuatl, living at the thirteenth, the highest, heaven, had four sons, the main gods, and in parallel created a human pair, Oxomoco and Cipactonal. Some authors refer to these humans as "archetypal"; López Austin (1980, I: 265) called them "humanity gods". They appeared before the creation of the heavens and before the establishment of the central realm and the constitution of a cosmic division of underworlds and earth. They also appeared prior to the existence of time and death; in fact they "invented" the time calendar and record keeping of celestial bodies' revolutions. This couple received divine orders to work and cultivate the lands, and also to weave and spin and use maize grains for divination and cures (*Historia de los mexicanos*, 25). This shows clearly that a woman was the first curer, but the representation in the *Borbonicus Codex* depicts both woman and man, Oxomoco and Cipactonal, in the act of consulting maize grains and soothsaying (Figure 5). From ancient times, the medical profession not was exclusive to any gender, but was practiced by both sexes, perhaps adjusted to different matrilineal and patrilineal patterns in different local contexts.

Medicine was conceived not exclusively as a human achievement, but as a

Figure 5 The primeval couple, Oxomoco and Cipactonal, inventing the art of medicine. The figure represents divination and soothsaying (*Borbonic Codex*, plate 22).

divine gift combined with the obligation to transform it into a human activity. It was never solely human, because it needed divine assistance.

MEDICINE MEN

In Ancient Mesoamerica there were many medical specialists. One very ancient clay figurine from Tlatilco, dressed in a jaguar skin, has been interpreted as a shaman. Some Olmec sculptures, including the Chalcatzingo reliefs, represent man as a jaguar acquiring immense powers, not only those of the animal, but also those owned by the night god (Gay, 1971: 54–55). These men were sorcerers and healers, and some of them were also Tezcatlipoca priests who took care of the preparation and administration of *tlilatl*, black water, a magic medicine useful in the most serious illnesses.

Tlaloc's relatives also played an important healing role. In addition to the priests charged with curing Tlaloc illnesses, there were some people selected by the rain god to be his adjutants. They came to this in two ways, either by being born in a Ce quiáhuitl (1-Rain) day or by being struck by lightning. The first one was regularly called *nahualli*, the second one *tlaciuhqui, tlamacaztli* or *granicero* (hailer). The Aztec informants to Sahagún said that *nahualli* might be good or bad, depending on whether they used their powers to benefit or harm people. A child born on that day, 1 Rain, would be provided with supernatural faculties. He would appear from and return four times inside his mother's womb and only come up the fifth time, symbolizing the fifth sun era. He could change his form into any kind of animal and behave according to his animal disguise. He knew the heavens and underworld regions and also knew how to protect and enhance some crops and harm others. He knew all kinds of illness prognoses and could act as a healer, but he could also become an owl-man (*tlacatecólotl*) and cause all kind of calamities and evils. Hailers behaved in a different way; a patient would recover only after he agreed to collaborate in the god's tasks of directing rain to the proper places and protecting harvests against hail and frosts. These people needed to learn everything the *nahualli* knew "naturally" and required an initiation time, learning the geography of magical and supernatural places, the control of the spirits and the use of sacred plants and curative substances.

Some ancient sources, like the *Florentine Codex*, (1974, XI: 37–38 and 55–57; López Austin, 1969), spoke about many different kinds of magicians and sorcerers, most of them harmful, like *tlacatecolotl*, a very dangerous *nahualli* associated with Tezcatlipoca who could convert himself into an owl, or *teyollocuani*, the heart devourer. Some could cure, some could protect common people from the malevolence of other sorcerers, and some could predict the future.

The most interesting medical figure was the so-called *ticitl (pl. titici)* in Central Mexico, *ilalix* among the Huastecs, *xurihca* in Michoacán, *h'ilojel* and *h'ilol* by the Tzeltal and Tzotzil respectively, *h men, ah cut* and *ah cun* among Mayan, Pokoman and Quiché. Jacinto de la Serna, a Spanish priest reporting some time after the Conquest, explained that the *ticitl* was more than a medical doctor, as his activities included being soothsayer, wise man (*tlamatini)* and sorcerer (Serna, I: 117). All the *titici* were healers following the professional activities of their parents (Viesca, 2001: 47). Both women and men could become healers, and the same term, *ticitl*, was employed indiscriminately to medical specialists and midwives (Viesca, 1984: 218). In big cities, as in Mexico-Tenochtitlan at the eve of the Conquest, *titici* lived in separate quarters (López Austin, 1973: 72–75). Data recovered from a Maya site at Verapaz describe two medical strata, separating those healers curing kings and the nobility from those treating lesser people. Curiously, those noble healers were called *ah cut*, like the Pokoman healers, and those curing common people were the *ahicom* (Miles, 1957: 768; Orellana, 1987: 60–61). This same situation might have been the case when, speaking about Martín de la Cruz, an indigenous doctor before the Conquest, Spanish authorities call him a *principal*, that is a high ranking person (Viesca, 1995: 480–498). These noble healers would have been educated at the *calmecac*, the same scholarly institution destined for nobility and priests, mentioned also as medical teaching centers.

The *titici* are reported both as good and bad physicians and as male and female practitioners. Sahagún's informants differentiated clearly between these health professionals, recording the ways each of them acted. Male physicians practiced wound treatment, fractures management, sutures and ear and nose plastic surgery. Women were responsible for midwifery, embryotomy, ulcer treatment, ocular surgery, topical applications and massage (Sahagún, 1959: III: 116–117, 124–125; Ruíz de Alarcón, 1987: 195). A deep knowledge of the clinical features of illnesses and of medicinal plants and other substances was imperative for good physicians, but they also needed to master the spirits' language, know tricks and magical formulae and to have at least a reasonable acquaintance with sorcery, to protect and cure their patients and reverse evil in some cases.

Some authors have mentioned the possible existence of specialists, separating midwives from medicine women, and surgeons, bone setters, from eye or ear illness specialists (Flores, 1888, I; Ocaranza, 1935; Gutiérrez, 1971: 10).

THERAPEUTICS

In Mesoamerican cultures, therapy was as diverse as the cultures themselves. This diversity increased because of growing social development just before the Spanish Conquest. We have a large repertory of local practices, as well as a common worldview from the Aztec (Mexica) Empire, heir of cultures two and a half millennia old, to relatively isolated tribal cultures dispersed along large territories. Therapeutics ranged from magical rituals to sophisticated medicinal preparations, but there was never a completely naturalistic view without reference to the supernatural.

RITUAL AND PRAYER

Many practices are recorded from the ethnic and political groups existing at the eve of the Conquest. Some prayers contained intriguing magical traits, as in the prayer directed to Tezcatlipoca in times of pestilence. As reported by Sahagún's informants, the suffering people prayed that god be benevolent to them, but at the same time insulted him, calling him homosexual among many other socially problematic names (*Florentine Codex*, VI: 2; Sahagún, 1959, II: 55). A friar in the early 17th century, in a Nahuatl zone near Mexico City, used a prayer identifying the curer with "The Lord of Transformations" and menacing the illness, specifically the pain, with his own strength and then with the accumulated strength of his predecessors and auxiliary spirits (Ruíz de Alarcón, 1953). The Ruíz de Alarcón collection of thirty-one conjures, coming from a Nahuatl village but maintaining a strong Indian tradition, constitutes first hand evidence about the structure and composition of these procedures.

Some preventive statements are easily ascribed to magical ideas. Pregnant women were ordered not to eat *tamales* to avoid having the child be intertwined in the placental membranes. The prohibition, also to pregnant women, against going out during an eclipse without a stone, preferentially obsidian, covering the abdomen, involves fear of being damaged by bad spirits.

Many magical procedures were mingled with other therapies. A good example is the prescription of *yolloxóchitl* (*Talauma mexicana*) infusion to strengthen the heart, a practice originally related to the flower's form, which in a lateral view looks like a heart cut through the middle. This magical idea was confirmed by its cardiotonic properties. In the *de la Cruz-Badianus Codex*, nearly a hundred magic based practices were identified (Viesca and de la Peña, 1974). Most of them were attributed to animal substances, perhaps because in modern medicine we have lost many of these types of medicines. Applying burned hair from the top of an animal's head to children that could not suck corresponds to the idea of transmitting *tonalli* properties, (Cruz, fo. 14r and 51v).

Maya medicine has been less studied than that of Nahuatl cultures, and the studies published until now have been mainly descriptive. We have evidence of rituals and prayers from the Maya area in the *Ritual of the Bacabs*, many of them containing references to gods and telling about the travel of medicines throughout the sick body, as in the case of the incantation against asthma. In it, illness was manipulated by a charmer with a stick and a fan over a stone or wood human figure representing the patient. At the same time, the magical words referring to the structure of the universe, some of the god's origins and the symbolic procedure to throw out asthma and conduct it to the Harmful Lady Palace, She Who Holds up the Earth's Opening, were uttered (*Ritual*, 23).

MATERIA MEDICA

Plants and their derivatives

Some 2,500 medicinal plants were recorded in Mesoamerican territories. Plants were recommended for magical, religious and natural reasons. *Yolloxóchitl* (*Talauma mexicana*) was prescribed for heart weakness, because of its magical identification with the heart, but it became a useful cardiotonic. This same flower was recommended as a treatment against madness, as a mental tonic, and as a diuretic (Viesca, 1986: 131).

In a similar way, the Maya employed ipecacuana roots to cure dysentery, explaining its effectiveness by means of the root color and consistency, comparable to the sick person's excrement, and related to Venus in its sun conjunction when a reddish halo is visible between both heavenly bodies.

Sometimes the plants were selected because of their relation with a god or goddess, as in Ortiz de Montellano's study of Tlaloc plants. The sacred *iyauhtli* (*Tagetes erecta*) had a role in deity festivities, but also had medical uses against intermittent cold fevers, thunderbolt fright or evil tendencies (1980: 287–314) (Figure 6). Other Tlaloc plants were aquatic, like lilies, or hallucinogenic, like

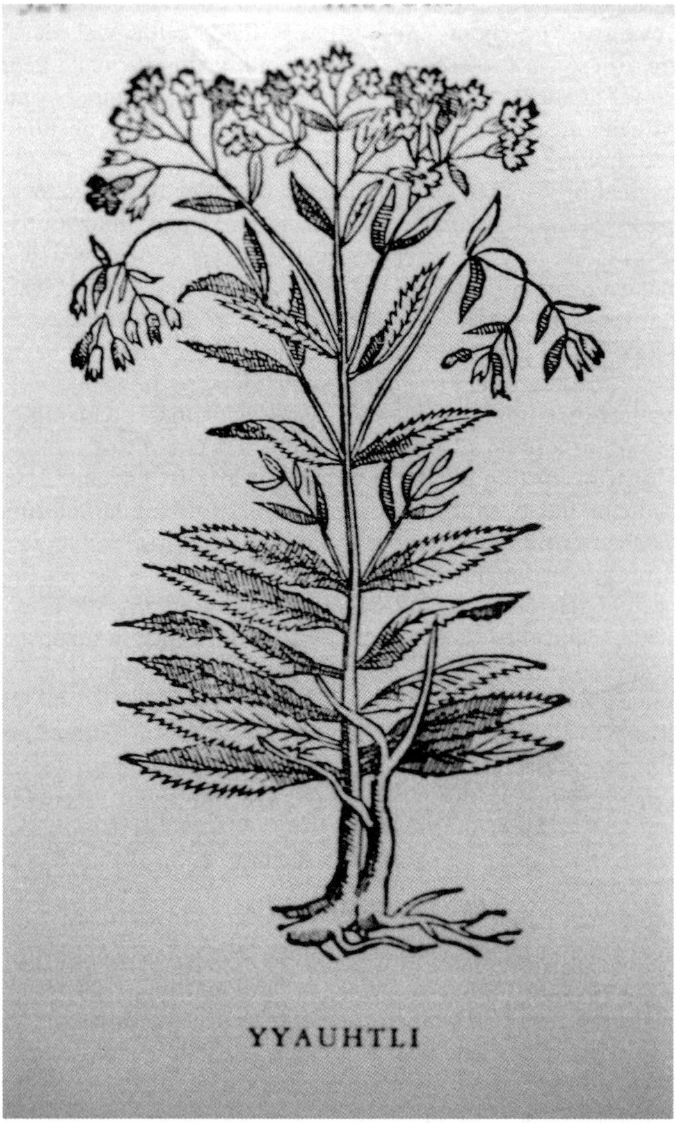

Figure 6 *Iyauhtli* (*Tagetes erecta*), a plant related to Tlaloc, as represented by Francisco Hernández, King Philip II's physician, *ca.* 1575.

ololiuhqui or enteogenic mushrooms. Datura was also depicted in the Rain God Temple at Tulum.

Pre-Hispanic doctors observed and recorded some somatic effects from plants; they used them to obtain sudorative, diuretic, laxative, purgative, or febrifuge actions, or to eliminate cold or heat, or expel *alahuac* and other pathogenic substances. There were cold plants, like *huitzquílitl* (*Cirsium* sp.), which diminished excessive heat and was useful against burns and itch; hot plants, like *cihuapahtli* (*Montanoa tomentosa*), useful to speed up delivery by building up heat inside the uterus; plants generating sweat, like *mecapahtli* (Mexican zarzaparrilla), prescribed against fever; and plants that expelled *aláhuac*, such as *tlatlancuaye* (*Iresine celosia*).

This list could be enlarged extensively. Medicinal materials were not single plant substances but composites of several parts and substances from plants, animals and minerals. The main purpose was to control the ill characteristics and substitute healthy ones. For example, in the *de la Cruz-Badianus Codex*, the treatment for child burns starts with *michihuauhtli* and red incense, both of them hot, followed with a mixture of *xiuhehecapahtli* (*Cassia occidentalis*), a hot plant, and cooling plants, such as *huitzitzilxóchitl* (*Salvia gesneraflora*) and *tlalahuehuetl* (*Taxodendron* sp.) (Viesca, 1997: 182; *De la Cruz*, fo. 61r).

Animals and animal substances

Medicines prepared from plants were the most numerous part of therapeutics. Animal and mineral substances were less abundant and usually given magical explanations for their actions. Some of them, like the *tlacuatzin* (*Opossum marsupialis*) tail, which was employed as a childbirth inductor and accelerator, have had their pharmacological properties experimentally tested with satisfactory results. Others, such as jaguar meat, reputed to strengthen the heart and cure mental illness, have not undergone any modern research. The same can be said about the little bird called *pito* or *oconénetl*, living in the Tlaxcala region in Central Mexico, whose hallucinogenic properties remain enigmatic (Muñoz Camargo, 1892: 134).

Other animal parts or substances were prescribed according to magical indications. For example, eagle wings employed to facilitate childbirth, or excrement for postpartum pain treatment, refer to the eagle's solar identity and its warrior-like comportment and the relationship between war behavior in men and childbirth as a correlative to it in women (*De la Cruz*, fo. 57r).

Puncture with a pointed eagle or puma bone in arthritis cases or boiled *huacton* (*Nycticorax* sp.), sparrow hawk, rabbit legs to cure a fractured knee, have to do with transmitting the animals' strength and flexibility to the sick (*De la Cruz*, ff. 36v, 47r). In a different way, leaving arthritis patients to be bitten by ants proceeds from the observation of the positive effects of what we now know as formic acid. A mixture of magical, calendric and medical theory is apparent in the use of a burned and desiccated *cuetzpalin* (small lizard) in the treatment of dysmenorrhea, relating its symbolic association with the small lizard calendar sign and sexual power, the penis' role in menstrual suspension and its extreme heat, which is useful in a hot illness (Viesca, 1986: 139).

Stones found in a bird's maw were extensively employed. At least twenty were mentioned in ancient sources coming from eagles, hawks, native doves, quails and some other little birds. These stones contrast with animal stomach stones or bezoars, used in European medicine from classical times. The careful observation of the animals' characteristics and of the effects obtained by their parts and products when used as medicaments is a good example of a pre-Hispanic medical achievement.

Mineral remedies

Every type of mineral was present in the prescription repertory. Gold was recommended for heart pain following the heart/sun/gold relationship (*De la Cruz*, fo. 28r). Some stones were considered useful to stop bleeding, like *eztetl*, identified as quartz; to cure tooth pain, like *iztac quetzaliztli*, nephrite; or to relieve heart pain, like *tetlahuitl*, a red stone. Rock crystal was thought to alleviate ocular heat and the ailments produced by heavy winds. *Quiauhteocuitlatl*, a special stone descended from heaven with a thunderbolt that remained concealed underground for a while, was used both in heart ailments and for those people frightened by thunder (Sahagún, XI: 7, 236 and 237). Among organic substances, amber was considered the forefathers' *tonalli*; it was prescribed to treat some illnesses related to *tonalli* loss or diminution like heart pain, or pathological fear, or to provide courage to those who had to cross a dangerous river (*De la Cruz*, ff. 28r, 53r, 56r).

Medicine was rich and complex, derived from a comprehensive knowledge of natural products and developed within a sophisticated medical theory.

SURGERY AND MIDWIFERY

Surgery was highly developed, even if it did not have any anatomical or pathological explanation. Wound management was also effective, incorporating haemostat procedures, among them compression and local application of spider webs and plants like *matlalxíhuitl* (*Commelina palida*) or jade stone (*éztetl*). They left wet maize tortillas until they were covered with a greenish coat coming from microscopic mushrooms, and applied them over suppurating wounds; this was an empirical treatment against infection, employing antibiotics without having the correspondent conceptual frame. In the same "antibiotic" way, maguey soap was applied extensively (Ortíz de Montellano, 1990: 184) (Figure 7).

Trepanation was a relatively common procedure in Oaxaca cultures from the Classical Period (Monte Albán IIIB) from the 5th century CE, but it was practiced at least seven hundred years before that. The first technique developed was drilling the bone; scraping and sawing came later. We do not have any ancient descriptions of trepanation, but the paleopathological evidence shows accurate surgical management and the survival of some individuals (Winter, 1985). In other Mesoamerican cultures it was practiced, but never with the

Figure 7 Breast abscesses were cared for with ointments prepared with two plants: *ixyayáhual* and *eeloquíltic*. *Florentine* Codex, X, 28.

frequency and ability of Zapotec surgeons. [Editor's note: See an extensive discussion of trepanation in Ruben Mendoza's article on South American medicine in this volume.]

The plastic surgical procedures for nose surgery were extraordinary. They reimplanted the nose and used very small stitches made from the pointed ends of maguey leaves. The *titici*, Sahagún's advisors, described the substitution of an artificial nose, but unfortunately they did not leave any detailed descriptions about the technique (*Florentine Codex*, X: 28). Parallel procedures were recommended for ear loss.

Another area where Mexican surgeons were really skillful was bone surgery. In addition to precise indications to immobilize fractured bones and the development of mud and feather devices supported by wood sticks, an astounding surgical procedure was described. When the femur, the long bone in the thigh, was fractured and did not fuse after being immobilized for a long time, it was necessary to reach the fractured surface surgically, clean it, scrape it and introduce a thin wood stick into the bone medular channel (Sahagún, 1959, III: 182). The procedure implies a very complicated surgical technique, involving skillful management of anatomical structures, bleeding and anesthesia or some effective pain control. It also implies good infection control, especially when the bone medular channel was manipulated. This kind of surgery was quite in advance of its time.

Midwifery was also a very developed specialty performed on postmenopausal women who were called *titici* in recognition of their medical role. They took care of pregnant women from the fifth or sixth month of pregnancy and were charged with caring for women on moral, psychological and physical levels. When delivery time arrived, the *ticitl* remained at the woman's side and touched

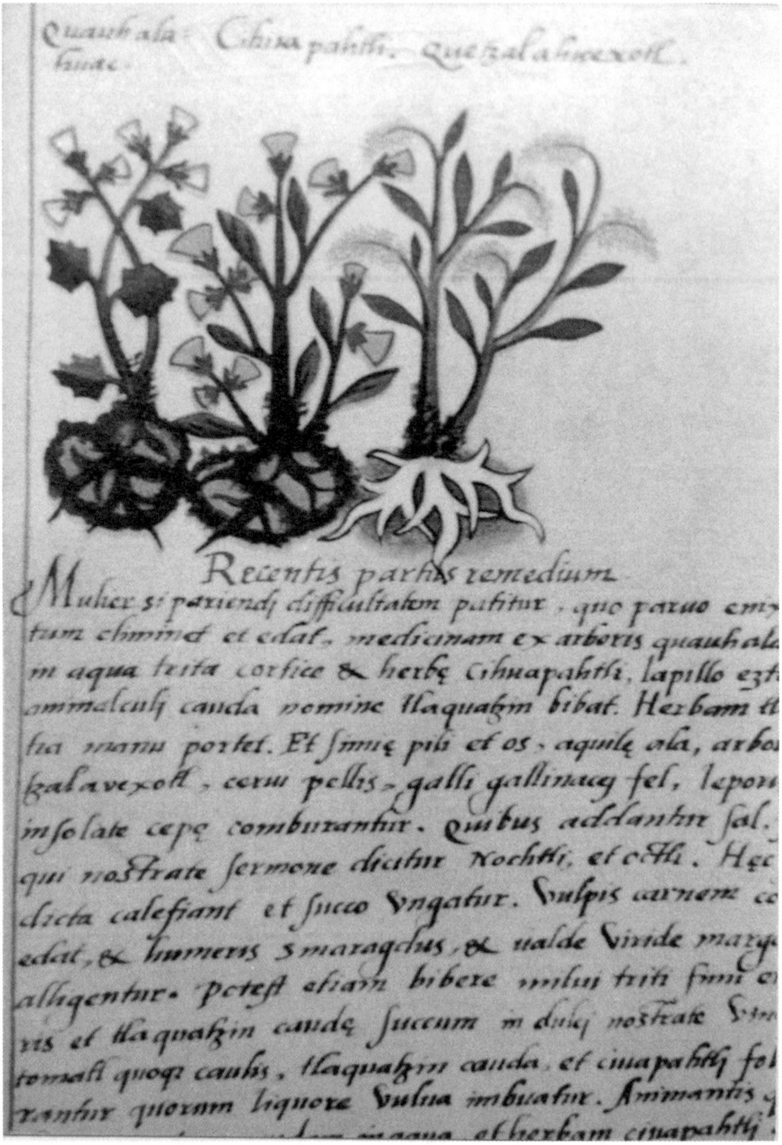

Figure 8 Remedies for puerperal women as depicted in the *de la Cruz-Badianus Manuscript*, fo. 56v. The plant represented in the central image is *cihuapahtli*, the most renowned uterine stimulant plant.

her belly, determining the foetal position and making changes if necessary. In all these circumstances, a *temazcal* bath was used, securing the protection of the female mother goddesses. They were also very skillful in oxytocic substance use, with precise knowledge about doses and contraindications to give *cihuapahtli* (*Montanoa tomentosa*), *tlacuatzin* tail or any other useful substance to induce labor. Finally, they were proficient with embryotomy when foetal death was confirmed. As was the rule all over the world at those times, mortality around childbirth was extremely high.

* * *

Mesoamerican medicine was an integrated system, closely related to and derived from the people's worldview and following a strict rationality in terms of intracultural parameters. It was provided with a general medical theory comprising magical, religious and natural aspects interpreted not only in terms of faith but also regarding physical characteristics and changes. All this knowledge and skill was used to control the main illnesses affecting them and to establish an adequate balance between people and their environments.

BIBLIOGRAPHY

Calepino de Motul, Ramón Arzápalo, ed. 3 vols. Mexico City, Mexico: DGAPA/Instituto de Investigaciones Antropológicas, Universidad Nacional Autónoma de México, 1995.

Codex Vaticanus Latinus 3738. Facsimile edition. In *Antigüedades de México*. Mexico City, Mexico: Secretaría de Hacienda y Crédito Público, 1964, vol. 3, pp. 7–313.

Codex Vaticanus 3773 (*Codex Vaticanus B*). Graz, Austria: Akademische Druck und Verlangsanstalt, 1972.

Codex Borgia. Facsimile edition. Mexico City, Mexico: Fondo de Cultura Económica, 1998.

Codex Fejervary Mayer. Facsimile edition. In *Antigüedades de México*. Mexico City, Mexico: Secretaría de Hacienda y Crédito Público, vol. 4, 1967, pp. 185–265.

Códice Dresden. Mexico City, Mexico: Fondo de Cultura Económica, 1989.

Códice Ramírez. Mexico City, Mexico: Editorial Innovación, 1979.

Códice Tudela. Facsimile edition. Madrid: Ediciones Cultura Hispánica, 1980.

Cruz, Martín de la. *Libellus de Medicinalibus Indorum Herbis (Cruz-Badianus Codex* or *Badianus Manuscript)*, E. Walcott Emart, ed. Baltimore: Johns Hopkins University Press, 1940; Facsimile edition and studies, E. Del Pozo, ed. Mexico City, Mexico: Instituto Mexicano del Seguro Social, 1964; 2nd edition, 1991.

Fernández del Castillo, Francisco. 'La cronología y la medicina náhuatl.' In *Antología de Escritos Histórico-Médicos del Dr. Francisco Fernández del Castillo*. Mexico City, Mexico: Universidad Nacional Autónoma de México, Departamento de Historia y Filosofía de la Medicina, 1984, vol. 1, pp. 61–84.

Florentine Codex (General History of the Things of New Spain), Charles Dibble and A. Anderson, eds. 11 vols. Santa Fe, New Mexico: School of American Research and Salt Lake City: University of Utah, 1950–1982.

Freidel, David, Linda Schele and Joy Parker. *Maya Cosmos*. New York: William Morrow, 1993.

Flores, Francisco. *Historia de la Medicina en México*. 3 vols. Mexico City, Mexico: Secretaría de Fomento, 1886–1888.

Foster, George. 'Relationships between Spanish and Spanish-American folk medicine.' *Journal of American Folklore* 66: 201–217, 1953.

Foster, George. 'On the origin of humoral medicine in Latin America.' *Medical Anthropology Quarterly* 1: 355–393, 1987.

Foster, George. *Hippocrates' Latin American Legacy*. Langhorne, Pennsylvania: Gordon and Breach, 1994.

Gay, Carlo. *Chalcazingo*. Graz, Austria: Druck und Verlagsanstalt, 1971.

Gutiérrez Vázquez, Samuel. *El Texoxotla Ticitl. Cirujano Mexica a Través del Tiempo*. Mexico City, Mexico: Academia Mexicana de Cirugía, 1971.

Hernández Francisco. *Historia Natural de la Nueva España*. Vols. 2 and 3 of *Obras Completas de Francisco Hernández*. Mexico City, Mexico: Universidad Nacional Autónoma de México, 1959.

'Historia de los mexicanos por sus pinturas.' In *Teogonía e Historia de los Mexicanos. Tres Opúsculos del Siglo XVI*, Angel María Garibay, ed. Mexico City, Mexico: Editorial Porrúa, 1965.

Landa, Diego de. *Relación de las Cosas de Yucatán*. Mexico City, Mexico: Editorial Porrúa, 1959.

León Portilla, Miguel. *La Filosofía Náhuatlo Estudiada en sus Fuentes*. Mexico City, Mexico: Universidad Nacional Autónoma de México, 1960.

Lombardo, Sonia and E. Nalda, eds. *Temas Mesoamericanos*. Mexico City, Mexico: Instituto Nacional de Antropología e Historia, Dirección General de Publicaciones del Consejo Nacional para la Cultura y las Artes, 1996.

López Austin, Alfredo. 'Cuarenta clases de magos.' *Estudios de Cultura Náhuatl* 8: 87–117, 1969.

López Austin, Alfredo. *Textos de Medicina Náhuatl*. Mexico City, Mexico: Secretaría de Educación Pública, 1971.

López Austin, Alfredo. *Cuerpo Humano e Ideología*. 2 vols. Mexico City, Mexico: Universidad Nacional Autónoma de México, 1980.

López Austin, Alfredo. *Tamoanchan y Tlalocan*. Mexico City, Mexico: Fondo de Cultura Económica, 1994.

Manzanilla, Linda and López Luján, Leonardo. *Historia Antigua de México*. 3 vols. Mexico City, Mexico: Universidad Nacional Autónoma de México, Consejo Nacional para la Cultura y las Artes: Miguel Angel Porrúa, 1994.

Martínez Ciortés, Fernando. *Las Ideas en la Medicina Náhuatl*. Mexico City, Mexico: Prensa Médica Mexicana, 1965.

Miles, S.W. 'The sixteenth-century Pokom Maya. A documentary analysis of structure and archeological setting.' *Transactions of the American Philosophical Society* 47: 731–781, 1957.

Montoliú, María. 'Cosmovisión y salud entre los mayas de Yucatán.' In *México Antiguo*, A. López Austin and C. Viesca, ed. Vol. 1, *Historia General de la Medicina en México*, De F. Martínez Cortés, ed. Mexico City, Mexico: Universidad Nacional Autónoma de México/Academia Nacional de Medicina, 1984, pp. 283–288.

Montoliú, María. 'Medicina Maya.' In *México Antiguo*, A. López Austin and C. Viesca, ed. Vol. 1, *Historia General de la Medicina en México*, De F. Martínez Cortés, ed. Mexico City, Mexico: Universidad Nacional Autónoma de México/Academia Nacional de Medicina, 1984, pp. 297–306.

Montoliú, María. *Cuando los Dioses Despertaron. Conceptos Cosmológicos de los Antiguos Mayas de Yucatán según el Chilam Balam de Chumayel*. Mexico City, Mexico: Universidad Nacional Autónoma de México, 1989.

Muñoz Camargo, Diego. *Historia de Tlaxcala*. Mexico City, Mexico: Oficina Tipográfica de la Sexcretaría de Fomento, 1892.

Ocaranza, Fernando. *Historia de la Medicina en México*. Mexico City, Mexico: Laboratorios Midy, 1935.

Orellana, Sandra. *Indian Medicine in Highland Guatemala*. Albuquerque: University of New Mexico Press, 1987.

Ortíz de Montellano, Bernard. 'Aztec medicine: empirical drug use.' *Ethnomedizin* 3: 249–271, 1974–1975.

Ortíz de Montellano, Bernard. 'Empirical Aztec medicine.' *Science* 188: 215–220, 1975.

Ortíz de Montellano, Bernard. 'Las yerbas de Tláloc.' *Estudios de Cultura Náhuatl* 14: 287–314, 1980.

Ortíz de Montellano, Bernard. *Aztec Medicine, Health and Nutrition*. New Brunswick and London: Rutgers University Press, 1990.

Recinos, Adrián. *Popol Vuh: Las Antiguas Historias del Quiché*. Mexico City, Mexico: Fondo de Cultura Económica, 1947.

Recinos, Adrián, ed. *Relación de las Ceremonias y Ritos y Población y Gobierno de los Indios de la Provincia de Michoacán, 1541*. Madrid: Editorial Aguilar, 1956.

Reilly, F. Kent III. 'Cosmos and rulership: The function of Olmec style symbols in Formative Period Mesoamerica.' *Visible Language* 24(1): 12–37, 1990.

Ritual of the Bacabs, A Book of Maya Incantations, Ralph Roys, trans. Norman: University of Oklahoma Press, 1965.

Ruíz de Alarcón, Hernando. *Tratado de las Supersticiones y Costum, Bres Gentilicias que oy Viven*

Entre los Indios Naturatles desta Nueva España. Mexico City, Mexico: Secretaría de Educación Pública, 1988.

Sahagún, Bernardino de. *Historia General de las Cosas de la Nueva España (Florentine Codex)*. 4 vols. Mexico City, Mexico: Editorial Porrúa, 1959.

Sahagún, Bernardino de. *Primeros Memoriales*. Facsimile edition. Norman: University of Oklahoma Press, 1993.

Schele, Linda. 'The Olmec Mountain and Tree of Creation in Mesoamerican cosmology.' In *The Olmec World: Ritual and Rulership*. Princeton, New Jersey: The Art Museum/Harry Abrams, 1996, pp. 105–117.

Segre, Enzo. 'El susto, una interpretación.' In *Simposio: Cuerpo y Alma en la Tradición Médico Filosófica*. Taxco, México. In press.

Sejourné, Laurette. *El Universo de Quetzalcóatl*. Mexico City, Mexico: Fondo de Cultura Económica, 1962.

Sepúlveda, Teresa. *La Medicina Entre los Purépechas Prehispánicos*. Mexico City, Mexico: Universidad Nacional Autónoma de México, 1988.

Serna, Jacinto de la. *Tratado de las Idolatrias, Supersticiones, Dioses, Ritos, Hechicerias y Otras Costumbres Gentilicas de las Razas Aborigenes de Mexico*. Mexico City, Mexico: Ediciones Fuente Cultural, 1953.

Viesca, Carlos and I. de la Peña. 'La magia en el Códice Badiano.' *Estudios de Cultura Náhuatl* 10(1): 267–301, 1974.

Viesca, Carlos. 'Los psicotrópicos y la medicina de los gobernantes entre los Aztecas.' *Estudios de Etnobotánica y Antropología Médica* 2: 121–135, 1977.

Viesca, Carlos. *México Antiguo*. Vol. 1 of *Historia General de la Medicina en México*, F. Martínez Cortés, ed. Mexico City, Mexico: Universidad Nacional Autónoma de México/Academia Nacional de Medicina, 1984.

Viesca, Carlos. 'El médico mexica.' In *México Antiguo*, A. López Austin and C. Viesca, coord. Vol. 1 of *Historia General de la Medicina en México*, F. Martínez Cortés, ed. Mexico City, Mexico: Universidad Nacional Autónoma de México/Academia Nacional de Medicina, 1984, pp. 217–230.

Viesca, Carlos. *Medicina Prehispánica de México*. Mexico City, Mexico: Panorama Editorial, 1986.

Viesca, Carlos. 'Y Martín de la Cruz, autor del Códdice de la Cruz-Badiano, era un médico tlatelolca de carne y hueso.' *Estudios de Cultura Náhuatl* 25: 480–498, 1995.

Viesca, Carlos. *Ticíotl. Conceptos Médicos de los Antiguos Mexicanos*. Mexico City, Mexico: Departamento de Historia y Filosofía de la Medicina, Universidad Nacional Autónoma de México, 1997.

Viesca, Carlos, A. Aranda and M. Ramos. 'El cuerpo y los signos calendáricos del tonalámatl entre los nahuas.' *Estudios de Cultura Náhuatl* 27: 143–158, 1997.

Viesca, Carlos. 'Curanderism in Mexico and Guatemala: its historical evolution from the sixteenth century to the nineteenth century.' In *Mesoamerican Healers*, B.R. Huber and A. Sandstrom, ed. Austin: University of Texas Press, 2001, pp. 47–65.

Villa Rojas, Alfonso. *The Maya of East Quintana Roo*. Washington, DC: Carnegie Institution, 1945.

Wagner, Elizabeth. 'Mitos de la creación y cosmografía de los mayas.' In *Los Mayas. Una Civilización Milenaria*, Nikolai Grube, ed. Köln: Könemann, 2000, pp. 280–293.

Winter, Marcus. 'La trepanación en Oaxaca.' In *México Antiguo*, A. López Austin and C. Viesca, coord. Vol. 1 of *Historia General de la Medicina en México*, F. Martínez Cortés, ed. Mexico City, Mexico: Universidad Nacional Autónoma de México/Academia Nacional de Medicina, 1984, pp. 59–65.

KAREN McCARTHY BROWN

HEALING RELATIONSHIPS IN THE AFRICAN CARIBBEAN

Healing traditions throughout the Caribbean share deep habits of the heart, mind, and body. These diverse but related cultures share ways of thinking, interpreting and reacting in the face of disease, suffering, or bad luck. People from West Africa and Western Central Africa are likely to have shared such common conceptions of life and attitudes toward living before individuals from these areas became enslaved (Mintz and Price, 1976: 45–56). Other similarities among African Caribbean healing traditions seem to have come out of the experiences of slavery itself. It is these broad commonalities, more than specific practices, that shaped what we now recognize as a distinctly African-Caribbean culture of healing. The similarities among healing practices found in former Caribbean slave colonies will be mapped out here, before the essay turns to ethnographic portraits of healing practices in Haiti and in the Haitian diaspora.

AFRICAN-BASED HEALING PRACTICES IN THE CARIBBEAN

Healers in the African-Caribbean traditions function as herbalists, psychologists, ethicists, and social workers, and they are also respected sacerdotes, acting as mediators between God and human beings. In the Caribbean, healers are almost always religious figures, and virtually all African-Caribbean religious practices, whether they are large public rituals or private one-on-one consultations, are mainly about healing. Traditional healers in the Caribbean are aware of Western medicine and usually have access to it, although that access is quite limited in some cases. These healers rarely try to compete with scientific medicine; in fact, they often recommend that their clients consult a medical doctor. These days medical doctors carry out their treatments alongside, although rarely in cooperation with, the spiritual treatments provided by traditional healers. Scientific medicine alone is inadequate because the problem the client presents to the healer is, more frequently than not, merely a signal of larger issues in that person's life. Most of the presenting problems brought to African Caribbean healers, problems of love, luck, disease, family life, money and so forth, eventually turn out to involve spiritual entities and/or spiritual issues. The ancestors and spirits are said to send problems – obstreperous

H. Selin (ed.), Medicine Across Cultures: History and Practice of Medicine in Non-Western Cultures, 285–303.

children, chains of bad luck, elusive pain and weakness, lumps and bumps that doctors cannot diagnose – into the lives of individuals to get their attention. The first thing the traditional healer must do is to find out what the real problem is. The African Caribbean healer, using one of the many forms of divination practiced in the area, diagnoses whether the problem involves other people. It is commonly understood that a strong emotion like jealousy can actually warp the life of its target. More frequently the problems are traced to neglected ancestors, unsatisfied spirits, or malevolent souls. Broken promises to the spirits, neglected commitments to feed and honor the ancestors, or malevolent spirits sent by an enemy can all result in problems for the living, even if the offences occurred several generations before the birth of the person currently suffering from the results of such dangerous behavior. The spirits can also bring problems, even quite serious problems, into the life of a person who is hesitant to heed the call to become a priest or healer. This type of shamanic illness is recognized throughout the Caribbean basin and far beyond.

There are widespread beliefs in the post-colonial black Caribbean that the conditions of health and the practices of healing necessarily involve more than the physical body and the immediate social environment of the individual with the complaint. The spirits and the ancestors are intimate presences, and they are active in the day-to-day life of every individual. Their primary role is protection, and only in the case of major relational blockages would these guardians allow trouble to enter the lives of the people they care for. The spiritual aspects of healing theory are easiest to understand by looking at the way the notion of self is configured in the African Caribbean.

Individuals in the Caribbean tend to practice a kind of dense sociality. They weave thick webs of relationships, webs that include extended family, as well as important contacts and close friends, both are which may be given honorary kinship titles: "brother", "sister", "auntie", and so forth. The purpose of this is to guarantee that the individual will not have to face single-handedly the frequent crises engendered by poverty, drought, hurricanes, and corrupt politics, elements that characterize life in former Caribbean slave colonies. Yet other human beings are not the only ones with a role responsible for creating an individual's social safety net. Ancestors and the spirits, with their capacity to warn, protect and empower, are also involved in weaving that net. At the same time, these same characters are almost as likely to create problems in a person's life as they are to solve problems for them. Virtually all the problems that African Caribbean healers treat can be traced to neglected or corrupted relationships within an individual's web of connection. Healing takes place through ritual adjustments of these relationships, both relations between the living and the dead and between human beings and spirits. Healing in the Caribbean is thus a thoroughly relational process. As a result, the problems and diseases – physical and social – that healers deal with are never attributed to characteristics essential to human beings, spirits, or ancestors. Instead, all problems reside in-between entities, in the relational space between them, and all healing takes place in the in-between as well.

Each individual is seen as a dynamic arena of diverse spiritual energies or tendencies, some of which were inherited from their ancestors. Yet, in the ordinary course of living, people also attract other spirit energies. Thus, both external connectedness and internal complexity contribute to an individual's identity and spiritual resources. The African-Caribbean view of self as a dynamic system of relationships allows for flexibility and conflict inside and outside of the person, and thus discourages rule-bound thinking in both arenas. The spirits, like the self and the community, are also dynamic relational constructs, more frequently referred to as forces or powers than as deities. African Caribbean people tend to attribute both physical and social problems to spiritual sources. Ancestors and spirits who have not been properly taken care of hector the offending individual in a variety of ways. Those who have been chosen by the spirits to be spiritual leaders often encounter the force of the spirit's desire through such problems. There are no rules allowing the healer to connect a presenting problem immediately with the deeper spiritual forces that caused it. Discerning those connections is half the work that the healer must do, and it lies at the heart of the process of diagnosis. Once the relevant connections are traced, often through repeated divination sessions (including dream analysis) then the healing rituals can begin. These include initiation for chosen spiritual leaders and, for less serious cases, the manufacture of charms and talismans, the prescription of prayer and other forms of spiritual discipline, along with cleansing and/or strengthening herbal baths. Herbal baths, that remove bad luck as well as raise good luck, are given routinely. More rare is the dramatic full-scale exorcism required in serious circumstances such as when a malevolent spirit or a restless and wandering soul has somehow attached itself to the client.

The manufacture of objects – charms, guards, *wanga*[1] – is a routine part of healing processes in the African-Caribbean. These objects represent troubled relationships. Once such a relational knot has gained a material presence it can be worked on ritualistically. For example, a wife unhappy in her marriage can cool down her husband's temper (or his attraction to other women), by writing his name a specific number of times on a piece of paper and immersing it in a bottle of heavily sugared ice water; this is to make him sweet and cool him down. Dolls can be manufactured from items of clothing belonging to individuals involved, and these can be blessed, prayed over, beaten into shape, bound in chains, or sprayed with alcohol, perfume, or holy water. Words – oaths, promises, or curses – can be recited while driving straight pins into candles. As the candle burns the words are activated over and over.

Wanga make material complex emotional, physical, and social situations and make them available for healing. These "anti-aesthetic" objects are related to Dahomean *bocio*[2] (Blier, 1995) and Kongo *minkisi*[3] (MacGaffey, 1994; Stout, 1995), and they are known by various names and in various forms throughout the African Caribbean. They are material objects that expel complex and highly charged emotional states (Blier, 1995: 95ff.), thus actualizing the possibility of dealing with those problematic feelings and perhaps even changing them. There is a series of less frequently used healing procedures that do just the opposite. These charms gather powers and different types of energy and introduce them into the body (e.g. herbal concoctions rubbed into small cuts in various places on the body – the head, upper arm, back of the knees, or the top of the feet). These "guards" can be thus secreted under the skin and are said to be able to signal a person when trouble is about to happen. Thus, in processes of ritual healing, what Margaret Lock (1993) calls the "mindful body" can be purged of destructive feelings or energies, and strength can be augmented in the same body. While discussing how memory was affected by the traumas of slavery, Roger Bastide (1960: 249–251) noted that memories can be carried in the body. The binding and tying characteristic of bocio and minkisi are reiterated in African Caribbean healing charms, where yards and yards of twine may be used to bind a wanga, and dolls can be tied together with copper wire or even chains and padlocks. While bocio and minkisi were being bound and tied before chattel slavery began, surely the depths of a kinesthetic memory of slavery are activated in the mindful body by the very process of manufacturing New World charms. Embodied meaning and memory are well understood in the African Caribbean and widely used in their healing arts.

In a culture where good and evil are not absolute opposites, but shifting places temporarily visited in the never ending search for survival of self and family, what I have here called the healing arts of the African Caribbean are not always considered beneficial by everyone involved. For example, a woman who wants to have a more faithful husband may try through spiritual venues to "tie him down", and he may not consider that a good thing. The stereotypes of Caribbean African-based religions (black magic, cannibalism and other forms of violence) are little more than fantasies of the popular media. Generally speaking the moral sensibility of the African Caribbean is highly refined but always contextual and never rigid. Morality and healing are closely connected, in that an unjust use of spiritual power, especially one that puts the success of an individual before the well being of the community, will almost certainly lead to punishment from the spirits.

HEALING AND HAITIAN VODOU: CASE STUDIES

Unlike a religion such as Cuban Santeria, which took shape in urban contexts, Haitian Vodou is a set of spiritual practices primarily shaped by the ecological, social, and spiritual accommodations of peasant farmers. Because of Haiti's successful revolution against France (1791–1804), a unique event in the history of trans-Atlantic slavery, those remaining in Haiti were able to restore, in many parts of the island nation, a type of farming similar to what many had known in Africa. (Two-thirds of the slaves in Haiti at the end of the revolution had been born in Africa.) This history accounts for the importance of the land – sometimes literally of dirt or earth – in Haitian Vodou ceremonies. Family cemeteries are the untouchable spiritual centers of family land worked for generations, land that should never be sold. The graves of the oldest male and female buried in the family cemetery are spiritually empowered places where a person can go to seek help for his or her family. By definition, the spirits connected to these graves, Bawon Samdi and Bwijit, refuse to lend their power

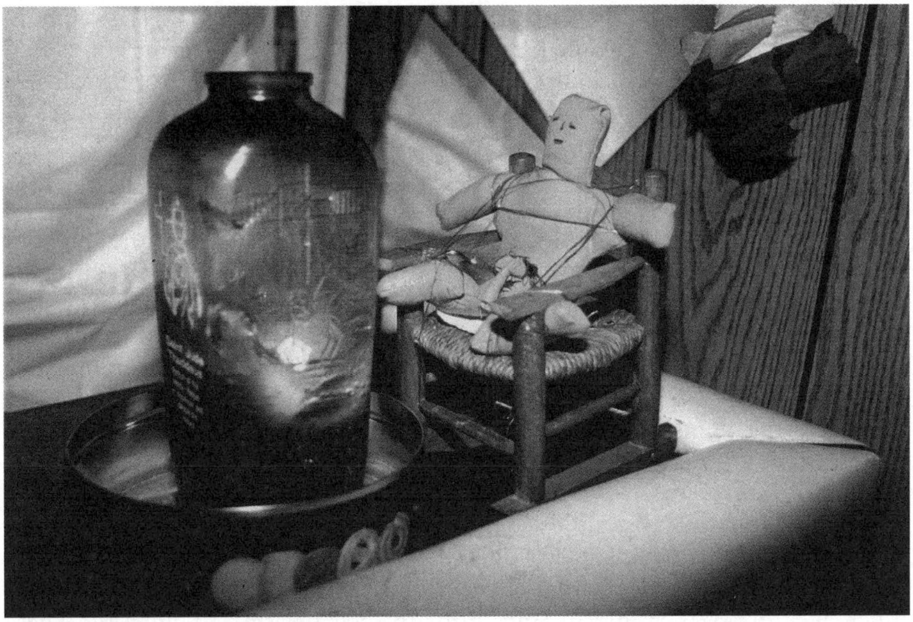

Figure 1 A *wanga* made for a woman who wants to "tie down" her husband. Photo by Martha Cooper.

to either side when it comes to disputes within the family. Some Vodou rituals must be carried out *pye pa te*, with bare feet on the earth. People need to be connected to the earth to be connected to the spirits. The fertility of the land, or the lack thereof, is also understood as a signifier of the mood of the spirits and ancestors. Natural disasters such as hurricanes are endured as punishment. Vodou is filled with metaphors derived from animals and plants.

Drought, poverty, political corruption, and disputes over land titles are among the reasons why large numbers of Haitians have, for several decades, been leaving the countryside and their extended families and moving to the cities. Vodou has been urbanized along with the migrating farmers. In the cities, it is the Vodou temple and the fictive kinship network it provides that compensate for the loss of the large rural family. The head of the temple is called "mother" or "father" and the initiates are addressed as "children of the house". The Vodou initiate owes service and loyalty to his or her Vodou parent after the pattern of filial piety owed all parents by their children. In turn, Vodou parents, like actual ones, owe their children protection, care and help in times of trouble. In certain circumstances this help is of a very tangible sort: food, a place to sleep, or assistance in finding work.

For the last half century political instability and the ill-conceived interventions of larger and more powerful nations have spawned Haitian migration patterns that led out of the island nation to the United States, and to some extent to Canada and France. Vodou has accompanied the émigrés. In the new diaspora, Vodou, following its familial model, functions to elaborate and strengthen the relational networks that make survival possible in these immigrant communities. Vodou and other religions of the African Caribbean, especially Santeria, thrive in urban centers around the United States. Streams of African spiritual sensibility, once separated and channeled by European colonial borders, are now reconverging in New York City, Miami, and Chicago. In places like these, African Americans are embracing a wide variety of Caribbean traditions, together with what might be called the Yoruba revival movement, a reclaiming of African religion, and particularly of the religious practices of the contemporary Yoruba. Although there are no trustworthy demographic data on these movements, it is highly likely that more than a million people within the greater New York area are involved in this constantly shifting yet always dynamic matrix of African and African Caribbean religious practices.

Among the hundreds of thousands of Haitians now living in the United States, there are many Vodou priests (*oungan-yo*) and priestesses (*manbo-yo*). Among them are healers who manage to make their living ministering to the many needs of a substantial community of Haitian immigrants, African Americans and a sprinkling of white clients from a variety of ethnic groups. They read cards, practice herbal medicine, manufacture charms and talismans, much as many of them did in Haiti prior to their decision to emigrate. These healers also provide their clients with access to the wisdom and power of the Vodou spirits, a configuration of African spirit entities loosely identified with Catholic saints, each of whom presides over a particular life domain such as childbearing and rearing, family roots, anger and assertion, humor and death. Access to the spirits usually comes through possession. Either at large Vodou "parties", or during intimate consultations between a *manbo* or *oungan* and a client, the spirits possess the healer and deliver messages to the client. It can also happen that such access is provided through the healer's interpretation of the client's dreams or visions.

In the language of Vodou there is a fundamental contrast between things that are bound or tied (*mare*) and things that have been opened (*ouvri*), let go (*lage*), or allowed to flow (*koule*). The point of healing is to open the conduits of relationship, to untie (*demare*) or loosen (*lage*) whatever it is that blocks the flow (*koule*) of reciprocity which is the nature and the fuel of all relationships. A Vodou healer, for example, once told a client that if she followed the prescribed cure being offered to her, "dlo kle va koule devan ou," (clear water will flow in front of you). In order to get things flowing among persons and between persons and spirits, gates have to be opened, pathways cleared, chains broken, blockages removed, and knots untied.

Vodou is a religion that can feed on paradox, and thus it is also true that a little tying is occasionally a good thing. For example, it can be put to good and productive use as routine love magic. Dolls are bound back to back with ribbon or string and then hung upside down in a tree in the local cemetery. A relationship needs to be dissolved. A bottle containing the name of a loved one is wrapped round and round in butcher's twine until it looks like a little mummy. In this case someone wants a lover to stay. And, when women are going to a Vodou ceremony but do not want to get possessed by the spirits, they sometimes put a knot in their hair or in the hems of their skirts. Also, new initiates into Vodou keep a scarf tied tightly around their heads for forty days. During the lengthy initiation period, the Vodou "mother", that is the one who initiated them, constantly reminds the new initiates: "mare tet-ou!" (Tie your head).

On a more somber note, chains and knots frequently are seen in Petwo (the "hot" pantheon of Vodou spirits) altar rooms, and even more so in spaces dedicated to spirits further toward "the left hand" such as those represented on the altars of secret societies. In secret society altar rooms, crosses may be planted in the floor and caught in a snare of massive, multiply knotted ropes that extend from floor to ceiling. I have seen old slave chains in such places on more than one occasion. The following story from my ethnographic notes illustrates the making of wanga and the dialectic of tying and binding. It also involves a well-known Brooklyn *manbo*, Mama Lola (Brown, 2001).

Case Study #1: The Wanga called Mare Djol

Sometime in the late fall of 1994, Mama Lola, a Vodou priestess of some repute, produced a *mare djol*, a charm called "tie the lips", for a client. This type of wanga can also be referred to as *mare bouch*, tie the mouth. Inside a milk bottle-sized terra cotta pot, Lola put what she identified as *sabilye* leaves, to foster forgetting. The name of this leaf puns on the Kreyol word to forget,

Figure 2 Mama Lola and a friend before her altar table in the basement of her Brooklyn, New York home.

bliye. She added salt, a common prophylactic in Vodou healing. This use of salt is a practice traceable to the Catholic tradition of putting salt under the tongue of a catechumen just before baptism, in order to chase away the devil. As a sweetener for the whole transaction Lola was about to be involved in, she added molasses. Lola finished by writing the name of an electronics company on nearby Long Island, and the name of a female employee of that company on a small piece of white paper and adding it to the mix. Next she draped the pot with a clean white cloth, noting at the time, "you tie the mouth with a white cloth." Then she wrapped yard after yard of white cotton string around the pot until it came to resemble a small, fat mummy. Having produced a wide loop with the last few feet of string, she suspended the *mare djol* from a heating pipe that crosses the ceiling of her small basement altar room. Whenever she thought of it in the course of a day's work, she would reach up and set the charm swinging.

This *mare djol* was created for a man having trouble with his supervisor at the electronics plant where he worked. The supervisor was jealous of him because he attracted favorable attention from those higher up in the firm. So she gossiped about Mama Lola's client in ways that damaged his reputation and, he feared, threatened his job. Mama Lola practices the "science of the concrete" as Levi Strauss once called it, in her healing work (1966). She objectifies complex relational situations. She literally turns them into objects which can then be worked with and even changed.

When Mama Lola made the *mare djol*, she was trying to clear the blockage in the relational network that sustained her client. Most immediately, there was a constriction in the relationship between him and his supervisor. This blockage that occasioned the client's trip to see Mama Lola was understood as a signal that other, more serious, blockages existed in his life. It was assumed that for Mama Lola's client to have a problem with his supervisor, the spirits who would otherwise be protecting him must be angry. If all had been well with them, they would have prevented any such trouble.

Because of this two-tier diagnosis, treatment for the problem proceeded on two levels as well. Mama Lola told her client what to do in his own home, what statues and images to buy, and how to feed and honor his ancestral and spiritual protectors. Then, turning to the more immediate symptom of the problem, Mama Lola made the *mare djol* to "tie the mouth" of the gossiping supervisor. Several months later, when her client told Mama Lola that his supervisor had been promoted and, at the same time, transferred to another branch of the company outside New York City, both he and Mama Lola pronounced the healing work successful. Note that the problem was not conceptualized as an essential flaw in either one of the persons involved, and in the end neither Mama Lola nor her client needed to get even with the offending supervisor (Brown, 1998).

Healing traditions such as this raise complex questions about ethics, questions to which there are no easy answers. The ethics of Vodou healing do, however, offer the intriguing possibility of an alternative to standard rule-bound, universalistic ethics. Systems such as Haitian Vodou assume that self-respect is integral to living a moral life. That is the starting point for an ethic that locates problems in the relations between people, rather than in the individuals themselves, and experiences goodness as a momentary achievement rather than a quality of being. In Haiti, moral absolutes become dangerous because they create situations in which the need to do good must be stronger than the need to survive. The moral challenge articulated by African Caribbean traditional religions such as Haitian Vodou is not a challenge to reach perfection, but rather one to find a flexible moral style, a collection of attitudes and practices that encourages the acceptance of conflict and suffering, the celebration of plurality in the group and within the person, and the search for the moral through whatever preserves life and enhances life energy.

While there is enormous herbal knowledge in Haiti, and particularly in rural Haiti, Haitians living in the United States appear to rely much more on medical doctors for medications than on Vodou healers. The herbs that healers in the New York immigrant communities use routinely are those used for baths. Yet as herbal knowledge appears to be diminishing in city life, the sophisticated psychological treatments people receive from Vodou traditional healers remain strong. In some ways, Vodou in New York functions as poor people's psychotherapy. Of course it is a different kind of psychotherapy. Much could be said about the differences between Western and African Caribbean views, but perhaps the most important are that Vodou psychotherapy focuses on the community as much or more than the individual, and it does not problematize sexuality or the body. In fact both are respected and celebrated. The third aspect that should be mentioned as fundamental to the differences between Vodou healing practices and those of psychiatrists and psychotherapists is that much of the therapy is not verbal but kinesthetic. The story of the soldier who was "hungry for family" reveals many things about the psychotherapeutic sophistication of Vodou healing practices.

Case Study #2: The Soldier Hungry for Family

Mama Lola can be discreet if that is what is called for. However, if there is nothing about a client's problem that needs protecting, she sometimes shares the details of cases. "I got somebody," she said. "I'm doing some work for him. He is in the army. Oh boy! That man so big! When he go downstairs, he got to go like this." With her usual refusal to reduce a story to words, Lola sprang up from her chair and hobbled across the kitchen floor imitating a tall man in fear of hitting his head on the ceiling. Lola continued,

> He from Virginia, but his mother in New York. She live in Brooklyn. But when he come he don't stay with her, he stay in hotel. I ask him: "Why you not stay with your mother? Let her take care of you?" He say, "Not me. I not dependent on nobody. What I need that for? The army is my family. I love the army and I going to stay in it until I die ... er ... until I retire." That what he say. I look at him and I think, "That funny," because I never meet nobody ... that's the *first* person I meet who say, "I love the army." So you know what I think? I think he just hungry for family.

As is the custom, the lieutenant had presented himself to Mama Lola with

Figure 3 *Wanga* hanging from the ceiling of Mama Lola's altar room. The largest is the *mare djol* described in Case Study #1. Photo by Martha Cooper.

an introduction from a mutual friend and a non-specific problem. He had "no luck." As she told me the story I could imagine the two of them in her cramped altar room tucked away in one corner of the basement, the one tiny window blocked by a heavy curtain. There were bunches of odorous herbs and dried leaves and a collection of baskets hanging from the ceiling; smoke-darkened color lithographs of the Catholic saints attached to the walls above altar tables, dense with tiny flickering flames; smooth dark stones sitting in oil baths; and bottles of rum, vodka and perfume, herbal brews and sweet syrups. She would have chosen to sit in the big armchair with the stuffing coming out of one arm, her face lighted from underneath by a candle stub that burned on one corner of the small table in front of her. The big army officer would be hunched on the tiny bench opposite her where clients routinely sat.

After a preliminary sign of the cross, Lola would have shuffled the cards and offered them to him to cut. When they were spread in front of her in four rows of eight, she would already have been in a concentrated state of mind. Some minutes of silence would follow and then Lola would have begun to tap the cards in an order determined by her inner vision and ask him questions, such as "You have some fighting in your house?" "You afraid of someone in your house?"

At this stage of the diagnostic divination process, the healer has a number of choices as to the direction of questions. The problems or generic bad luck she is faced with may manifest across a spectrum ranging from physical symptoms to social or spiritual crises. No matter how the problem is finally defined, it will always be diagnosed as due to a disruption in relationships. During diagnostic card readings, the client is free to answer "yes" or "no" to the healer's probing questions without prejudice. Thus the client is active in the diagnosis yet does not dictate the description of the problem. These details are echoes of the great West African divination systems Fa (Dahomean) and Ifa (Yoruba). When the questions and answers have gone on for some time, a joint definition of the problem will usually emerge. Most clients would acknowledge a significant difference between the problem they came with and the one finally defined via the dialectic of divination. The outsized lieutenant from Virginia was no exception, although it took him a while to realize this. After many questions and negative responses, the lieutenant finally acknowledged that he wanted to move to a higher rank in the army and one of his superiors was blocking his promotion. When she heard this Mama Lola was still not sure she had discovered the core of the problem.

> I read card for him, but I don't see nothing. You know some people, they really got bad luck and you just see it. But with him, I don't see nothing like that ... little thing, maybe, but no real big thing in his path ... nobody at work want to stop him. So, I say to him, "How long you feel you got bad luck?" He think ... he think ... he think, and he say, "Since I was a little boy." I say, "You can't have bad luck since you little boy!" Then I say, "What happened when you was a little boy," Oh, he tell me a story that was so sad ... so awful.

It seems that this troubled soldier had grown up in poverty in the Bahamas. A key event happened there, when he was no more than six or seven. One day his mother had nothing to give her children to eat. Neither did she have money to buy food. "He was so hungry ... so hungry," Lola said. The enterprising and desperate young boy went, without his mother's permission or knowledge, to the factory where his father worked to ask for money for food. The man, shamed in front of his co-workers by not being able to give the boy money, turned on the child with a terrible, hurtful truth that had been hidden until that moment. "Why you calling me Daddy?" he snapped, "I'm not your father." The young boy hid his hurt in anger and he reacted. He yelled at the man he had always thought to be his father, "You going to be sorry. Someday I going to be rich and I going to support you. I not going to be a hungry man like you!" The boy never said anything to his mother about this incident.

A few years later the boy found out who his real father was. There was a woman whom he had always addressed as "Auntie", yet he never knew if she was really a relative. One day as he walked by Auntie's front porch, a blonde woman sitting there with Auntie called him over. She took his chin in her hand, lifted his face for careful inspection and said to Auntie, "That brother of yours can't deny this one. Oh no!" The blonde woman turned out to be his biological father's current wife, someone he had met and married in Germany. That afternoon was the first time the young man heard the name of his "real" father. As an adult, Mama Lola's client went to Germany to search for his father, but he failed to make contact and said to Lola that he did not really care. She quoted him as saying, "What I need to find that man for? He not my father ... not really."

Some months after our first conversation about the soldier, Lola reported that deeper problems had been uncovered in the course of a subsequent card reading, and they were now starting to work on those. It seems that Auntie had quarreled with his mother shortly after the boy had accidentally encountered his stepmother. The now grown man had a dim memory of himself as a boy witnessing the burial of what Haitians who serve the spirits would call a wanga. The boy remembered seeing Auntie write his name several times on a piece of paper and bury it along with a tiny oil lamp beneath one of the trees in her courtyard. As a result of this revelation, Mama Lola realized the treatment she had to offer the soldier would be longer and more complicated. She was now clear that there had been many years of relational problems in the soldier's blood family, and, furthermore, the wanga buried under a tree which everyone knew was the dwelling place of a spirit guaranteed that the spirits themselves were now involved.

The treatment went on for some months, with hiatuses when the lieutenant was not able to be away from the army base in Virginia. In the course of things, Mama Lola got to know her client well, and when he came for treatments she offered him coffee and they sat for a while and talked at the upstairs kitchen table before descending to the basement where the proud soldier had to hunch his shoulders and squat down in order to squeeze himself into Lola's tiny, already crowded, altar room.

When I asked about him some time later, Lola reported a conversation she recently had with him:

> I talk to him real good. Oh ... I talk to him. I tell him to forgive his mother. She lie to him and tell him that other man his father. But she say that cause she ashame. So he got to forgive his mother and forgive that man who tell him, "I am not your father." Forgive him, cause he hurt and angry. I tell him, "Even he angry at you and your mother, that not your fault. You was just a child. No child responsible for that. Maybe at first your mother lie to him too. Maybe she say, "I pregnant for you," when it really not his baby. But ... maybe she hungry. Maybe she got to do that so he could feed her all that time she pregnant. Maybe he not know you was not his baby 'til after you born and then he hurt and angry. Don't judge her. Women gotta do all kinda' thing. I know ... cause when I got William [her third child and the last born in Haiti], I tell somebody "I pregnant for you," but it not true. What I going to do? I got to eat or that baby going to die. I tell him, "You got to love your mother ... do everything to see she happy before she dead ... she an old lady now ... send her money ... go see her." But you know who I hate? Who the bum? That man who go to ... [Germany]. He the bum! 'Cause maybe he find out she pregnant and maybe he think that going to be expensive. So he just run away. And you know what? He a racist. He got to go marry a white woman. Black woman not good enough. Ehhhh?

The course of treatment had been decided on. Healing, forgiveness, and reconnection were the desired ends (along with disconnection and moral censure for the biological father and his family), and Lola worked on these goals through her usual combination of prayers, herbal baths, and the making of wanga, as well as visits from the spirits who would possess Lola and through her tease the melancholy soldier, empathize with him, and when necessary deliver stern lectures.

Nearly a year later the name of Lola's sad and ungainly soldier friend came up in a conversation. When asked, "What ever happened to him?" "Did he get his promotion?" Mama Lola responded in a distracted fashion. She thought he did get it. She was not sure. Maybe he did. Maybe he did not. "But," and here she brightened up, "when he in New York now, he stay with his mother. His mother the only family he got. That man who raise him dead long time. He don't come to see me no more, but he okay now" (Brown, 1987).

The story of the soldier hungry for family shows the uses of empathy and intuition, necessary gifts for therapists, and the story of Ti-Klou further develops the psycho-dynamic potential of Vodou healing practices.

Case Study #3: Ti-Klou and the Milk Bath

Ti-Klou, "little nail", was diagnosed and treated by one of the spirits himself. Ogou is a warrior spirit with roots in Yoruba land and in Dahomey, where he is called Gu (Barnes, 1997). He is known as someone who is just and brave, yet he can also easily fall into a rage and even become self-destructive. This is the powerful spirit who showed up in the young manbo's altar room on the day Ti-Klou came to ask for help with her abusive husband. I was present at the time and will quote from field notes written on that occasion.

Ogou picked up a deck of cards and started to shuffle them. "Cut the cards," he directed, shoving them across the table to Ti-Klou, "and talk!" Ti-Klou cut the cards and somewhat reluctantly, explained she wanted "Papa" Ogou's help to rid herself of this man. Ogou listened but said nothing. After an uncomfortably long silence he said, "You gave him a lot of money. I can see that! Why did you do that?" Ti-Klou hung her head. "He needed it. He does not make much money on his job." "Look," said Ogou with a note of exasperation in his voice, "You sweat, you change your shirt. If someone does not sweat for you, you don't change his shirt." Ogou sat back in his chair and gave Ti-Klou a stern, but not unkind, look. "Okay, so you feel sad. Time passes. That goes away." "I'm going to make three baths for you. We will start next Tuesday ... It is no good to take a bath after the sun goes down. You've got to do it when the sun is out. Okay? ..."

Over the following three weeks, the healer prepared the baths and even made a protective perfume as Ogou said she should. For Ti-Klou's first bath, a quart of milk was heated and spices added to it, the dominant one being cinnamon. Ti-Klou was told to go into the bathroom, remove her clothes, and stand in the bathtub. She was directed to hold a lighted candle and pray. "What kind of praying?," asked Ti-Klou. "You know, talk ... say what you want," responded the priestess. Later, Ti-Klou was instructed in the way of applying the infusion to her skin. "This is a good luck bath. You have got to bring that luck up. So, you start from the foot." The *manbo* showed her how to take small handfuls of the liquid and spread it on her skin, always moving in the upward direction. "When you get to the top, you take a little in your hand – like this – and you put it on your face, and on the top of your head." Ti-Klou patted herself dry with a brand new towel she had been instructed to bring, and, as she was instructed, she left twenty-seven cents on the bathroom floor. As she left that day, the healer cautioned her: "You leave that on the skin for three days. Don't take no bath!"

When Ti-Klou returned the following week, she seemed depressed. I asked how she had been and she replied that she had spent most of the week in bed crying. Wrinkling her nose, she said the bath made her smell "like a sour baby". Ti-Klou's second bath given on the day of the week consecrated to Ogou, a spirit on intimate terms with righteous anger, stung her body. It was composed of alcohol and several different kinds of perfume. Later she told me that, on the first night after the bath, she awakened long before sunrise feeling very angry. In spite of it being the middle of the night, she telephoned her husband and argued with him for hours.

Ti-Klou's first two baths forced her into emotional maelstroms of sadness and anger, but the third and final one – concocted from rose petals, seven kinds of fruit, and champagne – soothed her. This bath was given on Thursday, the day of the week dedicated to Danbala, the serpent spirit with connections to deep wisdom distilled from generations of human experience (Brown, 1995).

Many a psychotherapist would envy the quick and effective "regression in the service of the ego" that the young Vodou priestess managed to orchestrate with these baths. Her success was not an accident. A good Vodou healer necessarily operates out of a refined sense of human emotion. Like a pool player, an accomplished healer knows how to slice into an emotional knot from just the right angle, so it will break into its constituent parts, which then can be handled one by one. Vodou healers are not masters of volumes of healing recipes. They are spiritually attuned persons who have learned to trust their own intuitive knowledge. In a very real sense the healer's knowledge is carried in her body and it is addressed to the mindful bodies of those who turn to her for help.

The cultures of former slave colonies in the African-Caribbean are considerably different from one another. These differences derive from the identity of the colonial power in charge and therefore the governmental, legal, culinary, religious, architectural, and social practices that formed the culture from the top down. The areas in Africa, or in the Caribbean, from which slave labor was drawn strongly influenced cultural structures extruded from the bottom up. While the slaves had little official power, it is important to remember that they were the great majority in most Caribbean colonial situations. Many Caribbean plantations produced enormous wealth during the 18th and 19th centuries, and therefore the wellbeing of the field laborers fluctuated in relation to the instability of the market and the unpredictability of the weather. During the period of chattel slavery, the desire for profit was enormous, relentless, and cruel. Profit justified sadistic masters and mistresses.

Contemporary African-Caribbean healing practices reflect both European and African influences, but they also clearly demonstrate influence from the traumas of plantation culture. Theirs are religious healing systems that give high priority to practices promoting both survival and resistance, sometimes even rebellion. For example, many Haitian herbalists have preserved knowledge of various poisons that could well have been used in the slave era. The wide variety of healers in the Caribbean Basin treat social wounds as well as physical ones. Yet, these days, the same healers frequently recommend medical doctors for the physical dimensions of their clients' illnesses. These traditional healing practices can be found throughout the Caribbean and their continuous practice configures a daily world charged with spiritual powers to be respected and nourished, much as friendships and blood ties need to be tended carefully.

Healing in the African Caribbean is not like Western allopathic medicine. The immediate goal is not the relief of painful or life threatening bodily conditions. In the African Caribbean, healing is about reconnecting a person to the powers that protect and sustain him or her; it is about eliminating whatever constricts the flow of life energy. Most African-Caribbean healers would say that even the best medical doctor could not cure a person whose real problem stems from the spirits.

NOTES

[1] A *wanga* is a magical charm or object used for healing, to achieve a wish, or for protection against evil. The Vodou spirits, the *loa*, bestow their essence and power into the *wanga*. *Wanga* are attached, worn, or kept in a sacred place. Each contains a collection of secret ingredients in a pouch or cloth square made of silk or satin. The *wanga* may be knotted together with string and cord or tied into a bundle with colorful ribbons.

[2] A *bocio* is a figural sculpture embedded and empowered with spiritually charged materials.

[3] *Minkisi* statues have the power to heal and to visit good or evil on a person. They acquire this through conjuring and consecration. They are composed of earths, ashes, herbs, leaves, and relics of the dead.

BIBLIOGRAPHY

Appardurai, Arjun. *Modernity at Large: Cultural Dimensions of Globalization.* Minneapolis: University of Minnesota Press, 1996.

Averill, Gage. *A Day for the Hunter, A Day for the Prey.* Chicago: University of Chicago Press, 1977.

Barber, Karin. 'How man makes god in West Africa: Yoruba attitudes towards the Orisa.' *Africa* 51: 724–744, 1981.

Barnes, Sandra T., ed. *Africa's Ogun: Old World and New.* 2nd expanded edition. Bloomington and Indianapolis: University of Indiana Press, 1997.

Bastide, Roger. *The African Religions of Brazil: Toward a Sociology of the Interpenetration of Civilizations.* Baltimore: Johns Hopkins University Press, 1960.

Bastide, Roger. *African Civilizations in the New World.* New York: Harper Torchbooks, 1971.

Bastide, Roger, Françoise Morin and François Raveau. *Les Haïtiens en France.* Paris: Mouton, 1974.

Bell, Madison Smartt. *Master of the Crossroads.* New York: Pantheon, 2000.

Benitez-Rojo, Antonio. *The Repeating Island: The Caribbean and the Postmodern Perspective.* Durham, North Carolina and London: Duke University Press, 1992.

Benoit, Max. 'Symbi-loi des eaux.' *Bulletin du Bureau d'Ethnologie* (Port-au-Prince) 3(20–22): 12–22, 1959.

Bhabha, Homi. *The Location of Culture.* London and New York: Routledge, 1994.

Blier, Suzanne Preston. *African Vodun: Art, Psychology and Power.* Chicago and London: University of Chicago Press, 1995.

Bourdieu, Pierre. *Outline of a Theory of Practice.* New York: Cambridge University Press, 1977.

Bourdieu, Pierre. *The Logic of Practice.* Palo Alto, California: Stanford University Press, 1980.

Bourguignon, Erika, ed. *Religion, Altered States of Consciousness and Social Change.* Columbus: Ohio State University Press, 1973.

Bourguignon, Erika. *Possession.* San Francisco: Chandler and Sharp, 1976.

Brown, David. 'Altared spaces: Afro-Cuban religions and the urban landscape.' In *Gods of the City: Religion and the American Urban Landscape*, Robert A. Orsi, ed. Bloomington: Indiana University Press, 1999, pp. 155–230.

Brown, Karen McCarthy. 'Olina and Erzulie: A woman and a goddess in Haitian Vodou.' *Anima* 5: 110–116, Spring 1979.

Brown, Karen McCarthy. 'Alourdes: A case study of moral leadership in Haitian Vodou.' In *Saints and Virtues*, John Stratton Hawley, ed. Berkeley and Los Angeles: University of California Press, 1987, pp. 144–167.

Brown, Karen McCarthy. 'Plenty confidence in myself: The initiation of a white woman scholar into Haitian Vodou.' *Journal of Feminist Studies in Religion* 3(1): 67–76, 1987.

Brown, Karen McCarthy. 'The power to heal: Reflections on women, religion and medicine.' In *Shaping New Vision: Gender and Values in American Culture*, Clarissa W. Atkinson, Constance H. Buchanan and Margaret R. Miles, eds. Ann Arbor, Michigan: UMI Research Press, 1987, pp. 123–141.

Brown, Karen McCarthy. 'Afro-Caribbean spirituality: A Haitian case study.' In *Healing and Restoring: Medicine and Health in the World's Religious Traditions*, Lawrence Sullivan, ed. New York: Macmillan, 1989, pp. 255–285.

Brown, Karen McCarthy. 'Writing about "The Other".' In *The Insider/Outsider Problem in the Study of Religion: A Reader*, Russell McCutcheon, ed. London and New York: Cassell, 1999, pp. 350–353. Originally published in *Chronicle of Higher Education*, April 15, 1992.

Brown, Karen McCarthy. 'The altar room: A dialogue with Mama Lola.' In *Sacred Arts of Haitian Vodou*, Donald Cosentino, ed. Los Angeles: UCLA Fowler Museum of Cultural History, 1995, pp. 226–239.

Brown, Karen McCarthy. *Tracing the Spirit: Ethnographic Essays on Haitian Art*. Davenport, Iowa: Davenport Museum of Art; distributed by Seattle: University of Washington Press, 1995.

Brown, Karen McCarthy. 'The moral force field of Haitian Vodou.' In *In Face of the Facts: Moral Inquiry in American Scholarship*, Richard Wightman Fox and Robert B. Westbrook, eds. New York: Cambridge University Press, 1998, pp. 181–200.

Brown, Karen McCarthy. 'Serving the spirits: The ritual economy of Haitian Vodou.' In *Sacred Arts of Haitian Vodou*, Donald J. Cosentino, ed. Los Angeles: UCLA Fowler Museum of Cultural History, 1995, pp. 204–225.

Brown, Karen McCarthy. 'Systematic remembering, systematic forgetting: Ogou in Haiti.' In *Africa's Ogun: Old World and New*, Sandra T. Barnes, ed. 2nd ed. Bloomington: Indiana University Press, 1997, pp. 65–89.

Brown, Karen McCarthy. 'Staying grounded in a high-rise building: Ecological dissonance and ritual accommodation in Haitian Vodou.' In *Gods of the City: Religion and the American Urban Landscape*, Robert A. Orsi, ed. Bloomington and Indianapolis: Indiana University Press, 1999, pp. 79–102.

Brown, Susan E. 'Love unites them and hunger separates them: Poor women in the Dominican Republic.' In *Toward an Anthropology of Women*, Rayna E. Reiter, ed. New York: Monthly Review Press, 1975, pp. 322–332.

Browning, Barbara. *Infectious Rhythm: Metaphors of Contagion and the Spread of African Culture*. New York: Routledge, 1998.

Chernoff, John Miller. *African Rhythm and African Sensibility: Aesthetics and Social Action in African Musical Idioms*. Chicago: University of Chicago Press, 1979.

Clifford, James and George E. Marcus. *Writing Culture: The Poetics and Politics of Ethnography*. Berkeley and Los Angeles: University of California Press, 1986.

Comhaire-Sylvain, Suzanne. *Les contes Haïtiens: Origine immédiate et extension en Amérique, Afrique, et Europe occidentale*. Vol. 1, *Manman d'Leau*. Wetteren, Belgium: De Meester, 1937.

Commaroff, Jean. *Body of Power, Spirit of Resistance: The Culture and History of a South African People*. Chicago: University of Chicago Press, 1985.

Cosentino, Donald J., ed. *Sacred Arts of Haitian Vodou*. Los Angeles: UCLA Fowler Museum of Cultural History, 1995.

Courlander, Harold. *The Drum and the Hoe: Life and Lore of the Haitian People*. Berkeley and Los Angeles: University of California Press, 1960.

Danticat, Edwidge. *Breath, Eyes, Memory*. New York: Vintage, 1994.

Dash, Michael. *Haiti and the United States: National Stereotypes and the Literary Imagination*. 2nd edition. New York: St. Martin's Press, 1997.

Davis, Wade. *Passage of Darkness: The Ethnobiology of the Haitian Zombie.* Chapel Hill: University of North Carolina Press, 1988.

Dayan, Joan. *Haiti, History and the God.* Berkeley and Los Angeles: University of California Press, 1995.

de Certeau, Michel. *The Practice of Everyday Life.* Berkeley and Los Angeles: University of California Press, 1984.

Deren, Maya. *Divine Horsemen: The Voodoo Gods of Haiti.* London and New York: Thames and Hudson, 1953; reprinted New Paltz, New York: McPherson, 1983.

Desmangles, Leslie G. *The Faces of the Gods: Vodou and Roman Catholicism in Haiti.* Chapel Hill: University of North Carolina Press, 1992.

Dunham, Katherine. *Island Possessed.* Chicago: University of Chicago Press, 1969.

Farmer, Paul. 'Bad blood, spoiled milk: Bodily fluids as moral barometers in rural Haiti.' *American Ethnologist* 15(1): 62–83, 1988.

Farmer, Paul. *AIDS and Accusation: Haiti and the Geography of Blame.* Berkeley and Los Angeles: University of California Press, 1992.

Farmer, Paul. *The Uses of Haiti.* Monroe, Maine: Common Courage Press, 1994.

Fick, Carolyn E. *The Making of Haiti: The Saint Domingue Revolution From Below.* Knoxville: University of Tennessee Press, 1990.

Fleurant, Gerdes. *Dancing Spirits, Rhythms and Rituals of Haitian Vodun, the Rada Rite.* Westport, Connecticut: Greenwood Press, 1996.

Foster, Charles R. and Albert Valdman, eds. *Haiti – Today and Tomorrow: An Interdisciplinary Study.* New York: University Press of America, 1984.

Gilroy, Paul. *The Black Atlantic: Modernity and Double Consciousness.* Cambridge, Massachusetts: Harvard University Press, 1993.

Gleason, Judith. *Oya, In Praise of an African Goddess.* San Francisco: Harper, 1992.

Gleason, Judith. 'Oya in the company of saints.' *Journal of the American Academy of Religion* 68(2): 265–292, 2000.

Glick Schiller, Nina and Georges Eugene Fouron. *Georges Woke Up Laughing: Long Distance Nationalism and the Search for Home.* Durham, North Carolina and London: Duke University Press, 2001.

Glisant, Edouard. *Caribbean Discourse: Selected Essays.* Charlottesville: University of Virginia Press, 1989.

Halbwachs, Maurice. *On Collective Memory.* Chicago: University of Chicago Press, 1992.

Heinl, Robert D. and Nancy G. Heinl. *Written in Blood: The Story of the Haitian People, 1492–1971.* Boston: Houghton Mifflin, 1978.

Herskovits, Melville J. *Dahomey: An Ancient West African Kingdom.* 2 vols. Evanston, Illinois: Northwestern University Press, 1967.

Herskovits, Melville J. *Life in a Haitian Valley.* Garden City, New York: Anchor Doubleday, 1971.

Hurbon, Laennec. *Dieu dans le Vodou Haïtien.* Paris: Payot, 1972.

Hurbon, Laennec. *Voodoo: Search for the Spirit.* New York: Harry N. Abrams, 1995.

Hurston, Zora Neale. *Tell My Horse.* Berkeley, California: Turtle Island, 1981.

James, C.L.R. *The Black Jacobins.* New York: Vintage, Random House, 1953.

Kiev, Ari. 'The psychotherapeutic value of spirit-possession in Haiti.' In *Trance and Possession States*, Raymond Prince, ed. Montreal: R.M. Bucke Memorial Society, 1968, pp. 143–148.

Laguerre, Michel S. 'Ticouloute and his kinfolk: The study of a Haitian extended family.' In *The Extended Family in Black Societies*, Dimitri B. Shimkin, Edith M. Shimkin and Dennis A. Frate, eds. Paris: Mouton, 1978, pp. 407–445.

Laguerre, Michel S. 'Haitian Americans.' In *Ethnicity and Medical Care*, Alan Harwood, ed. Cambridge, Massachusetts: Harvard University Press, 1981, pp. 172–210.

Laguerre, Michel S. *American Odyssey: Haitians in New York City.* Ithaca, New York: Cornell University Press, 1984.

Landes, Ruth. *The City of Women.* New York: Macmillan, 1947.

Laroche, Maximilien. *Bizango: Essai de mythologie haïtienne.* Quebec, Canada: GRELCA, 1997.

Larose, Serge. 'The meaning of Africa in Haitian Vodu.' In *Symbols and Sentiments: Cross-Cultural Studies in Symbolism*, Ioan Lewis, ed. New York: Academic Press, 1977, pp. 85–116.

Lawless, Robert. *Haiti's Bad Press.* Rochester, Vermont: Schenkman Books, 1992.

Levi-Strauss, Claude. *The Savage Mind.* Chicago: University of Chicago Press, 1966.

Lewis, Gordon K. *Main Currents in Caribbean Thought: The Historical Evolution of Caribbean Society in its Ideological Aspects, 1492–1900.* Baltimore: Johns Hopkins University Press, 1983.

Leyburn, James G. *The Haitian People.* New Haven, Connecticut: Yale University Press, 1966.

Lock, Margaret. 'Cultivating the body: Anthropology and epistemologies of bodily practice and knowledge.' *Annual Review of Anthropology* 22: 133–155, 1993.

Lopez-Springfield, Consuelo, ed. *Daughters of Caliban: Caribbean Women in the Twentieth Century.* Bloomington: Indiana University Press, 1997.

Lowenthal, Ira P. 'Ritual performance and religious experience: A service for the gods in Southern Haiti.' *Journal of Anthropological Research* 34(5). 392–414, 1978.

Lowenthal, Ira P. 'Labor, sexuality and the conjugal contract in rural Haiti.' In *Haiti – Today and Tomorrow: An Interdisciplinary Study*, Charles R. Foster and Albert Valdman, ed. New York: University Press of America, 1984, pp. 15–33.

Lowenthal, Ira P. *Marriage is 20, Children are 21: The Cultural Construction of Conjugality and Family in Rural Haiti.* Ph.D. dissertation, Johns Hopkins University, 1987.

MacGaffey, Wyatt. *Art and Healing of the Bakongo.* Bloomington: Indiana University Press, 1991.

MacGaffey, Wyatt and Michael D. Harris. *Astonishment and Power: Kongo Minkisi and the Art of Renee Stout.* Washington, DC: National Museum of African Art, 1994.

Marcus, George E. and Michael Fischer. *Anthropology as Cultural Critique: An Experimental Moment in the Human Sciences.* Chicago: University of Chicago Press, 1986.

Maximilien, Louis. *Le Vodou Haïtien: Rite Radas-Canzo.* 2nd edition. Port-au-Prince, Haiti: Imprimerie de l'État, 1945.

McAlister, Elizabeth. 'The Madonna of 115th Street revisited: Vodou and Haitian Catholicism in the age of transnationalism.' In *Gatherings in Diaspora: Religious Communities and the New Immigration*, R. Stephen Warner and Judith G. Wittner, eds. Philadelphia: Temple University Press, 1998, pp. 123–160.

McAlister, Elizabeth. 'Love, sex and gender embodied: The spirits of Haitian Vodou.' In *Love, Sex and Gender in the World Religions*, Joseph Runzo and Nancy M. Martin, eds. Oxford: One World, 2000, pp. 129–145.

McAlister, Elizabeth. '"The Jew" in the Haitian imagination: Pre-modern anti-Judaism in the postmodern Caribbean.' In *Black Zion: African American Religious Encounters with Judaism*, Yvonne Chireau and Nathaniel Deutsch, eds. New York: Oxford University Press, 2000, pp. 203–227.

McAlister, Elizabeth. *Rara!: Vodou, Power and Performance in Haiti and its Diaspora.* Berkeley and Los Angeles: University of California Press, 2002.

Metraux, Alfred. *Voodoo In Haiti.* New York: Schocken Books, 1972.

Mintz, Sidney and Richard Price. *The Birth of African-American Culture: An Anthropological Perspective.* Boston: Beacon Press, 1992.

Montilus, Guerin. *Dieux en diaspora: Les Loa Haïtiens et les Vaudou du Royaume d'Allada (Benin).* Niamey, Niger: CELHTO, 1988.

Murphy, Joseph M. and Mei-Mei Sanford, eds. *Osun Across the Waters: A Yoruba Goddess in Africa and The Americas.* Bloomington and Indianapolis: University of Indiana Press, 2001.

Murray, Gerald F. 'Population press, land tenure and Voodoo: The economics of Haitian peasant ritual.' In *Beyond the Myths of Culture: Essays in Cultural Materialism*, Eric B. Ross, ed. New York: Academic Press, 1980, pp. 295–321.

Murray, Gerald F. 'Women in perdition: Ritual fertility control in Haiti.' In *Culture, Natality and Family Planning*, John F. Marshall and Steven Polgar, eds. Chapel Hill: University of North Carolina Press, 1976, pp. 59–78.

Orsi, Robert A. *The Madonna of 115th Street: Faith and Community in Italian Harlem, 1880–1950.* New Haven: Yale University Press, 1985.

Orsi, Robert A. 'The religious boundaries of an inbetween people: Street feste and the problem of the dark-skinned other in Italian Harlem, 1920–1990.' *Amercan Quarterly* 44(3): 313–341, 1992.

Orsi, Robert A. *Thank You, St. Jude: Women's Devotion to the Patron Saint of Hopeless Causes.* New Haven, Connecticut: Yale University Press, 1996.

Price-Mars, Jean. *Ainsi parla l'oncle*. Ottawa: Léméac, 1973.
Reed, Ishmael. *Mumbo Jumbo*. New York: Scribner, 1972.
Rey, Terry. *Our Lady of Class Struggle: The Cult of the Virgin Mary in Haiti*. Trenton, New Jersey: Africa World Press, 1999.
Rigaud, Milo. *La Tradition Voudou et le Voudoo Haïtien: Son Temple, Ses Mystères, Sa Magie*. Paris: Éditions Niclaus, 1953.
Roach, Joseph. *Cities of the Dead: Circum-Atlantic Performance*. New York: Columbia University Press, 1996.
Rotberg, Robert. *Haiti: The Politics of Squalor*. Boston: Houghton Mifflin, 1971.
Salvatore, Nick. *We All Got History: The Memory Books of Amos Webber*. New York: Times Books, 1996.
Scott, James C. *Weapons of the Weak: Everyday Forms of Peasant Resistance*. New Haven, Connecticut: Yale University Press, 1985.
Scott, James C. *Domination and the Arts of Resistance: Hidden Transcripts*. New Haven, Connecticut: Yale University Press, 1990.
St. Juste, Laurore and Frère Enel Clerisme. *Présence polonaise en Haïti*. Port-au-Prince: n.p., 1983.
Stout, Renee. *Dear Robert I'll See You at the Crossroads*. Exhibition catalogue. Seattle and London: University of Washington Press, 1995.
Sutton, David E. *Memories Cast in Stone: The Relevance of the Past in Everyday Life*. New York: Oxford University Press, 1998.
Thoby-Marcelin, Philippe and Pierre Marcelin. *The Beast of the Haitian Hills*. San Francisco: City Lights Books, 1986.
Thompson, Robert Farris. *Flash of the Spirit: African and Afro-American Art and Philosophy*. New York: Random House, 1983.
Trouillot, Michel-Rolph. *Haiti, State Against Nation: The Origins and Legacy of Duvalierism*. New York: Monthly Review Press, 1990.
Trouillot, Michel-Rolph. *Silencing the Past: Power and the Production of History*. Boston: Beacon Press, 1995.
Wucker, Michele. *Why the Cocks Fight: Dominicans, Haitians and the Struggle for Hispaniola*. New York: Hill and Wang, 1999.
Wilcken, Lois. *The Drums of Vodou*. Tempe, Arizona: White Cliffs Media Company, 1992.
Wilentz, Amy. *The Rainy Season: Haiti Since Duvalier*. New York: Simon and Schuster, 1989.

SAMUEL KOTTEK

MEDICINE IN ANCIENT HEBREW AND JEWISH CULTURES

The Bible and the Talmud reflect the cultural setting of the ancient Jews. The Bible[1] covers a period of time extending over several hundred years. Since our approach is cultural rather than historical, we shall not consider the problem of the dating of the biblical texts. The Talmud is a huge corpus, basically legalistic but replete with "stories that happened", popular folklore and homiletic[2] discourse. The basic text, the *Mishnah*, was written down between the second century BCE and the early third century CE. The glosses and commentaries on this "Oral Law" were called *Gemarah* and are more commonly known as the Talmud. Two versions were edited, one in Palestine (the Jerusalem Talmud)[3] and one in Babylonia (the Babylonian Talmud). The latter version, completed about a century later than the former (around 500 CE), is far more extensive and much more widely studied in religious Jewish circles. Another source that cannot be overlooked, the *Midrash*, improperly translated as "Legends of the Jews", being rather a homiletic exegesis of the biblical narrative, extends well into the Middle Ages.

Secondary sources on this subject were often centered on one specific topic, such as *ẓaraʿat*, the so-called "biblical leprosy". A number of more comprehensive studies were written in German in the 19th and early 20th centuries (Wunderbar, 1850; Trusen, 1853; Ebstein, 1903). With one notable exception, these works were not translated into English. The exception is the epochal work of Julius Preuss, *Biblisch-talmudische Medizin.* It was translated and edited by Fred Rosner and published in 1978 and has been reprinted several times since then. Julius Preuss (1861–1913) was a medical practitioner without any religious background who became a meticulous scholar in biblical and talmudic literature and a strictly observant Jew. However, unlike many other authors of similar creed and training, he deliberately brushed aside any apologetic approach to the data studied. The well-known historian of medicine Karl Sudhoff (Leipzig) praised Preuss' "remarkable talent for historical and philological investigation". Although requiring some updating, Preuss' work remains the most comprehensive study to this day. Among the relevant studies written in English in more recent times, most are sketchy and/or superficial (Snowman, 1935; Smith, 1950).

H. Selin (ed.), Medicine Across Cultures: History and Practice of Medicine in Non-Western Cultures, 305–324.

One important caveat is in order. While speaking of medicine in ancient Jewish lore, we should make it clear that no purely medical work dating back to the biblical and talmudic periods has survived. The Bible and Talmud include no organized medical chapters, although some authors have argued that in the Bible the chapters on "biblical leprosy" are of medical relevance (Leviticus, chapters 13–14), and in the Talmud a "Small Treatise on Therapeutics" (b. Gittin 68b–70a) has been described (Freeman, 1999; Geller, 2000). We shall come back to these two exceptions later.

Contacts with, and influence of, other contemporary civilizations can be traced more easily in talmudic literature than in the biblical text. Ancient Egyptian, Mesopotamian, and later, Greco-Roman medical lore no doubt had an impact on Jewish lore. This impact is however very difficult to show with documentary evidence, for several reasons. The biblical and talmudic medical data are usually fragmentary and incidental, and the data cannot be dated with precision. In the case of biblical data, this is because of biblical criticism; for talmudic data it is because they are part of an oral tradition. We shall therefore omit such comparative aspects from this review, with a few exceptions.

We stated above that no medical work of old pertaining to Hebrew and Jewish culture has reached us. We might ask whether such medical literature in fact ever existed. There is actually a curious talmudic statement relevant to this question. It says (b. Pesaḥim 56a) that King Hezekiah (8th century BCE) hid the *Book of Medicines* and the Sages (of the talmudic period, i.e., centuries later) praised him for this deed, which is mentioned nowhere in the biblical text. Although no reason for this concealing is suggested, later (medieval) commentators offered two good reasons: (1) sick people used this book but forgot to pray to the Lord; or (2) the book contained magic, or astrological means, which were theologically problematic. At any rate, King Hezekiah's deed must not (should not) be considered a condemnation of medical practice, but rather a criticism of how medicine was practiced. As was accurately stated by Ben Sira (3rd century BCE), a Jewish moralist who was often cited by the talmudic Sages, medical practice was acceptable, and even praised, when (on condition that) both patient and physician entrusted the Lord to allow the healing process to succeed (Book of Ben Sira, chapter 38).

PHYSICIANS AND SURGEONS

Let us now consider how the Bible and Talmud viewed physicians and other medical and paramedical personnel.

In the Bible, the Healer is, in absolute terms, the Lord (Exodus 15: 26; Deuteronomy 32: 39). The Hebrew term *rofé* (physician, healer) is only used in the Pentateuch (the Five Books of Moses) in the context of the embalming of Jacob and of Joseph (Genesis 50: 2). However it is well known that physicians in Egypt, where Jacob's embalming took place, did not practice embalming. It seems, therefore, that the term was adapted to a pseudo-medical procedure (Preuss, 1978: 18). In the chapters devoted to "biblical leprosy" the term *rofé* appears nowhere. The priests acted as "sanitary officers" (diagnosis, isolation) but not as healers. They did not even pray (or rather, they were not supposed to pray) for the recovery of the "lepers". Their expertise was in pronouncing the fateful decision: "clean", or "unclean". Prophets, not priests, eventually practiced the art of healing. Thus Isaiah healed King Hezekiah from his skin disease (II Kings 20: 7). Elisha revived an apparently dead child (II Kings 4: 32–37); he also "healed" the insalubrious drinking waters of Jericho (II Kings 2: 19–22). The theme of the healing power of those inspired by the Lord's word was however less developed in the Hebrew Bible than in the Christian Gospels.

Physicians are mentioned in the context of King Asa's disease (II Chronicles 16: 12): the king is criticized for having sought their advice instead of entrusting the Lord. Commentators, however, remarked that Asa's attitude was not acceptable for a king; it was suitable for common people. Our approach is that the king should have *also* entrusted the Lord, and he should have prayed before seeking the help of the physicians. This would be in tune with Ben Sira's approach, stated above. Job calls his friends "physicians of no value" (Job 13: 4) and the prophet Jeremiah exclaims: "Is there no balm in Gilead? Is there no physician there?" (Jeremiah 8: 22). This signifies that there were physicians at that time and place. Medical anthropologists tend to attribute to the Bible a negative approach toward medicine and healers, the Lord being the *only* Healer (Avalos, 1995: 238–250). We would rather assert that the Bible does not feature much interest in the medical practice, the Lord being the *ultimate* Healer.

In the Talmud, the Aramaic term for physician is *assia* (close to the old Babylonian *a-su*). The assia treated wounds, injuries and even dental problems; it is preferable therefore to call him a healer rather than a physician. Nothing is known about the healers' training. The Midrash however mentions "a teacher who has completed teaching a student the art of healing" (Deuteronomy Rabba 6, 6). It should be stressed that there existed in talmudic times some kind of licensing by the Court. We are told that, "An expert physician who practiced with the licence of Court and caused some damage is not liable" (Tosefta Baba Kamma 9, 11). Obviously (as is indeed stated) he is liable if he caused damage intentionally. We might ask why the physician was not liable if he injured unintentionally. The answer is "for the public's sake" (Tosefta Gittin 4, 6). This means, according to several commentators, that without that provision nobody would be willing to practice medicine, and the public would be without physicians. But then, is this rule not a possible opening toward negligence in medical practice? The Talmudists did not overlook this question. The physician was not liable before human judges; his case was however deferred to the Heavenly Court (Tosefta Baba Kamma 6, 17).

Another category of personnel was what was later called the barber-surgeons, in Aramaic *umman* (Kottek, 2002: 275–279). It is clearly stated that this official was of much lower status than the healer. He is cited together with low-grade artisans; he could not be appointed an administrator or leader of the town council (Derekh Ereẓ Zuta 10, 2). His main office was bloodletting; he was therefore sometimes called *gara*ʿ (= Latin *minutor*). One of these officials was named (Abba Umna) and the Sages highly praised him for his commitment and his ethical behavior toward his patients (b. Ta'anit 21b).

No military medical personnel are mentioned in the Bible, although a number of war injuries to kings or generals are documented. There is no mention of specialists (who, according to Herodotus, were common in ancient Egypt). We are told, however, that among the officials appointed to serve in the Temple of Jerusalem, there was one, Ben-Aḥiya, "for the diseases of the bowels". The priests walked barefoot and ate much meat, which caused digestive problems. Ben Aḥiya is, however, not called *rofé*, and might have been just a folk-healer (m. Shekalim 5, 1). He was allotted a certain amount of money from the Temple treasury, so that he could order medications (*ibid.* 5, 2).

Midwives are mentioned several times in the Bible. We read (Exodus 1: 15–16) that the Egyptian ruler summoned the two Hebrew midwives, called Shifrah and Puah, and ordered them to kill all male newborns of the Israelites. This decree was not implemented by the midwives, for which they were praised. The Jewish historian Josephus offers a different version: "The king ... ordered ... that the labors of Hebrew women with child should be observed ... by the Egyptian midwives: for this office was ... to be performed by women who, as compatriots of the king, were not likely to transgress his will" (Antiquities II, 206–207 [H. St. J. Thackeray, trans.]). The midwife who assisted Tamar during her twin birth made a remark about one of the twins which led to the child being named Pereẓ (Genesis 38: 28–29). This shows the high standard of the midwife at that time: we are speaking about the son of Judah, the grandson of the Patriarch Jacob. In talmudic times, the midwife was called *ḥaya* (cf. m. Ḥullin 4, 3) or *ḥakhamah* (b. Eruvin 45a). The latter term actually means "the wise woman" (*cf.* French *sage-femme*).

The reader will hardly find it strange that thaumaturgy, i.e., the working of miracles, is not absent from talmudic lore. The zealous scholar Rabbi Ḥanina ben Dosa (1st century) was called by several famous colleagues (including his teachers) in order to pray for the sick members of their families. Once they came to fetch him for the sick son of Rabban Gamliel. Ḥanina retired to the upper floor and prayed for mercy. When he came back, he said, "You may go, for he (the sick child) is recovering". But they asked, "Say, are you a prophet?" He answered, "No, neither a prophet nor the offspring of a prophet, but I realize that when my prayer flows (freely) from my mouth, he (the sick person) is welcome; if not, he is rejected" (b. Berakhot 34b).

As a conclusion to this section let us quote from a talmudic saying: "No scholar should choose to live in a town where one cannot find (among ten things) a public bath, a public lavatory, a welfare fund, a physician (*rofé*) and a surgeon (*umman*)" (b. Sanhedrin 17b). For a learned person wishing to live in an organized environment, the presence of a physician (or a healer) and of a surgeon (or a bloodletter) was a must in talmudic times.

DISEASE AND DISEASES

In the Bible, being diseased (Hebrew *ḥoleh*) means essentially being diminished in strength (see Genesis 48: 1). The patriarch Jacob feels a weakness which announces death and accordingly summons his sons in order to bless them. Disease appears mainly in the Biblical text in the theological context of reward and punishment. "If you will diligently hearken to the voice of the Lord ... and keep all his statutes, I will put none of these diseases upon you, which I have brought upon Egypt; for I am the Lord that heals thee" (Exodus 15: 26). "And the Lord will take away from thee all sickness, and will put none of the evil diseases of Egypt ... upon thee ..." (Deuteronomy 7: 15; see also Exodus 23: 25). On the other hand, in the "Curses for Disobedience" it is stated that if the Children of Israel rebel against the Lord, they will be stricken by "great plagues, and of long continuance, and severe sicknesses, and of long continuance" (Deuteronomy 28: 59). A number of diseases are indeed mentioned in the chapter (also in Leviticus 26: 27–44), some of them difficult to differentiate. Moreover, many cases of individual punishment are reported in the Bible. Miriam, who has unduly criticized her brother Moses, is struck by *ẓaraʿat*, the paradigmatic skin disease. She is however soon relieved, owing to Moses' intervention (Numbers 12: 10). The same disease afflicts King Uzziah who has unduly performed an offering, which was the exclusive practice of priests (II Chronicles 26: 19–21). There are several other cases, not to mention the plagues that caused a striking number of casualties to the Children of Israel during their wandering in the desert, as retaliation for sinful behavior.

But disease may also afflict the righteous, as exemplified by the case of Job. His horrendous skin disease (Hebrew *sheḥin*) is central to the trial which he faces without giving up his faith in the Lord: "Shall we receive good at the hand of the Lord, and shall we not receive evil?" (Job 2: 10). The only answer to Job's question "Why?" (Or, as many patients still ask, "Why me?") is, "Hold your peace". There is no answer to the question (see Fohrer, 1963).

In the Talmud, a number of diseases are mentioned *en passant*; very few are described in detail. We will give more details later on about the epidemic disease called *askara* but will now concentrate on the way talmudic Sages discussed the question of what disease actually is. Their question was in fact centered on biblical exegesis: What kind of sickness will the Lord "take away from thee?" (Deuteronomy 7: 15). Two full folios of the two versions of the Talmud are devoted to this question (b. Baba Meẓia 107b and j. Shabbat 14, 3). The two versions are not identical, although they clearly stem from a common tradition. More precisely, the question addressed by the Sages should be formulated as follows: What may be considered the main cause of disease? In the Jerusalem Talmud it says: "99 die from ... and only one from Heaven". While trying to classify these causes of disease (cf. Kottek, 1985: 21–26) on the basis of the two versions put together, we may distinguish:

1. Magical aspects: the evil eye;
2. Environment: air (or wind), *sharav* (dry heat), cold;
3. Humors: bile, "secretions", blood;
4. Other causes: fever, obsession, carelessness, [abrupt] changes in the manner of living (or eating).

Such a heterogeneous list needs a few explanatory remarks. Two other items actually appear elsewhere in the Talmud. Blood (b. Baba Bathra 58b) is particularly stressed. "At the head of all [causes of] diseases am I, the blood". We know that in most ancient cultures (and through the Middle Ages), plethora, i.e., excess of blood in the body, was considered a foremost cause of disease, hence the regular practice of bloodletting. Change in life or eating style is mentioned four times in the Talmud. Three times it says "a change in habitual diet" (e.g. b. Sanhedrin 101a) and once "a change in habit" (b. Baba Bathra 146a). Three items appear in both Babylonian and Palestinian versions: the evil eye, bile, and "cold". This shows that the three basic categories, i.e., magic influences, popular beliefs and academic medicine, coexisted everywhere (in Babylonia and Palestine).

Two additional items deserve some comment. Obsession (Hebrew *ra'yon*) meant that whoever considered himself to be ill was indeed a sick person. Carelessness meant that whoever did not lead a healthy life was the main cause of his own ills. Magic, as well as astrology, was also practiced in Babylonia. The latter had little bearing on medical lore in the Talmud, with the exception of bloodletting (favorable vs. unfavorable days). Demons and spirits are also mentioned. How does a dog become rabid? One sage opines: "Female witches play with it". Another states, "An evil spirit dwells on it" (b. Yoma 83b). A state of transient mental confusion called *kordiakos* is considered by Mar Samuel, one of the foremost talmudic medical authorities, to be caused by a demon (b. Gittin 67b). Amulets were part of the accepted materia medica, although they were used cautiously. Only patent formulas (i.e., those which had been effective on three occasions) were legitimate. They should obviously not have contained any heathen reference. Incantations were also present in talmudic lore. However, no scenes of exorcism are recorded in the Talmud.

Epidemic diseases

Plagues are frequently mentioned in the Bible and in the Talmud. In the Bible they were inflicted on the Children of Israel during their wanderings in the desert as a punishment, with an impressive number of casualties. For instance, after they sinned with the daughters of Moab at Baal Peor, 24,000 perished (Numbers 25: 9). It should be made clear that the Hebrew term *dever* (plague) was used in ancient Hebrew writings for any disease which caused a great number of casualties in a short while. For instance, when King David incurred punishment for having unduly conducted a census of his people, 70,000 died in a very short time [one or three days, according to different readings of the text]. Here a detail of the story, with ethical implications, is in order: The king was presented with the choice among three forms of punishment; either seven [other version: three] years of famine, or three months of unsuccessful warfare, or else three days of pestilence (*dever*). The king chose the latter because he was not immune to that agent of death. David exclaimed, "Why! I have sinned ... but these sheep, what have they done?" (II Samuel 24: 17).

Another example is the so-called Plague of the Philistines (I Samuel 5–6). After they had taken the Holy Ark as a trophy of war, the Philistines were stricken by an epidemic disease called *ophalim*, usually translated as "swellings". Because mice (or rats?) were mentioned as a related scourge, and swellings could be seen as buboes, the diagnosis of bubonic plague has been raised. This can, however, hardly be proven (see Conrad, 1984).

In the Talmud, there is one epidemic disease called *askara*, which is mentioned several times. It struck children especially, often in night attacks, and it led to death by suffocation. It was the most terrible form of death. This seems to describe epidemics of diphtheria, although sub-glottic laryngitis could be included. The rather remarkable detail here is that one author of ancient Greco-Roman medicine called Aretaeus (2nd century) described diphtheria under the name *eschara* (used by Galen for a scab on a wound). We may thus have here an example of influence or symbiosis between Greco-Roman and talmudic medical lore (cf. Preuss, 1978: 157–160; Kottek, 1996: 2920–2922).

ANATOMY AND PHYSIOLOGY

In the Bible, a number of anatomical details are mentioned while describing wounds incurred in battle. Moreover, human defects, which invalidate the priests' fitness to perform their office at the Temple, are listed (see Leviticus 21: 17–21). On the other hand, human beauty is described in the Song of Songs with a luxury of (external) anatomical details. Impressive embryological descriptions are provided in Psalms (139: 15–16) and in the Book of Job (10: 8–12). The latter excursus reads: "Have You not poured me out as milk and curdled me like cheese? You have clothed me with skin and flesh, and fenced me with bones and sinews ..." The curdling metaphor can be found in Aristotle's *Generation of Animals* (739b22), the male semen fixing the female blood (see Needham, 1959: 50–51).

The body is often designated by two expressions. One is *basar ve-dam* (flesh and blood); the other is an association of *eẓem* (bone) and *basar* (flesh). When Adam first met his wife, he exclaimed: "This is now bone of my bones, flesh of my flesh" (Genesis 2: 23). Since they are the frame of the body, the bones are "essential", and the term *eẓem* (bone) also means "essence" and "self".

The soul is designated by three terms, all having a spiritual etiology (or at least relationship): *nefesh* may be seen as the vegetative soul, *ruaḥ* as the animal soul and *neshamah* as the intellect. Medieval commentators wanted to individualize the three principles of growth, movement and thought, which seems a somewhat overzealous attempt at finding a common denominator with Greek philosophy (see Carny, 1982).

In the Talmud, a wealth of anatomical terms may be found. It may be asked from where the Talmudists got their information. The answer is most probably from observation of wounded individuals and of corpses and certainly from analogies with animal anatomy observed at the slaughterhouse. But what did they do about autopsy? On the one hand, desecration of a corpse and/or taking profit from the dead were only allowed if this were helpful to save another sick patient (Jakobovits, 1959: 132–156). On the other hand, we are told (b. Bekhorot 45a) that the disciples of Rabbi Ishmael (2nd century) once "boiled" the corpse of a prostitute who had been executed by the (Roman) authorities, with the aim of counting the number of bones of the human body. The background of this "experiment" is relevant to the laws of purity/impurity. The corpse was boiled (Hebrew *shalaq*) in order to loosen the flesh from the bones, a technique sometimes used by the Crusaders to be able to bring back home the corpses of dead knights. The Talmudists considered that there were 248 bones in the human skeleton. The Mishnah (m. Oholot 1, 8) gives a full description. They counted 101 pairs of bones (51 for the legs, 39 for the arms and 11 ribs), and 46 bones which are not in pairs (spine, sternum).

It is stated elsewhere that once a physician called Thodos examined a box full of bones and asserted that they did not derive from one and the same skeleton (b. Nazir 52a). It says that Thodos came with a host of physicians. Were these physicians his pupils or just colleagues who wished to learn from his expertise? This question will remain unanswered.

Another appalling experiment allegedly performed in Alexandria under the auspices of Queen Cleopatra is recorded in the Talmud (b. Niddah 30b; also Tosefta Niddah 4, 17). Several handmaids had been sentenced to death. They were made pregnant and were put to death at certain times so that the physicians could examine the fetuses in order to determine when they were fully fashioned. The Rabbis had heard that the result of the experiment was that a fetus (male and female alike) was fully formed on the 41st day of pregnancy. These results are discussed at length in the Talmud, to the amazement of the historian of embryology Joseph Needham (1959: 65–66). While dealing with embryology, let us add here a midrashic story (Numbers Rabba 9, 43) illustrating the belief in psychic maternal influences on the fetus. An Arabian king came to consult Rabbi Aqiba (1st–2nd centuries) saying that he and his wife were both black but she had just given birth to a white baby. He wanted to know if he should kill her. The Rabbi asked if they had white statues in their bedroom; the king said they did. The Rabbi provided this interpretation: During intercourse, the queen had looked intently at these statues and bore a child like them! Did the clever Rabbi really believe this, or did he find the solution that the king wished to hear in order to absolve the queen?

HYGIENE

In the Bible dietary laws are imposed on the Children of Israel without any explicit reason and without any clearly stated relationship to health and disease (cf. Preuss, 1978: 501–506). The case of meat is of particular interest. Adam and Eve were apparently vegetarian: "You will eat the herb of the field" (Genesis 3: 18). To Noah and his offspring "every living moving creature will be food for you, even as the green herb I have given you all things" (Genesis 9: 3). During the time the Children of Israel were nomads (in the desert) they ate manna ("bread from Heaven", Exodus 16: 12). But soon they murmured against the Lord, asking for meat. When quails were offered, they overate and many of them died (Exodus 16: 13; Numbers 11: 31–34). Once having settled down in the Land, meat became part of the diet. "Because you desire to eat meat, you may eat meat to your heart's wish" (Deuteronomy 12: 20). This is not altogether left to one's "heart's wish", for soon after the Bible presents a list of permitted and prohibited animals (Deuteronomy 14: 3–21; also Leviticus 11: 1–31). Since Noah's time, and it is repeated several times, there has been a cardinal injunction not to eat blood, for "blood is life" (e.g. Deuteronomy 12: 23–25).

Wine, "the red blood of grapes" (Genesis 49: 11), is cited, together with bread, as a basic nutriment: "Come, eat of my bread, and drink of the wine I have mingled" (Proverbs 9: 5). Wine brings consolation to "those of heavy hearts" (Proverbs 31: 6). Nazirites vowed to abstain from wine (Numbers 6: 3–4), as did the Rekhabite tribe (Jeremiah 35: 6–10). Priests were not allowed to drink wine before and during their office (Leviticus 10: 9). Many other rules relating to hygiene, such as body cleanliness, housing, the weekly day of rest, burial, sexual activity, and more, are mentioned in the Bible (e.g. Grünwald, 1911: 175–188; Kottek, 1996: 37–65).

In the Talmud, hygienic data are extensively described. Water supply was of prime importance. Water was stored in cisterns and reservoirs; special officials insured their maintenance (b. Baba Bathra 91b). Laundry was considered a public health issue, as dirty clothes could cause various skin diseases (b. Nedarim 80b–81a). Wells and aqueducts were tended; the latter were not supposed to pass through cemeteries (b. Megillah 29a).

Public baths were one of the prerequisite facilities for a scholar choosing a dwelling place. They are described in the Talmud with a luxury of details describing the operations performed by the attendants. The terminology related to bathing, in both versions of the Talmud, is very rich in terms of Greek origin, showing clearly the Hellenistic influence in this domain (Kottek, 1996: 2915–2917). There are also aphorisms such as, "Whoever has bathed in warm water and has not been sprinkled thereafter with cold water is like iron that has been put into fire but not into cold water thereafter" (b. Shabbat 41a). Oil rub, massage and physical exercise are mentioned; these were often performed after bathing. Ritual bathing in "live waters" (Leviticus 11: 36) is only of relative hygienic relevance. It pertains more to theological categories such as the concepts of cleanliness/uncleanness, or in other terms purity/impurity. There were instructions that the hands be washed upon arising in the morning and before (and after) every meal which included bread.

Lavatories (called the house of stool, Hebrew *beit ha-kissé*) are described. There were mainly public privies; private ones seem to have been rare. People were advised not to delay going to the toilet. As one Sage put it: "As soon as your pot is warm, empty it out" (b. Berakhot 62b). The fact that this somewhat touchy topic was addressed in detail is quite remarkable (Kottek, 1995: 522–524).

Nutrition is often dealt with in talmudic lore. People were advised to eat moderately. "Eat one third, drink one third, and leave one third of the stomach empty" (b. Gittin 70a). It seems that most people ate only one real meal a day, and only people with leisure time could afford an elaborate cooked meal

(b. Ḥullin 84a). Therefore the Sabbath day, with its three traditional meals, was indeed festive.

Burial has a bearing on hygiene in sub-tropical regions. Corpses were as a rule buried the same day, unless there was some necessity "in honor of the dead" (m. Sanhedrin 6, 5). In mishnaic times the dead were laid to rest in sarcophagi, and the vault was watched for three days (m. Semaḥot 8, 1). From talmudic times to the present day, the conditions of burial were equal for rich and poor. An emptied catacomb could not be used for storage of any kind (even for straw or wood) or as a den for cattle (Kottek, 1995: 644–646).

SURGERY

As stated above, surgery was not, in ancient Jewish lore, isolated from practical medicine. The official called *umman* in Aramaic was hardly a surgeon, but rather a bloodletter; he may however have occasionally treated wounds or fractures. In the Bible, the treatment of wounds is mentioned metaphorically (e.g. Isaiah 1: 6; Ezekiel 16: 10).

In the Talmud, on the other hand, we find an impressive range of surgical interventions. Bloodletting is often mentioned (see Preuss, 1978: 248–257). The patient was instructed to drink immediately after bloodletting but was allowed to tarry a little before eating. This is expressed in the following aphorism: "Drink immediately! Eat up to [walking a distance of] half a mile" (b. Shabbat 129b). This well-developed topic is in tune with trans-culturally accepted rules, and need not be described in detail here.

Regarding wounds, the Talmudists knew that the healing process might be hypertrophic: "vegetation (Hebrew *ẓemaḥim*) growing over the scar" (b. Baba Kamma 85a). They also knew (empirically as it seems) that "sweet food is bad for the wound" (*ibid.*), possibly a reference to the difficult healing of wounds in diabetic patients. A scaly (?) scar (*gargutni*) was treated with a salve containing resin (galbanum?) and aloe (Jacobi, 1998–1999) or with squashed snails (b. Shabbat 77b).

Fractures and traumas are described in animals (cf. b. Ḥullin 77a). Regarding criteria for the evaluation of the healing of fractures, a Rabbi states, "This question I have raised before sages (philosophers?) and before physicians". Their advice was that the bone should be scratched (causing some issue of blood), and this would induce the closure of the fractured bone (*ibid.*). While considering a fractured skull, there is a mention of two membranes (upper and lower), as was commonly accepted in Greco-Roman medicine. Once a sheep fell off the roof and thereafter dragged its legs. One Rabbi diagnosed this as sciatica; the other said its spinal column had been severed. So they checked (i.e., they performed an autopsy) and found that the second opinion was the right one (b. Ḥullin 51a). This case story is paradigmatic of the experimental tendency of the Talmudists.

We shall briefly mention a few real operations mentioned in the Talmud. Circumcision, performed by special officials, will not be considered here. Splenectomy is briefly mentioned (b. Avodah Zarah 44a). It was allegedly performed on runners in order to increase their velocity. There is a similar statement in Pliny the Elder (11 [80], 205). *Atresia ani* (congenital absence of an anal opening) and its surgical treatment are recorded (b. Shabbat 134a): the membrane blocking the newborn's anus was cut open crosswise (not in a circle). Abbaié (4th century) described this technique in the name of his (step) mother, who had been (it appears) a midwife.

Plastic abdominal surgery is briefly hinted at. A well-known Rabbi was very stout. He was given a soporific drink and was brought to a room with marble (-covered) walls. His abdomen was opened and many baskets of fat were taken out (of his belly). No detail is given on who performed the operation (b. Baba Meẓia 83b). Pliny also mentioned this operation (11 [83], 213). Even more astonishing, the operation of hypospadias, a defect in which the ureter presents an opening below the penis, is described in detail. In order to close the unwanted hole on the ureter, they used an ant that would bite around the hole. Then its head was severed from its body, and when the head had decayed the hole would be closed. This technique is of particular interest (see b. Yebamot 76a), for it apparently refers to Indian lore. Indeed, Hindus are known for their ancient tradition of using ants for suturing wounds (Majno, 1975: 304–309). There seem to have been in this case a link between Hindu and Jewish medical lore, most probably through Babylonian mediation.

Cesarean section is another major operation allegedly performed even on living women. Embryotomy was performed whenever the mother's life was in danger during the birthing process. The life of the mother had priority over that of her fetus, although the fetus is considered to be alive from the time of impregnation. This talmudic rule is of particular interest, as it was opposed by the Church Fathers. Let us add that if the head of the child had come out, embryotomy was no longer allowed (m. Oholot 7, 6). A Jewish midwife, not a heathen one, was allowed to perform embryotomy (Tosefta Avodah Zarah 3, 4).

GYNECOLOGY AND OBSTETRICS

In the Bible the impressive event of giving birth is described quite often, factually as well as metaphorically. After the sin of Eve and Adam, the Lord addressed the woman as follows: "I will greatly multiply the pain of your childbearing; in sorrow you will bring forth children ..." (Genesis 3: 16). This translation, which is currently accepted, may be challenged: "pain" could be rendered by something like "labor", and "sorrow" by something like "anguish" or "discomfort". The prophets often used the birth metaphor, e.g. "Anguish and pains have seized on her, like a woman in travail" (Jeremiah 49: 24). "They will be afraid; pangs and pains will take hold of them, they will tremble like a woman in travail" (Isaiah 13: 8). "Like a woman with child who draws near the delivery is in pain and cries out in her pangs, so have we been before you, O Lord" (*ibid.* 26: 17). Midwives have already been mentioned above (in the section on Physicians and Surgeons).

In the Talmud, obstetrics is a widely documented topic (see Schapiro, 1904; Preuss, 1978: 381–431). Pregnancy extends over 271 to 273 days (b. Niddah

38a). The Talmudists knew that semen could remain fertile for three days after intercourse. "Most women give birth nine months after impregnation, some after seven months" (b. Yebamot 37a). In accordance with Hippocratic medicine, the Talmudists believed that an eight-month fetus could not survive (b. Yebamot 42a); however they observed that some did. These were – so they believed – babies "planned" to be born after seven months of pregnancy whose birth was delayed (j. Yebamot 4, 2).

Superfetation (the presence of two fetuses of different ages, not twins, in the uterus) was considered possible. One of the fetuses could therefore become a *fetus compressus* (Hebrew *sandal*, b. Niddah 45a). There were however contradictory opinions. One Sage affirms, "A woman who becomes pregnant does not conceive again thereafter" (b. Niddah 27a).

The role of the placenta was known. "There is no placenta (Aramaic *shilya*) without a fetus" (b. Niddah 26a). There was a custom of burying the placenta "in order that the newborn become warm" (b. Shabbat 129b), a procedure which existed in many cultures (Ploss, 1935–36, vol. II: 807–812).

Abortion could be caused by environmental hazards (climate, winds, water) or by fright. The priests at the Temple of Jerusalem prayed and fasted each Thursday with the aim that pregnant women would not miscarry (b. Taʿanit 27b). When a child died within the first month after birth, it was considered (regarding laws of burial and mourning) an aborted fetus (Hebrew *néfel*).

Cesarean section is mentioned in the Talmud several times. If a woman died during childbirth, everything was done, even on the Sabbath day, in order to save the child; her belly was opened and the child extracted (b. Arakhin 7a). Quite strikingly, the Codes (16th century) invalidated this ruling, arguing that "we are no more expert", which means that the Rabbis of the Renaissance period ruled that there was practically no hope to save the unborn child, and therefore the corpse of the mother could not be desecrated. This shows that Jewish Law has its own dynamics, in tune with the evolution of science and knowledge (see Shulḥan Arukh, Orah Ḥayyim 333). Scholars have had many lively discussions concerning whether Cesarean section was performed in talmudic times on living mothers. A male child born "through the wall" (Hebrew *yotzé dofen*) would not be circumcised on the eighth day, if this day were a Saturday (*Shabbat*), as he was actually not really "born". Such a child, if he were the first child of his parents, was not considered a firstborn legally. This was already stated in the Mishnah (m. Bekhorot 8, 2) but it is not known if Cesarean section on a living mother was performed at that period. This statement concerning a firstborn could therefore have been purely theoretical (cf. Preuss, 1978: 420–426).

Regarding gynecological data, there is a story of a sterilizing potion which was given to a woman who had had several difficult pregnancies. The composition of the potion is recorded: it included alum and garden crocus (b. Shabbat 110a). In order to avoid abortion, women were allowed to wear (even on the Sabbath day) an amulet, the "preserving stone" (Hebrew *even tequmah*), which was quite certainly aetites (eaglestone), as recorded in ancient Greek materia medica (see Kottek and Baader 2000: 92–93). Let us consider a case history, as there are many in the Talmud.

> It happened that a woman issued from her womb like red peelings. They [the husband, or the family] went and asked Abba [what should be done]. Abba asked the Sages, and the Sages asked the physicians. The latter answered: This woman has a wound [or a sore, Hebrew *makkah*] in her womb from which she issues like red peelings (b. Niddah 22b).

The sociological context is clear: families first consulted the bloodletter (Abba), who in turn consulted the Sages (who might have been confronted with similar cases). But the expert and final answer came from "the physicians", who added, "Put these peelings into water; if they dissolve, the woman is impure". This was the real question. Was this woman who issued "red peelings" permitted to have intercourse with her husband, or was it bloody so that she could not. This case may be seen as paradigmatic of how the historian is confronted with medical data within legalistic discussions in the Talmud.

DERMATOLOGY

Circumcision and "biblical leprosy" are very widely researched topics. The foreskin was (and still is) cut on the eighth day after birth, not for medical reasons, although perhaps with some positive medical consequences. The Bible offers no technical details whatsoever and the Talmud very few, although considering a number of accidents, complications, or contra-indications. The most striking case is a brief though clear reference to hemophilia and its genetic transmission by women while affecting male children (b. Yebamot 64b). Even a leper had to be circumcised (see Preuss, 1978: 240–248; 544–545).

Biblical leprosy (Hebrew *ẓara'at*), or as Preuss called it "Lepra legalis", is a paradigmatic chapter in cultural anthropology. An impressive number of symptoms are recorded (Leviticus chapter 13). They hardly combine to make any clear-cut diagnosis, although biased authors have put many syndromes forward. These attempts will not be discussed here. The important, fateful decision was to decide whether the individual was clean or unclean. The priests were in charge of taking this decision; in dubious cases, they would isolate the individual for one or even two weeks and then decide. They were also responsible for assessing the disappearance (the cure) of the "disease". In the time of the Mishnah, other knowledgeable persons were allowed to diagnose. However, a priest had to pronounce the verdict: clean or unclean (m. Nega'im 3, 1). It is also stated that the same priest had to follow the suspected "leprous" individual after eventual isolation in order to be able to evaluate the evolution of the lesion (Tosefta Nega'im 1, 15). There is not a single word about treatment of the condition, which was considered a punishment for a series of offenses, mainly for speaking ill (cf. Miriam's *ẓara'at*, Numbers 12: 9–13), but also for many others reasons (e.g. b. Arakhin 16a).

There is also an attempt at a natural etiology in the Midrash. It says that normally there is an equal amount of water and blood in the body. If blood exceeds water, the individual becomes leprous (Leviticus Rabba 15, 2). One Rabbi remarks that in Babylonia no cases of leprosy were reported, which he explains by some special dietetic habits and by bathing in the Euphrates (b. Kethubot 77b). (See Grmek, 1989: 152–176; Kaplan, 1993; Kahle, 1982).

Sheḥin is another skin disease which is mentioned several times in the Bible. One of the Ten Plagues in Egypt is called *sheḥin aba'bu'ot*, implying boils or blisters (Exodus 9: 8–10). Among the diseases mentioned in the "Curses for Disobedience", there is "sheḥin on knees and legs that cannot be healed, from the sole of your foot to the top of your head" (Deuteronomy 28: 35). King Hezekiah, who suffered from *sheḥin*, was said to be fatally ill. He nevertheless recovered, after he prayed for mercy and after the prophet had rubbed a cake of figs on the lesion (e.g. Isaiah 38: 13). *Sheḥin* was Job's primary disease. He was afflicted from foot to crown, he was disfigured, he suffered from acute itching (Job 2: 7–8), and his skin became black and "fell apart" (*ibid.* 30: 30). As was the case for *ẓara'at,* we cannot make any clear-cut diagnosis for *sheḥin*. Based on the root *shaḥan* (to be hot, parched), we have chosen the very general concept of "inflammation" as the fittest.

Another unidentified syndrome mentioned in the Talmud is *raatan*, which is one of the "24 types of leprosy" and is particularly harmful (painful?) during intercourse (b. Kethubot 77b). It was supposed to be contagious: "Beware of the flies of (i.e., coming from) someone suffering from *raatan*" (ibid.) [cf. Preuss, 1978: 323–353].

A number of other skin diseases are mentioned in the Bible and the Talmud including baldness. Youngsters derided the prophet Elisha for his baldness (2 Kings 2: 23). Esau, on the other hand, was born with hypertrichosis, an excessive amount of hair (Genesis 25: 25).

MENTAL DISEASES

In the Bible, several psychiatric syndromes are mentioned in the "Curses for Disobedience". "The Lord will smite you with madness (*shiga'on*), ... and astonishment of heart (*timahon levav*)" [Deuteronomy 28: 28]. "And you will be mad (*meshuga'*) from the sight of your eyes which you will see" [*ibid.* 34]. These terms are nowhere explained or described. "Astonishment of heart" is currently understood as confusion of mind.

King Saul was seized by an "evil spirit from the Lord" (Shy, 1982) which replaced the "spirit of the Lord" which had left him (I Samuel 16: 14). Saul had already earlier behaved strangely after meeting a company of prophets, and there it says, "the Lord gave him another heart" (*ibid.* 10: 9). As a matter of fact the term used here is not "gave him", but more exactly "overturned". Saul's mind was overturned and became different. We have in Saul's case a rare reference to a therapeutic measure, i.e., the calming effect of music on his fits of raving (*ibid.* 16: 15–16).

The Babylonian king Nebuchadnezzar is depicted in the Book of Daniel as stricken by a transitory psychic syndrome. For seven years, the king allegedly lived like a beast (hence the contemporary diagnosis of zooanthropy), after which he recovered and his kingdom was restored to him (Daniel 4: 22–24; 30–34). Ebstein (1901: 115–117) thought that this might well be considered a literary expansion of a transitory period of depression.

David once fled to the Philistine city of Gath, but he was recognized and brought before the king. Feigning insanity, he "scratched the doors of the gate and let his spittle run down on his beard" and was soon released (I Samuel 21: 11–16).

In the Talmud and Midrash, there are a number of relevant references. We would like to quote a meaningful comment on the feigned insanity of David. He once asked the Lord: "Everything you have created is fine, but a mad individual – what for? He runs about, people laugh at him, what kind of benefit can he have?". The Lord answered, "You complain about the value of a madman! You will experience for yourself its usefulness!"

In the Mishnah and Talmud, an insane person is called *shoteh* (from a root meaning "to roam about"). Such an individual "goes out alone at night, he spends the night in a cemetery, he tears his clothes" (b. Ḥagigah 3b). The Sages remark that even one of these characteristics might be enough if he acted otherwise strangely, while all three together might occur to a normal person for some special reason. The Talmudists also knew about the possibility of temporary mental confusion (see m. Niddah 2, 1). The case of the syndrome termed *kordiakos* (see m. Gittin 7, 1) will not be considered here (see Kottek, 1991: 47–49; Preuss, 1978: 320).

OTHER SPECIALTIES

Ear and throat

A popular maxim says that if you have earaches, use "liquid for dry and dry for liquid". If the ear is dry, use a liquid medication (urine from a goat was advocated), and if there is matter running out of the ear, use a dry sponge or a woolen wick (b. Avodah Zarah 28b). Presbyacusis (hearing loss) is mentioned in the Bible and Talmud (cf. b. Shabbat 152a), as are malformations of the ear (cf. Rosner, 1975).

Deafness was considered a serious defect. A deaf person was not considered intellectually fit; he could not testify in court (see b. Ḥagigah 2b). Deaf people were however protected by society: "You will not curse the deaf" (Leviticus 19: 14). "Open your mouth for the dumb" (Proverbs 31: 8).

Nose

Epistaxis (blood flowing from the nostrils) is mentioned. Amulets were among the treatments prescribed (b. Gittin 69a). Polypus (nasal polyp) was known for being a cause of a bad odor from the nose and/or the mouth. It was considered a valid reason for divorce (m. Kethubot 7, 10).

Eyes

A number of eye defects are discussed in the Talmud (see Preuss, 1978: 259–267). Diseases of the eyes (tearing, discharge of matter, pains, reddening) are recorded. They were, in talmudic times, considered dangerous, most probably a result of the frequent complications of (blennorrhagic) ophthalmia.

Blindness (Hebrew *ivver*, Aramaic *suma*) is one of the disabilities mentioned in the "Curses for Disobedience" (Deuteronomy 28: 28). Like the deaf, blind people were protected by the Bible: "You shall not ... put a stumbling block before the blind" (Leviticus 19: 14). "Cursed be he that makes the blind wander out of the way" (Deuteronomy 27: 18). According to a talmudic saying, a blind person is like a dead man (together with the poor, the childless and the leper) (b. Nedarim 64b). It is also stated that a blind individual is never really satiated; one cannot compare one who sees what he is eating with one who cannot (b. Yoma 74b). Among several causes of blindness mentioned in talmudic lore, let us mention traumatic blindness. It says that a blow to the brain (i.e., on the skull) may cause blindness (Tosefta Baba Kamma 9, 27). If such a case happened after one had beaten his slave, the latter would be set free (*ibid.*). Several famous Sages of the Talmud were blind. They are called "clear-sighted" (Aramaic *sagi nehora*), or the "light of the eyes" (Hebrew *meor einayyim*). Even the prince of the demons Ashmodai, meeting a blind man, put him back on the right path (b. Gittin 68b).

Cosmetic painting of the eyes existed in biblical times (cf. Ezekiel 23: 40), although it was considered an act of harlotry. In the Mishnah, the woman suspected of adultery "has painted her eyes for him, and therefore her eyes are protruding" (Tosefta Sotah 3, 3). In talmudic times, however, it had become a common custom. According to one talmudic authority, a woman might paint her own eyes, although (on the Sabbath day) dyeing was prohibited, but she could not do it for another woman (b. Shabbat 95a). The root *kaḥal* (cf. khol) was used for the application of medicines in the eye as well. The term collyrium (Aramaic *qilorin*, or *qilor*, cf. b. Shabbat 18a) also appears several times in the Talmud.

Cataract (?) is mentioned (Aramaic *barqa, barqit*) – a bright, shining spot in the eye (b. Shabbat 78a). If urine was considered remedial for earaches, spittle was (ubiquitously) used against eye diseases. Spittle of a firstborn or a fasting person was considered particularly efficient (b. Shabbat 108b).

THERAPEUTICS

Of the therapeutic remedies mentioned in the Bible we shall only mention mandrakes (Hebrew *dudaim*). Reuben gathered some dudaim for his mother Leah, who gave them to her sister Rachel (Genesis 30: 17). There is in this story no reason to conclude (as did many commentators) that mandrakes have aphrodisiac virtues. Talmudic authorities offered differing identifications for dudaim (b. Sanhedrin 99b). Be this as it may, the plant seems to have been offered for its fragrance (perfume?) rather than for any treatment of sterility (see Preuss, 1978: 462–463).

The link between vegetable medicines and diet, or in other words, plants used as food and medicine, is apparent in Ezekiel's messianic vision: "(The trees) will bring forth new fruit every month, because these waters have issued out of the sanctuary; and its fruit will be for food, and its leaf for medicine" (Ezekiel 47: 12). Two Talmudists, while offering an anagrammatic comment on the word medicine (Hebrew *terufah*), explained that the leaves would "untie the mouth". One says, "the upper mouth", i.e., they will cure the dumb; the other says, "the lower mouth" (or opening), i.e., they will cure sterility [tied wombs] (b. Menaḥot 98a).

In the Talmud, a drug is called *sam*, or *samma*, whether external or internal. A poison is a "deathly drug" (Hebrew *sam ha-mavvet*). If one applies one's learning correctly, a plant can be a medicine of life; if not, it becomes a deadly drug (b. Yoma 72b). A saying of Ben Sira (38: 3–4) is quoted in the Midrash (Genesis Rabba 10, 6): "The Lord has caused drugs (*sammim*) to sprout from the earth; with them the physician heals the wound".

A plaster, or emollient, called *samtar*, or *samtré*, was thought to be able to heal any wound (b. Baba Bathra 74b). Another poultice recommended by the physician Minjome (Benjamin) was made of wax and resin (b. Shabbat 133b). Cataplasms (poultices) with plants, or pressed fruit, or dough, are mentioned as well. Honey is given to someone who fainted (hypoglycemia?) and to someone who suffered from bulimia (b. Yoma 83b). The all-healing drug of ancient times called theriaca is mentioned in the Talmud; however its use was forbidden if prepared by heathens (j. Avodah Zarah, 2, 6).

A number of magic devices were part of popular lore and were quite common in Babylonian society. Amulets and incantations were used against evil spirits and demons. The Rabbis tolerated these practices as long as they did not entice people toward idolatry (see Veltri, 1997).

In ancient medicine, dietary rules played an important part in the prevention and management of disease. We shall quote here only a few among such rules mentioned in the Talmud, often in the form of brief statements or aphorisms. "Three things are harmful in large amounts, although beneficial in small quantities: yeast, salt and ... meekness" (b. Berakhot 34a). "After every meal, take some salt, and after every drink take some water, and no disease will befall you (b. Berakhot 40a). "Three things decrease the movement of the bowels, straighten the stature and give light to the eyes: (1) clean white bread; (2) fat goat meat and (3) old wine" (b. Pesaḥim 42a–b). Overeating was of course considered harmful: "Eat one third, drink one third, and leave one third (of the stomach) empty" (b. Gittin 70a). "If you consume your meal slowly you will prolong your life" (b. Berakhot 54b). Moreover, regularity regarding meals, in quantity/quality/time, was stressed as conducive to health. "Changes in one's (eating) habits lead to digestive ailments" (b. Sanhedrin 101a).

There are many relevant statements in the Talmud and Midrash regarding the nutritive value (and medical use) of all kinds of food, for instance meat, fish, eggs, oil, honey, fruit and vegetables. In the latter category, let us pick out garlic, of which it was said, "it satiates, it warms the body, it brightens the face, it increases semen, and kills intestinal worms" (b. Baba Kamma 82a). Among drinks, water and wine are prominent, together with milk. Wine is one among eight things which are considered beneficial in small quantities but harmful when excessive. The other seven things are warm water, sleep, intercourse, bloodletting, work, wealth and traveling (Abot de-Rabbi Nathan 37, 5). Eating habits and the impact of dietetics on health and disease are prominent chapters in talmudic literature and in ancient lore more generally (see Preuss, 1978: 553–575; Cooper, 1993: 1–52).

* * *

This outline of biblical and talmudic medical lore is indeed a fragmentary sketch aimed at giving a general impression of the data which may be gathered. Full chapters have been left out of our study, on topics such as injuries and malformations, dentistry (see Nobel, 1930), neurological disorders including epilepsy (see Kottek, 1988), child care (see Kottek, 1987: 452–471), and legal medicine (which included sexual laws and activity (see Preuss, 1978: 447–500). In the introduction to his work published in German in 1911, Julius Preuss wrote, "There does not exist a work [stemming] from Jewish antiquity devoted exclusively to medicine; nor even a compendium of natural history, such as that of Plinius" (trans. F. Rosner, 1978: 4). This statement induces me to remark that there is indeed some idiosyncrasy between the way medical lore was recorded in Pliny's *Natural History* and in the Talmud, in which popular lore was intermingled with "scientific" medicine, without even minimal critical discussion.

The oldest organized medical work written in Hebrew was the Book of Assaph the Jew, who probably lived in Palestine and/or Babylonia around the 6th or 7th century, but about which very little is known (Venetianer, 1915; Muntner, 1957). Assaph describes 123 drugs, clearly leaning, at least in part, on Dioscorides, the Greek pharmacologist. His [pseudo-] historical introduction and his Oath/Covenant for/with the students are of particular interest. But Assaph's work, although rather close to the talmudic period, was nevertheless beyond it.

We would like to close this study with a reference to the religious duty to visit the sick. "He who does not visit the sick is like one shedding blood" (b. Nedarim 40a). A number of rules, social, psychological and theological, are indicated in much detail (see Rosner, 1996: 2890–92). Only one of them will be recorded here. In case the sick person cannot bear visits, remain in the antechamber, ask whether you may help in some way, and listen to the patient's complaints. The main object of these visits is not to speak, but rather to listen.

NOTES

1 The Bible includes the five books of Moses, to which were appended the prophetic and hagiographic writings.

2 Homiletic refers to the arts of preaching and writing sermons.

3 In this chapter, we use the following abbreviations:
b. = Babylonian Talmud
j. = Jerusalem Talmud
m. = Mishnah

BIBLIOGRAPHY

Avalos, Hector. *Illness and Health Care in the Ancient Near East.* Atlanta, Georgia: Scholars Press, 1995.

Carny, P. 'The soul (nefesh) in the Bible: A psychosomatic unity.' *Koroth* 8(5–6): 122–133, 1982.
Conrad, Lawrence I. 'The biblical tradition for the plague of the Philistines.' *Journal of the American Oriental Society* 104: 181–187, 1984.
Cooper, John. *Eat and Be Satisfied: A Social History of Jewish Food.* Northvale, New Jersey: Jason Aronson, 1993.
Di Lella, Alexander A., ed. *The Hebrew Text of Sirach: A Text-Critical and Historical Study.* The Hague: Mouton, 1966.
Ebstein, Wilhelm. *Die Medizin im Alten Testament.* Stuttgart: Enke, 1901.
Fohrer, Georg. 'Man and disease according to the book of Job.' *Koroth* 9 [special issue]: 43–48, 1988.
Freeman, David. 'The Gittin "Book of Remedies".' *Korot* 13: 151–164, 1998–1999.
Geller, Mark J. 'An Akkadian vademecum in the Babylonian Talmud.' In *From Athens to Jerusalem: Medicine in Hellenized Jewish Lore and in Early Christian Literature*, Samuel Kottek and Manfred Horstmanshoff, eds. Rotterdam: Erasmus, 2000, pp. 13–32.
Grmek, Mirko D. *Diseases in the Ancient Greek World.* Baltimore: Johns Hopkins University Press, 1989, pp. 152–176.
Grünwald, Max. 'Die Hygiene der Bibel.' In *Die Hygiene der Juden*, Max Grünwald, ed. Dresden: Verlag der Historischen Abteilung der Internationalen Hygiene-Ausstellung, 1911, pp. 175–188.
Jacobi, Margaret. 'Mai gargutani? An obscure medical term in Baba Kamma 85a.' *Korot* 13: 165–170, 1998–1999.
Jakobovits, Immanuel. *Jewish Medical Ethics: A Comparative and Historical Study of the Jewish Religious Attitude to Medicine and its Practice.* New York: Bloch, 1959.
Josephus Flavius. *Josephus*, H. St. J. Thackeray *et al.*, trans. and ed. 9 volumes. Cambridge, Massachusetts: Harvard University Press; London: Heinemann, 1976–1980.
Kahle, Ehrhart. 'The problem of sara'at in the viewpoint of Samaritan exegesis.' *Koroth* 8(5–6): 48–56, 1982.
Kaplan, David L. 'Biblical leprosy: An anachronism whose time has come.' *Journal of the American Academy of Dermatology* 28(3): 507–510, 1993.
Kottek, Samuel. 'Concepts of disease in the Talmud.' *Koroth* 9(1–2): 7–33, 1985.
Kottek, Samuel. 'On children and childhood in ancient Jewish sources: A study in philological anthropology.' *Koroth* 9(5–6): 452–471, 1987.
Kottek, Samuel. 'Epilepsy in ancient Jewish sources.' *Israel Journal of Psychiatry* 25(1): 3–11, 1988.
Kottek, Samuel. 'Sur l'origine gréco-latine de certains termes médicaux utilisés dans le Talmud et le Midrash.' In *Le latin médical: La constitution d'un langage scientifique*, Guy Sabbah, ed. Saint Etienne: Publications de l'université, 1991, pp. 42–52.
Kottek, Samuel. 'Gems from the Talmud: Lavatories [Public health IV].' *Israel Journal of Medical Sciences* 31: 522–524, 1995.
Kottek, Samuel. 'Selected elements of talmudic medical terminology, with special consideration to Greco-Latin influences and sources.' In *Aufstieg und Niedergang der Roemischen Welt (ANRW)*, Teil II: Principat, Band 37.3, 1996, pp. 2912–2932.
Kottek, Samuel. 'Hygiene and health care in the Bible.' In *Health and Disease in the Holy Land: Studies in the History and Sociology of Medicine from Ancient Times to the Present*, Manfred Waserman and Samuel Kottek, eds. Lewiston, Maine: Edwin Mellen Press, 1996, pp. 37–65.
Kottek, Samuel and Baader, Gerhard. 'Talmudic and Greco-Roman data on pregnancy: A renewed examination.' In *From Athens to Jerusalem: Medicine in Hellenized Jewish Lore and in Early Christian Literature*, Samuel Kottek and Manfred Horstmanshoff, eds., Rotterdam: Erasmus, 2000, pp. 83–98.
Kottek, Samuel. 'The surgeon as depicted in talmudic literature.' In *Proceedings of the 37th International Congress on the History of Medicine*, Galveston, 2002, pp. 275–279.
Majno, Guido. *The Healing Hand: Man and Wound in the Ancient World.* Cambridge, Massachusetts: Harvard University Press, 1975; reprinted 1991.
Midrash Rabba. H. Freedman, trans. and ed. 3rd ed. 10 volumes. London and New York: Soncino Press, 1983.
Muntner, Suessman. *Introduction to the Book of Assaph the Physician.* Jerusalem: Geniza, 1957 [in Hebrew].

Needham, Joseph. *A History of Embryology*. New York: Abelard Shuman, 1959.
Nobel, Gabriel. *Zahnheilkunde und Grenzgebiete in Bibel und Talmud*. Leipzig: im Selbstverlag, 1930.
Pliny. *Natural History*. H. Rackham, trans. and ed. Cambridge, Massachusetts: Harvard University Press, 1967.
Ploss, Heinrich, Max Bartels and Paul Bartels. *Woman: An Historical, Gynaecological and Anthropological Compendium*. London: Heinemann, 1935–1936.
Preuss, Julius. *Biblisch-Talmudische Medizin*. Berlin: Karger, 1911.
Preuss, Julius. *Biblical and Talmudic Medicine*, F. Rosner trans. and ed. New York: Sanhedrin Press, 1978.
Rosner, Fred. 'Otology in the Bible and Talmud.' *Journal of Laryngology and Otology* 89(4): 397–404, 1975.
Rosner, Fred. 'Jewish medicine in the talmudic period.' *Aufstieg und Niedergang der Roemischen Welt (ANRW)*, Teil II: Principat, Band 37.3, 1996, pp. 2866–2894.
Schapiro, David. *Obstétrique des anciens Hébreux*. Paris: H. Champion, 1904.
Shy, Hadassah. 'Ruah ra'a (melancholy) as seen by medieval commentators and lexicographers.' *Koroth* 8(5–6): 94–105, 1982.
Smith, C. Raimer. *The Physician Examines the Bible*. New York: Philosophical Library, 1950.
Snowman, Jacob. *A Short History of Talmudic Medicine*. London: Staples Press, 1935.
Talmud [Hebrew-English edition of the Babylonian Talmud]. Isidore Epstein, ed. London: Soncino Press, 1985(?)–1989.
Trusen, J.P. *Die Sitten, Gebraueche und Krankheiten der alten Hebraeer*. Breslau: Korn, 1853.
Veltri, Guiseppe. *Magie und Halakha*. Tuebingen: Mohr, 1997.
Venetianer, L. *Asaf Judaeus: Der aelteste medizinische Schriftsteller in hebraeischer Sprache*. Budapest: Jahresbericht der Rabbinerschule, 1915.
Wunderbar, R.J. *Biblisch-Talmudische Medicin oder Pragmatische Darstellung der Arzneikunde der alten Israeliten*. Riga, 1850.
Zuckermandel, M., trans. and ed. *Tosephta, based on the Erfurt and Vienna codices*. Jerusalem: Wahrmann, 1970 [in Hebrew].

GUY ATTEWELL

ISLAMIC MEDICINES: PERSPECTIVES ON THE GREEK LEGACY IN THE HISTORY OF ISLAMIC MEDICAL TRADITIONS IN WEST ASIA

I have two aims in writing this survey of the history of Islamic medicine. The first is to unravel its historiography, to try to explain why certain aspects of Islamic medicine have attracted much attention while others have not. The second is to deal with specific themes pertaining to Islamic medical history that have been the subject of controversy or debate and that highlight broader historiographical trends or bring up issues of methodological importance.

Thus I shall not be so concerned with describing what Islamic medicine is, in all its variant forms, but rather with the enterprise of writing about it. Nevertheless, I will briefly describe some of the core principles associated with one of these traditions at the end of this section. I shall start with a broad discussion of how the historiography of Islamic medicine may be seen in the context of principally European scholarship on Islamic cultures and science. This will inform my examination of approaches to the study of the medical traditions associated with Galen of Pergamon (d. 199 CE) and the Prophet Muḥammad. I will focus on the association of Galenic medicine with the "golden age" and prophetic medicine with its "decline". In subsequent sections I shall revisit core themes in the history of Greco-Islamic medicine in which one can view the relationship between power, society and the medical tradition and question assumptions and interpretations about the role of Greek medical learning in Islamic medicine. These themes are the establishment of Greco-Islamic medicine in the context of the translation of ancient scientific and philosophical texts into Arabic, the development of the hospital, and the medical profession. Most treatments of the history of Islamic medicine deal in some way with the translation movement; such is its fundamental place in the development of sciences and philosophy in Islamic contexts. But only recently have its social, political and ideological dimensions been explored. We shall see that the development of the hospital in Islam is directly connected to these political currents. These are of great importance for understanding how the Greco-Islamic medical tradition became established in West Asia. In the discussion of the medical profession we will explore how physicians could gain

H. Selin (ed.), Medicine Across Cultures: History and Practice of Medicine in Non-Western Cultures, 325–350.

professional legitimacy through the composition of and their familiarity with medical texts, which will inform how one should interpret textual form and content in the context of a tradition of learning.

GRECO-ISLAMIC AND PROPHETIC MEDICINE – "GOLDEN AGE" AND "THE DECLINE"

The term "Islamic medicine" encompasses the traditions of medical theory, practice and literature that have been developed in Islamic cultural contexts and expressed most commonly in the languages of Islam, principally Arabic, Persian, Turkish and Urdu. In this article I shall be mainly dealing with two medical traditions that developed in Islamic contexts. One is a tradition that ultimately derives its authority, and many of its theoretical and practical components, from ancient Greek and Hellenist sources, which I shall refer to as "Greco-Islamic" medicine; the other encompasses traditions associated with the Prophet Muḥammad, or "prophetic medicine". Of the two, only prophetic medicine can be said to have essentially Islamic qualities, given its at least partial foundation on the sayings of the Prophet and his companions. Many works of prophetic medicine do however incorporate elements from Greco-Islamic medicine, as shall be pointed out below.

The dominant conceptual framework for understanding health and disease in the Greco-Islamic medical tradition is the theory of the four humours. The following brief summary of Greco-Islamic humoral physiology is based on the *Kitāb al-Malakī*, by 'Alī ibn al-'Abbās al-Majūsī (d. 994): All things are made from four elements (*ustuquṣṣāt* or *arkān*): fire (its qualities are hot and dry), water (cold and wet), air (moist and hot), and earth (dry and cold). Their relationship in a body is its temperament (*mizāj*), balanced if in equal proportions, or unbalanced if not. External factors may influence a body's temperament, i.e. environment, age, sex, profession, and lifestyle. In the body four humours (*akhlāṭ*) correspond to the elements: yellow bile/fire; blood/air; phlegm/water; black bile/earth. The imbalance of the humours causes illness; balance is health. Three spirits or pneumata (*arwāḥ*) support the functions of the body and nourish its innate heat, which gives life. These are natural (originating in the liver, to support nourishment and procreation), animal (in the heart, for life) and psychical (in the brain, for thought, movement and sense perception) (Ullmann, 1978: 55–69). The physician had to restore the patient to his natural humoral balance, usually by prescribing herbal, mineral or animal drugs of opposing quality to the patient's condition, e.g. a hot remedy for a cold complaint. Aside from an elaborate symptomology, the principal diagnostic tools of the *ṭabīb* (practitioner) were the analysis of urine (Arabic *baul*) and stool (*birāz*), and above all the art of reading the pulse (*nabḍ*).

Forms of Greco-Islamic medicine continue to be taught today in predominantly institutional settings in India and Pakistan, where the tradition is known as Unani tibb (Arabic *yūnānī*, meaning Ionian, i.e. Greek). Branches of the Hamdard foundation, a private teaching, research and clinical institution set up after Partition by the late Hakim Muhammad Said and his brother Hakim Abdul Hamid in Karachi and Delhi, and the Nizamia General Hospital (Unani) in Hyderabad (India) are very well-known as centers of Unani medicine throughout the subcontinent.

Most practitioners of Unani medicine are Muslim, but not exclusively by any means. Since at least the late 1950s many women have trained at some of the centers mentioned above as practitioners, although it is difficult to be sure of their role in the practice of this medical tradition historically, which seems to have been dominated by men. There have however been cases that have, exceptionally, remained in the historical record of successful practices run by women physicians (Conrad, 1995: 129). Although discussions of childbirth and women's diseases are common in Unani medical literature, a report by a British medical official in late 19th century Rajputana notes that obstetrics and women's diseases were almost exclusively in the hands of *dhais* (a class of midwives) (Hendley, 1900: 41). One should not however extrapolate from this isolated example about a complex socio-medical domain that may have varied tremendously in time and place.

The historiography of Islamic medicine that still finds its resonance today began to be written predominantly in the 19th century, concurrent with the expansion of colonial rule over Muslim peoples and the establishment of the Royal Asiatic Society of Bengal, the French Société asiatique and the Deutsche Morgenländische Gesellschaft (Said, 1978: 43). Greco-Islamic medical theory had been known in the Latin West since the Arabic-Latin translations began in the 11th century with Constantine the African at Salerno. These texts formed the basis of medical education until the 16th century in many European universities, such as Montpellier, Salerno and Salamanca (see Jacquart and Micheau, 1990; O'Boyle, 1999).

A number of scholars of Islamic civilization, of which medicine was one field of study, emerged in the 19th and 20th centuries; their knowledge, especially of Arabic and Persian, was applied to the editing, translating and study of Islamic manuscripts. Influential scholars on the history of Islamic science, philosophy and especially medicine were Ferdinand Wüstenfeld, Mauritz Steinschneider, Lucien Leclerc, Carl Brockelmann, Edward G. Browne and Paul de Koning. The enterprise of these scholars was text-based, and they focused on texts produced in the Abbasid period (750–1254), especially those composed from the mid-8th to the 11th centuries. They espoused, and no doubt the thrust of their work encouraged, the notion of a golden age of Islamic civilization, which then lapsed into decline by the 12th century. The philological emphasis of these scholars needs to be seen in the context of broader scholarly representations of Islam and the "Orient" (Said, 1978). The pioneering French philologist Ernest Renan's representation of science in Islam is clearly related to his notions of Western cultural and political superiority.[1] The famous German orientalist Ignaz Goldziher's influential account of an orthodox reaction against sciences in the 14th century derived from Greek sources has been shown to have more relevance to the politics of his day than to his purported topic of study (Dimitri Gutas, 1998: 173ff. on Goldziher, 1916). Equally some scholars saw the overriding value of the Arabic versions of Greek philosophical and scientific works in the Abbasid period in their functioning as a mere repository of classical knowledge, which was then transmitted to Europe and stimulated the European Renaissance.

The notion of the golden age of Islamic medicine has been applied to an age in which texts were produced, more specifically 8th and 9th century translations from Greek into Arabic, often via Syriac. During this time Arabic versions of Greek medical works by Galen, Hippocrates, Rufus of Ephesus, Paul of Aegina and others became accessible to the literate elite of cities such as Baghdad and Cairo. These texts formed the basis for medical education in humoral pathology and made possible the emergence of a medical profession grounded in these works. Historians have put forward many theories in order to account for the vast scope and dimensions of the translation movement in the 8th and 9th centuries to explain why patrons had invested so much in the production of texts from Greek, Persian and Indian cultures. These issues are clearly crucial for an understanding of the development of Greco-Islamic medical culture in West Asia, and they will be the subjects of discussion below. Nevertheless, the emphasis in the historiography, until the 1970s, on this form of medical knowledge in the Abbasid Caliphate (750–1258) tended to occlude the recognition of other medical traditions in the Islamic world as worthy of investigation and to set spatial and temporal limits on the study of Islamic medicine. There has been little serious attention paid to prophetic medicine, although this is beginning to change. Medical traditions associated with *Shi'as*, (*Ṭibb al-a'immā*, or medicine of the Imams) have yet, to my knowledge, to be analyzed historically, though Irmeli Perho does outline some of the works of this tradition (Perho, 1995: 62ff.), and one text has been translated into English (see Ibna Sabur, 1991). Neither am I aware of historical studies of medical practices associated with Sufi hospices (*khanqah*s) or shrines (*dargah*s). The reasons for this neglect by Western scholarship, I shall suggest, partly lie in the way that European and North American scholars have perceived Islamic culture.

One scholar whose ideas on early Islamic medical history have been influential on later generations is Edward G. Browne. For him "Arabian" medicine is at its core Galenic, and its worth is conceived in terms of its relation to the history of ideas in European classical and Renaissance scholarship. He quotes from E.T. Withington's *Medical History from the Earliest Times*:

> It was this people [i.e. Arabs/Muslims] who now took from the hands of unworthy successors of Galen and Hippocrates the flickering torch of Greek medicine. They failed to restore its ancient splendour, but they at least prevented its extinction, and they handed it back after five centuries burning more brightly than before (Browne, 2001: 67–68).

Browne did not agree with the overstated chronology, but he did concur with the sentiment that the importance of Arabian medicine lay in its role as a repository of Greek knowledge. Embedded in this argument is the question of originality; the basis of the viewpoint is that the Arabs did not creatively develop the Greek medical learning that they inherited. The validity of this statement is frequently debated in the historiography and is clothed in other terms, such as "value" or "contribution", and we will see examples of this below. The golden age of Islamic medicine in this view is the era of the Greek-Arabic translations, the works of the great physician Muḥammad ibn Zakariyya al-Rāzī (d. 935), and the famous synthesizers of Greek medicine al-Majūsī (d. 994) and Ibn Sīnā (d. 1037), to name the most prominent in the tradition. In Browne's account al-Rāzī is singled out as "probably the greatest and most original of all the Muslim physicians" (Browne, 2001: 44). Browne's main criterion in evaluating al-Rāzī is the approximation of his works to standards in the medical profession in Browne's day. Browne quotes the praise of the medical historian Max Neuburger on al-Rāzī's famous treatise on smallpox (translated by Greenhill, 1848) as the earliest monograph on the subject in the history of epidemiology (Browne, 2001, 47). Al-Rāzī's huge collection of case notes, comprising the work *al-Ḥāwī*, compiled by his students after his death, reveals him as "a clinical observer", and in that "he excels all his compeers" (Browne, 2001: 50). The tendency in the historiography of Islamic medicine in the 19th and at least the first half of the 20th century, of which Browne is typical, was prescriptive. It engaged with the "great" personalities of medicine in Islamic countries, where greatness was largely gauged by the extent to which their ideas, laid down in texts, conformed to an approach to health and disease based on natural causation.

A golden age inevitably needs a decline. European historians, particularly orientalists of the 19th century, have attributed the cause of the decline of scientific activity in Arabic to an orthodox Islamic reaction. A more succinct and unequivocal expression of the perceived fundamental opposition of Islam to rationality, science, or indeed a just political order, perhaps could not be found than in the writings of the influential philologist Ernest Renan, particularly in a lecture he gave at the Sorbonne in 1883, published as *L'islamisme et la science*. He sought to address what appeared anomalous to him – that scientific activity had flourished at one time in the Arab-Islamic world. His argument was that Islam "in reality, has always persecuted science and philosophy. It ended by wiping them out" (Renan, 1883: 16). In his periodization of the history of Islam, the golden age of learning, which lasted until the 12th century, happened *in spite of* Islam. Islam supported philosophy "because it could not prevent it; it could not prevent it because it lacked cohesion, was not much used to terrorize" (*peu outillé pour la terreur*) (Renan, 1883: 18). By contrast, after the 12th century, when the "Tartar races" adopted Islam, the story is one of rapacious and relentless persecution of the advocates of the ancient sciences.

The golden age/decline paradigm, as formulated in the works of Renan, the German Orientalist Ignaz Goldziher and others, has been quite persistent. Browne broadly espoused their interpretations of the rise of orthodoxy as the initial cause of the decline, which was accelerated by the Mongols. J.C. Bürgel presents similar views in an article in 1976. For Bürgel, however, orthodoxy and the Mongol invasions were not enough to explain the process, for, as he points out, some Mongol rulers did in fact patronize learning to some extent. He saw the supposedly increasing influence of Islamic mysticism as an additional factor, since it apparently had turned people away from worldly concerns and therefore the products of scientific activity (Bürgel, 1976: 54). Bürgel argues that the absence of a concept of science, as it is currently understood, explains in part why Islamic medicine did not go beyond humoral pathology. Felix Klein-Franke's work of 1982 on medical traditions in Islam adopts a slightly more nuanced approach, but is still working within a similar set of assumptions. He writes that medicine in Islam lacked a Paracelsus or a William Harvey, who took medicine in the West in a new direction, and that signs of its stagnation were already apparent in the 11th century (Klein-Franke, 1982: 2).

What these accounts share, especially the earlier ones, is a vision of Islam unconnected with social context or contingency and divorced from individual agency. They betray a misconception of how and under what circumstances the Greco-Islamic medical tradition evolved and its relation to the social, intellectual, economic and political context of society under the Abbasids, whose rule, for them, roughly coincides with the flourishing of Islamic culture.

J.C. Bürgel was one of the first to link the "decline and fall" of Greco-Islamic medicine to the emergence of prophetic medicine (Bürgel, 1967, 1976).[2] He argued that the orthodox reaction to the promotion of Greco-Islamic medicine in Abbasid times manifested itself in the emergence of a medical tradition whose authority derived from the Prophet. This is *ṭibb ul-nabī*, or *al-ṭibb al-nabawī*, (prophetic medicine), a literary tradition derived initially from sayings ascribed to the Prophet Muḥammad and his companions that were collected to form *hadīth* literature in the 8th and 9th centuries. There is little of directly medical content in the *Qur'ān*, except the use of honey (Arberry, 1989: Surah 16, *The Bee*: 265–266), the importance of cleanliness before prayer (Surahs 4, *Women*: 79; and 5, *The Table*: 100–101), and that no blame should be attached to the blind, lame or the sick (Surahs 24, *Light*: 360–361; 48, *Victory*: 533). However, the *hadīth* literature that reported the Prophet's and his companions' sayings on medicine compensated for this dearth of medical advice in the *Qur'ān* and in time became a specialized field of scholarship.

The *Ṣāḥīḥ* of al-Bukhārī (d. 870) is one of the most authoritative collections of *hadīth*, and in current discussions of this genre in the medical context reference is most frequently made to his *Kitāb al-ṭibb* (*Book of Medicine*). Al-Bukhārī's organizing principle was reliable authority in the transmission of the Prophet's sayings and not coherence of the contents of these traditions. In many of the *hadīths* the Prophet either sanctions or condones what must have been customary medical practices of the time. For example,

> [The Prophet] said: 'There is curative power in three things: a dose of honey, a stroke for a cupping glass, and a cautery iron from a fire; but cautery is forbidden to my people' (al-Bukhārī, 1908: vol. 4, 50).

The collections of *hadīth* by al-Bukhārī, Ibn Māja and others were the basis for the subsequent development of the genre of prophetic medicine from the 9th to the 16th century (Perho, 1995: 53ff.). The early works did not comment on the Prophet's sayings, and according to Perho's analysis for the most part did not contain accounts of medical theory, although this was a characteristic of some 13th and 14th century and later works, which we shall encounter below. The sanction of *'ruqā'* (incantation) in Bukhārī's *Ṣāḥīḥ* as a healing practice or to ward off the evil eye is for Bürgel an example of "magical healing" that later developed into "gross forms of religious sorcery", which involved amulets and other devices (Bürgel, 1976: 58–59). Prophetic medicine, along with astrology, alchemy and magic, are for Bürgel "the hot-houses of irrationalism", the perceived enemy to the development of rational thought (Bürgel, 1976: 54). Without any concern for context, Bürgel quotes from a 14th century text of al-Surramarrī (d. 1374), "one of its [prophetic medicine's] most fanatic defenders": 'The relation of prophetic medicine to the Galenic system[3] is correspondent to the relation of the latter to quackery' (Bürgel, 1976: 59).

Al-Surramarrī's views represented for Bürgel the "Islamic dethronement of Galen ... in favour of Bedouin quackery and superstition sanctified by religion" (Bürgel, 1976: 60).[4] This is clearly prescriptive history writing that reveals more about the author's own position relative to the data than either illuminating the complexity of the relationship between prophetic medicine and folk Arab medical practices or explaining why an age characterized by the recreation of Greek-derived medical knowledge in Arabic came to an end. That al-Surramarrī's teacher, Ibn Qayyim al-Jawziyya, had also used this phrasing was not commented upon by Bürgel, although it may be a matter of literary style and not necessarily a substantive comment. Furthermore, Bürgel's translation of the Arabic is misleading and at once revealing. He renders *ṭibbi l-'awāmmi*, which is literally "medicine of the people" or folk medicine, as "quackery", which is to introduce a value judgment that is not present in the original.[5]

One can perhaps assume that prophetic medical texts were written as a reaction among some scholarly Muslim circles to the status that Greco-Islamic medicine enjoyed, at least for some sections of the population, but this needs to be tested in each case. According to Felix Klein-Franke, the notion of there being any actual contest is not realistic. The name *al-ṭibb al-nabawī* is misleading, he writes,

> ... since it claims to be a medical system that is in competition with, and can replace, Greek-Hellenistic medicine. This then was the tendency of those orthodox circles in which the medical sayings of the Prophet were compiled (Klein-Franke, 1982: 8).

As Klein-Franke sees it, these claims are baseless, since prophetic medical texts are nothing other than compilations of incoherent (*unzusammenhängender*) statements of the Prophet, his companions and later theological authorities (*ibid.*). Klein-Franke's view still implies that contests were theoretically possible, although the character of the literature did not permit such contest. By contrast, Lawrence Conrad has pointed out that we may not be dealing with competition at all, since prophetic medicine "simply reflects an ongoing interest in medical issues among a religiously educated public familiar with medical chapters and hadith traditions" (Conrad, 1995: 125). He points out that elements in medical lore that went against Islamic tenets were reduced, but there was no fundamental opposition to the categories of Greco-Islamic medicine.

Fazlur Rahman gives a more thorough treatment of prophetic medicine. He is of the opinion that the notion of competition is the weakest argument for the origin of the genre, since the texts of prophetic medicine frequently cite the

Greeks as authorities as being learned in medicine (Rahman, 1989: 42–43). Hence there is no evidence of contest between the two traditions. This is apparent even in al-Surramarrī's text, which Bürgel used in his argument above. Rahman principally refers to the works of Muḥammad al-Dhahabī (d. 1348) and Ibn Qayyim al-Jawziyya (d. 1350) of Damascus, and Ibn al-Jawzī (d. 1200) of Baghdad. He believes that there were three main motivations for the rise of prophetic medicine. First, the authors aimed to spiritualize medicine, that is to take into account the power of faith and prayer in healing, elements that are not to be found in Greco-Islamic medicine. Second, the authors wished to make prescriptions more easily accessible to an educated public. Third, they intended to bring medicine more into the mainstream of Islamic faith (Rahman, 1989: 43–49). But it is not clear why these motivations should have been important at the times the texts were written. Regarding the availability of medical treatment, he quotes from the introduction of *Medical Benefits Made Accessible*, by the early 15th century writer Ibrāhīm ibn 'Abd al-Rahmān al-Azraq:

> When I saw that [today] there are very few people who concern themselves with medicine but those who seek help from it are many ... it appeared to deserve special devotion since no human being can avoid it ... (Rahman, 1989: 45).

But this statement cannot be taken at face value. Variations on this kind of justification were a common motif in Islamic medical literature, and may not have had any bearing on the situation prevailing at the time.[6] The main problem with Rahman's account is that the social and political spheres in which these authors were acting do not feature.

Given the number of the texts based on statements attributed to the Prophet, written in different times and places, it seems that a more differentiated approach to assessing the medical literature is necessary, an approach to the nature of this form of literature that is historical in that it seeks to situate the texts and their authors in specific time, space and context. Irmeli Perho has recently (1995) done a study in this direction. She analyzes the output of three scholars of religious sciences who were contemporaries in 13th and 14th century Damascus and wrote authoritative texts on prophetic medicine: Ibn Qayyim al-Jawziyya, Muḥammad al-Dhahabī and Ibn Mufliḥ. The texts by these theologians, Perho shows, give a fuller articulation of medical theory than other texts in the prophetic medical tradition had done before. The authors of these texts did not expunge the Greek authorities, Hippocrates, Galen and Plato, from their texts, as one might have expected had they wished to provide an alternative to Greek medicine. The Greeks remain authorities, and they are appropriated as such, even if they were considered pagans. Al-Dhahabī wrote, for example,

> This is what Hippocrates said and he [was] not one of the Faithful. I have already said that this Hippocrates was the Head of the Art of Medicine and a Leader. He was a Greek physician and their Father. He is held to be perfection in the Art of Medicine (Elgood, 1962a: 66).[7]

Perho, in addressing the question why prophetic medical literature developed, argues that traditionalist circles perceived the Mongol incursions as God's punishment for straying from Muḥammad's teachings and example. It was the task of these circles to influence policy and also to encourage people to adhere more closely to the *sharīʿa*, Islamic law (Perho, 1995: 144). In this context Perho identifies aspects of medical education and practice that would explain the interest of traditionalist scholars in medicine: many practitioners were Christians or Jews; medical studies involved an acquaintance with Greek philosophy; and Greek medical theory had an ethical element, in that the physician was supposed to give the patient advice on lifestyle and conduct (ibid.). All three authors whose works she studies were connected with Ibn Taimiyya (d. 728/1328), a Hanbali[8] theologian and scholar. He was politically active and played a role in negotiations with the Mongols.

Perho's analysis may, however, underestimate the significance of the Crusades – the other major threat to Mamluk society – in heightening the religious tensions of the time.[9] These tensions may also have prompted these authors to incorporate Greco-Islamic medicine and present medical learning derived from the Prophet in a more systematic fashion than previous writers had done.

Medical texts could also have been produced in response to professional demands in this climate. Perho implies that this could have been the case, since medicine could be taught at religious establishments (although it was not the norm) in early 14th century Damascus, and these accounts of prophetic medicine could have been used for this purpose (Perho, 1995: 48–49).

We have seen then that recent works have challenged the notion, proposed by earlier orientalists, of a contest between prophetic and Greco-Islamic medicine. Indeed, I would argue that the idea of contest in this context has its roots in earlier prescriptive history writing, which constructed the history of Greco-Islamic medicine into periods of golden age and decline and attributed this decline of ancient sciences to an orthodox reaction. In order to understand more accurately why texts of a certain medical tradition are written, and why they take a particular form, social, economic, political and cultural dynamics in the society that produced them have to be taken into account. While studies on the prophetic medical tradition or folk medical practices in Islamic contexts are scarce,[10] there is a great body of literature on Greco-Islamic medicine. In the following sections we will see how the translation movement, the functioning of early Islamic hospitals and the demands of the medical profession have been viewed by some historians as the motors for the emergence and development of Greco-Islamic medicine.

THE ESTABLISHMENT OF GRECO-ISLAMIC MEDICINE

The translation movement

Within the focus in the historical studies of Greco-Islamic medicine during the Abbasid caliphate, most work consisted of philological analysis and what may be termed intellectual history. Painstaking work has been done in producing bibliographies of Arabic medical and philosophical literature and in bringing out editions and translations of Arabic medical texts.[11] Central to these concerns is the translation movement, which saw the translation of mainly philosophical, astronomical, mathematical and medical texts from Greek, often via

Syriac, into Arabic on a vast scale over the duration from the late 8th to the end of the 9th century. It is within this period that we see the creation of Greco-Islamic medicine and the canonization of its texts. But it is only recently that historians of Islamic sciences have begun to ask questions of a more social-historical nature, which could elucidate the motivations for the translation movement and the circumstances under which it took place. The results of some of these treatments, Dimitri Gutas' *Greek Thought, Arabic Culture* in particular, reveal much of how Greco-Islamic medicine became established and its place among the elites of Abbasid society.

It is quite clear that a process that sees the translation of almost the entire corpus of Greek scientific and philosophical works, sustained over such a long time, was due to more than just the commitment of interested scholars. This is, however, the impression one gets from some accounts. According to Schacht and Meyerhof, the process of translation was initiated and carried out by Syrian Christians, and the "result was that of a fertilizing rain: hundreds of Muslim scholars eagerly began to study and to appropriate Greek learning ..." (1937: 7–8). Felix Klein-Franke has addressed the question of the receptivity to Greek sciences in Abbasid society to some extent. He suggests that the translation movement was only the outer manifestation of an intellectual (*geistigen*) process that began in the 8th century and which was based on two factors: an increasing openness to Hellenist culture and an increasingly critical attitude, noticeably in theology, literature and the sciences (Klein-Franke, 1982: 68). The first part of his argument is that through conquest Arabs had made territorial gains which landed them in possession of a large swathe of the ancient world, and they studied the intellectual and material culture of the peoples they had conquered. He cites the famous Arab philosopher, al-Kindī (d. after 870), who praises Aristotle and the nobility of the search for truth. The upturn of medicine in the 9th century, for Klein-Franke, is mainly attributable to the then current conviction of the superiority of Greek intellectual culture (1982: 69). There are numerous problems with these statements. First, they are not related to any particular social, political or historical context. It is not clear from his account why Arab peoples should have wished to engage with the cultures of the people over whom they had begun to rule. Nor is it self-evident that Greek thinkers should have been accepted as authorities. He goes some way to answering this last point in the second part of his argument: the early Abbasid caliphs sought to base Islamic belief on firm grounds by adopting a rationalist doctrine, the *mu'tazila* (1982, 69). The doctrine of *mu'tazila* was based on the unity of God and the createdness of the *Qur'ān*. Thus knowledge of God could be approached through reason. The Caliph al-Ma'mūn espoused this as official ideology, and instituted a *mihna* or inquisition against those who publicly asserted the revelation of the *Qur'ān* (Gutas, 1998: 77). The point being made here is that this rationalist programme, based on Greek sources, which was promoted by the highest authority in the Abbasid empire (the caliph) engendered an openness to non-Islamic and non-Arab learning. Klein-Franke, however, does not expand upon this point, to indicate why, for example, at this particular time such a doctrine would have served caliphal interests.

Lawrence Conrad takes this issue further and gives the discussion of the origins of the translation movement a significant social dimension (Conrad, 1993: 690ff.). He identifies three factors that stimulated the movement. First, in the early Abbasid period social and economic conditions were becoming increasingly favourable for scholarly enterprise, witnessed by the growth of a wealthy cosmopolitan urban elite in Baghdad. Second, he perceives a need among Muslims and Christians for access to the medical heritage of antiquity. Third, this material was relatively accessible, in the monastic libraries of Eastern Christianity, for example. Furthermore, he highlights the domination of religious controversy in elite circles at this time. By using Greek sources Muslims could defend their faith against Christian polemicists using *their* tools for disputation.

Even more of a threat was potentially posed by dualist Manichean doctrines, which were apparently current among intellectuals of the time. These held that there was no all-powerful divinity, and no *creatio ex nihilo*, and were therefore at fundamental variance with and a challenge to Islamic tenets. In this context Conrad points out that rational approaches in medical works could be applied to religious debate. They provided useful models for rational inference from empirical data, contained few statements that were offensive to monotheist sensibilities, and even could be used to argue for the existence of a creator, given the teleological element that often comes through in Galen's works – that Nature, for example, provided the parts of the body of man to fulfill specific functions.[12] Moreover, this Greek-based rationalist approach to religious authority was applicable to power struggles within Islam itself (Gutas, 1998: 99). The Caliph al-Ma'mūn's (r. 813–833) espousal of *muʿtazila* established a rational foundation for religious discussion that sought to disempower those traditionalists who asserted the literal authority of the *Qur'ān* and *ḥadith*, and in addition, as Klein-Franke suggested above, created a receptivity to ancient learning.

There remain, however, more profound social, political and ideological dimensions to account for this receptivity, which Dimitri Gutas has recently brought to attention. According to his analysis, although the religious motivations for the movement were present in the early Abbasid caliphate, the initial impetus for the translation of Persian Sanskrit and Greek texts came with al-Manṣūr (r. 754–785) and not with the caliphs Hārūn al-Rashīd (r. 786–809) and al-Ma'mūn, as is often portrayed in the secondary literature. Al-Manṣūr shifted the seat of caliphal power from Damascus and founded Baghdad in 762. In consonance with this eastward orientation, he sought to legitimize his authority among the predominantly Persian population of the region by adopting the Zoroastrian imperial ideology of former Sassanian rulers (226–642) and presented the Abbasid dynasty as their successor. This ideology embraced the concept of the recovery of knowledge, which had originally been part of the Avesta, the ancient scriptures of Zoroastrianism, but which, with Alexander of Macedon's conquests, had been dispersed and had to be "retranslated" into

Persian from Greek and Sanskrit. Al-Manṣūr thus inherited a tradition of translation and promoted it as a form of political ideology (Gutas, 1998: 48). In order to bolster the Abbasid claim to authority among the Zoroastrian Persians, the early caliphs appointed many of the Sassanian elite to prominent posts in the administration, including as court physicians. The Barmak family held high political positions between 750 and 803; the Nawbakhts were also prominent in court life as were the Bukhtīshū‘ as physicians. It is reported that when al-Manṣūr fell ill in 765 he summoned a physician from a prominent family of Nestorian Christian physicians, the Bukhtīshū‘, who practiced Galenic and Hippocratic medicine (Klein-Franke, 1982: 40). This family had an established practice in Jundishapur, a town not far from Baghdad, which had also been adopted as the royal residence of its Sassanian founder Shāpūr I (r. 241–271 CE). In the light of Gutas' study of al-Manṣūr's political maneuvres, it is tantalizing to see the political ramifications of Jurjis ibn Jibrīl ibn Bukhtīshū‘'s appointment as personal physician to the caliph: to underscore the caliph's claim to legitimate succession to Sassanian rule. The Bukhtīshū‘ line continued to be court physicians for successive caliphs in the 9th century.

Of all the caliphs associated in the historiography with the translation movement and the promotion of Greek sciences, al-Ma'mūn is the one whose endeavour has been emphasized by the historians. In almost all accounts of the translation movement and the history of Islamic medicine in this formative period, it is al-Ma'mūn who is credited with having made the translation of the Greek sciences systematic and institutionalized in the form of the *bayt al-ḥikma*. This institution has been variously referred to as an academy, translation centre and library that is conventionally thought to have been founded by him (Sourdel, 1962: 1141).[13] The sources for Dominique Sourdel's piece in the *Encyclopedia of Islam* on the *bayt al-ḥikma* are mainly the well-known 13th century historians and bibliographers, Ibn al-Qifṭī (d. 1227) and Ibn al-Nadīm (d. 990), and are therefore far from contemporary commentators. Sourdel also mentions references to the *bayt al-ḥikma* in the time of Hārūn al-Rashīd, predating al-Ma'mūn, and thus concludes that al-Ma'mūn "may have only given new impetus", which was nevertheless influential on Islamic culture. Sourdel informs us that the principal activity of the *bayt al-ḥikma* was the translation of philosophical and scientific works from Greek into Arabic. The most celebrated translator of Greek learning into Arabic is Ḥunain ibn Isḥāq (d. 873 or 877). He translated many of Galen's works, made his own compositions on medicine and ophthalmology, and has also been associated with the *bayt al-ḥikma* (Leclerc, 1876, vol. 1: 154; Meyerhof, 1926: 7; Sourdel, 1962: 1141; Conrad, 1993: 693).

Dimitri Gutas has recently cast doubt on some of these assumptions and the interpretations that have been made about the nature and function of this institution from available evidence. To start with, no date can be established for its foundation, so although the earliest reference to it is in the time of Hārūn al-Rashīd, it may have existed with the caliphs al-Manṣūr or al-Mahdī (r. 775–785).[14] Regarding its function, there are references in the sources to translation activity, but these are about work from *Persian* to Arabic, and there is nothing to suggest that there was any translation at the *bayt al-ḥikma* from Greek into Arabic (Gutas, 1998: 58–59). Similarly in the many accounts of the Greek-Arabic translation movement there is no reference to the *bayt al-ḥikma*. Neither does Ḥunain ibn Iṣḥāq mention it in his lengthy discussion of his translation work.[15] The name of the institution is in fact the Sassanian term for a library of poetic works of Persian historical lore written for the king (Gutas, 1998: 54). Although Gutas does point to some innovation in the time of al-Ma'mūn, (the caliph appointed astronomers and mathematicians to do research (not translation)), he argues that it was initially a library under al-Manṣūr, as part of the Abbasid administration with Sassanian precedent. Translations were made there in the time of Hārūn al-Rashīd, but these were on Sassanian history and culture. Sourdel writes that members of the Bānū l-Munajjim family were "translators" there (Sourdel, 1962: 1141). They were a Persian Zoroastrian family of the educated elite, associated with Sassanian royalty (Gutas, 1998: 128). This would provide further evidence to support Gutas' argument since, as he himself points out, they were introduced to the Abbasid court, initially as astrologers, by al-Manṣūr. Independently of the functioning of the *bayt al-ḥikma*, they were active in the translation movement. For instance, one of their line commissioned Ḥunain's *Risāla*, a bibliography of Galen's translated works (*ibid.*). Perhaps this can explain some of the confusion over the function of the *bayt al-ḥikma*. The main reason, as Gutas suggests, may be that by the time Ibn al-Nadīm was writing his biographical history the institution had assumed legendary qualities, which have continued to impress subsequent commentators.

What Gutas' study interestingly reveals is that the celebration of al-Ma'mūn in the historiography as the promoter par excellence of the Greek sciences in Islam is probably due to the success of his own propaganda in this regard. We have seen above that al-Ma'mūn took a firm grip to centralize authority on Islamic doctrine, but his promotion of Greek sciences also needs to be seen as an ideological tool in the context of his foreign policy (Gutas, 1998: 85). The greatest external threat to the caliphate during his rule was the Byzantines. He justified war against them on the grounds that they were infidels and also that they were inferior in learning to their own Greek descendants. They were therefore also inferior to the Muslims, who valued Greek sciences and studied them in the spirit of Islam, while the Byzantines had renounced this learning because of Christianity. To illustrate the influence of al-Ma'mūn's "philehellenism", Gutas quotes from a text by 'Alī ibn Riḍwān, who was well-known as the chief physician in Cairo during the rule of al-Mustanṣir (r. 036–1094) (and who will reappear below):

> Al-Ma'mūn revived [the teaching of medicine] by favoring the most excellent of men. But for that, all the sciences of the ancients, including medicine, logic, and philosophy, would have been forgotten, <just as they are forgotten> today in the lands in which they were most specifically cultivated, I mean Rome, Athens, the Byzantine provinces, and in many other lands (Gutas, 1998: 93).[16]

Under al-Manṣūr the caliphate was the heir to the Sassanians; under al-Ma'mūn Muslims were the heirs to Greek-Hellenic culture.

It should be pointed out that while the caliphs had given the initial impetus to the absorption and recreation of Greek sciences in Arabic, Baghdad's intellectual and financial elite, which cut across ethnic and religious lines, were also involved in some way in the commissioning, production and consumption of Greek-inspired works. Apart from the caliphs and their families, Gutas identifies three other groups of patrons: the courtiers, officials of the administration, and other scholars and scientists. One can get a sense of this range from Ḥunain's *Risāla* (Meyerhof, 1984, I: 713–720). For example, he mentions that he and others had done ten Arabic versions of Galen's works for 'Alī ibn Yahyā, a secretary to the caliph al-Mutawakkil and a military commander. According to Ḥunain's testimony, one of the most prominent patrons of Galenic works was the famous mathematician Abū Ja'far Muḥammad ibn Mūsā. Ḥunain writes that he and his nephew Ḥubaish had translated fifty Galenic works into Arabic for him. Max Meyerhof states that the family had given over a great amount of their fortune to collecting Greek manuscripts and having them translated (Meyerhof, 1926: 714). Ḥunain further mentions thirteen translations of Galen into Syriac that he had done for his former teacher and physician in the service of Hārūn al-Rashīd, Yūhannā ibn Māsawaih (777–857), native of Jundishapur.

In the politically and ideologically charged atmosphere of 9th century Baghdad, in which Greek learning had been imbued with such authority, being a patron was clearly a sign of status and prestige. However, there were also professional demands for translations that were being generated by the large scale involvement (within elite circles) with translations and the production of knowledge, as is illustrated by the patronage of Yūhannā ibn Māsawaih, although he himself was known to have translated Greek medical texts into Syriac.

Nonetheless, during the 9th century, translations began to give way to the composition of new and original works in Greco-Islamic medicine, although they were firmly rooted in the Galenic tradition. In this time the canon of Greco-Islamic medical texts was formed, and the profession of the *ṭabīb* (from then on a practitioner of Greco-Islamic medicine) took shape, aspects of which we will see below. Gutas' work convincingly reveals the nexus of political and cultural authority in Baghdad, which served as a model from Andalus to Central Asia (Gutas, 1998: 154). As we have seen, Greco-Islamic medicine's status was embedded in the fabric of elite Baghdadi society, and perhaps the case could be made that as Baghdad became the model for political and cultural spheres, so did Greco-Islamic medicine become the model medicine, because it also was disseminated in cities across North Africa and to Andalus.

In order to see how Greco-Islamic medicine became established in more concrete terms, we will turn now to the hospital, and how its role in early Islamic society has been variously interpreted.

Hospitals

One of the most obvious features that distinguishes Greco-Islamic medicine from medical practice in ancient Greece is the institution of the hospital, and it is frequently referred to in the historiography. One should be wary from the outset of superimposing our own conceptions of the hospital and its functions upon its operation in such different contexts. The word that denoted hospital in the first centuries of Islam is the Persian loanword *bīmāristān*. It literally means "place of the sick". This should inform us that the primary function of the hospital was to house and provide for the sick, not necessarily to cure.

These hospitals were, apparently, "well-equipped", and the "secular management" was selected according to "qualifications" (Hamarneh, 1982: 199). But in some senses Hamarneh is right to recognize the difference between the hospitals that arose in the first centuries of Islam and those of Eastern Christianity, which seem to have provided the inspiration. Deed records survive for some hospitals dating as far back as the 13th century that have allowed a reconstruction of their structures and functions, such as for the al-Manṣūrī hospital in Cairo (est. 1284). (See Dols, 1992; Conrad, 1995: 135–137). Judging by these accounts, they were large and complex institutions, with provisions for men and women and for different forms of medical treatment such as ophthalmology or bone setting. Neither were they static institutions. Dols has shown that many increasingly catered for patients considered insane (Dols, 1987: 388). Much debate in the historiography of the hospital in Islam has centered on its origins. It is in this context that we will see why they were patronized and how this patronage tied in with the establishment of Galenic medicine.

Prior to the studies by Michael Dols and Lawrence Conrad,[17] it was conventionally thought that the model of the Islamic hospital lay in the reputed medical center Jundishapur, which had been a royal residence for Sassanian kings.[18] The Bukhtīshū' family came from this place, not far from Baghdad. Another Eastern Christian physician family that was famous in early Abbasid times also came from Jundishapur, the Māsawaih. Scholarship had suggested that it housed an academy where astronomy, theology and medicine were taught. In addition there was supposed to have been a hospital, at which "treatment was based solely on scientific medicine". Thus, Jundishapur began to influence Islamic medicine in a systematic way with the establishment of the "first" hospital in Islam, attributed to Hārūn al-Rashīd, in Baghdad.

This prominent role ascribed to Jundishapur has been seriously called into question. A major methodological problem seems to be the lack of contemporary Persian sources that would support claims for its influence. Indeed, Conrad and Dols have highlighted the problems of using 13th century sources (Ibn abī Uṣaibī'ah (d. 1270) and al-Qifṭī are two of the most widely used) for reliable information on the status of Jundishapur (Dols, 1987: 367 n. 1). It appears that by the time of these commentators, Jundishapur had acquired a legendary status, and there is nothing to support the notion that a medical school had existed there. Nevertheless, there was doubtless a hospital in Jundishapur, as witnesses more contemporary than the Arab historians mentioned above testify. Dols (1987: 377) refers to 'Alī ibn Rabbān al-Ṭabarī, who mentions a hospital in Jundī-shāpūr in his famous work on Greco-Islamic and Indian medicine, *Firdaus al-ḥikma*, (written 850). Dols' contention is that nothing special can be claimed for this particular hospital, which needs to be seen in a tradition of Syrian Christian care for the sick in a Sassanian Persian context. It is perhaps the Sassanian element that is most relevant for our discussion here. Dols shows that the first reference to a *bīmāristān* in an Islamic context was directly connected not with Hārūn al-Rashīd, but with his wazir Yahyā ibn Khālid ibn Barmak, whose family we referred to above as prominent carriers of Zoroastrian Sassanian culture. Again we are in the realm of power politics and its role in the promotion of a certain form of medical treatment over others. The biographer Ibn al-Nadīm (d. 990) mentions him, and these passages are worth quoting.

> Translators of India and the Nabataeans. ... He [Mānkāh or Kānkāh al-Hindī] translated from the Indian language into Arabic. Ibn Dahn, al-Hindī, who administered the Bīmāristān (Hospital) of the Barmak family. He translated from the Indian language into Arabic ...
>
> ... Yahyā ibn Khālid ordered Mānkāh [Kānkāh] the Indian to translate it [the *Book of Sarsad*, i.e *Suśruta-samhitā*] at the hospital and to render it in the form of a compilation (Dols, 1987: 383) (Ibn al-Nadīm, vol. 2: 589–590, 710).

Here it is Indian medicine that is being promoted; the *Suśruta-samhitā* is one of the core texts of Āyurveda, an Indian medical system. This makes sense if we bear in mind Gutas' thesis that the early Abbasid caliphs had adopted the Zoroastrian Sassanian concept of the recovery of knowledge, which had been dispersed in the languages of different cultures of the region. The hospital could have been established between 750 and 803, when the Barmaks had power and influence. Conceivably it could have been founded under the aegis of al-Manṣūr, even if it came to be attributed to Hārūn al-Rashīd. Hārūn was known for his patronage of the ancient sciences, a factor that may have attracted this attribution.[19]

Significantly, in the sections quoted above, Ibn al-Nadīm made the explicit connection between translation and the hospital. Clearly one cannot infer from this that the connection had been, or became, customary, but it nevertheless suggests that the hospital did figure as a locus for the assembly of physicians and scholars, interaction with patrons and the production of textual knowledge, in addition to its role in providing for the sick.

To make a tentative case for the nexus of translation activity and the hospital, we could cite the examples of the famous Syrian Christian physicians, like the Bukhtīshū' and Yūhannā ibn Māsawaih, who occupied important posts at hospitals in Baghdad (Ullmann, 1970: 108–115). As we have seen, they were also involved as translators and patrons of translations. Similarly, at the end of the 10th century, physician-scholars who translated from Greek were affiliated with the 'Adudi hospital, founded in Baghdad in 982 (Ullmann, 1972: 73–74). What Dols' work does make clear is that the founding of hospitals, especially the early ones, was often granted by royal patronage, as an act of charity. Charity to the Muslim community accorded well with tenets in the *Qur'ān*. With the Barmakid precedent, founding hospitals became one of the duties of the ruler towards the people. This much is suggested by the letter that Dols quotes written by a governor of Khurasan, Ṭāhir ibn al-Ḥusain, to his son and governor to be. He wrote,

For sick Muslims you should set up hospitals [*dūrān*] to shelter them, people to attend them and doctors to treat their illnesses' (Dols, 1987: 382).[20]

Although some hospitals had been founded in the 9th century, especially in Baghdad, many more were established in the 10th, in Baghdad and around the Abbasid Empire (Dols, 1987: 388). Despite the increasing number of hospitals throughout the Islamic world, in the Ottoman empire[21] and in the Delhi Sultanate and Mughal empire in India,[22] they should not be viewed as being the primary source of medical relief, since their number and scale relative to the size of populations was still small. Hospitals in Islamic contexts were often thus symbols of the prestige of the ruler. They attracted influential physicians, like the Bukhtīshūʿ and al-Rāzī in Baghdad, and most importantly for our discussion of the Greek legacy in Islamic medicine, they gave official sanction to the form of medicine that these physicians practiced, the medicine that was based on the teachings of Galen and Hippocrates.

The medical profession

In the section on the translation movement above we mentioned that the professional needs of physicians stimulated the translation and production of texts, once the ruling elite had initiated the process. Texts were the cornerstone of an education in Greco-Islamic medicine (Leiser, 1983). The famous translator Ḥunain ibn Iṣḥāq compared what he knew of medical education in the school of Alexandria in late antiquity and in his time.

> [The members of the medical school in Alexandria] would gather every day to read and study one leading text among those [books by Galen], just as our contemporary Christian colleagues gather every day in place of teaching known as *skhole* for [the study of] a leading text by the ancients. As for the rest of the books, they used to read them individually – each one on his own, after having first practiced with those books which I mentioned – just as our colleagues today read the commentaries of the books by the ancients (Gutas, 1998: 15, quoting Bergsträsser, 1913: 18–19).

The significance of the text, whether it was Galen's, al-Majūsī's, or Ibn Sīnā's, whose works were often the basis of a medical education, becomes more apparent in the absence of formal means for examining the competence of would-be practitioners.

There is one genre of literature in Arabic that deals with an institution to guard against malpractice in the market place, including medical. This is the *ḥisba*, and the duties of inspection fell upon the *muḥtasib*, who was appointed by the Caliph or Imam. Texts on *ḥisba* survive from the 12th to at least the 14th century, from Aleppo to Andalusia (Meyerhof, 1984, XI: 119–134). Reference is frequently made to one such text by Ibn al-Ukhuwwā of Cairo, which devotes many sections to how the *muḥtasib* should test the knowledge of various medical practitioners, phlebotomists, oculists, physicians, surgeons and others (Ibn al-Ukhuwwā, 1938). The text presumes that the *muḥtasib* should himself be fully conversant with numerous medical texts, something which could however hardly be considered realistic for someone who does not have to be educated in medicine. This text prescribes the ideal state of inspection, and therefore cannot be interpreted as an institution for the regulation of medical practice. However, mastery of certain texts was a form of qualification. The teacher would issue the student an *ijāza*, or permission, to teach a certain book. An example of this is an *ijāza* that was given to a physician of Baghdad, Ibn Butlān, by his master, Abū l-Farāj ibn al-Ṭayyib, reported by the biographer Ibn al-Qifṭī,

> I saw the copy of an autographical note of Abū l-Farāj for him [Ibn Butlān] (in a copy of) the book "Fruits of argument", a commentary of Abū l-Farāj, in the following terms: 'Has read with me this book from beginning to end the excellent Shaikh Abū l-Ḥasan al-Mukhtār ibn al-Ḥasan [Ibn Butlān] – may Allah prolong his authority ... Written by 'Abdallāh ibn al-Ṭayyib' (Ibn al-Qifṭī, 314. Reproduced in Schacht and Meyerhof, 1937: 52).

Nevertheless, this was not necessarily a useful measure of a physician's learning, since *ijāza*s could easily be falsified. If studying with a teacher was perhaps the norm in Ḥunain's time, it was still possible to be self-taught. The famous Ibn Sīnā (d. 1037), whose work, *al-Qānūn fī l-ṭibb* (*Canon of Medicine*), the Islamic medical tradition has accorded the highest status since the 12th century, was himself an autodidact (Leiser, 1983: 51).

An interesting debate on the means of medical education by near contemporaries of Ibn Sīnā in 11th-century Cairo occurs in the context of what is probably one of the most well known polemics in Islamic medical literature. This polemic disputation between a physician from Baghdad, Ibn Butlān (d. after 455/1063), and the chief physician of Cairo, Ibn Riḍwān (d. 460/1067–8), is important for several reasons. First, more significant than the frame topic of the debate, which was whether the young of the bird was of a warmer and moister nature than the young of the chicken according to Aristotle and Galen, is what it reveals about the nature of the medical profession, as we shall see below. Second, for our purposes in examining aspects of the writing of the history of Islamic medicine, it is illustrative of two radically different approaches to the text in Islamic medical literature, exemplified on the one hand by Schacht and Meyerhof and on the other by Lawrence Conrad.

The polemic involves written exchanges between the two physicians, which were brought to light by the German orientalists Joseph Schacht and Max Meyerhof. They entitled their English translation of five treatises of the debate "The medico-philosophical controversy between Ibn Butlān of Baghdad and Ibn Riḍwān of Cairo: a contribution to the history of Greek learning among the Arabs" (1937). This title already suggests that the translators' interest in the exchange was primarily in the status of Greek authorities in Islamic medicine and that they were framing the discussion in terms of an intellectual contest. For them the *cause* of the controversy was the question in Greek physiology about the chicken and the bird, even if, in their view, the meaning of this controversy was "insignificant" (*ibid.*: 15). This approach presumably explains why their edition of the five treatises was only translated in part, and mostly summarized. They left out parts of the original treatises that are now apparently untraceable (Conrad, 1995b: 85 n. 7). In their analysis of Ibn Riḍwān's works *in toto* they commented that, although these works show that the author was well versed in Greek medicine, they also reveal that,

> [Ibn Riḍwān] was not an original thinker, but a mere exponent of Hippocrates' and Galen's thought without adding anything in the least original or of his own, except for his list of remedies which were unknown to the Greeks. He is the true representative of that scholastic turn of mind which prevailed in medicine and philosophy dating from the Hellenistic period (Schacht and Meyerhof, 1937: 29).

Clearly Schacht and Meyerhof were looking at these texts from the perspective of an intellectual tradition that views science in a linear manner that emphasizes originality and contribution. They do provide useful biographical information on the two protagonists, passages translated from the well-known medieval biographers and historians. But they do not probe into the questions of what was really at stake, and why Ibn Riḍwān and Ibn Butlān engaged in the diatribes and abuse against each other that characterize their treatises.

Ibn Riḍwān was of a humble background – his father was a baker – but, through single-minded ambition and rigorous self-teaching, he rose to the position of chief physician in Cairo under the caliph al-Mustanṣir (r. 427–87/1036–94). His adversary in the dispute, Ibn Butlān, was a Nestorian Christian physician who had been taught by the prominent Abū l-Farāj ibn al-Ṭayyib of Baghdad. Ibn Butlān left his native city to seek a position in Cairo. Upon his arrival he wrote the first of the treatises setting out his argument in support of Aristotle's position on the issue of chick physiology, against the opinion of a Cairene physician known to Ibn Riḍwān. He sent it to Ibn Riḍwān, and with the latter's dismissive reply, the dispute begins.

The social historian of Islamic medicine Lawrence Conrad has recently suggested a fundamentally different approach (Conrad, 1995b). Key terms in his analysis are "affirmation" and "confirmation". These terms are to be understood in the context of the socio-intellectual climate of Islamic societies at this time, more specifically how individual scholars, in any field of learning, situated themselves, and in turn came to be viewed, in relation to a living tradition. In the case of medicine, they can help us to understand what it meant to be a physician in Islamic society at this time and the crucial role of the text in this profession. Affirmation can be taken to refer to the assertion of competence and authority that the physician made through the text or in public debate. Ibn Butlān was moving to a new professional territory in which he had to stake his claim. The choice of subject matter of his first treatise, and the fashion in which he argued it, is certainly important. It is the domain of engagement, and from this one can gauge that Ibn Butlān thought it would earn him, if not favour, then respect and a high position. He engaged in a discourse in which he could display his acumen, replete with Aristotelian logical argumentation, references to Galen and Hippocrates etc. He sent the treatise to the chief physician, as the central authority of the medical profession in Cairo. As the debate gets heated other arenas in their exchange open up, in which personal attributes, competence and background are questioned and slighted. One theme for contention was medical education. Ibn Butlān had studied medicine with a teacher, and a renowned one, a point whose value he was at pains to point out, since he knew that Ibn Riḍwān had taught himself (Schacht and Meyerhof, 1937: 83–86).

The social dimensions of Ibn Butlān's affirmation have to be taken into account if one is to understand the nature of the texts that the two physicians wrote to each other. From Ibn Riḍwān's point of view Ibn Butlān's assertion was a threat to the professional order in which he, as chief physician, had most to lose. This explains why the dispute arose and was so bitter and protracted. It explains why Ibn Riḍwān felt it necessary to address one missive to the physicians of Cairo, his "brethren", enumerating all of Ibn Butlān's errors. Subsequently Ibn Butlān left Cairo frustrated. As we have mentioned above, the context for the significance of affirmation of authority in the medical profession is the absence of any fixed curriculum, universally accepted procedure for examination or body to regularize the practice of medicine.

Confirmation is a process that operated in parallel to the act of affirmation in medieval Islamic culture. It was the process that

> ... determined the scope and content of knowledge as living tradition – the body of knowledge actually known to scholars, discussed among them, and used in their own works (Conrad, 1995b: 98).

The process of confirmation addresses the problem of the accessibility of works, in any field of learning, due to the vastly increasing corpus of literature produced and the difficulties of distribution, and also, presumably, cost. The writings of scholars, or debates among them, in a certain field, confirmed whether a work or an idea would become part of an evolving tradition of learning, in our case a medical culture. This helps one to understand why, as Conrad illustrates, Ibn Sīnā's *al-Qānūn fī l-ṭibb* was not an authoritative work in the author's own lifetime (1995b: 98). In order to acquire this status it had to be confirmed in the socio-intellectual milieu of scholarly activity, which subsequently did happen. Once a text had been confirmed, it would likely be the source for acts of affirmation by other scholars in the field, since placing oneself in the living tradition was also to legitimize oneself.

* * *

We have seen then numerous ways in which the writing of the history of Islamic medicine has, until recent years, been focused on the Greek legacy in Islamic medicine. This is evident in the concentration of studies in the Abbasid period, whose renown for cultural efflorescence is very much related to perceptions in European scholarship of a renaissance associated with the appropriation of Greek scientific and philosophical learning by an Arabic educated public. I have shown how this scholarly emphasis has borne on accounts of the decline of Greco-Islamic medicine in relation to prophetic medicine and the notion of contest. The form that prophetic medical texts took in 13th and 14th century Damascus may have had more to do with a perceived need to raise awareness of the primacy of Islam in a society under siege from the Crusaders and the Mongols than with a desire to challenge the legitimacy of Greco-Islamic medicine per se. Similarly, the initial motivations for the translation movement, and the consequent emergence of Greco-Islamic medicine, cannot be explained in terms of a purely intellectual appreciation of Greek learning, but rather with reference to the complex political, religious and cultural dynamics of West Asian society. The motive for the establishment of the hospital in early Islam needs to be viewed primarily not through the medical benefits that it might have brought to society, but rather from the outset as a political act. It seems that its function was in part connected to the ideological stances of the ruling elites. Moreover, the hospital was a symbol of power in the public sphere, which served to consolidate the prestige of the medicine practised there – Greco-Islamic medicine. We have also seen that in the absence of formalized training and qualification in medicine, the Hippocratic and Galenic medical heritage was the vehicle by which physicians were able to assert professional legitimacy, especially at a time when al-Ma'mūn's philhellenism held sway in Cairo. On a general level, this episode reveals the dynamics at play in the evolution of a medical tradition, and the pivotal role that the text assumes in that process.

While there can be no doubt that the Abbasid era was a formative period for sciences and philosophy in Islam, it is clearly necessary to reassess the relevance of the term "golden age". One way of avoiding this clumsy and value-laden periodization is to localize the study of the sciences in Islam, as A.I. Sabra has called for (Sabra, 1996).

Another very important task in the history of Islamic medicine is to push the boundaries of research beyond the so-called "Arab world" and the conventional time period on which it is conducted, the 8th to 11th century. Studies in this direction are emerging. Some work has recently been done on Islamic medicine in the Ottoman Empire[23] and in South Asia,[24] and Khaled Fahmy has innovatively applied social historical perspectives to the history of medicine in 19th century Egypt.[25] It is also relatively rare that indigenous history writing is taken into account in English language scholarship.[26]

As has been suggested in this survey, the text, especially the medical text, has been the primary source for historical research. But recently different forms of literature and material have been located and have begun to be explored,[27] and the possibilities for going beyond the text into archival documentation, iconographic sources and material culture naturally increase the closer one approaches the present.

Equally, research in the history of Islamic medicine can benefit greatly from the input of the social sciences,[28] and also by the cross-fertilization with studies of medical anthropology.[29] Through these studies the world of medical practice can be opened up. This is of great importance, since underlying all the historical accounts of Greco-Islamic medicine is the assumption that Greco-Islamic medicine is a "system": a set of uniform and coherent theoretical principles and practices.[30] In this survey we have seen that one may justifiably speak of a Greco-Islamic medical tradition. However, viewing Greco-Islamic medicine as a system is questionable, given the possibilities over time and place for the influence of local cultures and local healing traditions. This problem can be addressed by focusing not on the tradition of learning, but on the relation of this learning to theory and practice in localized contexts.

NOTES

1 See Edward Said on Ernest Renan's ideas on the cultural and political superiority of the West (Said, 1978: 123ff.).

2 Prophetic medicine was only mentioned in passing in E.G. Browne (2001: 12–13); Cyril Elgood translated a text of prophetic medicine (1962a) that he attributed to Jalāl al-dīn al-Suyūṭī (d. 1505), but Perho shows that it should be attributed to al-Dhahabī (1995: 36–40). The notion of contest is found in Ullmann (1970: 185–189; 1978: 5). He says that prophetic medicine "was to counter Hellenistic Greek medicine, which for orthodoxy was suspect as being a science of heathen origin".

3 The word "system" is not present in the Arabic, which is *ṭibbi Jālīnūsa*, meaning "Galen's medicine" or "the medicine of Galen". The Arabic is from Ullmann (1970: 187), who quotes from Dieterich (1966: 119).

4 Although not as emphatically prescriptive, Ullmann (1970: 185) does describe prophetic medicine as *Aberglaube* (superstition).

5 See note 4 for the reference.

6 For example, the Andalusian physician Abū l-Qāsim al-Zahrāwī (d. after 1009) writes in the introduction to his famous work on surgery: "For the skilled practitioner of operative surgery is totally lacking in our time ... Therefore I decided to revive this art by expounding, elucidating, and epitomizing it in this treatise (Spink and Lewis, 1973: 2).

7 See note 3 above.

8 Adhering to one of the four schools of Islamic law, founded by Aḥmad ibn Ḥanbal (d. 855).

9 Gutas refers to the possible impact of the Crusades and Mongols on Mamluk society, and the traditionalism that Ibn Taimiya has been taken to represent in previous historiography (1998: 170–171). Cf. Goldziher (1916).

10 For an account of folk medicine in late 19th century Egypt from the perspective of a Western trained Egyptian doctor, see Walker, 1934. See also the anthropological studies of medical practices, especially Greco-Islamic medicine, in Iran by Good (1977), and Good and Good (1992).

11 Essential bibliographical works are Ullmann (1970) and Sezgin (1967–).

12 *Galen On the Usefulness of the Parts of the Body*, translated by Margaret Tallmadge May. Ithaca, New York: Cornell University Press, 1968.

13 No date is given for its foundation. This is the source for Ullmann (1978: 9) and Klein-Franke (1982: 70). Conrad (1995: 105) gives 832 as the date.

14 Gutas (1998: 55–56). Gutas refers to the problem of interpreting the Arabic preposition 'lī', as in 'lī Hārūn al-Rashīd'. It has been translated as "of", but he argues it should more correctly be "in the service of".

15 Klein-Franke remarks on this point that it is therefore possible that Ḥunain did his Greek and Syriac work independently of the *bayt al-ḥikma*, but he does not question the primary function of the institution in the translation movement (1982: 70).

16 Quoted from Ibn Riḍwān's *al-Kitāb al-nāfi' fī kayfiyyat ta'līm ṣinā'at al-ṭibb*, 107–108.

17 Dols also cites a typescript of an article by L. Conrad, "The institution of the hospital in medieval Islam: ideals and realities" (1986); Conrad (1993: 688).

18 The following account of the conventional interpretation of the role of Jundi-shapur and its hospital is from the *Encyclopedia of Islam*, revised edition, s.v. "Gondeshapur", by Aydın Sayılı.

19 Like his son, al-Ma'mūn, although not quite on the same scale, Hārūn did become renowned in later histories for his promotion of the Greek sciences. Gutas relates that the Arab chronicler Ibn Juljul (d. 994) reported Hārūn as a patron of the translation of certain texts that chronologically could only have been sponsored by one of his successors, al-Mu'tasim (1998: 123).

20 Dols quotes from *Sechster Band des Kitāb Bāghdād von Ahmed ibn Abī Ṭāhir Ṭaifūr*, Hans Keller, ed. and trans. Leipzig: Harrassowitz, 1908, p. 50, II. 14–15.

21 For hospitals in the Ottoman Empire see Terzioğlu (1996) and Evilyā Chelebi's account of hospitals in 17th century Egypt (Leiser and Dols, 1988).

22 According to chronicles (which may have the tendency to elaborate the munificence of the ruler), there were 70 hospitals in Delhi under the reign of Muhammad ibn Tughlaq (r. 1325–1352), and the pattern of establishing and adding to hospitals continued with Akbar (r. 1556–1605) in Delhi and Agra, and Aurangzeb (r. 1658–1707). In Hyderabad, Muhammad Quli Qutb Shah founded the Dar ush-Shifa in 1593 (Jaggi, 1977; Liebeskind, 1997: 50).

23 See for example Terzioğlu, 1996.

24 For general surveys of *Unani ṭibb* in South Asia see Jaggi (1977) and Liebeskind (1997); Rahman mentions more about *Unani ṭibb* than most surveys of Islamic medicine (1989). Quaiser (2001) takes the analysis of *Unani ṭibb* to new levels of sophistication in his study of debates on surgery in late 19th century colonial India.

25 Fahmy (2000).

26 Ahmed 'Issa Bey's work (1929) on hospitals is frequently referred to. Perho (1995) stands out for her summary approaches by historians of prophetic medicine writing in Arabic.

27 As some examples: Conrad (1993) makes use of a wide variety of sources, as does Dols (1992). Goitein (1971) and Motzkin (1970) draw on the Geniza documents. Savage-Smith (1995) explores legal literature on the question of dissection.

28 For example, see Rosenthal's (1979) study of the place of the physician in medieval Islamic society.

29 See, for example, Good and Good's (1992) excellent study of Greco-Islamic medicine in contemporary Iran.

30 Rahman does write that *Unani ṭibb* almost entirely absorbed Āyurveda in India, but refers only to pharmacology (1989: 75).

BIBLIOGRAPHY

Agha, Zouhir M. *Bibliography of Islamic Medicine and Pharmacy*. Leiden: E.J. Brill, 1983.

Ahmed 'Isa Bey. *Tārīkh al-bimāristānāt fī l-Islām*. Damascus, 1347/1929.

Alvarez Millan, Cristina. 'Graeco-Roman case histories and their influence on medieval Islamic clinical accounts.' *Social History of Medicine* 12(1): 19–43, 1999.

Arberry, Arthur J. *The Koran Interpreted*. London: Allen & Unwin, 1955, reprinted Oxford: Oxford University Press, 1989.

Bergsträsser, G. *Hunain ibn Ishak und seine Schule*. Leiden: E.J. Brill, 1913.

Brandenburg, Dietrich. *Die Ärzte des Propheten, Islam und Medizin*. Berlin: q. Verlags-GmBH, 1992.

Brandenburg, Dietrich. *Islamic Miniature Painting in Medical Manuscripts*. Basle: Editiones 'Roche', 1982.

Brockelmann, Carl. *Geschichte der Arabischen Literatur*. 2 vols., 2nd ed. Leiden: E.J. Brill, 1943–1949; *Supplement*. 3 vols. Leiden: E.J. Brill, 1937–1942.

Browne, Edward G. *Arabian Medicine*. Cambridge: Cambridge University Press, 1921, reprinted as *Islamic Medicine*, New Delhi: Goodword Books, 2001.

Bürgel, J. Christoph. 'Secular and religious features of medieval Arabic medicine.' In *Asian Medical Systems*, Charles Leslie, ed. Berkeley: University of California Press, 1976, pp. 44–62.

Bürgel, J. Christoph. 'Die wissenschafliche Medizin im Kräftefeld der islamischen Kultur.' *Bustan* 1: 9–19, 1967.

Chandpuri, 'Ali Kauthar. *Aṭibbā'-i 'ahd-i Mughalīya* [Physicians of the Mughal Period]. Karachi, 1960.

Conrad, Lawrence. 'The Arab-Islamic medical tradition.' In *The Western Medical Tradition, 800 BC to AD 1800*. London: Cambridge University Press, 1995a, pp. 93–138.

Conrad, Lawrence. 'Scholarship and social context: a medical case from the eleventh century Near East.' In *Knowledge and the Scholarly Medical Traditions*, Don Bates, ed. Cambridge: Cambridge University Press, 1995b, pp. 85–100.

Conrad, Lawrence. 'Arab-Islamic medicine.' In *Companion Encyclopedia of the History of Medicine*, W.F. Bynum and Roy Porter, eds. London: Routledge, 1993, pp. 676–727.

de Koning, Paul, trans. *Trois Traités d'Anatomie Arabes par Muhammed Ibn Zakariyya' al-Razi, 'Ali ibn al-Abbas et 'Ali ibn Sina*. Leiden: Brill, 1903.

Dietrich, Albert. *Medicinalia Arabica: Studien über arabische medizinische Handschriften in türkischen und syrischen Bibliotheken*. Abh. Gött. Dritte Folge, Nr. 66, Göttingen: Vandenhoeck u. Ruprecht, 1966.

Dols, Michael W. *The Black Death in the Middle East*. Princeton, New Jersey: Princeton University Press, 1977.

Dols, Michael W. 'Insanity and its treatment in Islamic society.' *Medical History* 31: 1–14, 1987a.

Dols, Michael W. 'The origins of the Islamic hospital: myth and reality.' *Bulletin of the History of Medicine* 61: 367–390, 1987b.

Dols, Michael W. 'Islam and medicine.' *History of Science* 26(4) no. 74: 417–425, 1988.

Dols, Michael W. *Majnūn – The Madman In Islamic Medieval Society*, Diana E. Immisch, ed. Oxford: Oxford University Press, 1992.

Eaton, Richard M. *Essays on Islam and Indian History*. New Delhi: Oxford University Press, 2000.

Elgood, Cyril. 'Medicine of the Prophet.' *Medical History* 6: 146–153, 1962a.

Elgood, Cyril. 'Tibb-ul-Nabbi or Medicine of the Prophet: Being a translation of two works of the same name.' *Osiris* 13: 33–192, 1962b.

Fahmy, Khaled. 'Medicine and power: towards a social history of medicine in nineteenth century Egypt.' *Cairo Papers in Social Science* 23(2): 1–46, Summer 2000.

Farooqi, M.I.H. *Medicinal Plants in the Traditions of Prophet Muhammad*. Lucknow: Sidrah Publishers, 1998.

Gallagher, Nancy E. *Medicine and Power in Tunisia, 1780–1900*. Cambridge: Cambridge University Press, 1983.

Goitein, S.D.F. 'The medical profession' and 'Druggists and pharmacists.' In *A Mediterranean Society: The Jewish Communities of the Arab World as Portrayed in the Documents of the Cairo Geniza*, II, S.D.F. Goitein, ed. Berkeley: University of California Press, 1971, pp. 240–272.

Goldziher, I. 'Stellung der alten islamischen Orthodoxie zu den antiken Wissenschaften.' *Abhandlungen der Königlich Preussischen Akademie der Wissenschaften*, Jahrgang 1915, Philosophisch-Historische Klasse, no. 8, Berlin: Verlag der Akademie, 1916.

Good, Byron. *The Heart of What's the Matter: The Structure of Medical Discourse in a Provincial Iranian Town*. Ph.D. dissertation, University of Chicago, 1977.

Good, Byron and Mary-Jo Del Vecchio Good. 'The comparative study of Greco-Islamic medicine: the integration of medical knowledge into local symbolic contexts.' In *Paths to Asian Medical Knowledge*, Charles Leslie and Allan Young, eds. Berkeley: University of California Press, 1992, pp. 257–271.

Greenhill, William Alexander. *A Treatise on the Small-pox and Measles, by Abu BecrMuhammad ibn Zacariya ar-Razi*. London: Sydenham Society, 1848.

Gutas, Dimitri. *Greek Thought, Arabic Culture*. London, New York: Routledge, 1998.

Hawting, G.R. 'The development of the biography of al-Harith ibn Kalada and the relationship between medicine and Islam.' In *The Islamic World from Classical to Modern Times, Essays in Honour of Bernard Lewis*, C.E. Bosworth *et al.*, eds. Princeton, New Jersey: Princeton University Press, 1989, pp. 127–140.

Hamarneh, Sami Khalaf. *Background of Yunani (Unani), Arabic and Islamic Medicine and Pharmacy*. Karachi: MAS Printers, 1997.

Hamarneh, Sami Khalaf. *Health Sciences in Early Islam*, 2 vols., Munawwar A. Anees, ed. Blanco, Texas: Noor Health Foundation and Zahra Publications, 1984.

Hamarneh, Sami Khalaf. 'Introduction to the history and philosophy of Arabic-Islamic medicine and allied sciences.' *Studies in the History of Medicine* V1(3): 195–213, 1982.

Hamarneh, Sami Khalaf. 'Medical education and practice in medieval Islam.' In *The History of Medical Education*, C.D. O'Malley, ed. Berkeley: University of California Press, 1970, pp. 39–71.

Hendley, Colonel Thomas Holbein. *General Medical History of Rajputana*. Calcutta: Office of the Superintendent of Government Printing, India, 1900.

Husain, S.A. 'Physicians of Hyderabad during Nizam IV, V, VI.' *Bulletin of the Indian Institute of History of Medicine* XV: 58–72, 1985.

Ibn abī Uṣaibi'ah. *'Uyūn al-anbā'*, A. Müller, ed. Cairo and Königsberg: al-Maṭba'a al-wahhaabiiya, 1882–1884.

Ibn Ḥabīb. *Mujtasar Fī l-Ṭibb (Compendio de Medicina)*, Camilo Álvarez de Morales and Fernando Girón Irueste, ed. and trans. Madrid: Consejo superior de investigaciones científicas Instituto de cooperación con el mundo Árabe, 1992.

Ibn al-Qifṭī, 'Alī ibn-Yūsuf. *Tārīkh al-ḥukamā'*, J. Lippert, ed. Leipzig: Dieterich'sche Verlagsbuchhandlung, 1903.

Ibn Qayyim al-Jawziyya. *Medicine of the Prophet*, Penelope Johnstone, trans. Cambridge: The Islamic Texts Society, 1998.

Ibn al-Ukhuwwa. *The ma'ālim al-qurba fī aḥkām al-ḥisba*, Reuben Levy, trans. London: Luzac, 1938.
Ibnā Sābūr and al-Ḥusain ibn Bistam. *Islamic Medical Wisdom – The Ṭibb al-a'immā*, Batool Ispahany, trans., Andrew J. Newman, ed. London: The Muhammadi Trust, 1991.
Isaacs, Haskell D. 'Arabic medical literature.' In *Religion, Learning and Science in the Abbasid Period*, M.J.L. Young *et al.*, eds. Cambridge: Cambridge University Press, 1990, pp. 342–363.
Jacquart, Danielle and Micheau, Françoise. *La médicine arabe et l'occident medièval.* Paris: Maisonneuve et Larose, 1990.
Jaggi, Om Prakash. *Medicine in Medieval India.* Delhi: Atma Ram & Sons, 1977.
Klein-Franke, Felix. *Vorlesungen über die Medizin im Islam.* Sudhoffs Archiv Zeitschrift für Wissenschaftsgeschichte, Beiheft 23, Wiesbaden: Franz Steiner Verlag, 1982.
Leclerc, Lucien. *Histoire de la médicine arabe.* Paris: Ernest Leroux, 1876.
Leiser, Gary. 'Medical education in Islamic lands from the seventh to the fourteenth century.' *Journal of the History of Medicine and Allied Sciences* 38: 48–76, 1983.
Leiser, Gary and Michael Dols. 'Evilyā Chelebi's description of medicine in seventeenth century Egypt.' *Sudhoffs Archiv* 72: 49–63, 1988.
Liebeskind, Claudia. 'Unani medicine of the Subcontinent.' In *Oriental Medicine: An Illustrated Guide to the Asian Arts of Healing*, Jan van Alpey and A. Aris, eds. Boston: Shambala, 1997, pp. 39–65.
Levey, Martin. 'Medical ethics of medieval Islam, with special reference to Al-Ruhawi's "Practical Ethics of the Physician."' *Transactions of the American Philosophical Society* 57(3): 1–100, 1967.
Levey, Martin. *Early Islamic Pharmacology.* Leiden: E.J. Brill, 1973.
Meyerhof, Max. *Studies in Medieval Arabic Medicine*, Penelope Johnstone, ed. London: Variorum Reprints, 1984.
Meyerhof, Max. 'Science and medicine.' In *The Legacy of Islam*, Thomas Arnold and Alfred Guillaume, eds. Oxford: Oxford University Press, 1931, pp. 311–355.
Motzkin, A.L. 'A thirteenth century Jewish physician in Jerusalem.' *The Muslim World* 60: 344–349, 1970.
Neuburger, Max. *Geschichte der Medizin.* Band 1, Stuttgart 1906, Band 2. Stuttgart, 1911.
O'Boyle, Cornelius. *The Art of Medicine: Medical Teaching at the University of Paris, 1250–1400.* Leiden: E.J. Brill, 1998.
Perho, Irmeli. 'The Prophet's medicine – a creation of the Muslim traditionalist scholars.' *Studia Orientalia* 74: 1–158, 1995.
Quaiser, Neshat. 'Politics, culture and colonialism: Unani's debate with doctory.' In *Health, Medicine and Empire: Perspectives on Colonial India*, Biswamoy Pati and Mark Harrison, eds. New Delhi: Orient Longman, 2001, pp. 317–355.
Qudratullah Husami, Syed. *Islam and Medical Sciences* (in 8 parts). Hyderabad: Fazil Academy, part 1, 1985, parts 2–8, 2000.
Renan, Ernest. *L'islamisme et la science.* Conférence faite à la Sorbonne, Paris. Paris: Calmann Lévy, 1883.
Rahman, Fazlur. *Health and Medicine in the Islamic Tradition: Change and Identity.* New York: Crossroad, 1989.
Rosenthal, Franz. 'The defense of medicine in the medieval Muslim world.' *Bulletin of the History of Medicine* 43, 1969, reprinted in Rosenthal, F. *Science and Medicine in Islam: A Collection of Essays.* Aldershot, UK and Brookfield, Vermont: Variorum, 1990, VIII, pp. 519–532.
Rosenthal, Franz. 'The physician in medieval Muslim society.' *Bulletin of the History of Medicine* 52, 1978, reprinted in Rosenthal F. *Science and Medicine in Islam: a collection of essays.* Aldershot, UK and Brookfield, Vermont: Variorum, 1990, pp. 475–491.
Sabra, A.I. 'Situating Arabic science: locality versus essence.' *Isis* 87: 654–670, 1996.
Sabra, A.I. 'The appropriation and subsequent naturalization of Greek science in medieval Islam: a preliminary statement.' *History of Science* 25: 223–243, 1987.
Said, Edward W. *Orientalism.* New York: Pantheon Books, 1978.
Said, Hakim Mohammed. 'Islamic medicine and the art of drug making: a historical perspective.' *Hamdard* 33(1): 43–57, 1990.
Savage-Smith, Emilie. 'The practice of surgery in Islamic lands: myth and reality.' *Social History of Medicine* 13(2): 307–321, 2000.

Savage-Smith, Emilie. 'Attitudes towards dissection in medieval Islam.' *Journal of the History of Medicine and Allied Sciences* 50: 67–110, 1995.

Schacht, Joseph and Max Meyerhof. *The Medico-Philosophical Controversy Between Ibn Butlan of Baghdad and Ibn Ridwan of Cairo*. Cairo: The Egyptian University, the Faculty of Arts Publication no. 13, 1937.

Sezgin, Fuat. *Geschichte des arabischen Shrifttums, 3. Medizin–Zoologie–Tierheilkunde bis ca. 430 H.* Leiden: E.J. Brill, 1970.

Sezgin, Fuat, ed. *Beiträge zur Geschichte der arabischen-islamischen Medizin. Aufsätze aus den Jahren 1819–1920.* 4 vols. Frankfurt: Institut für Geschichte der arabischen-islamischen Wissenschaften, 1980.

Spink, M.S. and G.L. Lewis. *Albucasis On Surgery and Instruments*. London: Wellcome Institute of the History of Medicine, 1973.

Strohmaier, Gotthardt. 'Ärztliche Ausbildung im islamischen Mittelalter.' *Klio* 61(2): 519–524, 1979.

Terzioğlu, Arslan. *Beiträge zur Geschichte der Türkisch-Islamischen Medizin, Wissenschaft und Technik*. Istanbul: Isis Verlag, 1996.

Ullman, Manfred. *Islamic Medicine*. Edinburgh: Edinburgh University Press, 1978.

Ullman, Manfred. *Die Medizin im Islam*. Handbuch der Orientalistik, Erste Abteilung: Der Nahe und Mittlere Osten, Ergänzungsband VI, Erster Abschnitt, B. Spuler hrsg. Leiden, Köln: E.J. Brill, 1970.

Van Ess, J. *Theologie und Gesellschaft im 2. Und 3. Jahrhundert Hidschra. Eine Geschichte des Religiösen Denkens im frühen Islam.* 6 vols. Berlin: W. de Gruyter, 1991–97.

Walker, John. *Folk Medicine in Modern Egypt: Being the Relevant Parts of the Ṭibb al-rukka or Old Wives' Medicine of ʿAbd al-Rahman Isma'il.* London: Luzac, 1934.

Wüstenfeld, Ferdinand. *Geschichte der arabischen Aerzte und Naturforscher, nach den Quellen bearbeitet*. Göttingen: Vandenhoeck und Ruprecht, 1840.

Zillurrahman, Syed. 'Unani medicine in India 1901–1947.' *Studies in History of Medicine and Science* 13(1): 97–112, 1994.

HUGH SHAPIRO

HOW DIFFERENT ARE WESTERN AND CHINESE MEDICINE? THE CASE OF NERVES

For 2,000 years, medicine in China was simply called "medicine" (*yi*). The word "Chinese medicine" (*zhongyi*) emerged in the mid-nineteenth century, a product of China's collision with European power, when "Occidental influence made it inevitable that what previously was thought universal be identified as native" (Sivin, 1987: 4 n1). The Chinese word "Western medicine" (*xiyi*) also dates from this period, even though medical ideas and physicians from Europe began reaching China more than two hundred years before the Opium Wars. From this new milieu of Sino-foreign juxtaposition emerged the question of how the differences between Chinese and Western medicine should be understood.

There are convincing answers to this question. In therapeutic aim, for example, Chinese and Western medicine exhibit distinct orientations, with the result that the two systems often succeed in different areas. The late Joseph Needham offered this formulation: the province of Western biomedicine lies in managing acute illness requiring interventions such as surgery and antibiotics, and includes a pharmacology that sometimes causes serious side effects. Chinese medicine, on the other hand, excels at treating chronic disorders and pain, with a pharmacology that visits fewer harmful effects on patients (Needham, 2000: 65). Farquhar contends that the wholly distinct approach of medicine in China challenges the very attempt at comparison, because

> Chinese medicine heals in a world of unceasing transformation. This condition of constant change, this fluidity of material form, stands in sharp contrast to a (modern Western) commonsense world of discrete entities characterized by fixed essences, which seem to be exhaustively describable in structural terms (Farquhar, 1994a: 24–25).

The contingency of transformation shaped bodily experience. In early China,

> The imagination of wind sprang from a concrete *experience* of oriented space and local place, a directly felt sense of seasonal drift and human moods. It bespoke an embodied apprehension of an ever shifting self in an ever changing universe. Personal breath could harmonize with cosmic breath, and habitually the two might be reasonably in phase. But the character of all winds, inner or outer, was that they always retained some chaotic contingency, the possibility of suddenly blowing in new and unexpected directions (Kuriyama, 1994: 37–38).

The lived experience of the self differed in unexpected ways, as did the

H. Selin (ed.), Medicine Across Cultures: History and Practice of Medicine in Non-Western Cultures, 351–372.

recognition of what constituted a person. We witness a "multiplicity" of medical bodies (Farquhar, 1994b), and a medical body that is ambiguously gendered (Furth, 1999) and regionally differentiated (Hanson, 2001). During the last two centuries, we also see the dramatic transformation of the archetypal human form (Elvin, 1985; Zito and Barlow, 1994; Honig and Hershatter, 1988).

Divergent understanding of the self also illustrates fundamental contrasts between the two medical traditions. From late antiquity, the imagination of the body diverged in ways so profound that even the outwardly symmetrical gestures of Chinese and Western physicians, such as pulse analysis, are in fact incommensurate (Kuriyama, 1999). Scholars argue that Chinese medicine cannot be understood in Western terms, that it must be approached from within its own epistemological vision (Sivin, 1987; Unschuld, 1985; Ishida, 1992). Research on texts unearthed from the formative period, and on gender, local culture, politics, and the encounter with Western biomedicine continues to transform our grasp of what we call Chinese medicine (Li, 2000; Harper, 1998; Lee, 1996; Bray, 1997; Wu, 1998; Cass, 1999; Chang, 1998; Andrews, 1996; Rogaski, 1996). In this article, I propose examining the contrasts between medicine in China and the West from a different perspective, from the idea of nerves. Before turning to this issue, however, let us first consider the idea of Chinese medicine since the invention of the term in the mid-nineteenth century.

CHINESE MEDICINE

The original comparison of what became known as Chinese and Western medicine first occurred in the wake of China's military and diplomatic defeat. This background shows through in the analysis of influential observers such as Zheng Guanying (1842–1922), who in 1892 humbly proposed that "Chinese medicine fails in the vacuity of its theory; Western medicine's problem is that it niggles too much over the tangible details" (Zheng, 1892: 22b).

Zheng Guanying judged Chinese medicine to be past its prime. But he argued that much could be salvaged from the teachings of late antiquity, especially from the *Inner Canon of the Yellow Emperor* (*Huangdi neijing*), a text comprising the most influential writings of medicine in China. Known also as the *Inner Canon* (*Neijing*), the work consists of two separate compilations, the *Basic Questions* (*Suwen*) and the *Divine Pivot* (*Lingshu*). The *Inner Canon* was compiled over a period of several centuries, probably between the first century BCE and the early first century CE, and drew together the work of many authors (Sivin, 1987: 5 n3; Keegan, 1988; Yamada, 1979; Sivin, 1993). Zheng also advised retaining texts from the imperial era, especially the masterworks of the Tang and Song Dynasties (7th to 13th centuries). He believed it was possible to rescue what was vital in China's medical legacy, combine it with new approaches from abroad, and do so with an unprejudiced eye, "not distinguishing between what is Chinese and what is foreign." Zheng warned that, in unskilled hands, Chinese medicine did more harm than good.

Even if Zheng's analysis was influenced by his times, his criticism of medical theory was hardly unique. Since the Han Dynasty (202 BCE-220 CE), medical scholars have been trying "to reconcile discrepant systems of correspondence," with the result that they have produced "increasingly comprehensive, complex schemata" (Sivin, 1995: 35). "Correlative cosmology" (Schwartz, 1985: 350–382) and the contingent and relative concepts such as *yinyang*, the two opposing, complementary forces of nature (Farquhar, 1994a: 217–220), and the Five Phases, *wuxing* (Sivin, 1987: 70–80), so central to medical thought, were employed precisely because they were part of a universal theory.[1] Driving the use of these concepts, in some measure, was an effort to integrate medicine into an acknowledged philosophical framework. By so doing, disorders and treatments could be explained to patients in recognizable terms. Part of the value of these themes was that they helped establish medicine as a system within a widely accepted worldview (Needham, 1956: 232–303). Tensions thus grew up between the framework of a universally applicable theory and what doctors could actually work with in clinical practice. In the medical texts written by physicians, for example, we encounter many things that do not actually fit into this scheme of *yinyang/wuxing* correspondences. However, practitioners were not blinded by these abstract and philosophical ideas, and they invented new ideas that represented a clinical experience not always consistent with the theory. The correlative system of *yinyang/wuxing* is only part of the medical vision.

The next generation of critics was less restrained. China of the early twentieth century witnessed creative ferment sometimes likened to the Weimar Republic in Germany. It was also a dark time: foreign incursions intensified, regional satraps eclipsed state power, civil strife sharpened, countryside and city were broadly militarized, and in some areas the population suffered pauperization. From this insecurity, exacerbated by a crisis in public health, there surfaced a notion that went much beyond the inferiority of the country's medicine: the body itself was innately defective.

The weakness of the nation, the argument went, began with the frailty of its people. The search for wealth and power thus linked national strength with personal hygiene and vigor. Examples of this discourse are abundant. We have Mao Zedong's 1917 call for the bookish to build steely bodies in *A Study of Physical Education* (Schram, 1976) and Yu Dafu's 1921 linking of sexual failure with national humiliation in the self-loathing story "Sinking" (Lau, 1981). And there were the morning exercises and reformist hygiene of the Nationalist Party's New Life Movement of the early 1930s and the Chinese Communist Party's veneration of purity, stamina, and androgynous robustness before and after the 1949 revolution.

Because weakness was perceived as both physical and national, affecting both the biological body and the body politic, the study of (Western) medicine attracted patriots hoping to ameliorate national ills by first healing individuals. Dr. Sun Yat-sen, China's first president and Father of the Country, initially trained in biomedicine, as did several of China's renowned twentieth-century writers, such as Lu Xun and Guo Moro. Lu Xun (1881–1936) abandoned medical training in Japan for writing, creating the character "Ah Q" (1921), a loathsome embodiment of human failings, a caricature of physical and moral imperfection (Lyell, 1990). The muscular statuary of early socialism in 1950s China did not merely mimic the motifs of Soviet art; its celebration of physical vitality repudiated the haunting image of Ah Q. This idea of weakness lay behind the early popularity of Bruce Lee. On screen Bruce Lee crushed Ah Q, and in "The Chinese Connection" (1972) he destroyed the idea of "the sick man of Asia" by forcing an arrogant adversary to eat a calligraphic work bearing those very words. In this discourse of public and private weakness, "traditional" medicine did not escape blame.

Writers working in a transitional vernacular disparaged Chinese medicine as not only failing to heal patients, but as actively killing them. "Medicine" (1919), another story by Lu Xun, vilifies a therapy that takes the body as a medicine, a pharmacology using material derived from the human body (Cooper and Sivin, 1993; Lyell, 1990). During the 1920s and 1930s, many forms of medicine fell under the writer's ax. In Lao She's (1899–1966) "A Brilliant Beginning", we find charlatans posing as physicians, injecting luckless patients with hypodermics full of tea (Jenner, 1970).

In 1928, the Ministry of Health, run by Western-style doctors and set up under the new Nationalist Government of Chiang Kai-shek, moved to abolish Chinese Medicine (Lei, 1999; Yip, 1995; Croizier, 1968). Practitioners of Chinese medicine organized themselves and overcame the threat. However, in the course of navigating the crisis, the physicians abandoned the name "Chinese medicine". In its place, they adopted "National medicine". Behind this striking shift in self-identification was the negotiation of state power by competing medical interests (Lei, 1999). What is more, given the diverse forces driving medical culture during the Republican era (1912–1949), and the "selective assimilation" of foreign practice (Andrews, 1996), it is difficult to generalize about the perception of what was then called Chinese medicine (Zhao, 1989). Western-trained and Western-styled physicians assailed Chinese medicine as amateur and dangerous, as did foreign doctors and other critics. But we know too that practitioners remained proud of the designation Chinese medicine, and that most patients in China would have had little exposure to foreign-based medicine, thus continuing to seek treatment from Chinese doctors. Chinese medicine became contested, particularly in the coastal cities (Rogaski, 1996), and it was not until the early 1950s that certain tensions that had accrued to the title were defused, after the establishment of the PRC and the restoration of the phrase (Hsu, 2000). Today, I would venture this generalization: many patients in the PRC, Taiwan, and Hong Kong consult physicians of different training and orientation, visiting both doctors of Western biomedicine and doctors of traditional Chinese medicine. This points to the difficulty of integrating the two systems of therapy.

In the West, the history of medicine for years had described medicine in China as backwards or fantastical. Positivist views of the progress of science conventionally depicted medicine as advancing from the dark ages to enlightenment. Chinese medicine was viewed as something akin to the medical medieval times, conceptually caught at a point of development where Western medicine had once been but had long since passed. Chinese medicine was but a curious artifact left behind by "medicine". English language exceptions to this attitude existed, but it was not until the 1960s that we witnessed a tectonic shift in this view. By the 1970s, scholars were reestablishing in the Western imagination the idea of Chinese medicine as a serious system of thought and practice. Since then, a sizeable literature has grown up devoted to assessing the efficacy of acupuncture and Chinese herbs in a biomedical environment.

ANATOMY

Critics also mocked Chinese medicine's anatomical system. It was here that disparagement was most harsh, and practitioners and scholars of biomedicine could point to physiology and gross anatomy as the basis for the impressive work of surgery. For China, we possess records of detailed forensic analysis from the third century BCE (McLeod and Yates, 1981), an official historical record of a dissection dated to 16 CE (Yamada, 1991), and in the Song period (11th century) we witness a proliferation of dissection and a sophisticated forensic medicine (McKnight, 1981, Needham, 2000: 175–200). However notable China's work in this field was, it is not the proper context for comparing the anatomical thinking of the two systems. With dissimilar concerns, physicians looked for and found different things.

In the words of the *Inner Canon*, "The subject of discourse, briefly put, is the free travel and inward and outward movement of the divine *ch'i* [*qi*]; it is not skin, flesh, sinews and bones" (Sivin, 1995: 14). Doctors accordingly fixed their gaze "on living persons rather than on lifeless corpses" (Kuriyama, 1995: 155).

> [In this] nonanatomical logic ... where generation and transformation are intrinsic to existence, fixity and stasis occur only as a result of concerted action and therefore demand explanation; motion and change are a given and seldom need be explained with reference to their causes. One consequence of this dynamic bias in Chinese medicine is that the body and its organs (i.e. anatomical structure) appear as merely contingent effects or by-products of physiological processes. Basing a pathology on anatomical structures and structural abnormalities would be like closing the barn door after the horse has fled (Farquhar, 1994a: 25–26, 92).

This anatomy is key to understanding how people fell ill.

> Breaching first the skin and pores, a disease sinks steadily, relentlessly inward – into the conduits, the sinews, the flesh, into the viscera, into the marrow of the bones. When it still hovers near the pores it can be expelled by poultices, say, or acupuncture; as it burrows deeper, medicines must be ingested; until eventually the disease infiltrates the marrow, where no cures can reach (Kuriyama, 1999: 164).

When Zheng Guanying took biomedicine to task in 1892 for getting bogged down with the "tangible details", I suspect that he was merely lending the elegance of symmetry to his criticism of Chinese medicine, for what impressed him above all about Western medicine was exactly its meticulous illustration of the internal body. His view was common. Anatomy and physiology struck late nineteenth-century scholars and physicians in China as a defining difference between the two systems. Western doctors viewed the body's interior as akin "to the mechanism of a watch, without opening it up for a detailed, careful analysis, its function cannot be understood, to the point of breaking it" (Zheng:

21b). The intellectual Tan Sitong (1865–1898), too, identified autopsy as the basis for the feats of foreign medicine (Tan, 1898: 403). Without anatomical tools such as "illustrations" and "wax figures", Tan stated, there was no way to know how the bones and joints fit together and how the blood vessels circulated around the body. Tan agreed that Chinese medical texts also discussed the organs in detail, but he concluded that the body could not be understood without cutting it open.

Both Tan's and Zheng's work represent a line of thought from *fin de siècle* China. In arguing for biomedicine's superiority, anatomy was often highlighted first, and then the idea of nerves was raised, to illustrate the fundamental contrast of approach and to suggest the clinical rewards of mapping the inside of the body in this new way. Revealingly, Tan asked how "the nerves spread throughout the body," for China had no tradition of nerves.

NERVES

Nerves did not exist in the medicine of traditional China. The nervous system as such entered medical discourse in the second half of the nineteenth century. The very word for nerves used today in China, *shenjing*, was translated only in the first decade of the twentieth century. And disagreement over how best to express the new idea lasted through the 1920s; into the 1930s, we see multiple expressions of the concept. Entire sections of the translated medical lexicon, not just nerves, remained fluid during the 1910s and 1920s. Physicians and translators had been grappling with a confusing array of translated medical language since the 1880s. However, for those attempting to craft a lexicon capable of sustaining the ideas of biomedicine, nerves presented an especially thorny case. Consider the deliberations of a group of eighteen doctors, officials, and educators, who gathered in Shanghai in 1916 to standardize the translated nomenclature of Western medicine.

Consisting mostly of men from China with a handful of Sinophone foreign physicians, the group moved alphabetically through long lists of terms (*Yixue mingci*, 1917). They debated scores of words (such as cartilage, endothelium, and intestine), discussing the merits of competing terms. Then, they voted on each item, deciding on a fixed Chinese translation of the foreign word. This method reflected linguistic transformations then occurring in several of China's cities, a movement towards simplifying literary language into an accessible vernacular (Chow, 1960). Choices for the word "aperture", for example, came down to the classical *kong* versus the colloquial *kai kou*. But most discussion focused on medicine, on physiology and anatomy (words that were also discussed at length). "Muscle", a distinctly Western notion (Kuriyama, 1999), presented a complicated case and required a full hour; "nerve" took two hours. "Few apt words exist" for its translation, opined a Dr. Tang, because "the function of nerves was not understood in ancient times" and thus we "lack an old style name" from which to work.

Yet there were many new names. Since the translation of nerve as "fine sinew" (*xijin*) in the early 1620s, physicians and authors of medical lexica had created more than a dozen Chinese neologisms for the idea. This first translation of nerve into Chinese came from a physician born near the Swiss German border, Johann Schreck (1576–1630, Terrenz), who entered China as a Jesuit in 1621. Schreck undertook the first translation of the Galenic vision of the body into Chinese. This event affords a context for thinking about how cross-cultural medical exchange actually occurred in the early years of contact between Europe and China.

Numerous obstacles militated against Schreck's translation reaching fruition. He first had to reach China, risking transoceanic travel and countenancing the deadly rivalries of European states in the sea lanes and, once in Asia, in the early colonial footholds. These conflicts were venomous, driven by politics, money, and religion. Then he needed to overcome the linguistic barrier, learning Chinese well enough to express meaningfully the ideas of his Latin anatomical texts. Conceptual differences between the medical systems often resisted translation. If an idea does not exist in the recipient language, how is it to be conveyed? Then, once in China, Schreck depended on officials who themselves were dependent on a system which could unpredictably influence the situation of guests. Finally, the target audience in China was not necessarily interested in foreign medicine. All this took place under the penumbra of deteriorating Ming dynastic power and an escalating anxiety regarding foreigners. Unease over foreigners grew not only from the dark memory of the Mongol conquest but also from an emerging power soon to be known as the "Manchus" menacing China from the north (Wakeman, 1985), the violence of coastal raids by pirates, and from a growing sensitivity to the desires of the Portuguese, Dutch, and English in the south. This was aggravated by horrors such as the mass murder of 20,000 Chinese in 1603 by the Spanish authorities in the Philippines (Spence, 1985: 51). The episode of Jesuit Schreck's translation is worth considering in some detail because it imparts a sense of the linguistic, conceptual, logistic, political, and ecclesiastic variables at play in these early intercultural encounters.

Before joining the Society of Jesus in 1611, Johann Schreck enjoyed a reputation as a gifted healer in the courts of Europe. Schreck also excelled at chemistry and astronomy; he knew Galileo in Padua and later corresponded with Kepler. He sailed for China in 1618 with some twenty others, including Adam Schall von Bell and Giacomo Rho. While waiting to enter China, in the Portuguese colony of Macao, Schreck studied the Chinese language for two years. When summoned into Ming China (1368–1644), it would be on the strength of his astronomy, and he would be put to work on the official calendar. But he remained interested in medicine. While in Macao, he practiced medicine and dissected the cadaver of a priest who had worked in Japan before the expulsion of foreign missionaries. Writing up the autopsy, the first performed by a Western doctor in the region, Schreck attributed the priest's damaged lungs to heavy smoking (Ma, 1993: 281).

Schreck entered China in 1621 but did not proceed directly to the Capital. Instead, he stayed in the southern city of Hangzhou, waiting for permission to proceed north towards the Emperor. In Hangzhou, he continued his study of

Chinese and began translating. He did so in private, on his own time, with no connection to official duties. In China, the polymathic Jesuits pursued a strategy of analogy: demonstrate the superiority of their civilization's science, and perhaps it will follow that the monarch and his supporters will become intrigued by the spiritual life of Europeans (Spence, 1985). As preparation for this style of proselytizing, before leaving Europe, Schreck and other Jesuits had assiduously gathered scientific books and instruments, such as the telescope, which was the first of its sort to reach China (Goodrich, 1976: 1282). Among the books in Adam Schall's possession was a 1597 anatomical text written in Latin by Casper Bauhin (1560–1624), a respected Swiss professor of anatomy, medicine, and Latin at Basel University. In 1622, in the Hangzhou home of a Chinese colleague, Li Zhizao, Schreck began interpreting the Latin text into spoken Chinese, dictating to a Chinese scribe.

Working from Bauhin's text, *Anatomica corporis virilis et muliebris historia*, Schreck and the scribe produced a manuscript in handwritten Chinese, organized into two volumes: 1) anatomy and physiology; and 2) perception, sensation, and movement. When called to Beijing, Schreck put aside the unpublished manuscript and turned his talent to astronomy, an area of importance at Court. Schreck died in 1630. Adam Schall laid in storage the draft that Schreck had titled *A Treatise on the Human Body* (*Renshen shuo*). Four years later, in Beijing in 1634, Schall encountered a man of letters from the highest echelon of China's scholarly world, Bi Gongchen, a "presented scholar" (*jinshi*). Schell asked Bi Gongchen to work the text into literary Chinese. According to Bi Gongchen, he endeavored to preserve both the concept and spirit of Schreck's labor; still, his work on the draft increased the text by one-half. Before publishing it in 1643, Bi found volume two shorter than the first. To balance the two parts, he appended the *Mnemonic Techniques of Western Countries* (*Xiguo jifa*) by Matteo Ricci, who died in China in 1610. Bi Gongchen titled the work, *Western Views of the Human Body, an Abbreviated Treatise* (*Taixi renshen shuogai*). One year after its publication, as the Ming Dynasty collapsed in 1644, rebel troops under the pretender Li Zicheng captured Bi Gongchen and murdered him (Ma, 1993: 282–286).

Schreck and Bi's *Taixi renshen shuogai* differs from the analysis of the body in China in fundamental ways, most conspicuously in its elevation of the importance of the brain. Quite different from the brain-centered view of the body in the West, scholars and practitioners of medicine in China viewed the brain with relative indifference. In the normal functioning of the body, the brain was viewed as a minor organ. In China's early medical writings, in the *Inner Canon*, the brain (*nao*) and medulla (*sui*) are formed just after the birth of the body (*Lingshu*, *Jingmai*, 10.1). Yet the very marginality of the brain generated ambiguities in these texts, a vagueness that moved the Yellow Emperor to question Qibo about the status of the *naosui* (brain and medulla).[2] "I have listened to the experts (*fangshi*), some view the *naosui* as a solid viscera (*zang*), some view it as a hollow viscera (*fu*). I wish to know what you think." In organizing his answer, Qibo grouped the brain and medulla with four other puzzling parts of the body: bone, circulatory tracts, gallbladder, and uterus (*gu*, *mai*, *dan*, *nüzu bao*). What these parts shared, Qibo explained, was that despite being *fu* (hollow viscera), they were located deep within the body, and thus shared the *yin* characteristics of the deeply placed five solid viscera (*wuzang*). These were the *qiheng zhi fu*, the "atypical hollow viscera" (*Suwen*, *Wuzang bielun*, 11.1). The incongruity of the *naosui*, along with other ambiguous sites, such as the gallbladder, demanded its own category. Clearly the *naosui* was not the brain of the translated Latin text, which celebrates it as "the center of consciousness".

In Schreck's work the brain is the site of memory and the tool by which one encounters the world. The medium through which the brain transmits and receives are the "fine sinews" (*xijin*), the nerves, which (Schreck, *Taixi*: *Xijin bu*)

> ... are uniformly distributed throughout the body, and are comprised of three types in all, being made up of skin, bone marrow, and flesh. Their function is to transmit the force of perception and movement. ... Body, brain, and skin, all contain the fine sinews, just as split wood exposes the tree's rings or just as the leaves of the russet pear tree contain a multiplicity of veins.

This choice of sinew (*jin*) reflects an understanding of nerves then witnessed in Europe. Robert Burton's (1577–1640) *The Anatomy of Melancholy*, for instance, widely published from the 1620s, speaks of nerve and sinew interchangeably. Etymologically, too, the word is apt. The Latin *nerv-us* means bow-string, tendon, sinew.

> However, within the fine sinew there is no space, it contains only *qi* and no blood ... the sinews (*jin*) themselves do not feel and do not move. Rather they depend wholly on the power and vitality of the spirit for sensation and movement. ... Therefore, when the body can neither feel nor move, this is due to a lack of *qi* which results in a lack of power.
>
> Therefore when a man's *jin* is cut, he loses the ability to move or if he is weighted down and bound, his four limbs fall numb, as if he is half dead. This is because the body's *qi* is blocked and cannot interconnect. This is irrefutable proof that the fine sinews contain *qi* but do not contain innate power.

This particular way of explaining nervous function, as inseparable from the circulation of the vital essence *qi*, was an intuition shared by subsequent, separate attempts at translating the idea of nerves. However, the *Taixi renshen shuogai* itself was forgotten, and we have no evidence that the elite few with access to this anatomical study viewed it as representing a competing or alternative system of diagnosis and healing. Several other Jesuits would attempt translations, but it was as astronomers that Schreck and his colleagues surviving him made an impact. The early Jesuit era that gave us the prefix Western (*xi*) did not give us Western medicine (*xiyi*).

The Jesuits had sailed to China with no military power backing them and gained entry only through official favor, circumstances totally different from the medical missionaries of the nineteenth century who roamed widely in the interior, treating patients, building clinics, hospitals, pharmacies, schools, and printing presses, all prerogatives written into Chinese law by treaties imposed after the Opium Wars, upheld and expanded by foreign armies. After the Jesuits, the next systemic effort to translate the anatomy and physiology of Western medicine came from Benjamin Hobson (1816–1873), a medical missionary out of England who arrived in Macao in 1839. In 1851, with the help of his collaborator Chen Xiutang, Hobson published *A New Theory of the Body* (*Quanti xinlun*), devoting one chapter to the brain and nervous system. In "The Brain Governs the Entire Body" (*Nao wei quanti zhi zhu*), Hobson translates the word nerve as had Schreck, building from "sinew" (*jin*), although Hobson almost certainly was unaware of the *Taixi*. To *jin* he added the words "brain" (*nao*) and "*qi*" (vital essence), creating *naoqi jin*. *Naoqi jin* can be read as "the sinew transmitting the brain's vital power", or more literally, the "brain-vital essence-sinew".

Qi, the life force, is experienced within the body and is the same substance that animates the heavens and pulses within the veins of the earth. Qi moves about the body in conduits, is stored and distributed by the kidney system (*shen*), and can be affected to a certain extent by interventions such as acupuncture. One is born with a finite amount of qi and when it is used up, one dies.

Hobson's translation proved more enduring than the words for nerve coined by other foreign physicians and translators through the end of the century. His vocabulary of nerves was picked up by serious thinkers, such as Yan Fu (1854–1921), the chief interpreter of Western thought during this epoch (Schwartz, 1964). Still, the appeal was limited, and by the 1916 meeting in Shanghai, the one to standardize the terms of biomedicine, Hobson's *naoqi jin* did not make the short list. It was all but forgotten.

Of the dozen neologisms for nerve fashioned since the 1620s, five made the short list at the 1916 meeting. After two hours of debate, eight voted for *naojing* (brain tract) and seven for *shenjing* (divine tract). The phrase that is used today for nerves, *shenjing*, lost the vote. The majority of eight voted for a term that did not appear in the medical texts or popular dictionaries of that period, a term that today no one remembers. My point is this: nerve was a novel idea, new to the degree that the word *naojing*, invented either on that day of August 7 or thereabouts, was perceived by a highly educated group as having more explanatory power than words that had been in circulation for decades, including Hobson's *naoqi jin*. In the first quarter of the twentieth century, the idea of nerves was still being digested by Chinese society.

The foreignness of the idea can also be seen in the long and tortuous history of its translation. These translations did not take hold because nerves are not particularly relevant to Chinese medicine. Moreover, the social context of the initial overture, first with the Jesuits and then the medical missionaries, proved inimical to stirring wide interest. But there is a deeper point. The history of nerves in the West is tied up with volition. The term chosen by the Greek scientist Herophilus (4th to 3rd century BCE) for the motor nerves, *prohairetika*, means "capable of choosing, purposive" (von Staden, 1989: 250–251). Action, volitional action, in the Western tradition, is a crucial defining feature of identity. People have volition and the action of nerves is inseparable from the exercise of will. If identity is not defined in this way, then the idea of nerves is not particularly germane or compelling. More than 300 years passed before the term for nerve was settled, because what was being introduced was a different vision of the self.

HEART, BRAIN AND VISCERA

In early Western thought, the heart was one of the perceived centers of the body, an idea that became a source of tension in the rise of a discourse of nerves in Europe. The head, too, lacks the somatic logic of the heart. Placing one's hand on the scalp divulges little of what lies beneath; we gain no hint as to the activity inside the skull. Compared to the heart, the brain *feels* inert, disconnected from sensation, emotion, and thought. And strongly felt emotions such as anger, terror, rage, and sorrow, were experienced "more keenly in the thorax than in the head," suggesting why certain functions were imagined as arising from the heart (Bynum, 1973: 457). This brings us to our next point: What did the idea of the brain and the nervous system displace (if anything) in Chinese medicine?

In his widely disseminated *Questions and Answers about Hygiene, in Nine Chapters*, Ding Fubao (1874–1952) states this position:

> Question: Chinese (*huaren*) assert that thought issues from the heart. Westerners assert that it issues from the brain. Of these two theories, which is correct?
>
> Answer: The Western theory is correct (Ding, 1901: 8).

Tan Sitong, in his 1898 address, contends that memory, sensory perception, and the ability to comprehend (*wu*) are controlled by the *nao* (brain) and not the *xin* (heart). In this line of discourse, seen in both Western and Chinese writings of the period, the nerves are then described as the vehicle of the brain's dominance. Yet the importance of the heart is not diminished. It is accorded a different, critical function; it moves the blood that nourishes the brain. In Tan's words, "Thought is assuredly only in the brain. However, that the brain is capable of thought depends completely on the heart's capacity to change the blood (*bianxue*) and nourish the brain (*yangnao*)" (Tan, 1898: 403). Brain replaces heart, but the heart feeds the brain.

Ding Fubao put a different nuance on the link between the two parts. "The brain governs the movement of thought, the heart governs the distribution of blood (*nao zhu yunsi, xin zhu faxue*)." So opens the chapter "On the Body" (*Lun quanti*) in his 1901 text. Despite arguing for cranial dominance, Ding establishes an equivalency of function between brain and heart. We see this in the grammatical symmetry of his formulation and in his choice of words. The phrase *yunsi* lends a physiological connotation to the notion of thinking, one commensurate with the action of the heart. The heart circulates blood; the brain circulates thought.

Despite setting up this dualism, Ding Fubao urges his reader to recognize the brain's supremacy. By analogizing the two parts, he aimed, I think, to preempt anticipated objections. He did this by providing a separate role for the heart, literally in the same breath as he argued the new vision of the brain. Ding Fubao possessed a keen sense of his audience; *Questions and Answers about Hygiene* would be published by four different presses, in four different parts of China. Over the next quarter century, Ding emerged as a leading medical publisher and a prolific translator of Western medical texts into Chinese, largely from Japanese. Less than two years prior to writing *Questions and Answers*, in 1899, at age twenty-five, Ding had fallen seriously ill; it was this bout with illness that stimulated his entry into medical writing. Ding's characterization of the heart as the once-perceived center of the body, a role that the brain inexorably must displace, echoes the reformist literature of his time. However, I think this notion is mistaken. I am speaking specifically of the idea that in the medicine of China the heart governed the body, as did the brain in Western medicine.

To say, however, that the heart did not enjoy a special status in Chinese thinking about the body, especially regarding the site of thought, would also be wrong. Authors such as Ding Fubao could have cited passages from the Confucian canon and China's vast medical corpus to argue the heart's ancient association with thought. As one traces the history of the heart, moreover, we see a gradual elevation of its significance, from the Song period on and especially during the Ming, when medical and philosophical writings on the heart become interconnected. What is more, there were multiple discourses on the heart in the scholarly medical tradition. One passage in the *Inner Canon*, for example, describes the heart as the sovereign (*junzhu zhi guan*). The sovereign heart has been explicated as heading up communication with other parts of the body, showing a formative relationship between the emerging medical corpus and the political and cultural unification of the Han. In this vision, we also see an irreducible interdependence of cosmos, state, and body (Sivin, 1995).

The *Inner Canon* also suggests that processes such as sensory perception were mediated neither by the heart nor by any single organ or system. To the contrary, a person's sensory encounter with the exterior world was mediated by distinct parts of the body, which from a Western point of view looks conspicuously decentralized.

Consider the Yellow Emperor's command of Qibo: "I wish to hear of the five *guan*" (*Lingshu*, *Wuyue wushi*, 37.2). Today, the five *guan* would be rendered as the five sensory faculties. We already see an element of this meaning in Qibo's response to the Emperor, which names the nose, eye, mouth, tongue, and ear. But absent from today's understanding of the five *guan* are vital clues about the nature of the body in Chinese medicine. Let us observe Qibo's answer, first without translating the term *guan*. Qibo expounds:

> The nose is the *guan* of the lung;
> The eye is *guan* of the liver;
> The mouth is the *guan* of the spleen;
> The tongue is the *guan* of the heart;
> The ear is the *guan* of the kidney.

We can surmise that the "seven head holes" (plus the tongue) have some sort of relationship with each of the five solid viscera (*wuzang*). This bond exemplifies the idea of correspondence so crucial in Chinese medicine. More specific to our discussion, the dynamic of the aperture-viscera connection describes perception, the process by which, for instance, the kidney hears through the ear. The ear is the opening through which the kidney senses outside sound.

Many centuries later the decentralized experience of perception is even more explicit, as articulated in the late fourteenth-century text, *Formulas for General Welfare* (*Puji fang, Fangmai yunqi zangfu, juan 1, Fangmai zonglun, wuchang dalun*):

> By way of the eye, the liver discerns the five colors;
> By way of the nose, the lungs inhale the five scents;
> By way of the mouth, the spleen tastes the five grains;
> By way of the tongue, the heart judges the five flavors;
> By way of the ear, the kidney distinguishes the five sounds.

The track of perception is utterly distinct from the route traveled in the centralizing and encephalocentric body of biomedicine. Sight and sound and the other senses are not mediated by the brain but by the five solid viscera. Stimuli from the exterior world reached the perceiving viscera in the deepest recesses of the body via the five mediating apertures (*wuguan*). The brain and nervous system did not displace the heart, they displaced the core of the body, the five solid viscera (*wuzang*).

We must then ask what the main sources behind the idea of the centralizing heart in nineteenth-century China were. We can be confident that learned figures such as Tan Sitong and Zheng Guanying were well-versed in philosophy, and from Ding Fubao's writings we know that he was conversant in the classics of Chinese medicine. But there was another source of this idea, imported into China by missionaries, educators, and translators. Let us briefly examine the ideas about the heart that were introduced, sometimes unconsciously, from Europe and the United States.

By stressing the primacy of the brain, foreign advocates of biomedicine argued for a single organ that unified the entire body. As they presented this idea, Western doctors critiqued the idea of the heart as the center. Foreign physicians had picked up on a certain discourse of the heart within China, not necessarily medical, but one providing a template against which to argue. The heart supplied a context for the idea of a centralizing system in the body. By making this argument, however, they inadvertently elevated the position of the heart in medical discussions. The foreign critique had another source, one stemming from a Western preoccupation with the heart.

Although Greek scientists had made revolutionary breakthroughs in the understanding of the nervous system in the third century BCE (von Staden, 1989), the idea of a brain and nerve-controlled body was for some time obscured by other visions. It was from the sixteenth century that the idea of nerves began gradually to gain acceptance again in Western thinking about the body's function. This anatomical renaissance, often associated with Andreas Vesalius (1514–1564) of Brussels, but occurring in different centers across Europe, witnessed not only the "rediscovery" of nerves but also the idea of the anatomical body, a concept that became the foundation for modern biomedicine. However, the heart continued to haunt the Western medical imagination.

Observe the weight of cardiocentrism on the brain research conducted at Oxford in the mid-1600s, which, among other breakthroughs, yielded the best images of the brain up to that time. The leader of this group, Thomas Willis (1621–1675), sought an anatomical explanation of humanity's superiority over animals (Martensen, 1993). The group located the answer in a "richer nerve supply...between heart and brain" (Bynum, 1973: 457, 459). In Willis' analysis, the transcendence of the human mind rests on its access, via the nervous system, to the creative resources of the heart. Purveyors of Western medicine in China understood, I think, the intuitive power of the heart as the vital center, because of the depth of this vision within the Western tradition. They tore down the idea preemptively.

Religion was also a factor. Most foreign doctors in nineteenth-century China were practicing missionaries. Whereas medical missionaries sometimes transmitted spiritual ideas through medical work and technical writings, not all foreign missionaries writing in the Chinese language about the body were doctors. Non-medical authors sometimes introduced a remarkable degree of ambiguity into topics such as the nervous system. "On the Human Body" (1876) illustrates this phenomenon. The unsigned essay published in the missionary journal, *The Chinese Scientific and Industrial Magazine* (*Gezhi huibian*), expounds on how the body moves:

> A person wishes to move a limb. From the brain and nerves (*naoqi jin*) this intention is transmitted to the muscles, and then the tendons either expand or contract and, following the person's intention, the limb moves. It is thus evident that the muscles of the body are entirely governed by the heart (*yi xin wei zhu*).

The passage embeds the heart in the idea of the nerves. To achieve this confusing locution, the author interchanges the words heart (*xin*) and intention (*renxin*), a term containing the word for heart. The text opens by describing nervous function but it ends with a characterization of the heart, not the nerves, as the vehicle of volition and will. Other missionary writings borrow the idea of the governing heart to make moral arguments, such as that the conscience rules the person.

This is the message of "The Heart Governs the Body" (*xin wei shen zhu*), written in 1872 by an American missionary educator, W.A.P. Martin (1827–1916). Martin, Director of the prestigious government school, the *Tongwen guan* (the Shanghai Arsenal), was an expert in the Chinese language who wrote literary Chinese, skillfully deploying images from philosophy, literature, and medicine. In this essay, Martin contends that with its irrational, intensely experienced emotions, the heart ruled humans, more so than will and fate. Writing in Chinese under his Chinese name, Ding Weilian, Martin offers several cases as proof that our very ability to exist is ruled by the conscience, by the heart. One is from France:

> A killer had been found guilty and imprisoned. Several physicians of the National Academy wished to borrow the prisoner in order to test the thought of man (*xinsi*), determining whether or not it has any bearing on life and death. With this goal in mind, they said to the prisoner: "Since you are facing the death penalty, today the sovereign has issued a special favor: you may be left intact, your head is to remain attached to your body. Instead your veins will be opened, waiting until the blood flows completely out and you die. Moreover, you will not suffer pain because a cloth will be wrapped around your head covering your eyes." Placing a blindfold over the condemned man's eyes, the doctors then set a basin of warm water beneath him, and only then did they produce a needle and gently prick his wrists. Then, drop-by-drop, they released water into the basin. The dripping water sounded just like the flowing of blood. The

condemned man lost his color. After some time his breathing grew short and shallow and he lost his strength. A doctor then spoke in faux *sotto voce*: "In the entire body, a person only has 20 *jin* of blood. Today we've already bled upwards of ten *jin*, already half!" Before he had finished speaking, the criminal suddenly collapsed onto the floor and died. However, his body in fact had suffered no injury. He died from fright.

The religiosity of foreign writing and the translated medical corpus was not lost on Chinese physicians and readers. It is hard to evaluate the influence of religious writings, especially when fused with technical ideas. In certain specific cases, we know that both sides of the message were transmitted, as with Tan Sitong who, in addition to immersing himself in Western ideas of science and the body, also became involved with Christianity, although he did not convert (Wright, 1994). However, the mixing of medical with spiritual messages is likely related to why many of the translated medical ideas initially did not take. In the case of nerves, the translations that eventually did take root in China were introduced from Japan.

JAPAN AND NERVOUSNESS

Shenjing, the word for nerve that eventually caught on in China, the term that everyone knows today, had been first translated into Chinese from Japanese in 1902. A Chinese scholar imported the word after a brief sojourn in Japan. The word *shenjing* was a translation of the Japanese word, *shinkei*, which in turn was a translation of the Western idea, nerve. (*Shinkei* and *shenjing* are written with the same Chinese characters). The transmission of aspects of Western biomedicine from Japan to China in the early twentieth century is an important history, but space does not permit even a skeletal synopsis. Instead, I will focus briefly on the history of the Japanese word for nerve, *shinkei*, which was coined in 1774 by a Japanese doctor trained in Chinese medicine. Sugita Gempaku (1733–1817) had come into possession of a post-Vesalian Dutch anatomical text in 1771. The finely engraved Western-style illustrations of the inner body stirred him. Wishing to put the images to empirical test he carried the book to an execution ground, to observe firsthand the dissection of a fresh cadaver. Sugita Gempaku watched the dissector, an elderly butcher, remove parts from the exposed innards. Using his hands, the dissector held them up organ by organ, announcing the name of each viscera, except for a few parts whose names he did not know. Looking back and forth from the text to the dissection, Sugita Gempaku became persuaded of the descriptive value of the anatomical images. In an extraordinary episode lasting four years, he translated the Dutch text into Japanese, together with like-minded physicians and scholars. The product, the *Kaitai shinsho* (1774), is seen as the seed of Western biomedicine in Japan.

The neologism *shinkei* (nerve) was the product of Sugita Gempaku's knowledge of the dominant medical theory of the time, from China, on the one hand, and a meticulous reading of a Dutch text on the other. Sakai argues that Sugita judged the Dutch term *zenuw*, nerve, as corresponding to the *keimyaku* (*jingmo*, in Chinese), the conduits of Chinese medicine in which substances such as blood and *qi* circulate. Sugita likened the Dutch term *zenuw-vogt*, nervous fluid, to *shinki* (*shenqi*, Chinese), the vital entities that move about the body in the *keimyaku* (*jingmo*) conduits. Joining *shin*(*ki*) with *kei*(*myaku*), Sugita formed the new word *shinkei* (Sakai, 1991: 99–157). (In Chinese this would be: *shen(qi)* joined with *jing(mo)* to form *shenjing*). Sugita Gempaku's work shared in common certain intuitions regarding the role of *qi* (*ki*, in Japanese) with translations of the idea of nerves in China.

However, Sugita Gempaku's neologism *shinkei* (*shenjing*) omits the word *qi* (*ki*). Also, there is no evidence that people in early twentieth-century China were aware of the history of the word, that they understood the word *shen* represented the phrase *shenqi*. (By itself, *shen* could mean a number of things, such as deity, spiritual, or the soul). The evidence in fact suggests the opposite. Members of the Shanghai group on medical terminology argued against the word *shenjing* (*shinkei*) precisely because the two-word phrase *shen jing* divulged no hint about the brain or the nervous system. As a compound, it seemed irrelevant. At least the term *naojing*, they reasoned, the term that won the day, contained the word for brain, *nao*. So why, we must ask, did the word *shenjing* eventually become adopted as the Chinese word for nerve? Part of the answer is that as the thousands of men who studied and trained in Japan in the early twentieth century returned to China and gained influence, the skills and material they acquired, such as Japan's analysis of biomedicine, became established. On the other hand, biomedicine in China drew on a diversity of sources. As an explanation, it is incomplete.

The word *shenjing* took root in China, I argue, because of the allied discourse of nervous weakness that was introduced from Japan at the same time. Nerves first became important not because of their reputed connection with cognition or perception or volition, or the fine points of physiology elucidated by the Jesuits and medical missionaries. Schreck and Hobson ably described the physiological capacity of nerves, but not their pathology. In their presentation, nerves were not things that wore out and caused sickness or debility; they animated and enabled human action. Yet it was the idea of nervous pathology coming from Japan that took hold first in China's medicine and then soon afterwards in the social imagination.

The particular form of nervous pathology that caught on was nervous weakness, or, neurasthenia (Kleinman, 1986; Lee, 1999; Shapiro, 2000). Neurasthenia, translated as *shinkei suijaku* in Japanese and then as *shenjing shuairuo* in Chinese, resonated with ideas of depletion that had sustained Chinese medicine since late antiquity. Despite its alien origins, neurasthenia would flourish in China as nowhere else in the world, long after the category was discarded in the countries of its origin and first popularity. The word for nerve, *shenjing*, entered China's medical and popular parlance through its association with the cognate idea of nervousness, which in its early twentieth-century form, showed affinity with age-old ideas about depletion.

* * *

We return to our original question: how different are Western and Chinese medicine? Since the time of Herophilus, nerves have figured centrally in Western thinking about the body, whereas in China the discourse of nerves is quite new. The notion of nerves would thus seem to be a prime example of the incommensurability of traditions. To be sure, the internal logic of Western and Chinese medicine is distinct in almost every respect, as is their analysis and treatment of disease. Yet the bold oppositions that characterize traditional Chinese medicine and Western biomedicine have also masked important points of resonance. I speak specifically of the thinking behind the idea of nervous weakness. Shared intuitions in China and in the West, regarding corporeal and emotional depletion, played a pivotal role in the entry of nerves into Chinese medicine and society. This resonant idea about depletion also helps explain the ubiquity and longevity in twentieth-century China of what was originally the foreign idea of nervousness. The dominant paradigm that currently posits the two traditions as incommensurate, while conceptually and methodologically transformative, also risks eclipsing our understanding of certain shared intuitions about the body.

ACKNOWLEDGMENTS

I am grateful to Shigehisa Kuriyama, Chang Che-chia, Marta Hanson, David Schlesinger, and Matthew D. Shapiro, for excellent and rigorous suggestions; to Nancy Hearst, for research aid; to the History Faculty of the National Taiwan Normal University, for generosity and encouragement during my stay there; and to Helaine Selin, for her work as scholar, editor, and project manager.

NOTES

1 Within China's system of thought, yin and yang are fundamental concepts. "Originally they were certainly connected with darkness and light, but gradually their meanings were greatly extended. Yin came to represent things or natures that are cold, dark, female, negative, passive and so on; while yang represented things or natures that are warm, light, male, positive, active, and so on. According to the Yin-yang school, ... the universe was balanced with the yin and yang forces, and all phenomena are the result of the interplay of these two opposite yet reciprocal forces. *Wu-xing* – earth, wood, metal, fire and water – at the very beginning apparently referred to material substances. They were later conceived in terms of their functional attributes. ... The Five Phases conception was developed, again by the Yin-yang school of Zhou Yan, to be the basis of a comprehensive theory for explaining change in the cosmos. The Five Phases were arranged in particular sequences of succession, either by "mutual conquest" or by "mutual production". Thus by mutual production wood produces fire, fire produces earth, earth produces metal, metal produces water, and water produces wood. By mutual conquest wood is conquered by metal, metal reduced by fire, fire extinguished by water, water blocked up by earth, and earth manipulated by implements made of wood.

Cosmologists of the late Zhou and Han eras used these sequences of the Five Phases to explain cosmic and historical evolution, as well as various seasonal, diurnal, and medical changes and rhythms. A good example is the explanation for the change of the seasons. Spring corresponds to wood, summer to fire, the third month of the summer to earth, autumn to metal, and winter to water. Thus the seasons change in the sequences of mutual production. ...

During the Han, the yin-yang and Five Phases theories were developed further. Dong Zhongshu of the Former Han incorporated the theory into the exposition of Confucian classics, thus extending the yin-yang and Five Phases mode of thinking into the realm of politics and social relations. By that time Han cosmologists could use this theory to build a complicated framework of correspondences between man, the state and the universe. They worked up a long list of things grouped by five and associated them with the *wu-xing*, including five planets, five seasons, five directions, five colors, five musical tones, five wise emperors, five viscera, five orifices, five animals, five conducts, five punishments, and still others. And all these were balanced with yin and yang." (Sun, Xiaochun, "Crossing the boundaries between heaven and man: astronomy in ancient China." In *Astronomy Across Cultures: The History of Non-Western Astronomy*, Helaine Selin, ed. Dordrecht, The Netherlands: Kluwer Academic Publishers, 2000, pp. 426–427).

[2] Parts of the *Inner Canon* are constructed as a series of dialogues.

BIBLIOGRAPHY

Andrews, Bridie. *The Making of Modern Chinese Medicine, 1895–1937*. Ph.D. dissertation. Cambridge: University of Cambridge, 1996.

Bates, Don, ed. *Knowledge and the Scholarly Medical Traditions*. Cambridge and New York: Cambridge University Press, 1995.

Benedict, Carol. *Bubonic Plague in Nineteenth-Century China*. Stanford, California: Stanford University Press, 1996.

Bray, Francesca. *Technology and Gender: Fabrics of Power in Late Imperial China*. Berkeley: University of California Press, 1997.

Bynum, William F. 'The anatomical method, natural theology, and the functions of the brain.' *Isis* 64: 445–468, 1973.

Cass, Victoria B. *Dangerous Women: Warriors, Grannies, and Geishas of the Ming*. Lanham, Maryland: Rowman and Littlefield, 1999.

Chang, Che-chia. *The Therapeutic Tug of War – The Imperial Physician-Patient Relationship in the Era of Empress Dowager Cixi (1874–1908)*. Ph.D. dissertation. Philadelphia: University of Pennsylvania, 1998.

Chang, Chia-feng. *Aspects of Smallpox and its Significance in Chinese History*. Ph.D. dissertation. London: School of Oriental and African Studies, University of London, 1996.

Chao, Yuan-ling. *Medicine and Society in Late Imperial China: A Study of Physicians in Suzhou*. Ph.D. dissertation. Los Angeles: University of California at Los Angeles, 1995.

Chow, Tse-tsung. *The May Fourth Movement: Intellectual Revolution in Modern China*. Cambridge, Massachusetts: Harvard University Press, 1960.

Chu, Pingyi. '*Shenti, linglun yu Tianzhu: Mingmo Qingchu xixue zhong de renti shengli zhishi*' [The body, the soul, and Catholicism: knowledge of the human body and physiology within Western learning during the late Ming and early Qing]. *Xin Shixue* [Taiwan] 7(2): 47–98, 1996.

Cooper, William C. and Nathan Sivin. 'Man as a medicine: pharmacological and ritual aspects of traditional therapy using drugs derived from the human body.' In *Chinese Science: Explorations of an Ancient Tradition*, Shigeru Nakayama and Nathan Sivin, eds. Cambridge, Massachusetts: MIT Press, 1973, pp. 203–272.

Croizier, Ralph C. *Traditional Medicine in Modern China: Science, Nationalism, and the Tensions of Cultural Change*. Cambridge, Massachusetts: Harvard University Press, 1968.

Cullen, Christopher. 'Patients and healers in late imperial China: Evidence from the *Jinpingmei*.' *History of Science* 31(2): 99–150, 1993.

Dikötter, Frank. *Sex, Culture and Modernity in China: Medical Science and the Construction of Sexual Identities in the Early Republican Period*. London: Hurst and Company, 1995.

Ding, Fubao. *Weishengxue Wenda Jiuzhang* [*Questions and Answers about Hygiene, in Nine Chapters*]. 4th ed., n.p., Shou yinlucong, 1901.

Dunstan, Helen. 'The late Ming epidemics: A preliminary survey.' *Ch'ing-shih wen-t'i* 3(3): 1–59, 1975.

Elvin, Mark. 'Tales of *shen* and *xin*: body-person and heart-mind in China during the last 150 years.' In *Fragments for a History of the Human Body*, pt. 2, Michel Feher, ed. New York: Zone, 1989, pp. 267–346.

Farquhar, Judith. *Knowing Practice: The Clinical Encounter of Chinese Medicine*. Boulder, Colorado: Westview Press, 1994a.

Farquhar, Judith. 'Multiplicity, point of view, and responsibility in traditional Chinese healing.' In

Body, Subject, and Power in China, Angela Zito and Tani E. Barlow, eds. Chicago and London: University of Chicago Press, 1994b, pp. 78–99.

Furth, Charlotte. 'Concepts of pregnancy, childbirth, and infancy in Ch'ing Dynasty China.' *Journal of Asian Studies* 46(1): 7–35, 1987.

Furth, Charlotte. *A Flourishing Yin: Gender in China's Medical History, 960–1665*. Berkeley: University of California Press, 1999.

Goldschmidt, Asaf. *The Transformations of Chinese Medicine during the Northern Song Dynasty (A.D. 960–1127): The Integration of Three Past Medical Approaches into a Comprehensive Medical System Following a Wave of Epidemics*. Ph.D. dissertation. Philadelphia: University of Pennsylvania, 1999.

Goodrich, L. Carrington and Chaoying Fang, eds. *Dictionary of Ming Biography, 1368–1644*. New York and London: Columbia University Press, 1976.

Grant, Joanna. 'Medical practice in the Ming dynasty – a practitioner's view: Evidence from Wang Ji's *Shishan yi'an*.' *Chinese Science* 15: 37–80, 1998.

Hanson, Marta. *Inventing a Tradition in Chinese Medicine: From Universal Canon to Local Medical Knowledge in South China, The Seventeenth to the Nineteenth Century*. Ph.D. dissertation. Philadelphia: University of Pennsylvania, 1997.

Hanson, Marta. 'Robust northerners and delicate southerners: the nineteenth-century invention of a southern medical tradition.' In *Innovation in Chinese Medicine*, Elisabeth Hsu, ed. Cambridge: Cambridge University Press, 2001, pp. 262–292.

Harper, Donald J. *Early Chinese Medical Literature: the Mawangdui Medical Manuscripts*. London and New York: Kegan Paul International, 1998.

Hinrichs, T.J. 'New geographies of Chinese medicine.' *Osiris* 13: 287–325, 1998.

Hobson, Benjamin and Chen Xiutang. *Quanti Xinlun* [A New Theory of the Body]. Hui ai Medical Office, n.p., 1851.

Honig, Emily and Gail Hershatter. *Personal Voices: Chinese Women in the 1980's*. Stanford, California: Stanford University Press, 1988.

Howard, Paul. *Opium Oppression in Qing China: Responses to a Social Problem, 1729–1906*. Ph.D. dissertation. Philadelphia: University of Pennsylvania. 1998.

Hsiung, Ping-chen. "Xie" and "li": a study of health and pathological problems concerning children's digestive tracts in late imperial China.' *Chungkuo wen hua yen chiu so hsüeh pao* 6: 129–161, 1997.

Hsu, Elisabeth. *The Transmission of Chinese Medicine*. Cambridge: Cambridge University Press, 2000.

Huangdi neijing zhangju suoyin. Taipei: Qiye shuju, 1987.

Hymes, Robert P. 'Not quite gentlemen? Doctors in Sung and Yuan.' *Chinese Science* 8: 9–76, 1987.

Ishida, Hidemi. *Chūgoku Igaku Shisōshi* [A History of Chinese Medical Thought]. Tokyo: Tokyo University Press, 1992.

Jenner, William. *Modern Chinese Stories*. Oxford: Oxford University Press, 1970.

Katz, Paul. *Demon Hordes and Burning Boats: The Cult of Marshall Wen in Late Imperial Chekiang*. Albany: State University of New York Press, 1995.

Keegan, David. *The 'Huang-ti nei-ching': The Structure of the Compilation; The Significance of the Structure*. Ph.D. dissertation. Berkeley: University of California, 1988.

Kleinman, Arthur. *Social Origins of Distress and Disease: Depression, Neurasthenia, and Pain in Modern China*. New Haven, Connecticut: Yale University Press, 1986.

Kleinman, Arthur. *Writing at the Margin: Discourse Between Anthropology and Medicine*. Berkeley: University of California Press, 1995.

Kuhn, Philip A. *Soulstealers: The Chinese Sorcery Scare of 1768*. Cambridge, Massachusetts: Harvard University Press, 1990.

Kuriyama, Shigehisa. 'Concepts of disease in East Asia.' In *The Cambridge World History of Human Disease*, Kenneth F. Kiple, ed. Cambridge: Cambridge University Press, 1993, pp. 52–59.

Kuriyama, Shigehisa. 'The imagination of the winds and the development of the Chinese conception of the body.' In *Body, Subject, and Power in China*, Angela Zito and Tani E. Barlow, eds. Chicago and London: University of Chicago Press, 1994, pp. 23–41.

Kuriyama, Shigehisa. 'Interpreting the history of bloodletting.' *Journal of the History of Medicine and Allied Sciences* 50(1): 11–46, 1995.

Kuriyama, Shigehisa. *The Expressiveness of the Body and the Divergence of Greek and Chinese Medicine*. New York: Zone Books, 1999.

Lau, Joseph S.M., C.T. Hsia and Leo Ou-fan Lee, eds. *Modern Chinese Stories and Novellas, 1919–1949*. New York: Columbia University Press, 1981, pp. 125–141.

Lee, Jen-der. 'Han-Tang zhi Jian Yishu Zhong de Shengchan zhi Dao' [Childbirth in Late Antiquity and Early Medieval China]. *Zhongyang yanjiu yuan Lishi yuyan yanjiu suo jikan* [Bulletin of the Institute of History and Philology, Academia Sinica, Taiwan] 67(3): 533–654, 1996.

Lee, Leo Ou-fan. *The Romantic Generation of Modern Chinese Writers*. Cambridge, Massachusetts: Harvard University Press, 1973.

Lee, Sing. 'Diagnosis postponed: *shenjing shuairuo* and the transformation of psychiatry in post-Mao China.' *Culture, Medicine and Psychiatry* 23(3): 349–81, 1999.

Lei, Sean H-L (Lei Hsiang-lin). *When Chinese Medicine Encountered the State: 1910–1949*. Ph.D. dissertation. Chicago: University of Chicago, 1999.

Leslie, Charles and Allan Young, eds. *Paths to Asian Medical Knowledge*. Berkeley: University of California Press, 1992.

Leung, Angela. 'Organized medicine in Ming-Qing China.' *Late Imperial China* 8(1): 134–165, 1987.

Li, Jianmin. *Si Sheng Zhi Yu: Zhou Qin Han Mai Xue Zhi Yuan Liu* [Mai and the Development of Medical Knowledge in Early China]. Taipei: Zhongyang yanjiu yuan Lishi yuyan yanjiu suo zhuankan 101 [Academia Sinica, Institute of History and Philology, Taiwan], 2000.

Lin, Fu-shih. *Chinese Shamans and Shamanism in the Chiang-nan Area during the Six Dynasties Period (3rd–6th century A.D.)*. Ph.D. dissertation. Princeton, New Jersey: Princeton University, 1994.

Lo, Vivienne. *The Influence of Western Han Nurturing Life Literature on the Development of Acumoxa Therapy*. Ph.D. dissertation. London: University of London, 1998.

Lyell, William A. *Diary of a Madman and Other Stories*. Honolulu: University of Hawaii Press, 1990.

Ma, Boying, Gao Xi and Hong Zongli. *Zhongwai Yixue Wenhua Jiaoliu Shi* [History of Sino-Foreign Intercultural Medical Exchange]. Shanghai: Wenhui chuban she, 1993.

Martensen, Robert L. *The Circles of Willis: Physiology, Culture, and the Formation of the 'Neurocentric' Body in England, 1640–1690*. Ph.D. dissertation. San Francisco: University of California, San Francisco, 1993.

McKnight, Brian E. *The Washing Away of Wrongs: Forensic Medicine in Thirteenth-Century China*. Ann Arbor: Center for Chinese Studies, University of Michigan, 1981.

McLeod, Katrina C.D. and Robin D.S. Yates. 'Forms of Ch'in law: An annotated translation of Feng-chen shih.' *Harvard Journal of Asiatic Studies* 41(1): 11–163, 1981.

Needham, Joseph. *Science and Civilisation in China*, vol. 2, *History of Scientific Thought*. Cambridge: Cambridge University Press, 1956.

Needham, Joseph and Lu Gwei-djen, *Science and Civilisation in China*, vol. 6, *Biology and Biological Technology*, pt. 6 *Medicine*, edited and introduced by Nathan Sivin. Cambridge: Cambridge University Press, 2000.

Porkert, Manfred. *The Theoretical Foundations of Chinese Medicine: Systems of Correspondence*. Cambridge, Massachusetts: MIT Press, 1974.

Rogaski, Ruth. *From Protecting the Body to Defending the Nation: The Emergence of Public Health in Tianjin, 1859–1953*. Ph.D. dissertation. New Haven, Connecticut: Yale University, 1996.

Sakai, Shizu. '*Kaitai shinsho to jutei Kaitai shinsho*' [Kaitai shinsho and the revised Kaitai shinsho]. In *Otsuki Gentaku no kenkyu*, Yogakushi kenkyukai [Research group on the history of Western medicine], ed. Kyoto: Shibunkaku Press, 1991, pp. 99–157.

Scheid, Volker. *Contemporary Chinese Medicine: Plurality and Synthesis*. Durham, North Carolina: Duke University Press, 2002.

Schipper, Kristopher. *Le Corps Taoiste*. Paris: Fayard, 1982.

Schram, Stuart R., ed. *The Political Thought of Mao Tse-tung*. New York: Praeger, 1976.

Schreck, Johann (Terrenz) and Bi Gongchen. *Taixi Renshen Shuogai* [*Western Views of the Human Body, an Abbreviated Treatise*], n.p., 1624–1643.

Schwartz, Benjamin I. *In Search of Wealth and Power: Yen Fu and the West*. Cambridge, Massachusetts: Harvard University Press, 1964.
Schwartz, Benjamin I. *The World of Thought in Ancient China*. Cambridge, Massachusetts: Harvard University Press, 1985.
Shapiro, Hugh. 'The puzzle of spermatorrhea in Republican China.' *Positions: East Asia Cultures Critique* 6(3): 551–595, 1998.
Shapiro, Hugh. 'Neurasthenia and the assimilation of nerves into China.' Presented at the Symposium on the History of Disease, Academia Sinica, Institute of History and Philology, Nankang, Taiwan, June 16–18, 2000.
Sivin, Nathan. *Traditional Medicine in Contemporary China. A Partial Translation of* Revised Outline of Chinese Medicine *(1972) with an Introductory Study on Change in Present-day and Early Medicine*. Ann Arbor: University of Michigan Center for Chinese Studies, 1987.
Sivin, Nathan. 'Huang ti nei ching.' In *Early Chinese Texts: Bibliographical Guide*, Michael Loewe, ed. Berkeley: University of California Press, 1993, pp. 196–215.
Sivin, Nathan. 'State, cosmos, and body in the last three centuries B.C.' *Harvard Journal of Asiatic Studies* 55(1): 5–37, 1995.
Sivin, Nathan. *Medicine, Philosophy and Religion in Ancient China, Researches and Reflections*. Aldershot, UK: Variorum, 1995.
Sivin, Nathan. 'Introduction.' In *Science and Civilisation in China*, vol. 6, *Biology and Biological Technology*, pt. 6, *Medicine*, by Joseph Needham and Lu Gwei-djen, edited and introduced by Nathan Sivin. Cambridge: Cambridge University Press, 2000, pp. 1–37.
Spence, Jonathan D. *The Memory Palace of Matteo Ricci*. New York: Viking Penguin, 1983, reprinted 1985.
Spence, Jonathan D. 'Opium smoking in Ch'ing China.' In *Conflict and Control in Late Imperial China*, Frederic Wakeman, Jr. and Carolyn Grants, eds. Berkeley: University of California Press, 1975, pp. 143–173.
Tan, Sitong. 'Lun quanti xue' [On anatomy; 1898]. In *Tan Sitong Quanji*, vol. 2 [Complete Works of Tan Sitong], Cai Shangsi and Fang Xing, eds. Beijing: Zhonghua shuju, 1981, reprinted 1990, pp. 403–405.
Unschuld, Paul U. *Medicine in China: A History of Ideas*. Berkeley: University of California Press, 1985.
Unschuld, Paul U. *Nanjing, the Classic of Difficult Issues, with Commentaries by Chinese and Japanese Authors from the Third Through the Twentieth Century*. Berkeley: University of California Press, 1986.
Van Gulik, Robert H. *Sexual Life in Ancient China: A Preliminary Survey of Chinese Sex and Society from ca. 1500 B.C. till 1644 A.D.* Leiden: E.J. Brill, 1961, reprinted 1974.
von Staden, Heinrich. *Herophilus, The Art of Medicine in Early Alexandria*. Cambridge: Cambridge University Press, 1989.
Wakeman, Frederic, Jr. *The Great Enterprise: The Manchu Reconstruction of Imperial Order in Seventeenth-Century China*. Berkeley: University of California Press, 1985.
Wile, Douglas. *Art of the Bedchamber: The Chinese Sexual Yoga Classics, including Women's Solo Meditation Texts*. Albany: State University of New York Press, 1992.
Wright, David. 'Tan Sitong and the ether reconsidered.' *Bulletin of the School of Oriental and African Studies* 57(3): 551–575, 1994.
Wu, Yi-Li. *Transmitted Secrets: The Doctors of the Lower Yangzi Region and Popular Gynecology in Late Imperial China*. Ph.D. dissertation. New Haven, Connecticut: Yale University, 1998.
Wu, Yiyi. 'A medical line of many masters: A prosopographical study of Liu Wansu and his disciples from the Jin to the early Ming.' *Chinese Science* 11: 36–65, 1994.
Xiong, Yuezhi. *Xixue Dongjian yu WanQing Shehui* [The Dissemination of Western Learning and Late Qing Society]. Shanghai: Shanghai People's Publishing, 1994, reprinted 1995.
Yamada, Keiji. 'The formation of the *Huang-ti nei-ching*.' *Acta Asiatica* 36: 67–89, 1979.
Yamada, Keiji. 'Anatometrics in ancient China.' *Chinese Science* 10: 39–52, 1991.
Yip, Ka-che. *Health and National Reconstruction in Nationalist China: The Development of Modern Health Services, 1928–1937*. Ann Arbor, Michigan: Association for Asian Studies, 1995.
'Yixue Mingci Shenchahui Diyici Kaihui Jilu' [Minutes of the first meeting of the group for the

review of medical terminology]. *Zhonghua Yixue Zazhi* [Chinese Medical Journal] 3(2): 30–60, 1917.

Zhao, Hongjun. *Jindai Zhongxiyi Lunzheng Shi* [History of the Controversies of Chinese and Western Medicine in Modern Times]. Anfei: Anhui keji chubanshe, 1989.

Zheng, Guanying. 'Yi Dao' [Medicine, 1892]. In *Sheng Shi Wei Yan* [Warnings to a Prosperous Age], by Zheng Guanying. Shanghai: Shanghai shuju, 1896, pp. 20a–23a.

Zito, Angela and Tani E. Barlow, eds. *Body, Subject, and Power in China.* Chicago and London: University of Chicago Press, 1994.

JIM B. TUCKER

RELIGION AND MEDICINE

This is the mistake some doctors make with their patients.
They try to produce health of body apart from health of soul.

Plato

Religion and medicine have been two pillars supporting human life throughout recorded history. Religion represented early humans' attempt to make sense of their world; it gave them an understanding and some feeling of control because of a belief in the power of ritual to effect changes in the world.

Medicine was an attempt to make changes in the natural course of physical illness. From the beginning, humans tried to understand the causes for illness and sought ways to alleviate it. The medical methods used followed from their understanding of the causes. Initially, the spiritual world was offered as an explanation for positive and negative events, including those related to physical health. Thus, just as severe weather or other calamities might be ascribed to an angry god or an evil spirit, spiritual causes were assumed for physical illness. Medical interventions therefore often involved religious rituals. Over time, however, the religious and medical spheres became more separate.

The interaction between religion and medicine has varied across cultures and time. In the West, the dominance of biomedicine in modern times has produced a strong movement away from recognizing any interplay between the two. Western medicine has been slow to recognize any effect on health by anything without a clearly concrete mechanism. While its failure to acknowledge the health effects of emotions and stress has changed to some extent over the last couple of decades, it continues to discount any effect that religion may have on health, beyond superficial ones. Other modes of thought have held religion and medicine in closer quarters, even as they have accepted non-spiritual causes of illnesses. Spirituality and physical health have been considered in concert at least, compared to Western medical thinking that has viewed spirituality as irrelevant. Because of this, Western medicine has been slow to accept recent evidence suggestive of the potential importance of religion on health.

H. Selin (ed.), Medicine Across Cultures: History and Practice of Medicine in Non-Western Cultures, 373–384.

In forms of medicine such as shamanism, illnesses are ascribed to factors such as evil spirits. Health and spirituality do not simply exist side by side; illness is an expression of activities in the spiritual realm. Thus, one person is both priest and physician, and illness is treated by spiritual means.

Shamanism is thought to have been practiced as far back as the stone age (Basilov, 1999) and may well have been practiced by all human groups in the early periods of human history. As defined by Basilov, shamanism is an early form of polytheism with the following key beliefs: 1) the surrounding world is inhabited by spirits who can influence people's lives; 2) there are general and reciprocal interconnections in nature that include human beings; 3) humans are equal to other forms of life; 4) human society is closely connected with the cosmos; 5) it is possible for a human to acquire some qualities of a spirit and visit other worlds; and 6) the aim of religious activity is to defend and make prosperous a small group of kinsmen.

In such a belief system, the shaman was thought to be able to establish direct contact with the spirit world by inducing a state of ecstasy, and in fact, he could only perform his duties with the assistance of his helping spirits. The shamans were, and in some places still are, viewed as having one foot in the spirit world and one in the physical. The ability to traverse from one to the other, via trance, enabled the shamans to gain some mastery in the spiritual plane and thus mastery of spiritual entities. As evil spirits were felt to be the cause of sickness and death, the shaman was the resource available to combat illness. He was at the very center of healing efforts. [Editor's note: for another discussion on shamanism, see Hultkrantz's chapter on The Relation between Medical States and Soul Beliefs among Tribal Peoples.]

As other concepts of health developed, specialization began to occur (Mauceri, 1986). The priest and physician were not necessarily the same person, even when medical traditions developed in concert with religious ones. Physical health was no longer purely subsumed under the spiritual realm, and the natural world was judged to have a large effect. Conceptualizations of the natural world developed differently in different places, and some were more inclusive of the effects of the spiritual world or the universe at large than others. And while some developed in concert with religious traditions, such as Ayurvedic medicine and Hinduism, they often placed greater emphasis on the effects of the natural world than the spiritual one.

ĀYURVEDA

In India, the practice of Vedic medicine began in the latter part of the second millennium BCE. *Veda*, "the knowledge", came from the Sanskrit liturgical literature of the *brāhmana* (hereditary priests), which involved primarily a magical and religious approach to illness and its treatment. Internal diseases were thought to have magical and demonic causes, while traumatic injuries such as broken bones were assumed to be due to external causes (Wujastyk, 1996).

The Vedic approach to religion came to be seen as mechanistic, with its emphasis on ritual incantations and prayers, and other religions developed with a greater emphasis on internalized values.

The Ayurvedic medicine that followed has been a significant part of Indian civilization for over 2000 years. In its view, the world consists of five elements: earth, water, fire, air, and space. These in turn make up the three humors that the body consists of: wind, bile, and phlegm. The Ayurvedic physician focuses on maintaining or restoring the equilibrium of these three, and this is done through practical measures involving herbal medicines or procedures that include interventions such as enemas and ointments. For the most part, the system focuses on a view of the natural world separate from magic emanating from the spiritual one. There have been exceptions to this, such as the idea that many childhood diseases represented attacks by demons; for the most part, however, the emphasis has been on a naturalistic view of health and illness. Ayurvedic medicine has continued to thrive in India to the current day, and in recent years it has expanded to the rest of Asia and to the West. It has evolved over the years, with a move toward standardized compound medications and away from the practice of preparing a medication specific to a particular patient, as the original texts dictated (Wujastyk, 1996).

While Ayurvedic medicine focused on the natural world, it did not discount the religious one. *Carakasamhitā*, the earliest of the three most authoritative texts of Ayurveda, covers subjects such as diagnosis and prognosis, but it also covers how the five sense organs relate to the five elements of the cosmos and how the gods and the ages of the worlds correspond to the emotions of a person and his or her phases in a single life cycle (Knipe, 1989). Thus, respect for the spiritual and the religious was inherent from the start in Ayurvedic traditions.

BUDDHISM

While Buddhism did not develop its own medical conceptualizations, Buddhist monasteries began medical treatment of the lay community by the middle of the third century BCE. Buddhist communities in India and Southeast Asia generally followed Ayurvedic principles, while those in China relied on Chinese medicine. Buddhist monasteries offered medical services, and Buddhism became associated with medicine (Kitagawa, 1989).

Buddhist thought, on one level, concurs with the Ayurvedic notion of disease being due to humoral disequilibrium. At the same time, the tenet of karma[1] that is so central in Buddhism must be considered in matters of physical illness. Karma is considered the root of all suffering and thus is a key cause of physical suffering. As explained by the ancient Buddhist master Nagasena,

> There are eight causes by which sufferings arise, by which many beings suffer pain. And what are the eight? Superabundance of wind, and of bile, and of phlegm, the union of these humours, variations in temperature, the avoiding of dissimilarities, external agency, and Karma. From each of these there are some sufferings that arise, and these are the eight causes by which many beings suffer pain (Rhys Davids, 1965).

Karma was considered one cause of disease, but the question of how much of disease can be ascribed to karma has been present in Buddhist thought since its beginnings (Kitagawa, 1989). The Buddha himself was said to have removed all effects of karma but still suffered physical maladies, so naturalistic causes were considered from the start.

Buddhism is nontheistic and has often accommodated various non-Buddhist beliefs in the communities to which it spread. These include beliefs in spirits and demons, such as ones that can cause disease. Some diseases are thought to have both naturalistic and demonic etiologies. For example, Buddhists in Sri Lanka have believed that menorraghia (abnormally profuse menstrual flow) can be caused by a natural excitement of heat or bile, or by the demon Sanni Yaka or Riri Yaka (Obeyesekere, 1969).

Such examples demonstrate how religious and spiritual beliefs have coexisted with medical beliefs and practices. While Buddhist teachings hold that causes of disease are either naturalistic or karmic, individuals have held both Buddhist and folk beliefs that led them to try various forms of faith healing.

THE CONNECTION OF SPIRIT AND BODY

All medical traditions have grown out of particular cultures and have been altered by the medical traditions of other cultures. For instance, Ayurvedic medicine was challenged by Unani (Greco-Arabic) medicine that entered India during the Muslim invasions of the first half of 6th century (Svoboda, 1996), and it subsequently had to deal with Western ideas that came with British colonization. In that sense, the medical traditions were cultural, just as the religious traditions were, and they often traveled together during cultural spread or retreat.

In the West, religious beliefs and biomedical methods developed separately. Biomedicine has been so successful in addressing many forms of illness that generally it is perceived as the only truth when it comes to physical illness. While many patients have religious beliefs, these beliefs have been considered irrelevant to their physical health. In fact, with its insistence on physical cause and effect, Western medicine precludes the possibility of the spiritual as a cause of illness.

Since Western scientific thought does not allow for the existence of the spiritual, there is nothing to connect an individual to the world other than direct physical connections and the need to follow general physical laws. This insistence on the physical world to the exclusion of all else has become unsatisfactory to increasing numbers of people. The emphasis on physical causes of disease has led to an increasingly technological approach to addressing illness, and while patients continue to want to take advantage of the advances that this technology has produced, more and more of them are dissatisfied when these approaches come at the cost of having treatment be purely disease-focused rather than patient-focused.

This has led to an increase of interest in complementary and alternative medicine. One recent study of 480 breast cancer patients found that 28% of them turned to alternative medicine after they were diagnosed (Burnstein *et al.*,

1999). The patients who did so had greater psychosocial distress and a poorer quality of life than those who did not, raising the possibility that patients at times seek out alternative medical approaches because they do not see their distress being addressed within the biomedical health care system.

In addition to wanting more of a personal approach to health care, many patients also want a more spiritual one. This is becoming a significant trend, and one recent poll reported that 63% of Americans wanted their physicians to discuss spirituality with them (McNichol, 1996).

Although this might seem to pose a problem for a field that does not acknowledge the existence of spirituality, the field is responding nonetheless. The number of American medical schools that provide coursework or lectures on spirituality has grown from four in 1992 to 79 in the year 2000; this represents more than half of the U.S. medical schools. The Association of American Colleges issued a Medical School Objectives report in 1999. One objective for medical students was that they demonstrate "the ability to elicit a spiritual history as well as an understanding that the spiritual dimension of people's lives is an avenue for compassionate care giving."

The Joint Commission on Accreditation of Healthcare Organizations also established Spiritual Assessment Standards in 1998 to address the need for a greater understanding of how spirituality affects patient care. In 1999, the American College of Physicians and the American Society of Internal Medicine issued a consensus report that suggested that physician care include consideration of the psychosocial, existential, or spiritual suffering of patients with serious illnesses.

While much of this movement is no doubt a response to the wishes of patients, it may also be due in part to some recent research looking at the interaction between religion and medicine. There is now increasing evidence that religion and spirituality are beneficial to both physical and spiritual health. More than 350 studies have now been conducted that look at the interaction between religious involvement and health, and many have found that individuals who are involved in religious activities tend to be in better health, have healthier lifestyles, and require fewer health services (Koenig, 2001). Religious involvement may have an impact on physical health that is comparable to cigarette smoking cessation (Koenig *et al.*, 1999) and may increase life expectancy by 7–14 years (Hummer *et al.*, 1999).

Studies have looked at specific illnesses such as cancer and hypertension in samples of Catholics, Protestants, Jews, Hindus, Buddhists, Parsis and Muslims, and religious involvement appears to be an epidemiologically protective factor (Dossey, 1999). While the mechanisms of this are hotly debated, the association between religiosity and physical health appears to be a genuine one. Some have argued that such an association can be easily explained by lifestyle effects such as churchgoers' being less likely to smoke cigarettes and abuse alcohol, thus attempting to discount spirituality as a possible force for improved health.

This association and its implications for treatment have made some people uncomfortable because of the blurring of the separation of medicine and religion. In the first of two articles by Sloan *et al.*, the authors acknowledge the large interest in issues regarding the intersection of spirituality and medicine, both among the public and among health care professionals. Yet they go on to state that, "When doctors depart from areas of established expertise to promote a non-medical agenda, they *abuse* their status as professionals. Thus, we question inquiries into a patient's spiritual life in the service of making recommendations that link religious practice with better health outcomes" (Sloan *et al.*, 1999: 666. Emphasis added). Implicit in this statement are the ideas that issues of a patient's spirituality constitute "a non-medical agenda" and that "making recommendations that link religious practice with better health outcomes" might constitute patient abuse. Thus, we see a strong push to maintain the separation of religion and medicine that has been such a strong part of Western biomedicine and that some patients are finding dissatisfactory.

In the second article, the authors express concern that emphasizing the health benefits of religion trivializes it, and they state, "We are concerned that attempts to obtain scientific evidence of the health benefits of religious activity and to use such activity instrumentally in achieving beneficial health outcomes not only are superficial but also suggest that the value of religion derives from its effects on health" (Sloan *et al.*, 2000: 1915–16). How these attempts would suggest that the value of religion derives from its effect on health is not explained.

While it is understandable that many in the biomedicine field would resist considering the health effects of religion, it is noteworthy that several of the authors of that piece are chaplains. The authors point out that "physicians are not trained to engage in in-depth conversations with their patients about their spiritual concerns." They go on to say that many facilities have chaplains or other community clergy who have such expertise, and that patients who seek spiritual support can be referred to them. While that is undoubtedly true, the question remains whether physicians need to be kept away from a holistic approach that includes addressing spiritual issues. Such thinking is part of the mechanistic approach of biomedicine that has been extremely successful in many ways but has also produced dissatisfaction in many patients. The evidence noted above suggests that, just as the effects of stress on health that were ignored for many years have been found to be important, the effects of the religious lives of patients may need to be included in assessments of health if we want even a mechanistic approach to be as complete as possible.

If convincing evidence is indeed found to support the health benefits of patients' addressing their spiritual concerns through religious practices, then medicine should at least show those practices enough respect to allow them to be appropriate for discussion.

While such discussion may raise the question of whether we are weakening the specialization of priests and physicians that has been common for centuries, recognizing the value of spiritual concerns and their potential importance in health is not the same as having physicians act as clergy. The 63% of Americans who want their physicians to discuss spirituality with them do not want their physicians to become their pastors. At the same time, they presumably do not like their physical health needs addressed in an environment that is completely isolated from their religious lives. Having medicine brush off spiritual concerns as someone else's area of expertise might lead patients to feel that their physical and spiritual needs are somehow incongruous when the research above suggests that their paths may be significantly linked. If physicians give patients the message that issues of spirituality should not be discussed, then they may be unable to help patients in their efforts to find a way to make sense of, and find peace with, their concerns about mortality and the various emotional reactions that arise during health crises. It would seem that if physicians can instead include spirituality in their concerns, then they might create an environment more conducive to good health, both physical and psychological.

INTERCESSORY PRAYER/DISTANT HEALING

Another area of research addressing the interaction of religion and medicine involves the effects that praying for others may have on them. In one study (Byrd, 1988), 393 patients admitted to a coronary care unit (CCU) were assigned to either an intercessory prayer (praying for others) group or a control group. In the prayer group, Christians outside of the hospital prayed for the patients, and the patients did not know if they were in the prayer group or the control group. It was found that the patients who were prayed for had a less severe hospital course, and multivariate analysis showed a highly significant difference between the prayed-for patients and the control patients.

Another study dealt with distant healing, defined as "a conscious, dedicated act of mentation attempting to benefit another person's physical or emotional well-being at a distance" (Sicher *et al.*, 1998: 356). It found that AIDS patients who underwent the distant healing (in a controlled, double-blind study) acquired significantly fewer new AIDS-defining illnesses than the control patients, had lower illness severity, and required significantly fewer doctor visits, hospitalizations, and days in the hospital.

There are different methods of distant healing, and LeShan (1975) developed one of the best known. In it, the healer meditates to achieve an altered state of consciousness in which he or she experiences a sense of oneness with the universe and feels an intense sense of love. The healer carries the patient into this state in the meditation (even though the patient is not physically present at the time), and in so doing is thought to produce healing. This method shows similarities to shamanistic healing in that the healer forms a connection with the universe through an altered state of consciousness. This modern form of healing differs from shamanism in that anyone can become a healer if he or she masters the exercises needed to develop the technique. It avoids specific gods and demons but still shows belief in the possibility that a special connection to the universe can be used for healing.

Another recent study involved 990 patients in a coronary care unit (Harris *et al.*, 1999) who were randomized to either an intercessory prayer group or a control group. As with the other studies, patients were not told which group they were assigned to, and in this particular study, they were not even told that they were going to be part of a study. The results of the study showed that while the lengths of the CCU and hospital stays did not differ between the two groups, the prayed-for group had significantly lower overall adverse outcomes.

Along with these studies, a recent review (Astin *et al.*, 2000) noted that there have been over 100 clinical trials of distant healing. When the authors limited their review to randomized, placebo-controlled, human studies published in peer-reviewed journals and involving clinical populations, they were still left with 23 studies meeting their inclusion criteria. Of these, 5 examined just prayer as the intervention, 11 looked at non-contact Therapeutic Touch, and 7 assessed other forms of distant healing. Thirteen of the 23 studies, or 57%, showed statistically significant treatment effects, while nine showed no effect and one showed a negative effect. Given these positive results, the authors concluded that the evidence thus far merits further study.

Since then, other studies have been published as well. One study looked at the effect of intercessory prayer on the pregnancy rates in women undergoing *in vitro* fertilization (Cha *et al.*, 2001). In it, 219 women in Seoul, Korea were randomly assigned so that half were prayed for by prayer groups of various Christian denominations in the United States, Canada, and Australia. The patients and their providers did not know that they were participating in a study. The main outcome measure was the clinical pregnancy rate in each group, and it was found that the prayed-for group had a significantly higher pregnancy rate of 50%, compared to 26% in the control group. Interestingly, the study was designed so that one prayer group prayed for each patient while another one prayed for the first group with the goal of increasing their prayer efficiency.

Another study at Duke University was performed because it was noted that patients undergoing cardiac procedures for unstable coronary syndromes have "emotional and spiritual distress that may promote procedural complications" (Krucoff *et al.*, 2001: 760). This study added four noetic therapies – stress relaxation, imagery, touch therapy, and prayer – to the treatment of patients undergoing acute coronary procedures. While the results of this pilot study did not reach statistical significance, patients in the treatment groups showed a 25–30% reduction in complications compared with the patients who received only standard treatment. The lowest complication rates were seen in patients receiving the prayer intervention. Those patients and the staff members treating them did not know if they were in the group receiving prayer or not. The praying was done by eight participating prayer groups in various parts of the world after they were given the patient's name, illness, and procedure being performed. The groups included several Christian ones, a Jewish one, and two Buddhist ones (one in France and one in Nepal), and they each prayed in their customary fashion. Because of the promising results, this pilot study has now been expanded to include several major medical centers in the United States.

In these various studies, prayer or distant-healing has been used as an adjunct to standard treatment. This is due in part, no doubt, to the fact that it would be unethical to withhold, for instance, a coronary procedure from a patient with an acute coronary condition, but it also shows that the researchers are not conceptualizing spiritual issues and health issues as completely separate. They are doing what patients have often done for hundreds of years: combining somatic treatments with religious ones in the hope that a better result can be obtained than with just one. In many of these studies, the somatic treatment remains the primary one, but the prayer may also be helpful. So, religion and medicine are combined where one can be complementary to the other.

These studies with human subjects have occurred on top of a foundation of studies with other organisms. These studies have looked at efforts by subjects to influence living systems with mental intention. There have been dozens of such studies, and they have looked at effects on various processes, including, among others, biochemical reactions in vitro, recovery rates of animals from anesthesia, growth rates of tumors, rates of wound healing in animals, and replication rates of microorganisms in test tubes (Dossey, 2000). At one point, out of 131 controlled studies that had been done, 56 of them produced results that were statistically significant to the point that the likelihood of their being due to chance was less than 1 in 100 (Dossey, 1993). So, where one or two positive studies would be expected by chance out of 131 studies, there were, in fact, 56 of them.

All of these studies raise the question of how prayer or distant healing may work. When an individual is praying for his or her own health, we may postulate a number of possible mechanisms for positive results: the placebo effect, the relaxation response noted in psychosomatic medicine, or the positive lifestyle changes that can accompany increased religious participation. While these possibilities have had trouble gaining acceptance from mainstream biomedicine, they can at least be incorporated into a slightly expanded understanding of influences on an individual's health.

When we move into the area of prayer at a distance about which the patient is unaware, however, we are walking on shakier ground. At the present time, biomedicine does not allow for the possibility that prayer at a distance could make a difference. If these studies continue to show positive effects, however, biomedicine might have to provide an explanation for how prayer works.

This goes against the notion that explanations for phenomena need to have consilience with generally accepted understandings of the world so that the pieces fit together to form a cohesive body of knowledge (Wilson, 1998). If prayer is found to have beneficial effects that cannot be explained by lifestyle issues or expectations or the like, then biomedicine and, in fact, Western science as a whole, could be shaken by a phenomenon that is at odds with its understanding of how the world works.

One issue of note is whether the improvements in health observed in these studies are the result of divine intervention or the direct result of the prayer itself. As the authors of one study said, "We have not proven that God answers prayer or that God even exists. It was intercessory prayer, not the existence of God, that was tested here" (Harris *et al.*, 1999: 2277). In other words, does God intervene to improve people's health, or is it the act of someone praying that makes a difference? Western biomedicine is not prepared to accept either option, but the issue may come down to one of a natural vs. a supernatural explanation (Levin, 1996). The naturalistic explanations would say that the act of prayer is the effective agent, and while they include processes that at this point are not currently understood or accepted, such as subtle energies or nonlocal mind, they are attempts to understand the effects in ways that use natural processes. These are very different from the idea of a supernatural intervention that is outside of natural processes. The naturalistic explanations are ones that could conceivably be incorporated into biomedicine at some point if the natural laws that they follow become understood. So, while they would dramatically change our understanding of the world, they could eventually become consilient with the rest of our knowledge. This would be very different from accepting that God is answering prayers. For now, of course, it is easiest for biomedicine to focus simply on the imperfections of the research, and clearly, the research has not been strong enough to necessitate a major shift in thinking. If it continues to accumulate, however, then mainstream biomedicine may eventually be forced to include prayer in the list of factors that influence health, and it would no doubt favor a naturalistic explanation over a supernatural one.

Other medical systems may have an advantage over biomedicine in incorporating studies such as these. While they have often run parallel to religious activities, their understandings have been complementary to, rather than in conflict with, religious understandings. Thus, effects of prayer may not necessarily be included in a humoral theory, but they do not clash with it.

These studies, ironically, can also be a challenge to religious thinking. Some have questioned them because they imply that God would show more love toward patients that were prayed for than toward other patients. Why would an all-loving God need prayers to produce improvements in patients' health? There is also the issue of to whom the prayers are directed. Most of the studies have used Christian prayer groups, though the recent Duke study also included Jewish and Buddhist ones. How might one religion explain the positive effects of prayers by members of a different religion with a completely different view of spiritual matters? Would negative results be viewed as examples of God's not answering prayers or as evidence of problems in the technique or character of those praying? What if members of one religion proved to be more effective in producing health effects than members of another religion? There are no easy answers to these questions, so we see that this research has the potential to challenge assumptions in religion as well as medicine and could lead to new understandings or insights in both areas.

In regards to medicine, there are currently the simultaneous trends of Western medicine continuing to spread throughout the world while there has been increasing interest in the West in other forms of medicine. This may be due in part to the exclusion of spirituality from the practice of medicine that has occurred along with a general loss of a human touch in many situations. Thus, incorporating a sense of spirituality in its practice is likely to be a task in the future for Western medicine. This certainly occurs in other places, even as biomedicine spreads. For instance, many people in India like to try Western medicine first because its medicines are considered "strong" in that the "hot" drugs are hotter and the "cold" drugs are colder than the Ayurvedic alternatives (Kakar, 1989). The patients do not give up their Ayurvedic views, much less their Hindu ones, but simply understand the medication effects in Ayurvedic terms.

The task everywhere will be to use effective, technological methods of treatment in a humane environment that can include spirituality. In this sense, religion and medicine may be joined in a way that they have not been in the West for a very long time. The physician will never become the priest, but with a greater sense of reverence, medicine can become a more spiritual endeavor that is effective and more fulfilling for patients and physicians alike.

NOTE

[1] *Karma* comes from the root *kri*, "to do". It literally means "deed", but it implies an entire cycle of cause and effects. Every human action – in thought, word, or deed – inevitably leads to good or bad results or consequences. These consequences become part of a person's makeup and shape what happens in the current life and the lives beyond.

BIBLIOGRAPHY

Astin, John, Elaine Harkness, and Edzard Ernst. 'The efficacy of "distant healing": a systematic review of randomized trials.' *Annals of Internal Medicine* 132: 903–910, 2000.

Basilov, V.N. 'Cosmos as everyday reality in shamanism: an attempt to formulate a more precise definition of shamanism.' In *Shamanic Cosmos: From India to the North Pole Star*, Romano Mastromattei and Antonio Rigopoulos, ed. Venice: Venetian Academy of Indian Studies, 1999, pp. 17–40.

Burstein, Harold, Shari Gelber, Edward Guadagnoli, and Jane Weeks. 'Use of alternative medicine by women with early-stage breast cancer.' *New England Journal of Medicine* 340(22): 1733–1739, 1999.

Byrd, Randolph. 'Positive therapeutic effects of intercessory prayer in a coronary care unit population.' *Southern Medical Journal* 81(7): 826–829, 1988.

Cha, Kwang, Daniel Wirth, and Rogerio Lobo. 'Does prayer influence the success of *in vitro* fertilization-embryo transfer?' *Journal of Reproductive Medicine* 46(9): 781–787, 2001.

Dossey, Larry. *Healing Words: The Power of Prayer and the Practice of Medicine*. San Francisco: Harper San Francisco, 1993.

Dossey, Larry. 'Do religion and spirituality matter in health?' *Alternative Therapies in Health and Medicine* 5(3): 16–18, 1999.

Dossey, Larry. 'Prayer and medical science: a commentary on the prayer study by Harris *et al.* and a response to critics.' *Archives of Internal Medicine* 160: 1735–1738, 2000.

Harris, William *et al.* 'A randomized, controlled trial of the effects of remote, intercessory prayer on outcomes in patients admitted to the coronary care unit.' *Archives of Internal Medicine* 159(19): 2273–2278, 1999.

Hummer, Robert, Richard Rogers, Charles Nam, and Christopher Ellison. 'Religious involvement and US adult mortality.' *Demography* 36: 273–285, 1999.

Kakar, Sudhir. 'Health and medicine in the living traditions of Hinduism.' In *Healing and Restoring: Health and Medicine in the World's Religious Traditions*, Lawrence Sullivan, ed. New York: Macmillan, 1989, pp. 111–126.

Kitagawa, Joseph. 'Buddhist medical history.' In *Healing and Restoring: Health and Medicine in the World's Religious Traditions*, Lawrence Sullivan, ed. New York: Macmillan, 1989, pp. 9–32.

Knipe, David. 'Hinduism and the tradition of Ayurveda.' In *Healing and Restoring: Health and Medicine in the World's Religious Traditions*, Lawrence Sullivan, ed. New York: Macmillan, 1989, pp. 89–109.

Koenig, Harold *et al.* 'Does religious attendance prolong survival?' *The Journal of Gerontology. Series A, Biological Sciences and Medical Sciences* 54: M370-M377, 1999.

Koenig, Harold, Michael McCullough, and David Larson. *Handbook of Religion and Health.* New York: Oxford University Press, 2001.

Krucoff, Mitchell *et al.* 'Integrative noetic therapies as adjuncts to percutaneous intervention during unstable coronary syndromes: Monitoring and Actualization of Noetic Training (MANTRA) feasibility pilot.' *American Heart Journal* 142(5): 760–767, 2001.

LeShan, Lawrence. *The Medium, the Mystic, and the Physicist: Toward a General Theory of the Paranormal.* New York: Ballantine, 1975.

Levin, Jeffrey. 'How prayer heals: a theoretical model.' *Alternative Therapies* 2(1): 66–73, 1996.

Maucceri, Joseph. *The Great Break: A Short History of the Separation of Medical Science from Religion.* Barrytown, New York: P.U.L.S.E., 1986.

McNichol T. 'The new faith in medicine.' *USA Weekend* April 5–7: 4, 1996.

Obeyesekere, Gananath. 'The ritual drama of the *Sanni* demons: collective representations of disease in Ceylon.' *Comparative Studies in Society and History* 11(2): 174–216, 1969.

Plato. *Laches and Charmides*, Rosamond Sprague, ed. Indianapolis: Bobbs-Merrill, 1973, p. 63.

Rhys Davids, T.W., trans. *The Questions of King Milinda*, Delhi: Motilal Banarsidass, 1965, pp. 191–192.

Sicher Fred, Elisabeth Targ, Dan Moore, and Helene Smith. 'A randomized double-blind study of the effect of distant healing in a population with advanced AIDS. Report of a small scale study.' *Western Journal of Medicine* 169(6): 356–363, 1998.

Sloan R.P., E. Bagiella, and T. Powell. 'Religion, spirituality, and medicine.' *Lancet* 353: 664–667, 1999.

Sloan, Richard *et al.* 'Should physicians prescribe religious activities?' *New England Journal of Medicine* 342(25): 1913–1916, 2000.

Svoboda, Robert. 'Theory and practice of ayurvedic medicine.' In *Oriental Medicine*, Jan Alphen, and Anthony Aris, eds. Boston: Shambhala Publications, 1996, pp. 67–97.

Wujastyk, Dominik. 'Medicine in India.' In *Oriental Medicine*, Jan Alphen, and Anthony Aris, eds. Boston: Shambhala Publications, 1996, pp. 19–37.

Wilson, Edward. *Consilience: The Unity of Knowledge.* New York: Alfred A. Knopf, 1998.

ÅKE HULTKRANTZ

THE RELATION BETWEEN MEDICAL STATES AND SOUL BELIEFS AMONG TRIBAL PEOPLES

In most medical handbooks, early and "primitive" notions of diseases are described as entirely physical pains, which can only be removed through some kind of external interference. The archaeological discoveries of Palaeolithic brain surgery (trepanation) are the oldest testimonies of clinical operations that can be dated. However, there is a great deal of evidence of other medical methods from the large range of tribal peoples throughout the globe which may be very ancient. Examples include the establishment of taboos (for instance, against performing certain actions, mentioning certain names), the laying on of hands, and seeking the causes of disease and their removal, transference to other beings, or destruction.

In most of these ceremonies, some sort of ailing person appeared, such as an old man or a wise old woman, or perhaps a member of the family such as an experienced grandmother, or a particular specialist. There are many names for this specialist: in Africa the "witchdoctor", in Europe the "doctor" or "practitioner", in Northern Asia and America the "seer" or "shaman", and in general the "healer". The term "medicine man" is often used now to denote medical persons among tribal peoples anywhere, but it was originally only used for native doctors in French North America. The role of this knowledgeable assistant is quite important and has been so for many centuries. Some authors might say that the medicine man (if we now choose to use this term more generally for the professional doctor) worked out the medical pattern among tribal peoples. This is a good guess, but our sources do not go back far enough in time to verify it. Certainly such a modern specific term as shaman was unknown to Westerners until the seventeenth century when it became a fashionable word for Siberian (or more adequately Tungusian) native doctors (see Eliade, 1964: 495ff.). However, everything seems to point to the fact that it was the interior, soul experiences of the shaman that modelled the fanciful landscape and the dramatic dreams of the shamanic therapist.

In this chapter it is my intention to discuss this therapist and demonstrate how his extraordinary psychic states have a direct bearing on soul beliefs and medical procedures.

H. Selin (ed.), Medicine Across Cultures: History and Practice of Medicine in Non-Western Cultures, 385–395.

THE SHAMAN

Shaman is today a general term for tribal doctors who cure the sick by some sort of supernatural method. We can leave aside at first what the term "supernatural" stands for – since to many Westerners it simply means a deception or swindle that preferably should be avoided. There is, as far as I can see, no reason to bind the shaman to a philosophy which he or she does not necessarily embrace. My main reason for retaining the concept supernatural is that from the outset it points to another reality beyond our natural world. Some scholars have asserted that the idea of two worlds is a Christian idea, but I do not share this perception. When I lived with a North American Shoshoni Indian medicine man/shaman I learned that the distinction between the natural and supernatural world is very much alive to these Indians, and I later experienced that differentiating between the two worlds was a real conviction among other tribes as well. Although many of my academic colleagues do not accept this opinion, I think that such a differentiation is realistic in view of the religious conceptions of the peoples concerned. If they believe in at least one realm of the dead, how can they imagine it? It is evidently not part of our everyday world; it is part of another order of things, possible to reach in dreams and trance experiences but usually not attainable within the bounds of this world.

All doctors do not have the ability to make journeys to this other world, but some have it, and Western scholars usually call those who do "shamans". In other words, there are therapists who are supposed to be able to transgress the boundaries of the natural world which a Western doctor is unable to do. It is interesting to note how many American hospitals install American Indian shamans as colleagues of the academically trained doctors to serve at the sickbeds of their Indian patients. The reasons of the hospital director are of course not that they expect any miraculous cures (which the Indians might hope for), but that they wish to provide consolation and security for the ill person in an unaccustomed environment. It is for this same reason that hospitals employ priests and rabbis to tend to their Western patients.

The characteristic trait of the shaman, in comparison with other doctors, is that he may transfer himself to another world. Eliade describes how the shaman can, in trance, visit the world above, the divine world, or the world below, the underworld or the deep sea (Eliade, 1950: 299, 302; 1964: 192ff., 200ff., 259; 1987: 205). The journeys are often imitated, unless they take place in a deep trance. The trance as such is a clear sign that it is another kind of existence that the shaman enters. The detailed accounts of the ecstatic journey and its wonders constitute a travelogue that we could almost call mythological if it were not for the fact that the shaman is not a divine being. However, on his cosmic journeys, he appears almost as such a being.

Let us dwell a while on the meaning of the shamanic role. Many scholars, particularly in America, use the word shaman as a label for any native doctor. This is not correct (cf. Hultkrantz, 1999: 2f.). It is meaningless if we extend the concept "shaman" so that it refers to other dimensions from those contained in the Siberian word. The word "shaman" is anchored in the prefix *sha*, which refers to thinking and to knowledge. Shamans are thus people who own a secret wisdom – the knowledge of what is beyond the world that we all know. A general practitioner may fall into a light trance when he is curing. A true shaman can himself produce a deep trance, often displayed in an apparent loss of consciousness – although his real consciousness is contained in his interior dream.

In the following I shall use "shaman" in the technical sense. Recently the French social anthropologist Roberte Hamayon (1998, see Hamayon, 1993, 1995) accused me as well as other scholars of making ecstasy or trance (which I claim are equivalent phenomena) necessary ingredients in the shamanic phenomenology (Hultkrantz, 1993, 1997, 1998). She has conducted fieldwork among Buryat and Mongol tribes and has, she claims, scarcely ever found trance experiences in their ceremonialism (see her large tome on the latter, 1990). This is a remarkable statement, but it is in line with what we know of, for instance, the Buryat from earlier times. If then Buryat shamans do not know ecstatic states in their professional activities it would, in this author's opinion, be better not to call themselves shamans. However, it is, as Madame Hamayon says herself, extremely difficult to decide, in several cases, if ecstasy has occurred or not (Hamayon, 1998: 176f.). It should also be observed that many shamanic tasks, for instance offering ceremonies associated with the shaman's duties, do not involve his falling into a trance.

We thus arrive at the following conclusions: many medicine men may operate as shamans, and all shamans may be called medicine men (or doctors etc.), but few medicine men can be called shamans, according to the functional understanding that I will propose.

Hitherto we have discussed shamans *quo* medical experts. However, shaman is a wider concept than that. Shamans appear as intermediaries between humans and the powers, for instance the supernatural masters of the animals. They are psychopomps or guides for the departed to the world beyond, diviners and soothsayers and ceremonial leaders in matters that are close to their shamanic activity. In some areas, for instance among the Tartars of the thirteenth century, shamans were political leaders, which is however a secondary, atypical function. What is it then that links together all the shamanic functions? One obviously important trait is the shaman's concentration on his interior life. However, he is no sacred hermit who meditates for his own good; he is a helper for the individuals of his society (or sometimes even for people from other societies). He puts his mental powers at their disposal. And, importantly, he is a mediator between man's world and the spiritual world, because the world of trance and dreams is the supernatural world.

We cannot say whether the shaman was man's first link to this supernatural world. What happened in mankind's early existence is hidden from us. But it is certain that shamanism is a very early phenomenon in the history of man. Its existence over wide parts of the world hints that it is very old, and that it has been so widespread is primarily due to the ancient hunters (and their Arctic followers, the mounted nomads).

For us, the ways of internalising dreams and trance experiences is a problem of particular importance.

SOUL CONCEPTIONS AND SHAMANISM

We have seen how shamanic experiences are dramatised on an inner plane, in the consciousness of the shaman, and imitated in ritual actions which sometimes supplant the original narratives of the experiences, as for instance among the Coast Salish of the North American West Coast (Jilek, 1982; *cf.* the interpretation in Hultkrantz, 1992: 65ff.). Here the doctors act out the journey to the land of the dead where they catch the patient – who has already arrived there – and bring him/her back to the land of the living. Such "plays" are however exceptional, even if they also occur in Siberia.

Since the shamanic healer works with inner experiences, they reflect the possibility that he will realise them. On this point he/she may have built up the formal structure of shamanic realisations. We should however understand here that both sexes, depending on cultural preferences, might appear in shamanic roles. I mean simply that the intricate system of soul ideas is construed on the shaman's understanding of soul ideas. This is in line with later scholarly understanding of the origins of man's conception of the soul. Earlier it was considered that the ideas of the soul arose among any gifted thinkers, as imagined by, for instance, Paul Radin (1937; see especially 275ff.). A more realistic point of view was presented to me by Ralph Linton (whose chief work *The Study of Man* [1936], nevertheless does not contain anything considering early speculations about the soul). In a meeting with Linton at the time when I was writing my book on American Indian soul conceptions in the early 1950s, he paid attention to the "double", as he said, in these soul beliefs: a soul that is able to leave the body in dreams and trance and thus establishes its own independence from the body (see Hultkrantz, 1984: 30). At the time of our conversation I was not quite convinced when Linton postulated that it was thus the shaman who had been the originator of the soul dualism; indeed, Linton made it clear that the shaman was the only person who could see this double and therefore had created this concept (Hultkrantz, 1953: 40f.). My position then was that there were also other persons who could see the double, whether it belonged to oneself or to others.

My further investigations changed my view. While we can endorse the view that experiences of a double self are not limited to the shaman, the search for data shows that the intensity of double experiences is at its height whenever shamanism occurs. This was brought out in a doctoral thesis by my former disciple, Ivar Paulson (1958: 332ff.; see also Hultkrantz, 1984: 31). The area of his investigation was Northern Eurasia, which is the apex of shamanism. Paulson's findings tally with my own (later) conclusions from North America.

Linton was an original thinker. But he was not the first to deal with the belief in doubles, or what we would call "multiple souls". Basically he drew on the unitary soul concept that belongs to European history and still exists today as part of Christian preaching. However, in most parts of the world the pre-

Christian plurality of souls has been dominant. It is possible to state that what we call "soul" among the Indians – and I daresay among most tribal peoples – corresponds to a scale of spiritual potencies that are inherent to or belong to man. They are mostly born with the child, or develop with the growth of man (as the *Kinderseele*, or "children's soul" in northern Asia). Most of these souls represent specific potencies of man: the life power, the breathing power, the motion power, the will power etc. Sometimes they are all united as one potency (or soul), the "body-soul" – just like Wundt's *Körperseele*. Of course, in many tribes the different souls have their own indigenous names. Several authors have noted this, and a few of them have tried to understand the relation between the souls. Thus, Wilhelm Wundt, a psychologist, found a distinction between souls bound to the body and souls of the breath and the psyche (Wundt, 1910: 78ff.). His efforts to demonstrate a division of souls was carried on by Ernst Arbman, who may be the most convincing troubleshooter in the field of soul beliefs (Arbman, 1927).

Arbman proceeds from the fact that in "primitive" thinking the souls really stand in a dualistic relationship to each other. He makes a distinction between the souls that operate when a person is awake and those that are at work when he is in a dream trance and coma. When awake the individual depends on his body souls – all the small organ souls that grant him the force of breathing, moving, or grant him will, understanding, knowledge and so on. The latter may be summarised as ego-souls. When a person is asleep, in a trance or in a coma, the free-soul takes over, and this is "soul dualism", according to Arbman.

HUMAN CONSCIOUSNESS AND SOUL BELIEFS

Most philosophers derive human discussions of the nature of consciousness from theoretical formulations in Classic Antiquity (in particular Greece) and Vedic India. Behind this idea lurks, it seems to me, the conviction that only high cultures can produce reasonably critical individuals. There is indisputable evidence of such thinking in simple cultures and tribal societies without advanced education. Scholars like Paul Radin (1937) and Ralph Linton (see Hultkrantz, 1984) suggested that the shamans were the thinkers in older and newer tribal societies, but my own field research with the Shoshoni points in a partly different direction: sometimes there is an identity between the two categories, but at other times laymen may be greater thinkers than shamans.

Whatever the fact of the matter, in practically every society there are persons who discuss ontological questions from their own prerequisites (as we do from ours). In connection with an extensive investigation of North American Indian soul beliefs I met with quite a few examples of their opinions of man's consciousness proceeding from ideas of the ego.

As just said, the characteristic trait of the body-soul or body-souls is that they are functioning only when man is conscious and active. The whole system is weakened when man's consciousness tends to disappear, as is the case when the person is asleep or dreaming, or, even more, when he is dizzy or feverish, not to say mad and crazy. Such states imply a dysfunction of the body-souls. In their place the free-soul appears, an ethereal picture of man himself as you remember him in your memory and your dreams of him. That is, an observing outsider may see the free soul of another person – perhaps in a dream, or observed as a changed creature, an *eidolon* according to the old Greek beliefs. It is this alternation, or should we say opposition, between two soul constellations, those of the conscious and those of the unconscious man, that are characteristic of most archaic soul systems.

Oppositions demand mediations. Since soul dualism is rarely, in the actual cases, as consequent as it has been drawn up in this general outline, we find that occasionally some of the body-souls may jump in to take over the role of the free-soul (I have called the resulting compound soul "the psychological free-soul"). The breath-soul is one possible bridge (*cf.* English expressions such as give up the breath or ghost); another is the ego-soul, the soul through which you think and which often appears as a free-soul and thereby promotes the development towards a unitary soul. I see the ego-soul as the soul that endows man with consciousness. The Naskapi-Montagnais Indians in the Labrador area consider that their mind-soul stands for intellect and comprehension, whereas the Navajo speak about a soul that expresses mind, reason and awareness. There are examples from other Native American cultures as well. J.N.B. Hewitt, an Iroquois, writes that these Indians believe that the mind-soul could "see itself, take cognizance of itself, know itself"; hence it represented the faculty of consciousness (Hultkrantz, 1953: 211; 1997: 78).

There is other information on American Indian tribes which confirms this view that the ego-soul is the primary repository of the consciousness which thus appears as an operative factor in people's self-understanding. An investigation of the thinking patterns of other tribal populations would probably give the same results.

When on the other hand the person is in a twilight state the body-souls more or less momentarily fall away, and instead there appears man's outer form in the guise of a spirit. As pointed out, this spectre is technically known as the free-soul, the soul that looks like man himself and may move outside the body. Indeed, it may be loosened completely from the body. It is as a free-soul that a sick person may go missing or be snatched away (by spirits) to the world of the dead, and it is as a free-soul that the shaman may send his own soul to the spiritual realms. In the characteristic phases it means that the shaman lies unconscious while his free soul transcends to the spiritual world. This model of alternating souls we can call "soul dualism". It is usually, but not always, associated with deep trance. Where there is no trance it may happen, as pointed out above, that the trance is supplanted by a ritual performance which imitates the happenings in the shamanic action.

The shaman initiates his soul journey when he needs information from the high spirits or needs their aid to bring back the souls, whether they have been captured or gone astray. The person who is feverish or unconscious is in a great number of cases supposed to have entered the road to the spiritual realm: this is then an automatic process – the patient feels as if he were drawn to the land of the dead. The shaman can visit this land and return from it, unless he is unable to defeat the dead who easily commence a fight with him and his attendant helping spirits. The ill person can only return to the living world with the help of the shaman. His soul is then restored to his body, usually by being pressed into the fontanel, the soft, membrane-covered space at the back of a baby's skull. The journey of the soul to the land of the dead and back again is a favourite theme in North American mythology. The journey of the soul to the beyond follows the same trail as that of the shaman (Hultkrantz, 1957: 229ff.), and the restitution of life to the sick person shows a pattern which is common to shamanism and the American Orpheus tale (Hultkrantz, 1957: 240ff.). As I tried to show in *The North American Indian Orpheus Tradition* (1957), the Orpheus motif in America is built on the shamanic flight to the other world in order to save a human life. The shaman is thus the doctor *par preference* in North American mythology.

THE CURING JOURNEY OF THE SHAMAN

The shaman may heal his patients in many different ways, depending on different traditions and the shaman's own experiences and abilities. In most shamanic areas we find two major manners of shamanic healing, seeking up the fleeing soul (usually the free-soul), and dislodging the disease object or spirit that has taken the patient's body in possession. The latter state may reach its climax in what we can call psychological possession: the patient's own ego is discharged and supplanted by the intruding spirit's own ego. Possession in this sense is common in the southern parts of Central Asia, China, Tibet, the Himalayas and Polynesia. However, this type of possession is not characteristic of normal types of shamanism.

Scholars have discussed the two dominating types of shamanic recovery since Lowie published his classical essay on disease types (1924). Proceeding from his historical understanding of the relation between "intrusion" and "soul loss" (Lowie, 1925 and 1934) Lowie originally presumed that intrusion was the older idea, whereas soul loss had been imposed later on. Another anthropologist, Clements, suggested that, "all cases of soul loss are historically connected" (Clements, 1932: 237) and assigned both soul loss and object intrusion to the Palaeolithic (*idem*: 241). A more careful investigation leads to the conclusion that we can scarcely express any definite opinion about the age of different disease forms. Intrusion and soul loss are both ancient. The same may be said about the two healing methods, which originally corresponded to the catching of the soul and the removal of intruding objects or spirits; they are both most probably Palaeolithic.

We need not concern ourselves here with the shaman's curing of intrusion cases, since this subject is not the topic for this article. Perhaps it should be said, however, that the shaman frequently makes use of soul-flight methods in his cures, just as patients suffering from soul loss may be healed with methods taken from the intrusion of objects. It is difficult to know why this is so. Perhaps the reason is that doctors used to make use of both methods for different purposes and therefore did not see any great gap between them.

However, the most important method of healing a patient is for the shaman to seek the place where the seriously sick person is held. In some cases the shaman does not see his patient until he meets him in the realm of the dead; in other cases (and very often in the Orpheus Tradition) he follows the diseased person's tracks. We have excellent descriptions of the paths the deceased move along among, for instance, the Ojibway Indians who live close to Lake Superior. There is a very interesting travel journal by the German, Johann Georg Kohl, who visited the island La Pointe in 1855, which was then inhabited by the Ojibway (Kohl 1859, I: 287–301). Other information on the same tribe may be found in later anthropological papers, such as Jones, 1916: 382–383, 388, and Jenness, 1935: 18–19, 109–110. The Ojibway display a pattern of a story that is widespread in North America, from the Plateau in the Northwest to the Woodland Indians in the Southeast, and even beyond these limits. Since it is so widely known it is interesting to follow its general contents.

According to the Ojibway legend, in several versions, the people who are seriously ill or close to death travel the road of souls. They travel a long and perilous route to the spirit land in the west where the night of our world is day and our day is night. After a while the nearly dead encounter a dreadful watchdog, which can be dangerous. If in life the traveller has been a tormentor of dogs the monster dog will devour him. If not, he can continue his journey. On the side of the trail he finds strawberries – some say raspberries or blueberries – that are tasty and delicious. However, they are dangerous: anyone who eats them cannot return to life. The tale also tells that the next obstacle on the trail is a river spanned by two logs that move alternately up and down. The reason for this, according to those who know it, is that the bridge in reality is a gigantic snake. Most people pay a small amount of tobacco for the passage and may safely pass over. Others however, such as sorcerers and small children, fall into the stormy river and thus go missing. It is said that they turn into crayfish that live in the river. Next the wanderer meets a barrier of fire which he has to leap across. If he manages to succeed he is forced to stand up to a man carrying a knife. According to witnesses, this man "extracts the brain of the soul (the seat of the shadow) and stores it in his dwelling" (Jenness, 1935: 109). This means that his memory of the past is obliterated and he has no wish to return to his previous life. At last the wanderer arrives at the wigwam of Grandmother Earth, a very venerated divinity (and also grandmother of the culture hero and his brother), and the home of the dead people is close by.

We see here how difficult and almost impossible the walk to the other world was. In fact, only the medicine men, the shamans, knew how to go there and safely return home again. It was less certain that the shaman could save the soul of a patient whom he had to rescue from the other world. An informant from the Ojibway told Jenness that his father, who was a medicine man, once "followed the soul of a young girl to the land of the dead and brought it back in his hands. But it was as light as wool, and slipped through his fingers as he was inserting it through the girl's forehead. So he returned it to Nanibush [the culture hero], and the girl did not recover" (Jenness, 1935: 110). The speaker said that he had in vain tried to reach first his daughter, then later on his son, in the realm of the dead. However, his two children remained there. On the last occasion, "For half a day my body lay unconscious in the house while my soul fared forth to the land of the dead. But again I failed in my quest" (*loc.cit.*). It was apparently impossible to act like a shaman; you had to be one.

Similar narratives about the shaman's success in retrieving lost souls and the impossibility of the laity to perform the same feats have been recorded among Northern Siberian tribes, as a cursory reading of Siberian shamanic records reveals. In all these cases it is the shamanic trance that opens the trail to the other world. As noted above, on the basis of these traditions the Orpheus myths and legends have been created among the American Indians for certain and possibly in the classical Greek myth (see Hultkrantz, 1957).

* * *

Of all the medical performers in tribal societies the shamans are the most interesting and appealing. They are the mediators between the two worlds, our common daily world and the dream world of another existence. (I would prefer not to use the word sacred here, Eliade's favourite word, since the "other" world includes so many non-sacred beings, like evil goblins, evil dwarfs, tricksters, awful ghosts – and the devil!). In this article many interesting phases of shamanic activities lying outside the medical associations have not been dealt with, in particular the important initiation rites which are more frequent in Siberia than in America. Nor have the artificial inductions of the shamanic trance been discussed. The emphasis lay on shamanic curing, in which the details vary from tribe to tribe and legend to legend. The details on for instance the shaman's journeys to retrieve souls vary very much, but the variations fit within the general framework that has been given here.

The most important results of this study refer to the ways in which shamans *qua* ecstatics try to overcome the most critical diseases of their tribesmen. The disease is dangerous. The shaman finds that his patient is more or less on the road to the realm of the dead which is the individual's final destination. How can one change this destructive course? Perhaps some prohibitive means have some effect, but the best thing is for the shaman to catch up with the fleeing soul (free-soul in most cases). The shaman walks or runs on the road used by the sick person and knows how to avoid all the pitfalls on the road that threaten common people. He travels on in a canoe to the beach of the dead on the other side of the big river/ocean or swims down to the abode of the dead at the bottom of the water. Or he climbs a tall tree, mostly identified as the world tree, in order to come into contact with the gods and spirits in the superior world. This is what happens for instance among the Altaic shamans in Siberia, who have to pass nine heavens before they reach the Supreme Being, Ulgen (see Harva, 1938: 557; on the world tree, see Harva, 1922–23; Hultkrantz, 1996).

Shamans tell us that the chief of the dead, the Supreme Being, Grandmother Earth, the culture hero or his brother, or the Prince of the Dead – the titles of the dignitaries vary with the different cultures – may agree with the shaman to set the lost soul free (if the latter wants to, because life over there might be very happy). However, the majority of the dead are often described as unwilling to let the recently received soul go, and our sources often tell of a fight between the dead and the shaman. In most of these fights the shaman is depicted as being victorious.

As hinted earlier, the shaman may sometimes replace the ecstatic journey with a ritual drama representing his soul journey. Such performances are known from both Siberia and North America. Whereas in the ordinary soul flights the shaman may be totally absorbed in his trance, cut off from the world around him, the dramatically active "shaman" is certainly inspired, but he is aware of the world around him. At the most we can state that he is in a "light trance" that allows him the insight into two worlds at the same time.

BIBLIOGRAPHY

Arbman, Ernst. 'Untersuchungen zur primitiven Seelenvorstellung mit besonderer Rücksicht auf Indien.' *Le Monde Oriental* 20: 85–226, 1927 and 21: 1–185, 1927.

Clements, Forrest E. *Primitive Concepts of Disease*. University of California Publications in American Archaeology and Ethnology, vol. 32, no. 2. Berkeley: University of California Press, 1932.

Eliade, Mircea. 'Shamanism.' In *Forgotten Religions*, V. Ferm, ed. New York: The Philosophical Library, 1950, pp. 297–308.

Eliade, Mircea. *Shamanism: Archaic Techniques of Ecstasy*. Bollingen Series 76. New York: Pantheon Books, 1964.

Eliade, Mircea. 'Shamanism: an overview.' In *The Encyclopedia of Religion*, vol. 13, Mircea Eliade, ed. New York: Macmillan, 1987, pp. 203–208.

Hamayon, Roberte. *Le Chasse à l'Âme: Esquisse d'une Théorie du Chamanisme Sibérien*. Nanterre: Société d'Ethnologie, 1990.

Hamayon, Roberte. 'Are trance, ecstasy, and similar concepts appropriate in the study of shamanism?' *Shaman* 1(2): 3–25, 1993.

Hamayon, Roberte. 'Pour en finir avec la transe et l'extase dans l'étude du chamanisme.' *Études Mongoles et Sibériennes* 26: 155–190, 1995.

Hamayon, Roberte. 'Ecstasy or the east-dreamt Siberian shaman.' In *Tribal Epistemologies: Essays in the Philosophy of Anthropology*, Helmut Wautischer, ed. Aldershot, England: Ashgate Publishing Company, 1998, pp. 175–187.

Harva, Uno [Holmberg-Harva.] *Der Baum des Lebens*. Helsinki: Suomalainen Tiedeakatemia, 1922–1923.

Harva, Uno [Holmberg-Harva.] *Die religiösen Vorstellungen der altaischen Völker*. Helsinki: Suomalainen Tiedeakatemia, 1938, pp. 1–634.

Hultkrantz, Åke. *Conceptions of the Soul among North American Indians: A Study in Religious Ethnology*. Stockholm: The Ethnographical Museum of Sweden, 1953.

Hultkrantz, Åke. *The North American Indian Orpheus Tradition: A Contribution to Comparative Religion*. Stockholm: The Ethnographical Museum of Sweden, 1957.

Hultkrantz, Åke. 'Shamanism and soul ideology.' In *Shamanism in Eurasia*, vol.1, Mihály Hoppál, ed. Göttingen: Edition Herodot, 1984, pp. 28–36.

Hultkrantz, Åke. *Shamanic Healing and Ritual Drama*. New York: Crossroad, 1992.

Hultkrantz, Åke. 'Introductory remarks on the study of shamanism.' *Shaman* 1(1): 3–14, 1993.

Hultkrantz, Åke. 'A new look at the world pillar in Arctic and sub-Arctic religions.' In *Shamanism and Northern Ecology*, Juha Pentikäinen, ed.. Berlin and New York: Mouton de Gruyter, 1996, pp. 31–49.

Hultkrantz, Åke. 'Some points of view on ecstatic shamanism, with particular reference to American Indians.' *Shaman* 5(1): 35–46, 1997.

Hultkrantz, Åke. 'The meaning of ecstasy in shamanism.' In *Tribal Epistemologies*, Helmut Wautischer, ed. Aldershot: Ashgate, 1998, pp. 163–173.

Hultkrantz, Åke. 'The specific character of North American shamanism.' *European Review of North American Studies* 13(2): 1–9, 1999.

Jenness, Diamond. *The Ojibwa Indians of Parry Island, Their Social and Religious Life*. Bulletin 78. Ottawa: National Museum of Canada, 1935.

Jilek, Wolfgang G. *Indian Healing: Shamanic Ceremonialism in the Pacific Northwest Today*. Surrey, British Columbia: Hancock House Publishers, 1982.

Jones, William. 'Ojibwa tales from the North Shore of Lake Superior.' *Journal of American Folk-Lore* 29: 368–391, 1916.

Kohl, Johann Georg. *Kitschi-Gami oder Erzählungen vom Obern See*. 2 volumes. Bremen: C. Schünemann: 1859. Reprinted Graz: Akademische Druck-und Verlagsanstalt, 1970.

Linton, Ralph. *The Study of Man*. New York and London: D. Appleton-Century Co., 1859.

Lowie, Robert H. 'On the historical connection between certain Old World and New World beliefs.' In *Proceedings of the 21st International Congress of Americanists*. Gothenburg, 1925. Nendeln, Liechtenstein: Kraus Reprint, 1968, pp. 546–549.

Lowie, Robert H. 'Religious ideas and practices of the Eurasiatic and North American areas.' In *Essays Presented to C.G. Seligman*, E.E. Evans-Pritchard, *et al.*, eds. London: K. Paul, Trench, Trubner & Co., 1934, pp. 183–188.

Paulson, Ivar. *Die Primitiven Seelenvorstellungen der nordeurasischen Völker*. Stockholm: The Ethnographical Museum of Sweden, 1958.

Radin, Paul. *Primitive Religion*. New York: Dover Publications, 1937.

Wautischer, Helmut, ed. *Tribal Epistemologies: Essays in the Philosophy of Anthropology*. Aldershot, England: Ashgate Publishing Company, 1998.

Wundt, Wilhelm. 'Völkerpsychologie.' *Mythus und Religion*, vol. IV: 1. Leipzig: A. Kröner, 1910.

INDEX

Science Across Cultures

1. H. Selin and S. Xiaochun (eds.): *Astronomy Across Cultures.* The History of Non-Western Astronomy. 2000 ISBN 0-7923-6363-9
2. H. Selin and U. D'Ambrosio (eds.): *Mathematics Across Cultures.* The History of Non-Western Mathematics. 2000 ISBN 0-7923-6481-3
3. H. Selin and H. Shapiro (eds.): *Medicine Across Cultures.* The History of Non-Western Medicine. 2003 ISBN 1-4020-1166-0
4. H. Selin (ed.): *Nature Across Cultures.* Views of Nature and the Environment in Non-Western Cultures. 2003 ISBN 1-4020-1235-7

KLUWER ACADEMIC PUBLISHERS – DORDRECHT / BOSTON / LONDON